# PULP & PAPER DICTIONARY

## Also published by Miller Freeman for the pulp and paper industry:

Paper and Paperboard Manufacturing & Converting Fundamentals
*by James E. Kline*
Pulp Technology and Treatment for Paper
*by James d'A. Clark*
An Introduction to Paper Industry Instrumentation
*by John R. Lavigne*
Instrumentation Applications for the Pulp and Paper Industry
*by John R. Lavigne*
International Glossary of Technical Terms for the Pulp and Paper Industry
*edited by Paul D. Van Derveer and Leonard E. Haas*
Modern Mechanical Pulping
*edited by Ken L. Patrick*
Paper Coating Trends
*edited by Ken L. Patrick*
New Trends and Developments in Papermaking
*edited by Kelly Ferguson*
Environmental Solutions
*edited By Kelly Ferguson*
New Maintenance Strategies
*edited by Ken L. Patrick*
Modern Paper Finishing
*edited by John C. W. Evans*
Paper Recycling
*edited by Ken L. Patrick*
Bleaching Technology
*edited by Ken L. Patrick*
Issues in Paper Distribution
*edited by Robert B. Galin*
Mastering Management
*by Roberta Bhasin*
Pulp and Paper North American Factbook
Lockwood-Post's Directory
Pulp and Paper Mill Map
International Pulp & Paper Directory
Pulp & Paper International Factbook
PPI Map of the European Pulp and Paper Industry
PPI Map of the Asian Pacific Rim's Pulp & Paper Mills

# PULP & PAPER DICTIONARY

by John R. Lavigne

technical editing by Ken L. Patrick

Miller Freeman Books
San Francisco

*To the pulp and paper industry*
*in appreciation for providing me with*
*the opportunity to be associated*
*with people who have shared and made the*
*pursuit of my career so enjoyable.*

Published by Miller Freeman Books, 600 Harrison Street, San Francisco, California 94107
Publishers of *Pulp & Paper* and *PPI* magazines
A member of the United Newspapers Group



Cover Design: Brad Greene

Library of Congress Catalog Card Number: 93-79787
ISBN 0-87930-303-4

Printed in the United States of America
93 94 95 96 97 5 4 3 2 1

# *Contents*

# *Editor's Preface to the Second Edition*

The first edition of this dictionary was published in 1986. During the seven years since then, the pulp and paper industry has changed dramatically in almost every process segment, particularly in the environmental area, including pulping and bleaching. These waves of change have been more sweeping than probably anytime in the history of papermaking.

A virtual renaissance has occurred in the bleach plant as mills rapidly moved away from chlorine-based bleaching to chlorine dioxide and then non-chlorine technologies. The paper mill, too, has evolved significantly with more paper machines being started up or retrofitted with gap forming technology, high impulse presses, and single tier drying.

Process control advances have continued to escalate, and the industry has the rudiments of artificial intelligence in place. As non-chlorine bleaching has finally opened the door to total effluent closure, and as the fiber recycling race continues to escalate, wet end paper machine chemistry has become more important than ever. With extremely high organic loads now moving into the chemical-energy recovery cycle, new recovery loop technologies are being developed and implemented.

Reflecting these many changes, this dictionary has been significantly updated, with new information added to many of the existing terms, along with deletions of some information that has become obsolete. Altogether, almost 800 new terms have been included.

The *Pulp & Paper Dictionary* is designed primarily for persons acquainted with the pulp and paper industry, either directly on the manufacturing side, or indirectly as a supplier, consultant, student, or analyst. Selection of terms and specifically the wording and structure of their definitions assume, to some degree, a basic understanding of papermaking fundamentals. However, those outside of the industry may also find this book to be quite useful due to its overall nontechnical tone, and particularly after reading the introductory overview beginning on page xvi. This section provides a brief history of papermaking and outlines the various processes in use today.

All terms are listed alphabetically, except that some compound nouns are grouped in order after their respective base words, and therefore may not be alphabetical in relation to other terms that follow. For example, there are 27 terms between *acid* and *acidic*, all of which are compound nouns built from the base word *acid*. These begin with *acid alum* and end with *acid wash*, and about half would follow rather than precede *acidic* if normally alphabetized. The author and editor feel that the value of listing related words in this manner outweighs the convenience of a routine alphabetical arrangement.

Common industry abbreviations and acronyms are included in alphabetical order, with a *see* reference to the full, spelled-out term, where there is one. We believe that this is the way that most users of dictionaries look for these abbreviations and acronyms, rather than scanning through stacked terms at the beginning of each letter chapter. Most of the important North American industry associations and related agencies are also included.

Terms chosen for inclusion in this dictionary are from the author's own listing compiled during many years of involvement with the industry and from various personal contacts and reference sources, including the *International Glossary of Technical Terms for the Pulp and*

*Paper Industry* (Miller Freeman Inc.). The definitions are specific to the paper industry, i.e., the terms are described only as they apply to the manufacture, conversion, or sale of paper products. Chemicals, for example, are defined by their roles in the pulping and papermaking processes and not by their compositions and characteristics (although formulas are shown for major chemicals used in the industry). Likewise, such terms as abrade and abrasion are described relative to their effects on paper and papermaking equipment.

American spellings are used for all terms. Definitions do not show pronunciations of the terms and do not include etymologies. Also, inflected forms (including plurals of nouns), syllabication, and examples of usage are felt to be beyond the scope and purpose of this dictionary and are not included. However, this information is typically available in most standard English language dictionaries.

Where multiple definitions for a term are given, the order of appearance reflects the commonness of use. The first description is generally the most widely used, with subsequent definitions showing increasingly less popular usage or secondary applications. In most cases, measurement terminology is in U.S. standard (not metric) and is spelled out (no abbreviations used, such as tpd, $ft^2$, hp, rpm, etc.). With few exceptions involving major equipment, trade names are not noted.

Cross-references are shown in all capital letters within a definition and are usually preceded by *see* or *see also.* These cross-references refer the reader either to another more commonly used term or other related terms whose definitions broaden and expand the meaning of the main entry. Cross-references are listed alphabetically and not necessarily in order of pertinence.

Some terms and definitions blend colloquial and even regional expressions with established technical terminology and common meaning. This reflects the varied makeup and development of an industry that has matured in a hodgepodge fashion over a relatively long period of time. Certain terms may not be used in the same way in various regions of the U.S. and Canada, and particularly in other parts of the world. Definitions for some of the terms may also vary widely from area to area. Thus, this dictionary is not intended to be a definitive source for the use and meaning of such terms, but rather to record their applications according to established practice and convention over time.

As manufacturing trends develop and new technology is implemented, basic vocabulary for the industry will continue to evolve accordingly. Therefore, future editions of this dictionary will incorporate more new terms and show revisions of existing terms and their definitions, as necessary.

Ken L. Patrick
Editor in Chief, *Pulp & Paper*

# Preface

I was born in a small mill town in the foothills of the Adirondack Mountains, literally in the shadows of the pulp and paper industry. The huge wood pile and the mill in which my father was employed were clearly visible from my bedroom window.

When my father took me to the mill on frequent visits during the early 1930s, I was bewildered by some of the quaint and peculiar terms and expressions used by the employees in and around the mill. As I became older and spent more time at the mill, not only as a visitor but also as a part-time employee unloading pulpwood, operating digesters and as a bleach plant tester, mixing lime in cooking liquor makeup, working on the pulp machine and the paper machine, as a tester in the paper mill, and on the maintenance crew, my bewilderment slowly changed to curiosity. I began to inquire as to the meanings of these expressions whenever I encountered them and found that some of the definitions were indeed expressive and colorful; those I mentally noted.

By the time I left for college to pursue a career in the pulp and paper industry, my curiosity developed into a definite interest. I found myself compiling a list of any new words and statements I encountered that were peculiar to the pulp and paper industry. I have attempted to maintain that list during my continual association with the industry.

While doing the literature research for my previous books, I mentioned this compilation to Charles C. Baake, a vice-president with the publisher. He suggested that it be further enhanced by researching the literature in other related areas of common interest and be made available as a dictionary. Without his persistent encouragement, which was the guiding force throughout, this dictionary would not have become a reality.

Subsequently, Miller Freeman Publications and the Foxboro Company agreed to collaborate in the compilation and publication of the dictionary, and I was requested to work on the project. An extensive literature search was conducted to compile a compendium of unique words and phrases commonly used in the pulp and paper industry. Other industries were also researched to determine if they used peculiar terms and expressions that would be of interest to the pulp and paper industry. These industries were printing, graphic arts, logging, lumbering, box and container manufacturing, converting, water and waste treatment, environmental, and power. The findings were incorporated into this dictionary.

The original intent was to make this book a basic reference not only for people with technical and nontechnical knowledge in the pulp and paper industry, but also for anyone with an interest in or a need for a convenient source of information about the terminology commonly used in the pulp, paper, and some associated industries. Although the words and phrases selected may have a number of definitions, depending on their use, only those descriptions and explanations have been included that are peculiar to the pulp, paper and allied industries. Where applicable and considered unique, common equipment nomenclature has also been included.

This compilation cannot be technically considered a dictionary in the strictest sense, but it is more than just another glossary. Being a compilation of colloquial and technical terms which have been used to a greater or lesser extent within the pulp and paper industry throughout its period of development, it would be an impossibility to cover the field completely. Therefore, the authority for the definitions stems from the sources from which they were derived, some of which are listed in the appendix, and must not be construed as carrying any legal implications.

Although this dictionary contains those words, names, phrases, and expressions which have been adopted by the pulp and paper industry for its particular use, the interpretation and use of these definitions must be the responsibility of the user, from whom many of them have originated. However, it is hoped that the pulp and paper industry will accept this reference as one that was developed by and for them, and any incompleteness or inconsistencies, when encountered, will be resolved by them in future editions by notifying the publisher, editor, or author.

The author would like to express his appreciation and gratitude to all his past and present acquaintances in the pulp and paper industry, employees of The Foxboro Company and Miller Freeman Publications, who, perhaps unknowingly, have encouraged, assisted and provided the initiative to assume and complete this project.

Special thanks go to Roberta A. Tierney for typing the manuscript, to Frank M. Todaro for his proofreading efforts, and with particular indebtedness to John B. Prendergast and Judith Brown for coordinating the many activities involved in this endeavor.

John R. Lavigne, P.E.

# Introduction

Pulp and paper manufacturing is one of the oldest major industries in the world, deeply imbued with historical romance and tradition. To briefly recount the events that have transpired throughout its evolution will provide a better appreciation of the sources and definitions of the language, words, terms, expressions, and phrases used by the industry. A general description of typical processes will also help in identifying where they are used.

## HISTORY

The instinctive desire to perpetuate achievements has produced the vast supply of knowledge available today. In the early days of human history, memories of noteworthy achievements and great deeds were kept alive by oral transmission from generation to generation. However, this technique was obviously fraught with inaccuracies and misinterpretations.

As civilization continued to develop, a more permanent and accurate record was required in order for learning to advance. For example, our not-too-distant ancestors began using stones placed in piles to convey information, with odd-shaped rocks and boulders commemorating great deeds and solemn covenants. This eventually led to the use of obelisks, which were hewn shafts of stone on which hieroglyphics and crude pictures of people, birds, animals, and plants were carved. Some of these still exist and can be traced as far back as 4000 B.C. in Egypt. As most of the record keeping was of a religious nature, it naturally followed that the process eventually showed up on stone tablets and panels on temple walls. Biblical history notes that the Ten Commandments, for example, were graven on stone tablets in 1491 B.C.

Next came the process of writing on soft clay cakes, which were enclosed in soft clay envelopes and baked, followed by the use of six- and eight-sided prisms made from terra cotta.

With time, other miscellaneous materials emerged as record-keeping mediums: plates of various metals, including gold and silver; skins, bones and intestines of animals; ivory; bark, leaves, and tablets made from wood; bamboo strips, and silk.

### Papyrus and Parchment

As civilization evolved from a primitive to a more sophisticated society, less-crude writing mediums were needed. In the search for new materials, interest in substances of vegetable origin led to the making of a crude type of "paper" from the papyrus plant in Egypt. Although these sheets have been dated as early as 3000 B.C., historically the invention of papyrus has been credited to 2400 B.C. The name "paper" was derived from papyrus, which is a reed-like plant growing along the Nile River. It was converted into a sheet by cutting it into thin slices and then pressing them together. Papyrus became the most widely used writing material of its time and continued to be up until the 12th Century A.D., when it spread into Italy.

The manufacture of parchment as a writing material followed the making of papyrus, sometime around 500 B.C. The skins of sheep and goats were employed in the making of parchment by steeping them in pits impregnated with lime, followed by stretching them on frames where their thickness was reduced by paring and scraping with sharp instruments. Its use as a writing material continued until the 15th Century. Most of the material known today as parchment paper is made by passing webs of paper through acid baths. Vellum, sometimes considered a type of parchment, was made in a similar manner from the skins of young calves or stillborn lambs. There is evidence that the Hebrews wrote on skins of animals as early as 1085 B.C.

## Paper

The invention of paper as we know it today is credited to Ts'ai-Lun, a Chinese official in the court of Ho Ti, emperor of Cathay around 105 A.D. It was first produced at Lei-Yang, China, from fermented and beaten mulberry bark fibers. The fibers were suspended in water and scooped out with a screen made of bamboo splinters tied with horse hair. This method of making paper was kept secret from the rest of the world for several hundred years until the Arab armies captured Samarkand, a city in western China, around 704 A.D. The Arabs learned the art from captured Chinese papermakers who were taken to Baghdad to make paper. It was at the end of that century that these papermakers began to use linen rags as a source of fibers for making paper. The art spread to all Arab domains and passed into Europe with the invasion of the Muslims. The first paper made in Europe was produced at a mill in Toledo, Spain, in 1085 A.D. The papermaking art was spread further by the Crusaders who visited Palestine and Syria during the 12th Century.

The art of papermaking was introduced into France in the late 1100s and in Italy in the early 1200s, where watermarks first appeared on paper. Paper mills were reported in Germany as early as 1336. As the knowledge of papermaking spread through Austria in 1356, Holland in 1428, Switzerland in 1433, Russia in 1576, and the Scandinavian countries in 1630, the use of paper slowly replaced the use of parchment and vellum to a great extent, even though it was still generally made by hand.

Papermaking finally came to the North American continent in 1690 with the building of a mill by William Rittenhouse and William Bradford at Wissahickon Creek in Germantown, Pennsylvania. Individual sheets of paper were made from linen rags by hand, a method which prevailed until the invention of the paper machine.

The first paper machine was invented by a Frenchman, Nicholas-Louis Robert, in 1799. The machine formed the fibers on an endless wire screen belt to produce paper in a continuous sheet. Due to lack of support, the machine was not fully developed until the English brothers Henry and Sealy Fourdrinier developed it into a practical device in 1810. It became one of the greatest single events in the history of paper manufacturing, and has become commonly known as the Fourdrinier.

About 1805 in England, Joseph Bramah invented a cylinder-type paper machine, which consisted of a rotating, screen-covered cylinder partially submerged in a vat of pulp. It picked up pulp from the vat and formed a continuous sheet of paper which was then pressed and dried. John Dickinson actually took out a patent on this machine and has been given credit for the idea. The first paper machine used in America was of this type and appeared in 1817 at Philadelphia.

Today, paper is manufactured on huge, expensive machines of tremendous weights and large widths. Some extend several stories in height for a distance longer than a city block. They are normally operated 24 hours a day, 7 days a week, throughout the year, only stopping for an occasional cleaning, repair, or a possible short pause for a change in the type of paper being made.

## PULP AND PAPERMAKING PROCESS

The manufacture of pulp and paper employs many different and varied operations which, historically, have been primarily batch or semi-batch types. For many years, the trend has been toward continuous operation; today this method is by far the most common one used. Paper is basically comprised of a mat of fibers. Since its invention, many different fibers have been used for its manufacture. These have included the bast fibers of flax and mulberry, the stalks of bamboo and other grasses, various leaf fibers, cottonseed hair, wool, asbestos, and the woody fibers of trees. Cotton and linen rags and straw have also been used. Many research projects have been conducted on the possible use of glass fibers, rayon, nylon, and other synthetics to impart unique properties to the sheet of paper. Today, wood is the primary raw material from which most paper pulp is made.

The actual process begins in the woodlands where trees, designated as pulpwood, are cut into prescribed lengths. The logs are either hauled directly from the forests to the mill woodyards or, during certain times of the year, the bark is removed by peeling it before shipment to the mill. Also, during certain times of the year, the logs can be chemically debarked. A strip of bark is peeled off around the base of the tree, and the exposed area is painted with a special chemical solution. The tree begins to die and by the following year when it is cut, the bark is loose enough so that it can be easily removed during the harvesting operation. Also, logs are often reduced to chip form before being hauled to the mill woodyard for storage.

## WOOD PREPARATION

In the mill woodyards, the unbarked logs are fed into a giant revolving drum barker in which their bark is stripped away as they tumble against each other and the steel-channeled wall of the drum. Bark is also removed by moving the logs past streams of high-pressure water in a hydraulic-type barker, or by sets of mechanical knives. Depending on the type of pulping process they are going to be used in, the debarked logs from the barker and/or the woodlands are either sent directly to the process or to chippers where, by dropping against a revolving disc with heavy, sharp knives set at an angle, they are reduced to small chips approximately one-half to three-quarters of an inch in size. These chips and the unscreened chips from the woodlands are conveyed to vibrating screens. Oversized chips are removed, or removed and sent to a rechipper and returned for another pass through the screens. Usually, the undesirable fine chips and sawdust that are removed at the screens are burned in the power boiler as fuel.

The screened chips are then either transported by air conveyors to outside chip storage piles or belt-conveyed to huge tanks or silos for storage, according to the type or species. They are then transported to smaller storage bins located in the pulp mill, usually directly over the digesters in which the wood is to be processed.

## PULP MANUFACTURE

Before wood can be made into paper, paperboard, plastics, and other products, it must be reduced to its basic components to form pulp.

Wood is made up primarily of cellulose fibers bound together with lignin, a glue-like binder, plus sugars, gums, resins, and mineral salts in lesser quantities. The objective of pulp manufacturing is to separate the wood into fibers and other components, remove the undesirable components, and provide a means for treating the fiber to produce a suitable pulp for the paper mill.

The conversion of wood into wood fibers is usually carried out by one or a combination of three general types of processes, namely: mechanical, chemical, and semichemical, depending on the type of wood used and the requirements of the end product.

### Mechanical Pulping

The simplest pulping method is generally referred to as mechanical pulping. This process was invented in Germany by Friedrich Keller in 1844 and was introduced into the United States at Curtisville, Massachusetts, by the Pagenstecher brothers in 1866. It differs from other pulping methods in that the reduction of wood to fibers is essentially a physical operation in which the fibers are actually pulled away from each other by the application of some type of mechanical force. The three most prevalent forms of mechanical pulping are: groundwood or stone groundwood, refiner groundwood or refiner mechanical, and thermomechanical.

*Groundwood Mechanical Pulping.* In mills using the groundwood mechanical pulping method, debarked logs from the barking drums or from the woodlands are transported to the grinders. The grinders consist of huge, rough-faced grinding stones driven by an electric motor or a hydraulic turbine. The stone is enclosed in a housing having pockets. The logs are placed in these pockets, and their sides are pressed against the revolving grindstone by the use of hydraulic pistons or gravity. In this way, the fibers are literally torn from the log.

The resulting pulp is screened to remove slivers, knots, and oversized pieces, which are sent through a refiner to be broken up and returned to the screens. After washing, the cleaned suspension is sent to the paper mill to be used for papermaking, as shown in the simplified flow diagram of a typical groundwood mill (Figure 1).

Since lignin is not removed, paper made with pulp produced by this method does not generally maintain its brightness and strength as long as papers made from chemical pulps. However, groundwood paper has certain desirable qualities for various printing processes. Paper on which newspapers are printed consists largely of groundwood pulp, as do the lower grades of tablet paper and other papers which do not require unusual strength or

**Figure 1.** Simplified flow diagram of a typical groundwood mill.

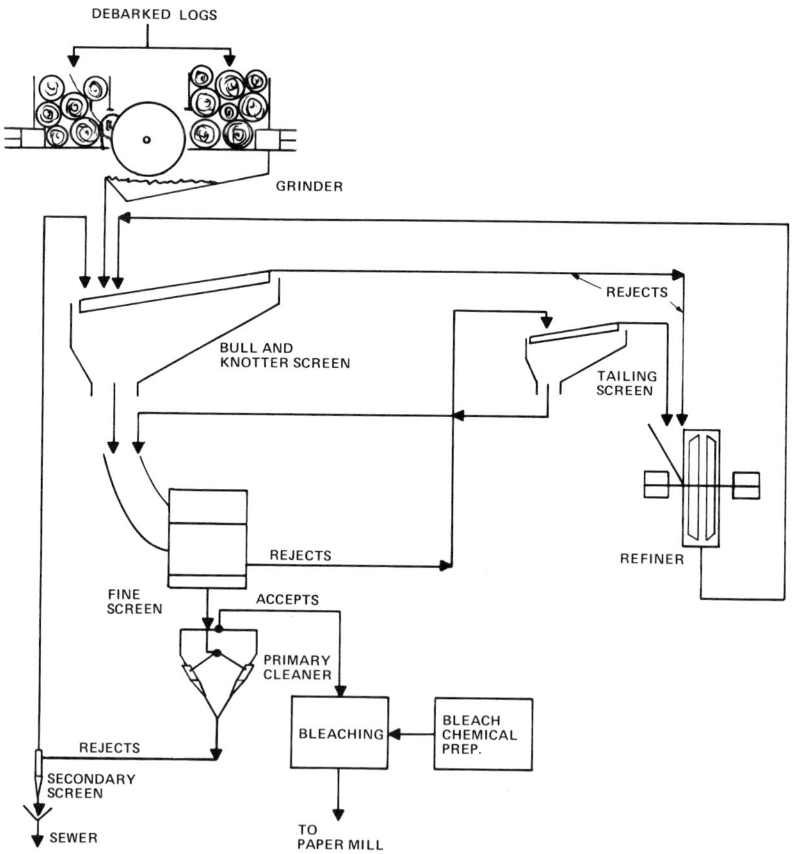

are intended to be used only for a short time. With the development of highly successful bleaching methods, groundwood pulp has now become one of the important raw materials for making papers such as those used for color printing in magazines, as well as other quality papers.

*Refiner Mechanical Pulping.* First used in the 1960s, this method involves the physical separation of fibers after the logs are reduced to chip form. A number of techniques have been developed to accomplish this by using refiners of various designs. The fundamental operation of refiners is based on the passage of chips between two disc-shaped rough surfaces moving in opposite directions and in close proximity to each other. The number of refiners used varies, depending on the operating practices of the mill. A common configuration consisting of primary, secondary, tertiary, and reject refiner stages is illustrated in Figure 2.

**Figure 2.** Block diagram of a typical refiner mechanical pulping process.

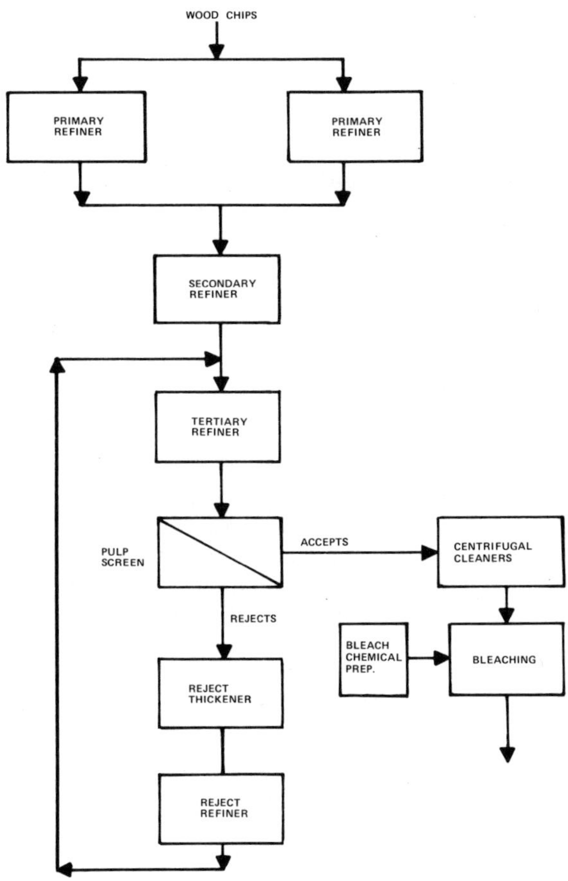

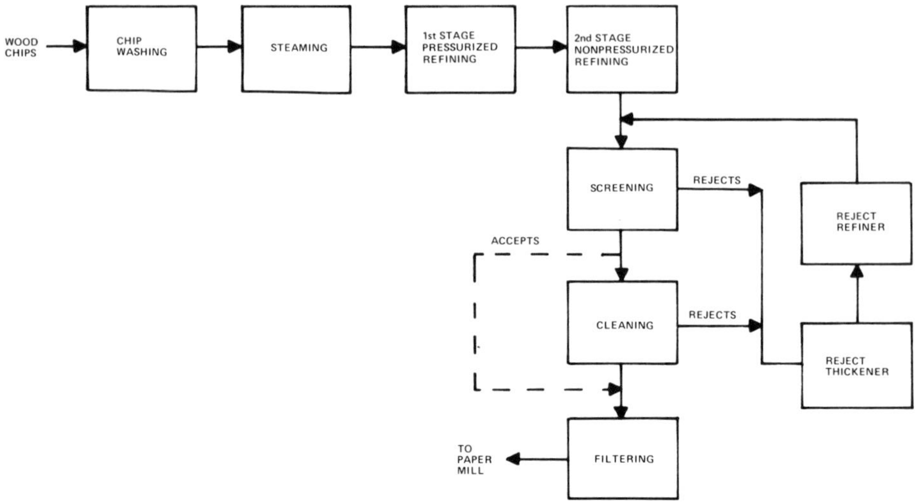


**Figure 3.** Block diagram of the thermomechanical pulping process.

*Thermomechanical Pulping.* This process, originating in the early 1930s, differs from the refiner mechanical pulping process in that, with the thermomechanical pulping system, chips are preheated and then subjected to mechanical forces in a disc refiner under pressure. The chips are subjected to repeated compressions and stress relaxations in the gap between the opposing bars and grooves. A second-stage mechanical treatment is performed on the pulp from the pressurized refiner in a nonpressurized, atmospheric-type refiner located downstream, as shown in Figure 3.

## Chemical Pulping

In these types of processes, the chips are cooked in a chemical solution under pressure until the fibers are separated by dissolution of their lignin binder and fall apart with very little or no mechanical action. There are several chemical pulping processes using one of several chemical solutions, depending on the type of wood used and the kind of paper desired. The most well-known are the soda, sulfite, and sulfate processes.

*Soda Pulping Process.* The first chemical pulping process was invented in 1851 by Hugh Burgess in England. The process used a caustic soda solution to reduce wood shavings and chips to their individual fiber components by cooking them in pressurized vessels. Today, it is hardly ever used, being largely replaced by the sulfite and sulfate processes.

*Sulfite Pulping Process.* This common pulping process uses mainly softwood species such as spruce, hemlock, and fir. It was invented by Benjamin C. Tilghman of Philadelphia in 1867. The chip cooking liquor consists of sulfurous acid and a salt of this acid produced by burning sulfur to sulfur dioxide. The desired cooking liquor is formed by the absorption of the sulfur dioxide gas by the hydroxide of a base chemical. Formerly, calcium was used, as shown in Figure 4, but due to problems with chemical recovery, this never became feasible. Newer mills are using recoverable bases such as magnesium, sodium, and ammonium. Figure 5 illustrates the cyclic nature of chemical recovery in a typical magnesium-base sulfite pulping process. It is similar to the sulfate process in configuration.

Pulp and spent sulfite liquor are discharged into a blow tank from which the digester contents, diluted with additional weak liquor, are pumped to stock washers. The first-stage washer filtrate enters the weak red liquor storage tank. The liquor is concentrated in multiple-effect evaporators and transferred to the strong liquor tank. The strong liquor is then fired in a recovery furnace. The combustion products of the sulfur and the magnesium in the liquor are discharged from the furnace in the gas stream as sulfur dioxide and solid particles of magnesium hydroxide. The sulfur dioxide is recovered by reaction with the magnesium hydroxide to produce a

**Figure 4.** Flow diagram of a typical calcium-base sulfite pulping process.

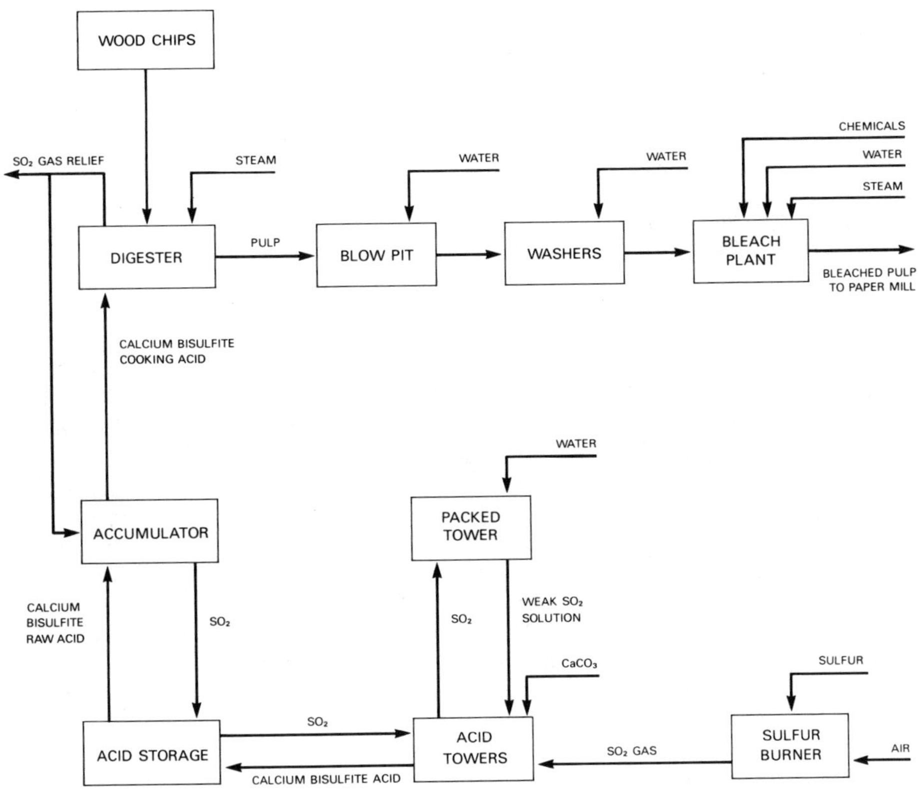

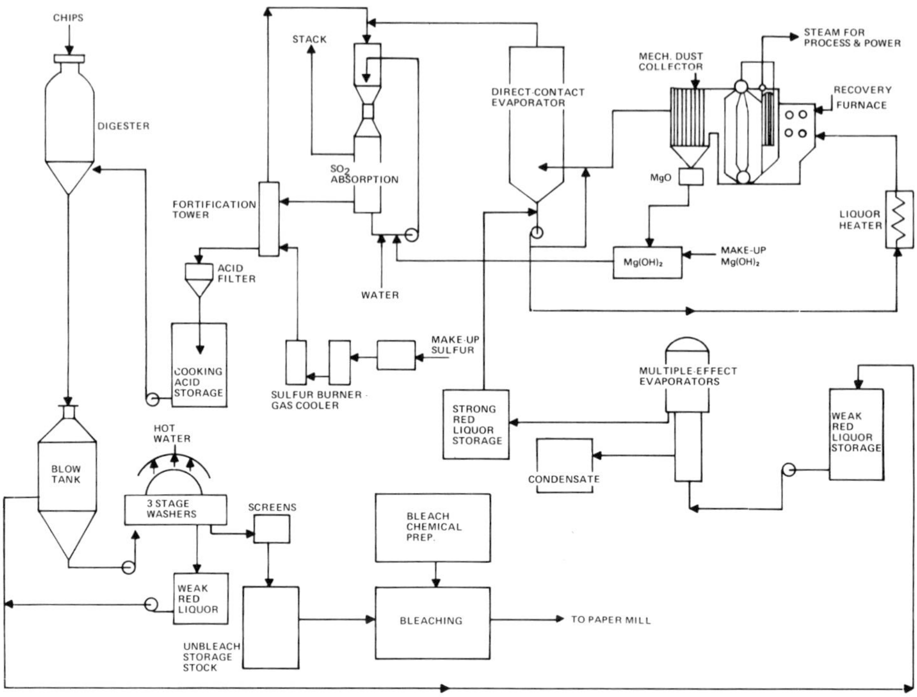


**Figure 5.** Flow diagram of a typical magnesium-base sulfite pulping process.

magnesium bisulfite acid in an absorption system. This acid is passed through a fortification, or bisulfiting, system and fortified with makeup sulfur dioxide. The finished cooking acid is filtered and placed in storage for reuse in the digester.

The unbleached pulp from the washers is screened, bleached, and sent to the paper mill. A typical ammonium-base sulfite pulping process flow diagram is shown in Figure 6.

Sulfite pulp is used in the manufacture of bond, writing, high-grade book, and other fine papers.

*Sulfate Pulping Process.* The most common and important chemical process is the sulfate or kraft process, invented by Carl F. Dahl in Germany about 1884. Practically any species of wood can be used. The cooking chemical is a solution of sodium hydroxide and sodium sulfide. As shown in Figure 7, the chips are boiled under pressure in a strong solution of sodium hydroxide and sodium sulfide. After cooking, the liquor, which is separated from the pulp suspension at the washing stage, is concentrated in multiple-effect, steam-heated evaporators. Sodium sulfate is added and the mixture is burned in the furnace. The molten smelt issuing from the bottom of the furnace is dissolved in water to form green liquor, which is causticized by adding slaked lime in the causticizing area. The sodium sulfide-caustic soda

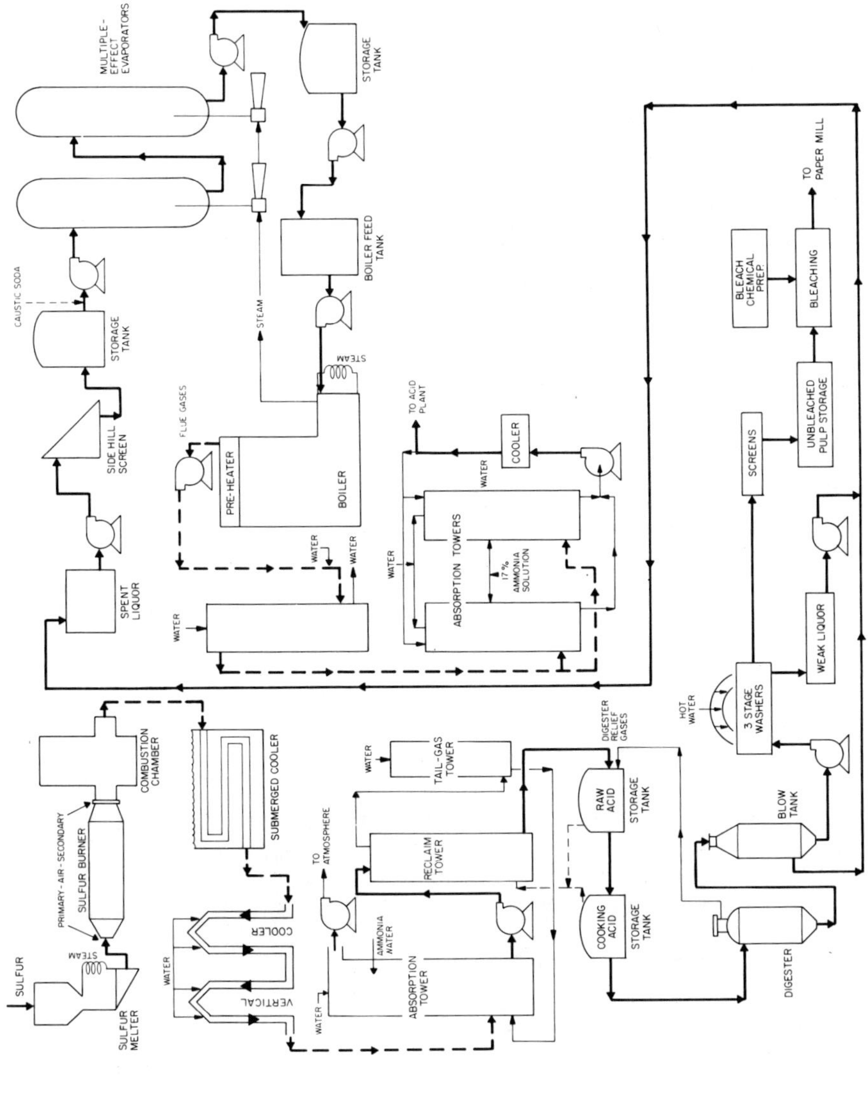


**Figure 6.** Flow diagram of a typical ammonium-base sulfite pulping process.

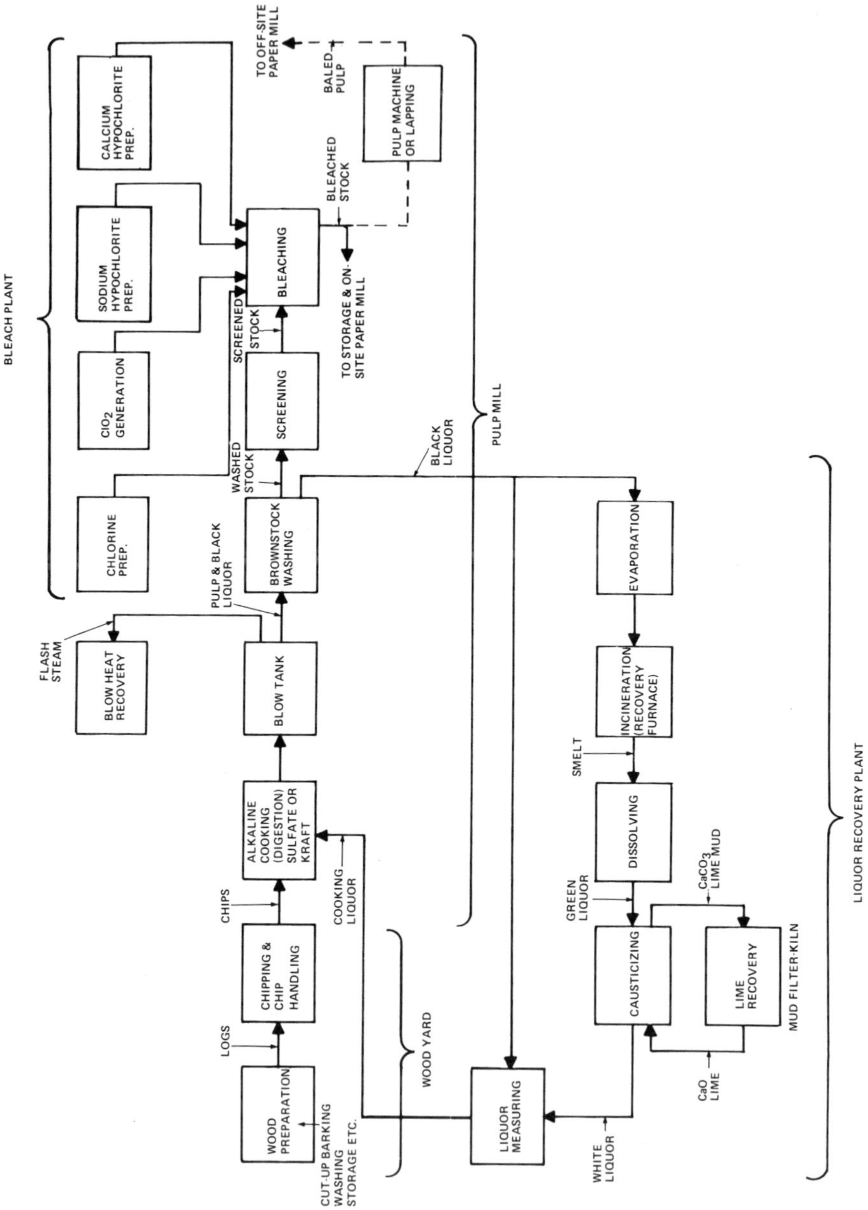


**Figure 7.** Block diagram of the sulfate pulping process.

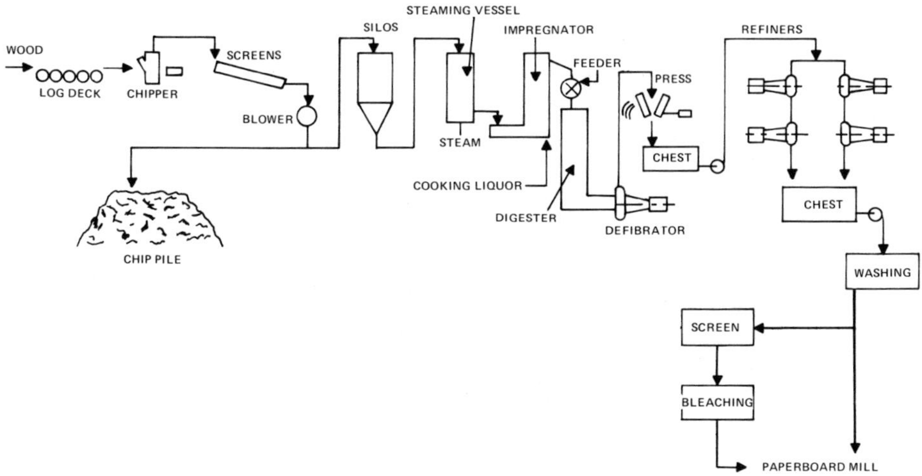


**Figure 8.** Simplified flow diagram of a typical NSSC pulping process.

solution, which is now called white liquor, is sent to the digester area for reuse in cooking new chips.

The washed brown stock is screened, bleached, and either pumped directly to an integral on-site paper mill or processed for shipment to other paper mills or pulp processing plants.

## Semichemical Pulping

Another way of making wood pulp, dating back to 1874, combines mechanical and chemical methods and is called the semichemical process. In semichemical pulping, the wood is given only a mild chemical treatment, softening or removing just enough lignin to loosen the fibers, but not enough to separate them. Separation is then effected by mechanical means. The process was developed particularly for the pulping of hardwoods and has many variations. Three of the important methods are the neutral sulfite, cold soda, and chemigroundwood process. Semichemical pulp produces stiff, resilient products and is used in making corrugated paperboard, egg cartons, and similar items.

*Neutral Sulfite Semichemical Pulping Process.* Commonly referred to as the NSSC process, this method uses a sodium sulfite liquor buffered with sodium bicarbonate to cook the chips. Although the design configuration variations are many, a typical process for the production of corrugating paperboard pulps is shown in Figure 8.

The process begins with the heating of screened chips in a steaming vessel, from which they enter a screw conveyor where they are impregnated with cooking liquor. After cooking in a digester, the chips are broken up by an impeller or defibrator under full pressure. The coarse pulp produced is

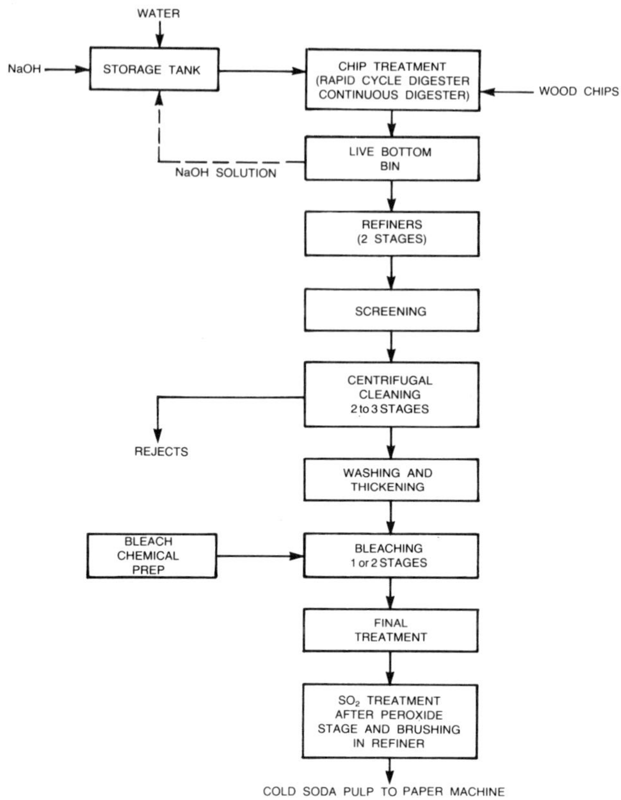


**Figure 9.** Major steps in the cold soda pulping process.

compressed by a disc or screw press, diluted, and sent to the refiners. After washing on filters or screw presses, the refined stock is sent to the paperboard mill. In the case of bleached grades, the pulp is screened and bleached, and is then sent to the paperboard mill.

*Cold Soda Pulping Process.* (Also referred to as the "chemical-mechanical process" because there is no major change in the lignin.) In the cold soda process, developed between 1919 and 1930, chips are treated with cold sodium hydroxide using hydrostatic, mechanical, or atmospheric pressure. Like the NSSC process, the cold soda process entails many variations in design. However, the major steps are shown in Figure 9.

After treatment, the liquor is usually collected, fortified by fresh sodium hydroxide solution, and used again. The pulp is refined in a two- or three-stage disc refining operation, screened, centrifugally cleaned, washed, and thickened. When the bleached pulp is used for printing papers or boards, one- or two-stage, high-density bleaching is applied. After the final treatment, the pulp is ready to be used on the paper machine. In some cases, unbleached cold soda is used, eliminating the bleaching operation.

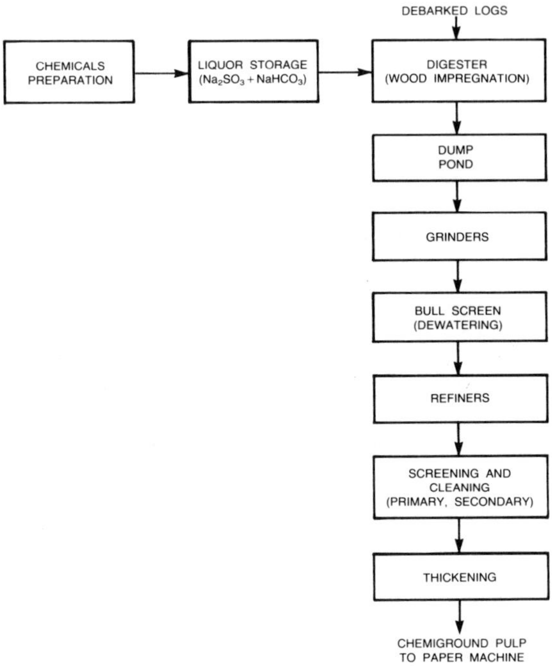


**Figure 10.** Major steps in the chemigroundwood process.

*Chemigroundwood Pulping Process.* Another process, introduced in 1890 but not fully developed until the late 1940s, is called the chemigroundwood process. In this process (Figure 10), 4-ft logs are cooked in digesters with a mixture of sodium sulfite and sodium bicarbonate. The digester is evacuated for a short period. Following evacuation, the digester is filled with heated liquor. After treatment under pressure, the logs are dumped in a pond and ground on conventional wood grinders. The stock is then directed through a bull screen, refiners, primary and secondary screens, thickeners, and then to the paper mill.

## PAPER MANUFACTURE

Pulp, produced by any of the foregoing or miscellaneous processes, is made into paper in a mill that is at the same location as the pulp mill (referred to as an integrated mill), or it is dried and shipped to a paper mill located remotely from the pulp mill. Paper mill layouts vary somewhat, based on the pulps used and grades of paper produced. Figure 11 shows the basic operations carried on in a typical paper mill. They are generally divided into stock preparation, paper machine, and finishing operations. The stock preparations can be further divided into stock proportioning or blending and mechanical treatment. Paper machine operations can also be subdivided into wet end and dry end.

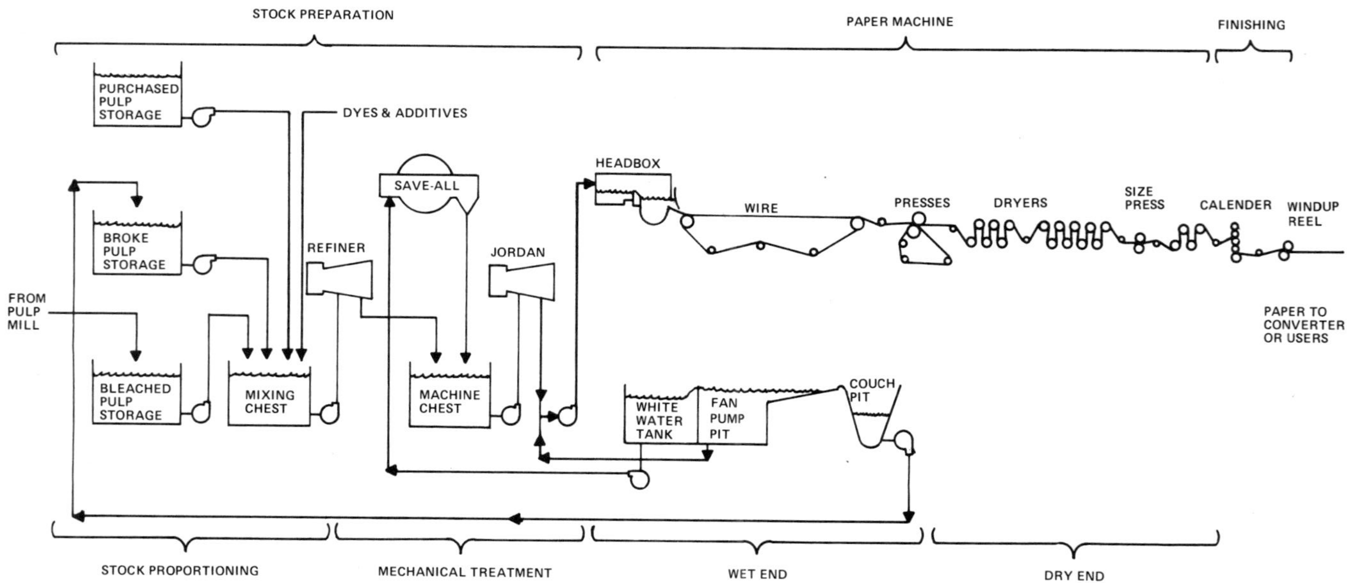


**Figure 11.** Simplified flow diagram of a typical paper mill.

## Stock Preparation

Most pulp cannot be used for papermaking as it comes directly from the pulp mill. To obtain final desired qualities, pulp having different characteristics (sometimes purchased from another pulp mill) may be blended with it, and dyes and special additives are also included to achieve the specific color and physical properties of the sheet. These operations are usually referred to as stock proportioning or stock blending.

To impart mechanical strength to the final sheet, the pulp is refined in a variety of machines, typical of which are the refiners, jordans, and beaters. Basically, this operation consists of passing the pulp repeatedly between sharp moving bars that cut and abrade the fibers. The sets of bars can be adjusted to turn out various lengths of fibers with rougher or smoother edges. This improves fiber-to-fiber bonding, making it more uniform, more dense, less porous, or more transparent, depending on the kind of paper to be made.

Before going to the paper machine, the resulting pulp is screened and, in some cases, cleaned by passing through centrifugal-type cleaners to remove heavy particles of dirt that have not as yet been removed.

## Paper Machine

*Wet End.* The most common major component of the wet end portion of the paper machine on which the paper is formed is a fourdrinier, consisting mainly of a continuous fine screen, called a wire, on which the pulp suspension is spread. The wet end section varies in widths in accordance with the size of the machine and moves at different speeds as determined by the type of paper being produced. Most of the water drains at the top end of the wire to form a mat of fibers. The wire then passes over a series of vacuum suction boxes which draw more water from the wet mat through the wire. The wet paper leaves the fourdrinier machine at a consistency of about 20% (20% fiber and additives, 80% water).

Paperboard, which consists of several layers of paper, is usually made on a cylinder-type machine instead of a running flat screen. Pulp is pumped to several vats. In each vat, a cylinder covered with a fine wire screen turns at the same speed as the rest of the machine. Water passes from the vat through the screen and is removed from the center of the cylinder, leaving a mat of paper on the wire screen. A continuous felt blanket in contact with all cylinders picks up layers of paper, forming a laminated, layered board. A good grade of pulp is frequently used in the end cylinders to give the outside of the board a good appearance. The second sheet is often made of a slightly inferior grade, while the inner layers are normally reclaimed paper of poor grade.

*Dry End.* After leaving the wet end section of the paper machine, the wet end paper is sent to the presses where it is supported by endless woven or synthetic loops called felts. The paper on top of the felts is then passed between heavy press rolls to press out as much water as possible. The paper leaves the press section at approximately 35% consistency on conventional press sections. The rest of the water is then evaporated on steam-heated rolls located in the dryer section. Endless felts again carry the paper through and press it against steam-heated rolls on opposite sides.

## Finishing

The dried paper then passes on to the finishing stage of the process. Most papers go through one or more additional processes, one of which is calendering. This process consists of ironing the paper between heavy, polished steel rollers, giving it a much smoother surface. Some paper is wound in large rolls as it comes from the calenders. These are later rewound and cut into smaller rolls or sheets as required by the user. Some papers are produced specifically for further processing by converter plants which make envelopes, milk containers, grocery bags, paper cups, and many other consumer products.

Paperboard is used for corrugated cartons, folding boxes for frozen foods, and many other items. Large quantities of pulp for some types of paperboard are made up of repulped wastepaper from which ink and other impurities have been removed.

# A

**AAW:** *See* ACTIVE-ALKALI-TO-WOOD-RATIO.

**A-FLUTE:** The corrugated section of a paperboard box that is approximately 3/16-inch thick; specifically, the triangular shaped fold(s) of this dimension in the corrugating medium, averaging 34-36 per linear foot. Corrugations of other dimensions and counts per foot have differing designations. *See* B-FLUTE, C-FLUTE, and E-FLUTE.

**A-LIFT:** When cutting paper, the maximum number of sheets placed at one time under the knife of the cutting machine.

**A-SIZE PAPERS:** Paper that has been trimmed such that the ratio of length to the width is equivalent to the ratio between the diagonal and the side of a square.

**ABACA:** Fiber, also known as *manila hemp* or *manila fiber*, that is prepared from the outer sheath of the stems of a variety of plantain and used to manufacture such products as rope papers, tea bags, etc.

**ABACA PULP:** A pulp made from abaca fiber in both wet and dry form.

**ABERRATION:** Image imperfection produced by an optical system.

**ABIETIC ACIDS:** An isomeric form of resin acid found in wood, such as spruce and pine.

**ABRADE:** To scrape off or wear away paper and paperboard as well as the equipment used in its manufacture and conversion into various finished products. *See* ABRASION.

**ABRASION:** The process of wearing out and/or scuffing paper product surfaces when sheets contact and rub against each other, or against equipment in the various stages of production and finishing. *See* SCUFFING. Also refers to the wear of papermaking and converting equipment, e.g., that caused on forming fabrics by abrasive filler clays and additives in the paper stock, and that caused to printing press components, particularly plates, by the finished paper as it moves through the process at high speeds.

**ABRASION INDEX:** A measure of the relative abrasiveness of substances such as clays and other materials. It is an indication of the degree a material will abrade or "wear out" other substances. *See also* ABRASION NUMBER.

**ABRASION NUMBER:** The relative abrasiveness of clays and similar minerals. It is generally determined by noting the loss in weight of a piece of fourdrinier machine wire (forming fabric). In North America, the apparatus currently in almost universal use for this purpose is the Valley Abrasion Tester which reports values in abrasion number. *See also* ABRASION INDEX.

**ABRASION RESISTANCE:** The ability of a paper product to withstand abrasion. It is measured by determining the degree and rate that a sample loses weight under the specific rubbing action of an abrading substance, such as an eraser or various abrasive papers.

**ABRASIVE BACKING PAPERS:** Well-sized kraft papers used as backing for sand, aluminum oxide, silicon carbide, garnet, and other abrasive materials used for sanding and smoothing of wood, metals, etc. Glue is normally used as the bonding agent.

**ABRASIVE FIBER:** A chemically treated fiber used in the manufacture of base sheets for abrasive grit used on sanding equipment. It is very hard and strong but maintains relatively high elasticity after the vulcanization type of treatment.

**ABRASIVENESS:** A characteristic of finished paper that indicates its potential for abrading other materials as it moves in contact with them. The degree of abrasion and wear is generally related to additives in the paper stock, as well as any "external" coatings, rather than the fibers themselves.

**ABRASIVE PAPERS:** Heavy, strong, hard-sized rope or kraft paper, coated on one side with glue and some abrasive material, such as flint, garnet, aluminum oxide, or silicon carbide, in varying degrees of fineness to suit the work to be done. For wet use, the adhesive may be a varnish-resin compound. It is sold under various trade names as sandpaper, flint paper, emery paper, garnet paper, carborundum paper, etc. *See* ABRASIVES.

**ABRASIVES:** (1) *See* ABRASIVE PAPERS. Refers to the entire class of these types of papers. (2) A reference to abrading materials such as flint, garnet, aluminum oxide, silicone carbide, etc. that are used to coat abrasive papers along with glue, and to make up a substance comprising the gritty particles in a pulpwood grindstone.

**ABSOLUTE ALARM:** An alarm caused by detection of a process variable of equipment operation that has exceeded a set of prescribed high or low limit conditions.

**ABSOLUTE HUMIDITY:** A measure of water vapor in each cubic foot or other unit volume of a gas or gases, whether isolated or as part of the atmosphere, and expressed in various units such as dew point, grains of water per pound of air, or pounds of water per million standard cubic feet. Sometimes used interchangeably with *vapor pressure* due to known relationship between absolute pressure and the pressure of water vapor in air. *See also* RELATIVE HUMIDITY.

**ABSOLUTE PRESSURE:** A measurement of pressure relative to a perfect vacuum. It is also the sum of the gauge pressure and the atmospheric pressure.

**ABSOLUTE TEMPERATURE:** Temperature measured from absolute zero on the absolute scale. *See* ABSOLUTE ZERO TEMPERATURE.

**ABSOLUTE ZERO TEMPERATURE:** The zero point on the absolute temperature scale; the hypothetical point of total absence of heat and molecular motion, theoretically equivalent to –273.18°C or –459.72°F.

**ABSORB:** *See* ABSORBENT.

**ABSORBENCY:** That property of pulp, paper and its constituents and products which permits the entrainment and retention of other materials they contact, such as other liquid, gaseous, and solid substances.

**ABSORBENT:** A material that takes up moisture or liquid when a physical or chemical change occurs during the process, e.g., calcium chloride (solid absorbent) and sulfuric acid (liquid absorbent).

**ABSORBENT PAPERS:** Papers having the specific property of absorbing water, ink, and other liquids. They are characterized by being soft, loosely felted, unsized, and spongy, and are known functionally as blotting paper, filter paper, toweling, matrix paper, and impregnation paper for resins and vegetable parchment. *See also* BIBULOUS PAPER.

**ABSORPTION:** The taking up of one substance by another.

**ABSORPTION TOWER:** A process vessel commonly used in the production of calcium bisulfite cooking acid in which sulfur dioxide ($SO_2$) is absorbed by water and reacted with limestone. *See also* ACID TOWER.

**ABSORPTIVE CAPACITY:** A measure of the ability of pulp, paper, and paperboard to imbibe or take up liquids with which they are in contact. *See* ABSORBENCY.

**ABSORPTIVITY:** The ratio of incident radiant energy absorbed by the surface of a body at a given temperature to that absorbed by a *black body* at the same temperature.

**A.C.:** *See* ALTERNATING CURRENT.

**ACCELERATED AGING TEST:** An evaluation of paper, products, and raw materials in the pulp and papermaking process where an artificial fast method of aging is used to determine the effects of time on their chemical and/or physical properties in order to predict useful life span.

**ACCELERATING AGENT:** A material used to hasten a particular act or process. It is often used in tests to determine the effect of time on chemical and/or physical properties or processes. Heat, light, moisture, and certain chemicals are among the most common ones.

**ACCEPTANCE SAMPLING:** The method of obtaining a representative portion or quantity of pulp, paper, paperboard, raw materials, and process materials for specifications testing. *See* ACCEPTANCE TEST.

**ACCEPTANCE TEST:** A laboratory or on-site quality control evaluation made on pulp, paper, paperboard, raw materials, and process materials to

determine whether they meet prescribed specifications establishing their satisfaction for use.

**ACCEPTED CHIPS:** Chips screened for a specific size and thickness then sent to storage bins in the pulp mill for subsequent processing in digesters or mechanical pulping equipment. Those chips not accepted are called "rejected" chips and are recycled for rechipping or disposal by burning in power boilers and by other means.

**ACCEPTED STOCK:** Pulp stock that is sent on for further processing in the papermaking operation after passing through screens or cleaners. The stock not "accepted" is called "rejected" stock and is recycled for recleaning and/or disposal.

**ACCEPTS:** That portion of material (chips, stock, water, etc.) sent on for further processing by some type of screening, cleaning, classification, etc. That portion of material not "accepted" is recycled for further processing or disposal. *See* REJECT REFINING, REJECTS, and REJECTS REFINER.

**ACCESSIBILITY:** That fraction of a cellulose system whose intercellular structure is such that it allows quick penetration and subsequent reaction to chemicals. It can be measured by subjecting samples to standard test reagents under specified conditions of reaction.

**ACCORDION PLEAT:** A method of multi-folding a sheet of paper so that each successive fold is parallel and in opposite direction.

**ACCOUNT:** Term used by pulp and paper sales organizations to designate their clients.

**ACCOUNT BOOK PAPER:** Paper used primarily for pen and ink record purposes. It is usually made from cotton fiber, bleached chemical pulp, or a mixture and possesses the physical properties of good strength, tearing resistance, erasability, water and ink resistance, uniformity of surface, color, and smoothness. *See also* LEDGER PAPER.

**ACCUMULATOR:** Pressure vessels in a pulp mill in which cooking acid or liquor is held in storage during fortification by $SO_2$ from digester relief gas or by the addition of liquid $SO_2$ in the sulfite pulping process.

**ACCUMULATOR ACID:** That acid held in an accumulator in the sulfite pulping process.

**ACETALDEHYDE:** A by-product chemical formed by the fermentation of sugars when alcohol is produced from spent sulfite cooking liquor.

**ACETALS:** The various compounds formed when aldehydes combine with alcohols under the higher temperatures existing in spent sulfite cooking liquors.

**ACETATE:** A term used to describe salts of acetic acid. It is commonly used in reference to the viscose solution resulting from the dissolving of pulp in acetic acid, acetic anhydride, and sulfuric acid.

**ACETATE FIBER:** A synthetic fiber extruded from a viscose solution made by dissolving a highly purified quality pulp in acetic acid, acetic anhydride, and sulfuric acid.

**ACETATE PULP:** A highly purified quality pulp made especially to be dissolved in acetic acid, acetic anhydride, and sulfuric acid to make acetate rayon.

**ACETATE RAYON:** A viscose solution made by dissolving a highly purified quality pulp in acetic acid. It is also used to refer to the synthetic fiber extruded from it, called acetate fiber or cellulose acetate.

**ACETIC ACID:** An acid formed from the acetyl groups of carbohydrates by acid hydrolysis during the cooking of wood in the sulfite process. It is also the acid used to make cellulose acetate.

**ACETIC ANHYDRIDE:** A chemical used as a dilutent in the manufacture of cellulose acetate or acetate rayon.

**ACETONE:** A volatile, fragrant liquid ketone formed during a sulfite-type cook and present in the spent liquor.

**ACETYLATION:** The treating of cellulose fibers with acetic acid, acetic anhydride, and sulfuric acid to form cellulose acetate.

**ACETYL GROUPS:** Types of softwood cellulose that are removed by a neutral or alkaline precook prior to a sulfite cook.

**ACHROMATIC COLOR:** (1) The color obtained in paper and paperboard by using dyes that will cause it to refract white light without breaking it up into its component colors. (2) Possessing no hue or without color as related to black or white.

**ACID:** Generally, any of various typically water-soluble and sour compounds used in pulp and paper mills capable of reacting with a base to form a salt. It frequently refers to the cooking liquors used in the sulfite pulp mills.

**ACID ALUM:** $Al_2(SO_4)_3 \cdot 14H_2O$, $Al_2(SO_4)_3 \cdot 18H_20$ or a mixture of these hydrates. Papermaking chemicals commonly used for precipitating the rosin size onto pulp to impart water resistant properties to the paper. It characteristically produces low pH in solution, sometimes caused by residual acid from which it is manufactured.

**ACID DYES:** Synthetic organic dyes and pigments originally derived from aniline which are in the form of sodium salts of dye acids and are used for surface coloring. They are set onto the fiber by the addition of size and alum to the paper furnish because they have little affinity to cellulose. They are characteristically very soluble and have a good fastness to light.

**ACID DYESTUFF:** Acid-base compounds made from acid dyes. These dyes are usually in the form of the sodium salts of acid dyestuff. Sometimes used interchangeably with and meaning the same as *acid dyes.*

**ACID FILLING PUMP:** A specially designed pump, usually a single stage, double-suction, direct-connected, centrifugal type, used in the sulfite pulping process to pump hot accumulator acid into the chip-filled digester, develop a hydrostatic head in the digester, and pre-circulate the hot accumulator acid through the digester.

**ACID FREE:** A term describing paper and paperboard that can be used with anything that may be tarnished or otherwise harmed by acid. It also refers to pulp and any other materials and substances used in the pulp and papermaking process to which acid would be harmful to itself or to anything with which it is used or comes in contact.

**ACID FREE PAPER:** Type of paper that does not contain any acid substances that may contaminate or affect any acid sensitive materials with which they may come in contact. *See also* ALKALINE PAPERMAKING and ANTIRUST PAPER.

**ACID GRINDING:** The addition of acid to a pulpwood grinder shower, or the impregnation of logs with acid and subsequently grinding to increase pulp strength and reduce specific grinding energy. Recently, instead of acid, caustic soda is used. *See* ALKALINE GRINDING.

**ACID INSOLUBLE ASH:** That portion of the ash resulting from the combustion of paper that will not dissolve in a fixed concentration of sulfuric acid ($H_2SO_4$) under specified conditions, and is an indication of the acid resistance of additives and fillers used to make up the sheet.

**ACID INSOLUBLE LIGNIN:** The lignin portion of pulp that will not dissolve in sulfuric acid ($H_2SO_4$) of fixed concentration (72 percent) under specified conditions.

**ACID MAKER:** The sulfite pulp mill employee responsible for the operation of the acid plant where the cooking acid is manufactured. He or she keeps the strength, quality, and supply to the cooking operation consistent.

**ACID MAKING:** In the pulp and paper industry, this term refers to all the operations involved in the manufacture and refortification of cooking chemicals used in the sulfite pulping process.

**ACID NUMBER:** The amount of free acids in fats, oils, waxes, and resins expressed as number of milligrams of potassium hydroxide (KOH) required to neutralize one gram of the material being tested.

**ACID PLANT:** The area of a sulfite pulp mill where cooking chemicals are prepared and refortified.

**ACID PULPING:** The conversion of chips to pulp by cooking in acid sulfite liquor under pressure.

**ACID RESISTANCE:** The ability of paper and paperboard or any other materials and substances used in the pulp and papermaking process to be used or come in contact with acid or anything containing acid without causing any harmful effects to itself.

**ACID RESISTANCE TEST:** A test to determine to what degree paper or paperboard will resist acid solution penetration and/or attack.

**ACID RESISTANT PAPER:** Type of paper manufactured with special ingredients or colored with dyes not affected by acidic conditions.

**ACID SIZE:** A rosin size used to impart water resistant characteristics to paper and paperboard. It is made up largely of unsaponified, unemulsified rosin.

**ACID STABLE SIZE:** A wax-type paper sizing material that remains in a fluid state when used in acidic papermaking process. *See also* WAX EMULSION and WAX SIZE.

**ACID SULFITE:** Cooking liquors made by the absorption of $SO_2$ gas by alkaline-base solutions of calcium (Ca), sodium (Na), ammonium ($NH_4$), or magnesium (Mg).

**ACID SULFITE DISSOLVING PULP:** Bleached sulfite pulp of high alpha-cellulose content used in the manufacture of rayon, cellophane, cellulose acetates, cellulose nitrates, cellulose esters, cellulose ethers, etc.

**ACID SULFITE PROCESS:** A pulp manufacturing process using a mixture of bisulfite salt and excess sulfurous acid as the cooking chemical for the digestion of wood to pulp. Also, its associated operations of manufacturing, recovery, and refortification of these cooking chemicals, which can be calcium-, ammonia-, magnesium-, or sodium-based.

**ACID SULFITE PULPING:** *See* ACID SULFITE PROCESS.

**ACID TOWER:** A pulp mill process vessel commonly used in the production of calcium bisulfite cooking acid in which sulfur dioxide ($SO_2$) is absorbed by water and reacted with limestone. *See also* ABSORPTION TOWER.

**ACID TOWERS:** Tall metal tanks used in the acid plant of a sulfite pulp mill to make up cooking liquor by absorbing $SO_2$ in an excess of Ca, Na, $NH_4$, or Mg alkaline-base solution by passing the gas countercurrently upward through a downfalling solution over packing material.

**ACID TREATMENT:** The conditioning of paper, pulp, board, or any other materials in the pulp and papermaking process which causes a reaction to acid that produces beneficial characteristics in the final product.

**ACID WASH:** The conditioning of paper, pulp, board, or any other materials in the pulp and papermaking process in order to remove any undesirable caustic-type contaminants by reacting acid with them and then washing the acid away. This is usually with water, which results in a product free from these undesirable caustic components.

**ACIDIFICATION:** The act of making or converting anything in the pulp, paper, and board-making process into an acid or acidic conditions.

**ACIDITY:** The condition in paper which results in an acid solution (caused by the concentration of hydrogen ions exceeding those of hydroxyl ions) when the paper is treated or extracted with water.

**ACID PROOF BRICKS or ACID PROOF LINING:** Special bricks of such composition as to make them resistant to acid solutions. They are used as lining in sulfite digesters to protect the metal body from the corrosive effect of acid-base cooking liquors. The inner lining of sulfite digesters made of acidproof brick to protect the metal digester body.

**ACID PROOF PAPER:** Paper that is not affected physically or chemically by acid and is used on or with acid-containing materials.

**ACOUSTICAL ABSORPTION COEFFICIENT:** That degree of intensity of incident sound absorbed by a material, such as acoustical paperboards, measured at a specific frequency. Also referred to as *sound absorption coefficient.*

**ACOUSTICAL BOARD:** A paperboard used primarily as a wall covering to absorb sound. It also has good thermal resistance. It is made in various thicknesses by gluing together thinner boards made from raw materials such as wood, straw, or bagasse. *See also* ACOUSTICAL INSULATION.

**ACOUSTICAL INSULATION:** Sound absorbing pulp and board that can also be used for insulating against heat loss. *See* ACOUSTICAL BOARD.

**ACRE FOOT:** A term used in measuring the volume of water, equal to the quantity of water required to cover one acre one foot in depth, or 43,500 cubic feet. It is used in reference to holding ponds for effluent from pulp and paper mills.

**ACROSS MACHINE DIRECTION:** *See* CROSS DIRECTION.

**ACROSS THE GRAIN:** *See* CROSS GRAIN.

**ACRYLIC BINDERS:** A binding material of acrylic base that provides high bonding strength for coatings used on either dry or wet paper and boards.

**ACRYLIC COATINGS:** Paper and board coatings using acrylic-base binders to provide high bond strength, water resistance, and light stability.

**ACRYLIC DISPERSIONS:** Acrylic-base suspensions used to make up binders and coatings for paper and board. They provide good bonding strength, water resistance, and light stability.

**ACRYLIC FIBER:** Generic name for a manufactured fiber in which the fiber-forming substance is any long chain polymer composed at least 85 percent by weight of acrylonitrile units. It is used in a broad range of apparel and pile fabrics, blankets, carpets, and industrial applications.

**ACRYLICS:** Group of binders, coatings, and adhesives formulated to deform for good contact with surfaces of pigment and cellulose fiber surfaces. They provide high bond strength on dry paper or boards, as well as on wet products. They also have good water resistant properties, with an absence of odor on dried coating and excellent light stability.

**ACRYLIC LATEXES:** Polymer and copolymer particles dispersed in water and used as coatings on board to provide good water resistance, high gloss ink and varnish hold-out, and good glueability.

**ACS:** *See* AMERICAN CHEMICAL SOCIETY.

**ACTINIC RAYS:** Type of light rays that will cause paper and plastic film chemical components to undergo changes.

**ACTION PAPER:** Paper formulated to make copies by creating internal blue-purple images in response to pressure applied to it which ruptures chemically filled cells blended with the pulp fiber when the sheet was made up. It eliminates the need for carbon paper. Also known as *carbonless paper. See also* NO CARBON REQUIRED (NCR).

**ACTIVATED CARBON:** A highly adsorbent powdered or granular carbon usually made by carbonization and activation, but also made as a byproduct of the recovery process in a sulfate or kraft pulp mill. It is used in mills chiefly for purifying by adsorption.

**ACTIVATED SLUDGE:** The settled solids after treatment of pulp and paper mill effluent with high-purity oxygen. The solids are collected at the bottom of a clarifier tank after mixing with oxygen in an aeration tank. Part of the sludge is recycled back to the aeration tank to maintain high solids concentrations and efficient treatment.

**ACTIVATED SLUDGE PROCESS:** The treatment of pulp and paper mill effluent with high-purity oxygen instead of air to oxygenate the biological mass. *See* ACTIVATED SLUDGE.

**ACTIVE ALKALI:** The caustic soda (NaOH) and sodium sulfide ($Na_2S$) content in sulfate cooking liquor expressed as $Na_20$. In the soda process, it is the caustic soda (NaOH) expressed as $Na_2O$.

**ACTIVE-ALKALI-TO-WOOD-RATIO (AAW):** The relationship between the caustic soda and sodium sulfide in fresh cooking liquor to the amount of wood being cooked with it. It is calculated as weight per ton of oven dry chips.

**ACTIVE CHLORINE:** That portion of chlorine in chemical compounds available to do useful work in the chlorination of mill water supply and in the bleaching of pulp. *See also* AVAILABLE CHLORINE.

**ACTIVE SULFUR:** Sulfur and sulfur compounds in the sulfate pulping and chemical recovery process that are converted to sodium sulfide.

**ACTIVITY:** The active alkali content divided by the total alkali in alkaline cooking liquors, all expressed as $Na_2O$ and calculated on a percentage basis. For sulfate cooking liquor the formula is as follows:

$$\left(\frac{NaOH + Na_2S}{NaOH + Na_2CO_3 + Na_2S + \frac{1}{2}Na_2SO_3} \times 100\right)\%$$

For soda cooking liquor:

$$\left(\frac{NaOH}{NaOH + Na_2CO_3} \times 100\right)\%$$

**ACTUAL WEIGHT:** The absolute or actual weight of a ream or reams of paper, or a package or packages of paperboard, as compared with nominal weight, which refers to the "billing rate" per 1,000-sheet reams regardless of whether or not they weigh a stated amount (within certain tolerances). *See also* NOMINAL WEIGHT.

**A.D.** *See* AIR DRY.

**AD MIXTURE:** Something added by mixing, which is very common in the stock proportioning phase of papermaking. It is essentially a blending or mixing of dyes, chemicals, and additives with a combination of various types and species of pulp fiber.

**ADDING MACHINE PAPER:** A sized paper with a fuzz- and lint-free surface of medium finish and absorbency used on adding and tabulating machines. It is of writing and book paper quality and made in various size rolls to fit the machine on which it is to be used.

**ADDITION POLYMERS:** Long chain link chemical compounds that make up the constituents of wood, pulp, chemicals, and additives used in the paper industry.

**ADDITIVES:** Chemicals or any other material added to pulp stock slurry to impart special physical and visual properties to the paper sheet or board made from it. *See* PAPER ADDITIVES.

**ADDRESS LABEL PAPER:** Paper made from chemical pulp or a mixture of mechanical and chemical pulp with fairly good strength used to make labels for magazines, periodicals, etc. Dyes used to color it are of the acid- and alkali-resistant types to prevent discoloration by the adhesives used.

**ADHESIVE DEMAND:** The adhesive required to achieve specific properties of paper and paperboard binders in coating as determined by wax *pick test.*

**ADHESIVE FELT:** A heavy gray paper made from inferior low-grade stock and used as a backing to stiffen leather or leather-type products such as are found in briefcases, billfolds, and cap visors.

**ADHESIVE GLASSINE TAPE:** Gummed glassine paper made up in narrow-width, small-diameter rolls and used to mend tears in printed paper sheets such as currency, books, documents, etc.

**ADHESIVE MIGRATION:** The intrusion of surface adhesive coating of a sheet into the base structure. It is considered a defect when the intrusion is beyond desired limits.

**ADHESIVE PAPER:** Paper coated with an adhesive, the type depending on specific end uses (cold or hot setting, liquid activated, pressure sensitive, etc.). Also refers to paper that is to contain an adhesive coating. *See also* GUMMED PAPER, GUMMING PAPER.

**ADHESIVE RATIO (AR):** The amount of adhesive required to bind pigment to a predetermined coating. Strength varies with the type of pigment and is expressed as the ratio in parts by weight of adhesive required to bond 100 parts of pigment to some determined coating strength.

**ADHESIVE TAPES:** Paper- and/or cloth-based tapes coated with an adhesive. They are generally divided into two categories: (1) gummed tapes activated by water or by an organic solvent; and (2) pressure sensitive tapes that are normally tacky, and are based on natural or synthetic rubber in combination with resin and plasticizers.

**ADHESIVE TENSION:** The attraction of liquid molecules for fiber as in surface coating for a paper sheet.

**ADHESIVES:** Substances used in the paper industry to add to or coat materials and which can be activated by pressure, heat, water, other solvents, or other means so they will adhere to each other. They are commonly used in coatings, beaters, paper machine wet ends, board manufacture, and converting operations.

**ADIABATIC CONDITIONS:** A description of processing conditions under which the system is perfectly insulated so that no heat is exchanged with the surroundings. It is a papermaking objective continually pursued but seldom achieved.

**ADJUSTABLE SPEED:** A term used to describe pulp and papermaking equipment, the speed of which can be varied gradually over a considerable range, but when adjusted remains practically unaffected by load. For example, a shunt motor is designed for a considerable range of speed variations, and it is used to drive adjustable speed pumps and other pulp and paper mill machinery.

**ADJUSTING COLOR:** A dye or additive that is used in combination to achieve a desired shade of color of a sheet of paper.

**ADSORB:** *See* ADSORBENT.

**ADSORBABLE ORGANIC HALIDES (AOX):** Refers to a test that combines all types of *chlorinated organics* into a single value number. The *AOX* test has become a popular and fairly accurate measure of a bleached pulp mill's environmental performance.

**ADSORBATE:** A substance that is taken up by pulp, paper, board, or other pulp and papermaking materials, such as water, ink, additives, coatings, etc.

**ADSORBENCY:** That property of a material that allows other substances such as gases and fluids to form a coating on its surface.

**ADSORBENT:** A material that takes up moisture and liquid without any physical or chemical changes taking place in the process. Activated bauxite, alumina, charcoal, and silica gels are good examples. Also referred to as *sorbent.*

**ADSORPTION:** The coating of substances by thin films of gases or fluids with which they come in contact.

**ADSORPTIVE CAPACITY:** The degree to which pulp, paper, board, and other materials take up moisture without any physical or chemical changes to basic structure or composition.

**ADSORPTIVITY:** Having the ability to take up moisture and liquid without physical or chemical changes.

**ADVANCED WASTE TREATMENT:** The processing of pulp and paper mill wastes beyond primary mechanical and secondary biological treatment. Advanced treatment at the tertiary level (mechanical and chemical) is used generally to meet existing or anticipated regulations and conditions that are particularly stringent or demanding. These methods of treatment may also be used to render an effluent reusable in some way. *See* TERTIARY TREATMENT.

**ADVERTISING STICKERS:** Printed or lithographed paper in small sizes used for advertising, with one side coated with a gum-type adhesive for sticking purposes.

**AERATE:** To carry on the process of *aeration*, i.e., creating intimate contact between air and a liquid.

**AERATION:** The bringing about of intimate contact between air and a liquid by spraying the liquid in the air, bubbling air through the liquid (diffused aeration), or by agitation of the liquid to promote surface absorption of air (mechanical aeration).

**AERATION BASIN:** A depressed area or pond area where air is introduced into liquid, as in a paper mill secondary effluent water waste treatment system. Also referred to as *aeration lagoon*.

**AERATION LAGOON:** *See* AERATION BASIN.

**AERATION TANK:** A vessel for carrying out the process of aeration.

**AERATOR:** The device that disperses air during the aeration process of mill effluents.

**AEROBIC:** Living or active only in the presence of oxygen, as applied to organisms that cause slime buildup in pulp stock storage tanks in mills.

**AEROBIC PROCESS:** A biological process that takes place in an atmosphere where oxygen exists.

**AERODYNAMIC DRY FORMING:** A paper and paperboard sheet manufacturing process using the principle of introducing fibers into a high velocity air stream which deposits the fiber on a moving wire (forming fabric) and recirculates all of the air.

**AERODYNAMIC FORMERS:** Paper and paperboard manufacturing machines using the principle of a high velocity air stream to carry the fibers and deposit them on a wire (forming fabric) to form a sheet.

**AERODYNAMIC FORMING:** A method of producing dry formed paper by introducing fiber into a recirculated high velocity air stream, which deposits it onto a moving wire (forming fabric) and recirculates the air.

**AEROSOL DEVELOPMENT:** The process of producing a fine suspension of solids or liquid particles in gas as in smoke and stack gas from power and recovery boilers.

**AFC:** *See* AMERICAN FOREST COUNCIL.

**AFFINITY:** The attractive force of pulp and papermaking materials that causes them to combine and to remain in combination with each other. Affinity is a desirable characteristic in dyes and additives during the sheet formation stage.

**AFTER DRYER:** A dryer or dryer section located after the coating operation on a paper machine.

**AGAINST THE GRAIN:** The paper sheet dimension that is parallel to the side-to-side direction of the machine on which it is made. *See* CROSS GRAIN.

**AGALITE:** A gray, hydrated magnesium silicate used as a natural filler in papermaking. *See* ASBESTINE.

**AGATE MARBLE PAPER:** Paper printed with blended color bands to simulate agate, or be agate-like in appearance. It is made primarily from chemical pulp and most commonly used as end leaves in bookmaking.

**AGAVE:** An aloe plant sometimes used as a source of long papermaking fibers similar to sisal hemp. *See* ALOE.

**AGE:** To condition paper and pulp in such a manner as to approximate the length of time it takes to diminish certain physical properties such as strength and brightness. Also refers to the alteration and general deterioration of paper and board characteristics during a period of time. *See also* AGING.

**AGENT:** Something that produces or is capable of producing an effect. It is commonly used when referring to dyes and additives in papermaking. It also refers to chemicals in other parts of the process, such as oxidizing and reducing agents.

**AGGLOMERATE:** The reaction to forces in dry pigment coating that restricts dispersion of particles. The reaction is strong in such pigments as titanium dioxide and zinc oxide, but is weak in clays and talcs. Agglomeration is the actual exertion of these forces.

**AGGREGATE:** The result of the exertion of agglomeration forces in dry pigment coatings. Aggregates are classes of dry pigment coatings that are characterized by strong agglomeration forces. *See also* AGGLOMERATE.

**AGING:** Irreversible alteration, generally deterioration, of the characteristics of paper or board produced in the course of time, such as folding endurance, color, etc. It is commonly associated with the diminishing of brightness of pulp and paper with age or the developing of a yellow color. Aging is sometimes referred to as *brightness reversion* or *yellowing.*

**AGING RESISTANCE:** The ability of pulp or paper to maintain certain physical properties, such as strength and brightness, over long periods of time. Sometimes referred to as *stability* or *permanence.*

**AGING TEST:** An evaluation of products and raw materials in the pulp and papermaking process based on an accelerated artificial aging procedure to determine the extent that physical and chemical properties will change with time.

**AGITATE:** To keep the contents of a storage tank in motion and well mixed by mechanical means.

**AGITATION:** The process of agitating by use of mechanical, pneumatic, or other means.

**AGITATOR:** Mechanical or other equipment provided in or on storage chests to keep contents in motion and well mixed. Agitators are generally classified as internal or external.

**AICHE:** *See* AMERICAN INSTITUTE OF CHEMICAL ENGINEERS.

**AIR:** That medium which normally exists in the atmosphere and theoretically consists (by weight) of 23 percent oxygen and 77 percent nitrogen; or (by volume) 20.7 percent oxygen and 79.3 percent nitrogen. One pound at 62°F (barometer at 30 inches) occupies 13 cubic feet, and 56 cubic feet at this temperature contains 1 pound of oxygen.

**AIR BALANCE:** Ratio of supply to exhaust in any pulp and paper mill building, unit process, or air system.

**AIR BELL:** A blister defect on the wire (forming fabric) side of a sheet, caused by air that is not actually within the stock but rather trapped between the sheet and the felt during drying. *See also* BLISTER.

**AIR BLADE:** A metal blade used in a method of coating paper that extends across the paper machine. It contains a slot which distributes air so that it removes excess coating and smooths the surface. *See also* AIR DOCTOR and AIR KNIFE COATER.

**AIR BLOWER:** A mechanical device that moves air into desired areas such as in air dryers. It is similar to air fans in housings except higher volumes of air are handled at higher velocities.

**AIR BRUSH:** A device used for coating a sheet of paper in which air pressure is used to spray coating onto the web. In the printing industry, it refers to a pencil-like apparatus for spraying liquid colors by means of compressed air. It is used to obtain tone or graduated tonal effects in artwork.

**AIR BRUSH COATER:** A type of paper coating equipment which uses the principle of air pressure to atomize the coating mixture and spray it on the web.

**AIR BRUSH COATING:** A method of coating paper on one side using pressurized air. The coating is distributed by a thin, flat jet of air coming from a slot in a metal blade extending across the paper machine. Sometimes called *air knife coating.*

**AIR BUBBLES:** Light spots in a sheet caused by bubbles of air that form in the stock on a fourdrinier wire (forming fabric) and are entrained in the stock as it cascades in trays and channels. The term also refers to bubbles of air formed in the coating operation and the defective imprint that is caused by the bubble bursting. *See also* BLOW, BUBBLES, FOAM MARKS.

**AIR CAP:** The enclosed area of an air dryer that contains the heated air through which the paper sheet is carried.

**AIR-CONDITIONER:** Units used to control temperature and humidity in paper testing laboratories, storage rooms, control centers, and other areas in pulp and paper mills.

**AIR CONTENT:** The quantity of air, usually expressed in percentage, that exists in any air-containing material such as wood, pulp, pulp slurries, paper, process water, chemicals, additives, fillers, dyes, etc.

**AIR COOLED:** Anything whose temperature has been reduced by the use of lower temperature air such as paper, fluffed pulp, etc., prior to further treatment or handling.

**AIR COOLING:** The process of making something air cooled.

**AIR CURTAIN:** A blanket of temperature-controlled air that is created and maintained between mill areas, usually across an entrance to prevent loss of heat from the area or building structure.

**AIR CUSHION HEADBOX:** A pressurized type of paper machine headbox that has a cushion of pressurized air called an air pad above the stock pond level. *See also* AIR PAD and PRESSURIZED HEADBOX.

**AIR DECKLE:** The untrimmed feather side edges of a sheet of paper made on the fourdrinier or cylinder wire (forming fabric) by means of a jet of air. *See also* JET DECKLES.

**AIR DOCTOR:** A long, thin jet of air acting as a flexible doctor blade which meters off the controlled excess of previously applied coating on a paper coater, leaving the correct weight of coating on the sheet. *See also* AIR BLADE.

**AIR DOCTOR MOTTLE:** A paper coating defect caused by excessively high solids or viscosity, or poor adjustment of the doctor.

**AIR DRIED:** Reference to pulp and paper when dried artificially with the use of heated air in appropriate type dryers.

**AIR DRIED BOARD:** Paperboards which have been dried with air on festoon and other types of air dryers after being made on wet-type board machines.

**AIR DRIED PAPER:** Paper that has been artificially dried by contact with ambient or heated air, either in suspension or over some type of skeleton dryer.

**AIR DRIED WRITINGS:** Writing papers that have been dried by contact with heated air in festoon or other types of air dryers after wet sheet formation on the paper machine.

**AIR DRY (A.D.):** Refers to the weight of moisture-free pulp or paper plus 10 percent moisture based on a traditional assumption that this amount of moisture exists when the pulp or paper comes into equilibrium with the atmosphere, which in actuality is dependent on the conditions of the atmosphere to which it is exposed. Air dried weight is determined by dividing the oven dried weight by a factor of 0.9.

**AIR DRYER:** Pulp and paper drying machine that dries a pulp or paper sheet by contact with air over skeleton dryers.

**AIR DRYING:** Using air to dry a sheet of pulp or paper or other papermaking material.

**AIR ENTRAINMENT:** The trapping of air in white water or pulp stock slurries caused by mechanical agitation in the process, e.g., by equipment in motion such as pump impellers and agitators. It is the primary cause of foam which impedes operation of screens, filters, cleaners, wires, and other process equipment.

**AIR EXHAUST:** The outlet side of any pulp and papermaking process or machinery discharging air into the surrounding atmosphere.

**AIR FELT:** A process of forming a felt mat on a screen by drawing an air suspension of fibers through it.

**AIR FILTER:** Device used to remove impurities from air before it is used in the process and other pulp and paper mill equipment.

**AIR FILTRATION PAPER:** A type of paper used to remove undesirable suspended particulates from an air stream prior to being used in the process, machinery, or equipment where they may be harmful.

**AIR FLOAT CONVEYOR:** A moving blanket of air used to transport material such as paper, pulp, or chips from one location to another, or through a piece of process equipment such as a dryer.

**AIR FLOTATION:** A method of refining clay. A dry process where finely ground clay particles are fed into an air current under conditions such that particles of requisite fineness are transported by air to suitable collecting chambers, whereas the coarser lumps and/or the oversize mineral impurities are either separately rejected or else fed back into a pulverizer and reduced to the desired particle fineness.

**AIR HEATER:** Equipment used to preheat air before being used in the pulp and papermaking process. This can be done by the use of steam coils or by the use of hot gases such as those from power and recovery boilers with tubular-type heat exchangers.

**AIR HOOD:** A canopy located over the dryer section of a paper machine. It is used to confine and control vapor-laden air rising from the dryer section so that the heat may be recovered. This is done by using that heat to heat air entering the machine room. The "open" type hood is usually a canopy over the top of the dryers with side panels extending down to the journals of the top dryers. The "closed" type hood has side panels that extend to the paper machine room floor, and they may also extend into the basement.

**AIR IMPACTOR:** A metal blade used in one method of coating paper. It extends across the paper machine and has a slot to allow the distribution of air in such a manner that it spreads and smooths the coating material without removing any of it.

**AIR IMPINGEMENT:** A method of heat transfer in the paper industry using impinging air jets to provide high heat transfer rates, such as in the drying of lightweight paper grade of tissue and toweling. The sheet is usually dried on a steam-heated cylinder (*yankee dryer*) with drying rates augmented by the use of a round or slot-shaped, high velocity impinging hot air jet.

**AIR INTAKE:** The inlet side of any pulp and papermaking machinery using air for processing or in its operation.

**AIR JET:** Round or slot-shaped air supply ports designed to increase the velocity of an air stream where desirable in the pulp and papermaking process, such as in the drying of lightweight papers. It is augmented by the use of impinging hot air jets to provide high heat transfer rates.

**AIR KNIFE:** *See* AIR BLADE.

**AIR KNIFE COATER:** Type of coating equipment that operates on the "air knife" principle using a thin, flat jet of air for removing excess coating immediately after it is applied to the sheet. *See also* AIR BLADE and AIR DOCTOR.

**AIR KNIFE COATING:** A paper coating process in which a large amount of coating material is applied to the web and the excess is removed in a doctor blade fashion by a thin, flat jet of air leaving a smooth film covering the sheet. *See also* AIR BRUSH COATING.

**AIR KNIFE MARK:** A paper coating defect caused during the coating operation with an air-knife coater, producing areas that are more transparent than the rest of the sheet. They are called *blade streaks* when occurring in narrow areas running in the machine direction. *See also* BLADE SCRATCH.

**AIR LAID:** A sheet made by fibers that have been mechanically separated and carried by air to form the basic fibrous web, which is then bonded and dried. Also called *dry laid*. The sheets are classified as *nonwovens*. Air laid forming is the process of making an air laid sheet. Also known as *dry laid forming*.

**AIR LOAD:** The pressure of air used in pneumatically operated process equipment.

**AIR LOADED TENSION DEVICE:** A device that applies tension to a piece of equipment and is driven by air. It is used to control tension of the paper sheet or felt on a paper machine.

**AIR MAIL PAPER:** Lightweight, high-opacity writing paper intended to be used for letters, flyers, circulars, and other printed matter transported by airlines.

**AIR MAKEUP:** Additional air that is introduced in large quantities into the paper machine room, the basement, in back of the machine, in the drive aisle below the fan mezzanine, or into adjacent buildings from which it can find its way into the machine room in order to attain proper air balance in the paper mill when the regulator air supply units cannot keep up.

**AIR NOZZLE:** A means of directing a concentrated stream of air in a certain direction or on a spot or area.

**AIR PAD:** That portion above the stock in a pressurized headbox that is filled with air and whose pressure is controlled to maintain a desired total head and level in the headbox. *See also* AIR CUSHION HEADBOX and PRESSURIZED HEADBOX.

**AIR PERMEABILITY:** A paper sheet characteristic that denotes its ability to allow air to pass through its fiber formation. It is measured by subjecting a specified sample to standard pressure differences across its boundaries under known conditions of temperature and relative humidity and determining the rate of air flow. Sometimes incorrectly referred to as *vapor permeability* and *porosity*.

**AIR POLLUTION:** The contamination of the atmosphere around a pulp or paper mill by exhausting process and machinery gases, vapors, and particulate materials into the air.

**AIR PORT:** The covered opening of the pressurized air section of a paper machine headbox; also used as a cleaning port.

**AIR QUALITY:** The purity of air used in the pulp and papermaking process. Also refers to the level of pollutants in outside air.

**AIR SHEAR BURST:** A paper roll defect in the form of an irregular separation or rupture of the paper web that normally extends for some length in the machine direction. It is normally located 5–10 inches from the edge of the roll.

**AIR VENT:** An opening in machinery and buildings through which air can flow in or out for cooling and heating purposes.

**AIRBORNE:** A method of supporting or transporting material by use of a cushion of air, such as in the moving of a paper sheet through an air dryer or the blowing of fluffed pulp through an air dryer.

**AKD:** *See* ALKYLKEYTENE DIMER.

**ALABASTER:** A paper-loading material, anhydrite calcium sulfate ($CaSO_4$), that contains practically no water of crystallization. Also called *annaline. See also* ANHYDRATE, GYPSUM, PEARL FILLER, and TISSUE FILLER.

**ALABASTER PAPER:** Paper with the appearance of alabaster, achieved by coating it with lead acetate crystals.

**ALARM:** A visible or audible signal alerting mill personnel that a process variable or mill equipment has exceeded prescribed limit conditions.

**ALBUM BOARD:** A thick, heavy cover paper used in making photographic albums. It is usually made in solid black or gray color.

**ALBUM PAPER:** A heavy black, gray, or other solid color cover paper used in photographic albums. It has a soft surface that will not wrinkle or cockle when photographs are pasted or glued on it, and when wet with such adhesives, album paper will not "bleed" contaminants or discolorants onto the photographs.

**ALCELL PROCESS:** *See* SOLVENT PULPING.

**ALCOHOL:** A by-product chemical made by fermenting spent sulfite liquors with torula yeast. It also can be made from wood directly by hydrolyzation followed by neutralization and the addition of a nutrient together with fermentation with torula yeast.

**ALCOHOL RESISTING:** A common reference used to describe dried paper coatings that are insoluble in alcohol.

**ALDEHYDE GROUPS:** Classes of resins used to impart wet strength properties to paper and paperboard.

**ALDER:** A broadleaf hardwood tree used as a source of fiber for making soda pulp.

**ALDONIC ACID:** An acid formed by a carbohydrate reaction during a sulfite cook, involving oxidation by bisulfite ions.

**ALGAE:** Microorganic plant life that forms in paper mill water supplies and causes problems in the water treatment plant by clogging filters, screens, showers, etc. They may impart tastes and odors which are carried through to the finished paper. They also tend to color the water and cause discoloration of pulp when water is used for bleached pulp washing.

**ALGICIDES:** Chemicals introduced into the paper mill water supply to discourage the growth of algae.

**ALGINATES:** Colloidal substance made from marine-grown algae and used as stabilizers and emulsifiers in the paper industry.

**ALIGNING PAPER:** A stable paper made from cotton fiber and chemical pulp, having a smooth surface, good printing qualities, erasability, and formation, used as a base for printed lines to make such products as *graph papers.* It also refers to a special extensible typewriter paper with multi-purpose applications, but most particularly for photographic reproduction of various documents (can be stretched to obtain uniform right hand margins).

**ALIGNMENT:** The relationship of the printing on a form to the edges of the paper and to the writing mechanism.

**ALIZARIN DYES:** Acid-type dyes used as colorants for aqueous mineral coatings and prepared by precipitating fatty acid complexes of their metallic (other than alkali metals) salts on a substrate of aluminum hydrate. Also called *anthraquinone dyes.*

**ALKAFIDE PROCESS:** A variation of the kraft process using hydrosulfide (sulfide-type cooking chemical). It was developed for continuous cooking using a two-body digester for a two-stage cook.

**ALKALI:** Sodium-base chemicals such as those found in sulfate pulping liquors; sodium hydroxide, sodium sulfide, and sodium sulfite. Also used in reference to sodium hydroxide (NaOH) and its solutions.

**ALKALI CELLULOSE:** A long chain compound resulting from the reaction of sodium hydroxide with cellulose occurring during the cooking of chips with alkali cooking liquors or during the manufacture of viscose from pulp.

**ALKALI CONSUMPTION:** The total amount of alkali used up during the cooking phase of wood. It is determined by calculating the total alkali in fresh cooking liquor and subtracting the calculated total alkali in the spent liquor. It is usually expressed in quantity consumed per ton of *oven dry (O.D.)* pulp produced.

**ALKALI EXTRACTION:** The second stage in a pulp bleaching sequence where the first stage is "chlorination" (in which chlorine or chlorine dioxide is added and allowed to react with the pulp slurry). The resulting chlorinated fiber residuals and other alkali-soluble constituents are then dissolved in the second or "alkali" extraction stage.

**ALKALI LIGNINS:** Precipitates that form under certain conditions of countercurrent continuous cooking when too low an alkalinity exists at the transition point between prehydrolysis and kraft cooking stages.

**ALKALI RESISTANCE:** The ability of paper and paperboard or other materials and substances used in the pulp and papermaking process to resist damage when used or brought in contact with alkali or anything containing alkali. *See also* ALKALI STAINING RESISTANCE.

**ALKALI REACTIVE BINDERS:** Paper and paperboard additives used as binders that require alkali conditions to become set up or established within the sheet.

**ALKALI RESISTANT CELLULOSE:** The fraction of cellulosic material in pulp that will not dissolve in alkali solutions of fixed concentration under specified conditions.

**ALKALI SOLUBILITY:** The fraction of cellulosic material that is soluble in alkali solutions of fixed concentration under specific conditions of testing .

**ALKALI SOLUBILITY OF PULP:** The fraction of cellulosic material in pulp that is soluble in alkali solutions of a fixed concentration under specific conditions.

**ALKALI STAINING RESISTANCE:** The ability of paper and paperboard to remain undamaged by contact with alkali and alkali substances. *See also* ALKALI RESISTANCE.

**ALKALINE AGENTS:** Chemical substances having markedly basic properties, i.e., NaOH and $NH_4OH$, usually added for the purpose of raising pH.

**ALKALINE BALANCE:** The proper ratio between sodium hydroxide (NaOH) and sodium sulfide ($Na_2S$) in sulfate liquor.

**ALKALINE COOKING LIQUOR:** Cooking chemical solution made up of sodium-based compounds such as NaOH, $Na_2S$, $Na_2CO_3$, and $Na_2S_2O_3$. Also referred to as *kraft pulping liquor.*

**ALKALINE FILLERS:** Types of fillers used in paper furnishes which result in alkaline conditions when exposed to water, and also will react with any acids present. A common alkaline filler is calcium carbonate, both natural ground and precipitated forms.

**ALKALINE GRINDING:** Addition of caustic soda to the showers of a log grinder, or impregnation of logs with caustic soda and subsequent grinding to increase pulp strength and reduce specific grinding energy.

**ALKALINE PAPERMAKING:** The manufacture of paper on a paper machine under alkaline conditions by the use of wet end additives (fillers) such as calcium carbonate ($CaCO_3$) and *neutral size.* Paper made under these conditions are used particularly in special applications–documents, books, maps (where aging resistance is important), milk carton stock, cigarette papers–but operating advantages offered by the process have now prompted almost all fine paper mills (especially in Europe) to convert to the process from acid systems.

**ALKALINE PEELING:** The reaction and removal of carbohydrates at lower temperatures (60-160°C) during a kraft cook.

**ALKALINE PROCESS:** A chemical process that reduces fibrous raw materials to their individual fiber components, or the treatment of pulp and paper produced from it by the use of alkaline-based chemicals.

**ALKALINE PULP:** Pulp made by the cooking of chips with alkaline-based chemicals. Sulfate and soda pulps are examples.

**ALKALINE PULPING:** The process of cooking chips with alkaline-based chemicals.

**ALKALINE SIZE:** *See* NEUTRAL SIZE.

**ALKALINE SIZING:** Making a paper or board sheet water-resistant by the addition of an *alkaline size* material before or at the time it is formed on the machine.

**ALKALINE SULFITE:** Refers to a pulping process in which the cooking liquor consists of sodium sulfite, sodium sulfide, and small quantities of sodium hydroxide or carbonate as the active cooking chemicals.

**ALKALINITY:** The amount of free alkali present in pulps and other materials.

**ALKALIPROOF:** A descriptive reference made to dyes and additives used in papermaking that will not change color under alkaline conditions. It is also used with reference to papers dyed with colors that resist the action of alkali.

**ALKALIPROOF PAPER:** A paper, either white or colored, that does not discolor when in contact with alkaline materials such as soap. Careful selection of fibers and coloring matters is necessary, but no particular strength requirements need be met. Many book papers are sufficiently alkaliproof, and glassine and waxed papers are also satisfactory.

**ALKALIPROOF SOAPBOX BOARD:** A type of board made to withstand the discoloration caused by alkaline substances in soap and soap products, and used to make packaging materials for them.

**ALKALI-TO-WOOD-RATIO (AWR):** The relationship between total alkali added to a pulp digester, plus the moisture in the wood and the bone-dry wood fed or loaded into the digester.

**ALKALIZATION:** The addition of sodium hydroxide or other alkaline materials to protein binders in paper and paperboard coating for the purpose of deflocculation.

**ALKEYNL SUCCINIC ANHYDRIDE (ASA):** A synthetic, internal paper sizing agent used extensively in alkaline papermaking systems.

**ALKYD GROUP:** A group of resin derivatives used in making shellac, lacquers, emulsions, and adhesives used in the paper industry.

**ALKYD RESINS:** Types of synthetic resins described as oil-modified polyester resins. As surface coatings they provide tough, versatile, and uniform finishes. They are widely used in paints and printing inks.

**ALKYLATION:** The process of modifying starch to be used as an adhesive by treatment with alkali.

**ALKYLKETENE DIMER (AKD):** A synthetic sizing agent used for internal sizing of paper, specifically in neutral or alkaline papermaking systems.

**ALL-RAG PAPER:** Nominally, paper made entirely from rag pulp, but may contain a small, incidental amount of nonrag fibers.

**ALLIGATOR:** A tug-boat used to tow pulp wood log booms on lakes and equipped with winches so that it can navigate through shallow water.

**ALLIGATOR IMITATION PAPER:** A cover paper embossed to resemble alligator leather. It is used for the manufacture of book and manuscript covers, wallets, and large envelopes.

**ALLOWABLE CUT:** The quantity of pulpwood that should be removed from a woodland in which good forest management is being practiced to achieve a sustained and substantial yield.

**ALLOY:** A metallic material made up of a metal mixed with a more valuable metal to give durability, anticorrosiveness, or some other desired quality. Examples are stainless steel, Hastelloy, Stellite, etc. It is commonly used in pulp and papermaking equipment, particularly in corrosive environments such as the bleach plant.

**ALOE:** A plant having long sisal-hemplike fibers sometimes used in the manufacture of paper. *See* AGAVE.

**ALONG MACHINE DIRECTION:** *See* MACHINE DIRECTION.

**ALPHA CELLULOSE:** The portion of pulp or other cellulosic materials that will not dissolve in 17.5 percent strength caustic solution at 20°C.

**ALPHA PRINTING PAPER:** Printing paper made from esparto pulp. It is also referred to as *esparto paper* when it includes writing and other papers made from esparto pulp.

**ALPHA PROTEIN:** The alpha-portion fraction of soybean protein used in making an adhesive that is subsequently used in sizing and coating of paper with and without casein.

**ALPHA PULP:** A specially processed chemical pulp of high alpha-cellulose content. It was developed especially for the manufacture of specialty papers such as photographic, overlay, printed design paper for laminates, resin-saturating papers, and gasket papers. It is also referred to as *dissolving pulp* when used to make products that involve a "chemical conversion" of the cellulose fiber into a different physical form or to some derivative, e.g., pulps for the manufacture of cellophane, textile rayon, cellulose acetates, cellulose nitrates, etc.

**ALPHA WRITING PAPER:** Writing paper made from esparto pulp. It is also referred to as *esparto paper* when it includes printing and other papers made from esparto pulp.

**ALPHA YIELD:** The percentage of alpha pulp remaining after it has been treated, purified, and removed from whole pulp.

**ALTERNATING CURRENT (A.C.):** Electrical current that reverses direction at regular intervals, as compared with *direct current (D.C.)*, which flows unidirectionally.

**ALUM:** A papermaking chemical, $Al_2(SO_4)_3 \cdot 14H_2O$, $Al_2(SO_4)_3 \cdot 18H_2O$ or a mixture of these hydrates, commonly used for precipitating rosin size onto the pulp fibers to impart water resistant properties (when used for water treatment) to the paper made from it. Also called *aluminum sulfate* or *papermaker's alum*.

**ALUM SPOTS:** Blemishes in a paper sheet resulting from alum crystals that did not dissolve in the paper furnish. These crystals are generally crushed and fall from the sheet in a powdery form during subsequent finishing operations.

**ALUMINA HYDRATE PIGMENTS:** Types of alumina coating pigments prepared by hydrolysis of sodium aluminate liquors that are completely free of organic material and are available in a variety of size ranges below 5 microns down to about 0.5 microns. They are also referred to more specifically as *alumina trihydrate pigments*, which is the type used in paper coating.

**ALUMINAS:** Paper coating pigments used as titanium extenders, having particle uniformity and no grit. They improve the brightness of paper.

**ALUMINA TRIHYDRATE PIGMENTS:** Hydrated alumina pigments used in paper coatings. *See* ALUMINA HYDRATE PIGMENTS.

**ALUMINUM:** A light metal used in around a pulp and paper mill for structural and equipment material.

**ALUMINUM CHLORIDE:** In solution, it is used to reduce sulfate ions when melamine formaldehyde resins are employed for producing wet strength in the manufacture of paper.

**ALUMINUM COATED BOARD:** A paperboard, boxboard, or bristol-type board made with a coating of aluminum powder and an aqueous sizing material, and used for show cards, tags, folders, decorative cartons, etc.

**ALUMINUM COATED PAPER:** A paper coated with aluminum and made by incorporating flake aluminum powder with casein or another aqueous-sizing vehicle or with lacquers in organic solution to form a coatable composition, and applied by brush or other coating means.

**ALUMINUM COMPOUNDS:** Chemicals and additives used in pulp and paper manufacturing that are made from or have an aluminum base.

**ALUMINUM DUSTED PAPER:** Similar to and manufactured the same as *aluminum coated paper*, but has the appearance of aluminum-bronzed powdered coating. It is usually calendered or embossed after coating.

**ALUMINUM FOIL:** Aluminum or aluminum alloy formed into a sheet and used for wrapping and packaging purposes.

**ALUMINUM OXIDE ADHESIVE:** Adhesive coatings for paper made with aluminum oxide powder.

**ALUMINUM PAPER:** A wrapping-type paper made by mixing aluminum powder into the furnish or by coating the sheet with aluminum powder. It is used for wrapping food products, tobacco, etc. When used in general label work, it is referred to as *silver label paper*.

**ALUMINUM SILICATE:** A chemical additive used as a pigment in papermaking derived from natural deposits of such materials as china clay, feldspar, etc.

**ALUMINUM SULFATE:** A chemical used as a coagulant in the treatment and clarification of paper mill source water. *See* ALUM.

**ALUNDUM:** Abrasive grit used in ceramic pulp stones for grinders.

**A/M STATION (AUTOMATIC/MANUAL):** A device that permits an automatically controlled system to be run manually by the operator.

**AMBIENT AIR:** Air that exists external to the process, around the pulp and paper manufacturing equipment, and out and around the mill buildings.

**AMBIENT PRESSURE:** The pressure of the air or medium outside and surrounding the process, pulp and paper manufacturing equipment, and around mill buildings.

**AMBIENT TEMPERATURE:** The temperature of the air or medium outside and surrounding pulp and papermaking process and equipment.

**AMERICAN CHEMICAL SOCIETY (ACS):** Headquarters: Washington, D.C., (202) 872-4600.

**AMERICAN FOREST COUNCIL (AFC):** Headquarters: Washington, D.C., (202) 463-2455. Merged in 1993 with the American Paper Institute Inc. (API) and the National Forest Products Assn. (NFPA).

**AMERICAN INSTITUTE OF CHEMICAL ENGINEERS (AICHE):** Headquarters: New York, NY, (212)705-7338.

**AMERICAN PAPER INSTITUTE INC. (API):** Headquarters: New York, NY, (212) 340-0600. Merged in 1993 with the National Forest Products Assn. (NFPA) and American Forest Council (AFC).

**AMERICAN PULPWOOD ASSOCIATION INC. (APA):** Headquarters: Washington, D.C., (202) 347-2900.

**AMINES:** Types of resins widely used in liquid form in adhesives for plywood and lumber core gluing and laminating, in textile and paper treating, and in surface coatings that provide marproof finishes. It also improves wet strength, rub resistance, dry tensile strength, and bursting strength of paper. Also referred to as *amino resins*.

**AMINO RESINS:** Liquid resins used in papermaking to improve wet strength, rub resistance, dry tensile strength, and bursting strength of paper. *See also* AMINES.

**AMMONIA:** A gas ($NH_3$) used to manufacture ammonia base sulfite cooking acid by putting it into solution and bubbling sulfur dioxide through it.

**AMMONIA BASE LIQUOR:** A sulfite cooking acid made by using ammonia gas in solution and reacting it with sulfur dioxide.

**AMMONIUM BISULFITE:** The chemical name for sulfite cooking acid ($NH_4HSO_3$). It is the result of the chemical reaction of ammonia and sulfur dioxide in solution.

**AMMONIUM COMPOUNDS:** Paper and pulpmaking chemicals and additives made from and having an ammonium base.

**AMMONIUM HYDROXIDE:** A solution of ammonia in water ($NH_4OH$). It is typically used to manufacture ammonia-base sulfite cooking liquor by absorption of sulfur dioxide.

**AMMONIUM SULFATE:** A chemical, $(NH_4)_2SO_4$, used as a coagulating agent in the treatment of river water before use by a pulp and paper mill.

**AMMUNITION PAPER:** The type of paper used in the manufacture of ammunition, such as cartridge paper that forms the tube section of a shotgun shell, and basewad paper, a dense paper used in the base of the shell. The paper is usually made on a cylinder machine from rag, flax, chemical wood pulps, or a combination of high-grade, reclaimed paper stock.

**AMORPHOUS:** Refers to non-crystalline structure of pulp and papermaking materials.

**AMORPHOUS CELLULOSE:** That part of the molecular structure of cellulose that is noncrystalline and will not produce a coherent diffraction pattern when subjected to X-rays.

**AMORPHOUS MATERIAL:** Pulping and papermaking material that is of a non-crystalline nature.

**AMORPHOUS REGIONS:** Portions in cellulosic fibers which, on the basis of X-ray diffraction or other techniques, do not show evidence of crystalline structures.

**AMPHOTERIC:** A material used in paper coatings containing both positive and negative charge. It usually exhibits cationic characteristics in acid solutions and anionic characteristics in alkaline solutions.

**AMYLASE:** Enzymes used in the paper mill to accelerate the hydrolysis of starch when preparing it to be used as an additive in the manufacture of the paper sheet, and is now being used as a pretreatment or combined stage (with ozone and/or various peroxygens) in non-chlorine bleaching sequences.

**AMYLOPECTIN:** One of two molecular configurations existing in starch used in papermaking. It is comprised of 1-6 linkage polymers and makes up 75 percent of the starch. The remaining portion is *amylose.*

**AMYLOSE:** One of two molecular configurations existing in starch used in papermaking. It is comprised of 1-4 linkage polymers and makes up 25 percent of the starch. The remaining portion is *amylopectin.*

**ANAEROBIC PROCESS:** A process which takes place only within an atmosphere free of oxygen. It usually refers to biological processes.

**ANALYSIS:** The methodical investigation of a problem and the separation of the problem into smaller related units for further detailed study, or the determination and identification of components of a substance or material by subjecting it to chemical or physical tests, usually in a laboratory.

**ANALYTICAL FILTER PAPER:** A type of low ash content paper that has a high filtering rate and does not allow fine precipitates or particles to pass when slurries are poured through it. It is used in chemical laboratories for analytical testing, in process filtering equipment, to remove suspended materials from process fluids, and in automobiles and products for the home such as coffee makers and air conditioners. *See also* ASHLESS, ASHLESS FILTER PAPER.

**ANATASE:** A term used to denote a grade of titanium dioxide paper coating pigment that is water dispersible.

**ANCHOR BOLT:** A type of bolt that holds pulp and papermaking equipment to the floor or another mounting base.

**ANCHORING:** The process of attaching pulp and papermaking equipment to the floor or other mounting base, such as fastening with anchor bolts.

**ANEMOMETER:** An instrument for measuring and indicating the force or speed of wind, commonly used at a boiler house and sometimes used to measure speed of air flows or air streams in the paper mill.

**ANGIOSPERMS:** Types of plants that produce seeds in a closed biological vessel, as distinguished from *gymnosperms*. Among those used as a source of papermaking pulp are hardwood trees such as oak and maple, bagasse, and esparto.

**ANGLE:** The type of paper that when converted (e.g., into envelopes, stationery, etc.), the "cut" is made at an angle to the machine direction.

**ANGLE CUT:** Any paper that is so cut that its edges do not parallel the machine direction or general fiber alignment of the sheet.

**ANGLE GRINDING:** The grinding of wood where the log is pressed against the stone at an angle, i.e., with the axis of the stone in the same plane but not parallel, in order to produce stock with lower freeness and higher strength.

**ANGLE OF CONTACT:** The angle formed by a liquid or surface coating spread on paper when the surface tension of the liquid is greater than the adhesive tension for the fiber.

**ANGLE SPLICE:** A splice made on paper at an angle to the sides of the sheet. *See also* SPLICE.

**ANGLE VALVE:** A type of valve used in mills for installations in pipe lines to adjust horizontal flow in the side and vertical flow at the bottom.

**ANGULAR:** *See* ANGLE.

**ANHYDRATE:** Anhydrous form of either natural or calcined calcium sulfate ($CaSO_4$) derived from ground gypsum. It is used as a paper loading material. Also referred to as *alabaster, annaline, gypsum, pearl filler,* and *tissue filler.*

**ANILINE:** A coal tar derivative from which the first synthetic dyestuffs used in papermaking were made.

**ANILINE DYES:** Currently, a broad range of synthetic organic dyes and pigments, so named because the first such dyes were made from aniline as derived from coal tar. They are classified according to the degree of brightness or fastness in light, such as basic dyes that have extreme brightness but are not fast to light, acid dyes that are less brilliant but are faster to light, and direct dyes that have the least brilliance but are fastest to light.

**ANILINE PRINTING:** A rotary letter press printing process using ink made of *aniline dyes* and pigments (mixed with a binder) that dries primarily by evaporation due to the solvent vehicle (with rapid evaporation properties) used. The process is generally used for printing on media such as paper, paperboard, laminations, films, and foils. It is commonly known as *flexographic printing* because of the rubber plates and rolls employed in the process.

**ANIMAL FIBERS:** Papermaking fibers obtained from animal sources such as wool and other animal hairs.

**ANIMAL GLUE:** Organic colloids of complex protein structure obtained from animal materials, such as bones and hides from the meat packing and tanning industries. It is used for gumming, tub sizing, and adhesive for paper coating and general adhesive. *See also* BONE GLUE.

**ANIMAL PARCHMENT:** Writing material made from the skins of various animals.

**ANIMAL SIZE:** A type of size used in papermaking made from gelatinous material obtained from animal hides. Also referred to as *gelatin* and *glue.*

**ANIMAL SIZED:** Paper which has been beater- or tub-sized with animal sizes (glue or gelatin).

**ANION EXCHANGE:** The ionization of synthetic surface active agents when dissolved in water.

**ANION EXCHANGER:** A synthetic surface active agent which is capable of ionizing when dissolved in water.

**ANIONIC:** (1) A negative electrical charged ionic particle. Anionic materials in liquid, when subjected to electric potential, migrate to positive pole or anode. Anionic surfactants are commonly used as polymerization emulsifiers and stabilizers in paper coatings. (2) Also refers to materials in papermaking furnish that interfere with the retention and performance of chemical additives, such as pulping and bleaching residues, recycled fiber, mechanical pulps, fillers, anti-scalants and pitch dispersants. Sometimes referred to as *anionic trash.*

**ANIONIC TRASH:** *See* ANIONIC.

**ANNALINE:** *See* ALABASTER, ANHYDRATE.

**ANNOUNCEMENTS:** Paper used to make up business cards, greeting cards, wedding invitations, and other business and social stationery. They may be plain or decorated and made with matching envelopes from the same paper. *See* WEDDINGS.

**ANNUAL CROP FIBERS:** Pulp and papermaking fibers derived from nonwoody plant species, and only considered suitable when harvested on a seasonable basis (usually only one growth period per year).

**ANNUAL RING:** Annual increment of wood as it appears in a cross section of a tree, representing a layer of wood laid down in given years, or its annual growth.

**ANTACID:** Acid free paper and paperboard.

**ANTACID MANILA PAPER:** Acid free paper made from manila rope or sulfate pulp so that it is suitable for one of its more common uses as insulation on wire or cable. It is referred to as *cable paper* when used as such. It must also be free of foreign materials and have high dielectric strength and a lower power factor.

**ANTHRAQUINONE (AQ):** A quinoid compound added to alkaline pulping to improve yield and to increase the delignification rate.

**ANTHRAQUINONE DYES:** *See* ALIZARINE DYES.

**ANTIBLOCKING AGENTS:** Antisticking substances used to prevent the adherence of paper to itself and other materials. They are applied as a coating over the completed sheet or in the coating makeup operation.

**ANTICHLOR:** Chemical substance used to remove residual bleaching chemical after the chlorination, hypochlorite, and chlorine dioxide bleaching operations in a pulp bleaching sequence. Sulfur dioxide ($SO_2$), sodium bisulfite ($NaHSO_3$), and sodium thiosulfate are some of the substances commonly used.

**ANTIFALSIFICATION PAPER:** A grade of paper on which any printing, typing, writing, etc. cannot be fraudulently altered by eradicators, mechanical erasures, etc. without noticeably affecting the surface design and/or uncovering or making otherwise visible a "hidden warning." This paper is usually used to make forms having a negotiable value, such as bank checks, tickets, postal money orders, certificates, etc. Also referred to as *safety type paper.*

**ANTIFOAM:** A material which, when included in a liquid composition, retards or inhibits the development of foam. Note: this is not the same as *defoamer.*

**ANTIFOAMING AGENT:** An additive that prevents, inhibits, or reduces the tendency of a liquid to foam. *See* DEFOAMER.

**ANTIMONY OXIDE:** A non-opaquing, low-reflectance synergist that permits the formulation of deeper, darker colors with less pigment in paper. When used with halogens, it acts as a flame retardant in paper, plastic, and textile production.

**ANTIOXIDANTS:** Agents that retard the action of oxygen in ink drying oils and other substances subject to oxidation.

**ANTIQUE BOARD:** A board made with a rough-finished surface design resulting from large, long, and narrow hill-and-valley-type felt marks, randomly aligned and not running in the machine direction.

**ANTIQUE BOOK PAPER:** A rough, antique finish, cream-colored, short-fibered paper made from chemical or a mixture of chemical and mechanical pulp in the specific bulk and weight ranges to be used primarily in books and novels.

**ANTIQUE BRISTOL:** A type of cardboard made with a rough, antique finish.

**ANTIQUE COVER PAPER:** A rough, antique finish paper made to be used for outer covers of pamphlets, brochures, booklets, etc.

**ANTIQUE EGGSHELL PAPER:** A paper having a finished surface design resulting from small hill-and-valley-type felt marks, randomly aligned and not running in the machine direction.

**ANTIQUE FINISH:** A type of rough paper and paperboard surface finish design, primarily the result of felt marks. To produce this effect, wet press felts with long naps are run with little pressure, followed by the use of low pressure on the machine calender.

**ANTIQUE GLAZED PAPER:** A two-sided finished paper, with a rough antique finish on one surface and a glossy-type finish on the other surface.

**ANTIQUE LAID BOND:** A bond writing and printing paper made with a parallel-ribbed surface formed by embossing with the wires of the dandy roll at the wet end of the paper machine. The wires are laid side-by-side on the outer surface of the roll.

**ANTIQUE PAPER:** Any paper with a rough surface design resulting from the use of long-napped felts on the wet press using little pressure on the presses and calenders.

**ANTIRUST PAPER:** Paper used for packaging metal and metallic equipment and made specifically without any ingredients that will cause discoloration by enhancing oxidation of the contained materials. Also referred to as *acid free paper* and *antitarnish paper.*

**ANTISCALANTS:** Chemicals added to the pulp and paper process slurries to minimize the formation of scales on pipe line and manufacturing equipment surfaces.

**ANTISKINNING AGENTS:** Materials that retard the skin formation on the surface of a drying oil or printing ink.

**ANTISTATIC AGENTS:** Substances that eliminate or reduce static electricity on paper. These agents can be incorporated in the paper or applied to the surface, in either case the purpose is to prevent the paper from attracting "dust" particles or other sheets of paper.

**ANTITARNISH BOARD:** An acid/alkali and reducible sulfur compound-free paperboard used to make containers for packing silverware, aluminum goods, leaded glass, hardware, razor blades, coins, and other metallic articles to prevent rust and discoloration.

**ANTITARNISH PAPER:** A term originally applied to higher weight tissues used for wrapping silverware, but now used for all papers so prepared that they will not rust or discolor razor blades, needles, silverware, etc. Various fibers are used and weights of paper made; the chief requirements are freedom from acidity and alkalinity as well as reducible sulfur compounds. Also referred to as *rust-preventive paper, silver tissue paper,* and *tarnishproof paper.*

**ANTITARNISH TISSUES:** Same as *antitarnish paper* but referring to the lower weight tissues used for wrapping silverware, steel tools, cutlery, and other polished steel hardware.

**AOX:** *See* ADSORBABLE ORGANIC HALIDES.

**APA:** *See* AMERICAN PULPWOOD ASSOCIATION INC.

**API:** *See* AMERICAN PAPER INSTITUTE INC.

**APPARENT DENSITY:** The weight (mass) per unit volume of a sheet of paper obtained by dividing the basis weight by the caliper (thickness). The resultant numerical value is dependent upon the definition of the ream. *See also* APPARENT SPECIFIC GRAVITY, APPARENT SPECIFIC VOLUME, BASIS WEIGHT, and POUNDS PER POINT.

**APPARENT SPECIFIC GRAVITY:** Specific gravity is the ratio of the weight of the paper to that of an equal volume of water. The specific gravity of paper cannot be measured in the usual way (by dipping it into the water to obtain its volume by displacement). Therefore, the weight (mass) of paper is expressed as *apparent density* or simply *density.*

**APPARENT SPECIFIC VOLUME:** The volume per unit weight (mass) of a sheet of paper, calculated by dividing the caliper (thickness) by the basis weight, thus functionally the inverse of *apparent density.*

**APPEARANCE:** The visual characteristics of paper, taken as a whole. Elements of appearance include color hue and intensity, brightness, surface features, and "purity" (cleanliness).

**APPLE WRAPS:** Tissue-type papers made from chemical or a mixture of chemical and mechanical pulps and treated with special oils to help retard decay in fruits when used as a wrapping (usually with individual pieces). The oils possess no color, taste, or odor. Generally referred to as *fruit wrapping paper.*

**APPLICATOR:** The device on a paper coater that applies the coating material to the sheet.

**APPLICATOR ROLL:** The roll on a reverse roll coater that runs in reverse direction to the web travel and applies the coating onto the sheet.

**APPRENTICE:** A mill employee assigned to work with a skilled laborer in order to learn a trade.

**APPROACH FLOW:** The portion of liquid feed flow in a fourdrinier type of paper machine from the discharge of the fan pump to the slice. On a cylinder machine, it refers to the portion of liquid feed from discharge of the mixing box or centrifugal screen to the making board or conventional vats, or to the entrance into the web-forming zone.

**APPROACH FLOW SYSTEM:** On a fourdrinier machine, all equipment between the fan pump and the slice of the headbox. This includes pipeline, cleaners, screens, distributor, and headbox. The system delivers and transforms the stock as uniformly as possible and eliminates flow disturbances. On a cylinder machine, it includes the pipeline, flow distributor, nozzles, and mixing chamber. It proportions stock to the mixing chamber feed nozzles and mixes stock before passing it on to the forming zone.

**APPURTENANCES:** Machinery, appliances, and auxiliary structures attached to a main structure, but not considered an integral part thereof, for the purpose of enabling it to function.

**APRICOT PAPER:** A type of *fruit wrapping* paper used specifically to package apricots; generally having a yellow-orange color.

**APRON:** A rigid metal plate attached to the headbox of a fourdrinier paper machine (between the sideplates) with its forward edge located just above the breast roll centerline. Drainage of stock as it first contacts the forming fabric can be controlled by adjusting the apron in the machine direction. Not all aprons are adjustable. Control of drainage with fixed aprons is done by adjusting the slice lip, which meters the flow against the apron to the fourdrinier wire (forming fabric).

**APRON BOARD:** A rigid metal plate attached to the paper machine headbox and extending over the fourdrinier wire (forming zone) from the breast roll to the slice lip.

**APRON CONVEYORS:** Material movers made up of a series of slats of wood, steel, or occasionally other materials, 2 to 12 feet in width, mounted on two or more roller-type chains in a continuous belt with sprockets or rolls at the ends.

**AQ:** *See* ANTHRAQUINONE.

**AQUA AMMONIA:** Liquid chemical sometimes used to manufacture ammonia-base sulfite cooking acid by bubbling sulfur dioxide through it.

**AQUATINT:** A copper or steel printing plate etching process (intaglio) which produces an India ink, sepia, or watercolor drawing effect on the printing press.

**AQUATONE PRINTING:** A method of printing using a photographically sensitized zinc plate. The special gelatin coating on the plate produces particularly fine, high quality halftones. An image (of bichromate) is transposed and light-fixed onto the coating by development in alcohol and subsequent heating.

**AQUEOUS:** A term used to refer to pulp and papermaking materials made from, with, or by water-like pulp stock slurries, chemicals, dyes, additives, etc.

**AR:** *See* ADHESIVE RATIO.

**ARABINOSE:** A five-carbon sugar that occurs in hemicelluloses.

**ARAGONITE:** A crystal-type calcium carbonate paper coating pigment that provides high opacity and covering power.

**ARBISO:** A type of bisulfite pulp made by the arbiso process, a single-stage sodium bisulfite cook starting at pH 4 with a maximum temperature of 155°C.

**ARCHING:** A common problem with pulp, chips, and other material storage bins in pulp and paper mills when the formation of an arch or bridge prevents flow through the bottom opening(s) to the discharge mechanism. The arch is normally broken up by means of poking, vibrating, agitating, or blasts of air or steam. Also known as *bridging.*

**ARMATURE PAPER:** A very tough and durable rope-fiber paper made up of very long fibers with the appearance and strength characteristics similar to vellum. It was first made in Japan and now used primarily for insulation in electrical armatures.

**ARSENICAL PAPER:** As its name implies, a paper containing arsenic. Formerly used in wallpapers and flypaper.

**ART BOARD:** A stiff, multi-layered board sheet lined on one or both sides with white or colored paper. It is used primarily to make posters for advertising purposes. *See also* ART POSTER BOARD.

**ART COVER:** A cover paper suitably decorated to be used for holiday and social greeting cards, formal invitations, announcements, etc.

**ART PAPER:** An English term applied to a coated paper or board. In America, it may mean a coated paper or a drawing paper with a high finish and smooth surface.

**ART PARCHMENT:** A vellum-like, heavily-sized paper used for official documents and certificates. It is made from chemical pulp, cotton fiber, or a combination of both.

**ART POSTER BOARD:** A rigid board made of several layers. It is lined on one or both sides with white or colored paper and used for advertising posters, show cards, campaign posters, article display advertising, game cutouts, etc. Also referred to as *art board.*

**ART VEGETABLE PARCHMENT:** A type of paper with an antique vellum appearance, manufactured from vegetable parchment and used for official documents, certificates, diplomas, etc.

**ARTICULATING PAPER:** A paper used by dentists to determine and adjust the "bite" of human teeth. It is impregnated with a relatively soft vegetable wax to receive and "record" bite information. Nontoxic colorants contained in the wax are easily transferred to teeth and various synthetic dental materials.

**ARTIFICIAL FILLERS:** Those types of inorganic substances, such as ground or precipitated calcium carbonate, titanium dioxide pigment, zinc sulfide pigments, etc., that have been made or prepared for use as loading materials in the manufacture of paper. Sometimes referred to as *manufactured fillers.*

**ARTIFICIAL INTELLIGENCE:** The use of computers to complement, and imitate to some degree, the human reasoning process. In the paper industry it is being progressively used for decision support purposes. Its various branches include *expert systems, fuzzy logic,* and *neural networks.*

**ARTIFICIAL LEATHER:** An imitation leather made from a variety of wood pulp, rag, or rope furnishes.

**ARTIFICIAL LEATHER PAPER:** The base paper for *artificial leather.* It is made from a combination of wood and non-wood fibers and sometimes contains a coating of latex, or is impregnated with it.

**ARTIFICIAL PARCHMENT:** A greaseproof sulfite paper with a parchment finish used primarily for the manufacture of greeting cards.

**ARTIFICIAL PULPSTONE:** Pulpwood grinding stones made from a composition of artificial binding material and abrasives, as compared with grinding stones made from natural materials.

**ARTISTS ILLUSTRATION BOARD:** A high rag content paperboard, usually made of pasted layers, solid or filled, and designed principally for pencil or pen sketching or water coloring. The rigid board is relatively warp resistant.

**ARTISTS PAPER:** A superior type of drawing paper used by artists and made with a close weave.

**ASA:** *See* ALKEYNL SUCCINIC ANHYDRIDE.

**ASAM PROCESS:** *See* SOLVENT PULPING.

**ASBESTINE:** A fibrous magnesium silicate mineral paper filler having physical properties similar to both talc and asbestos. A principal use is blotting-type papers and some boards. Also referred to as *agalite.*

**ASBESTOS:** Fibrous, smooth, nontubular silicate materials existing as natural deposits which can be mechanically separated and used as fillers in the paper industry, specifically where good heat-resistant properties are required in the product. There are several varieties of asbestos. The magnesium silicate variety is white, green, gray, or a combination of these colors, and it is called "chrysolite." The iron silicate is blue in color and is called "crocidolite." The iron magnesium silicate is brown, and it is referred to as "amosite."

**ASBESTOS BOARD:** A paperboard made of asbestos fibers, fillers, and binding materials. It is used to make heat-resistant building material.

**ASBESTOS CEMENT BOARD:** A pressure-formed, fire-resistant board made from a mixture of asbestos fibers and cement, or other suitable materials, and used to manufacture building materials. Sometimes called *asbestos lumber.*

**ASBESTOS DIAPHRAGM PAPER:** A type of paper containing high-quality asbestos fibers that is used generally to make electrolytic cell separation membranes.

**ASBESTOS ELECTRICAL INSULATION PAPER:** A white asbestos fiber-base paper made specifically for high temperature, heat-resistant insulations. It is sometimes treated with plastic or mixed with organic fillers to improve its strength.

**ASBESTOS FELT:** A thick mat-type sheet made from a combination of asbestos fibers and additional material. Used in locations requiring heat-resistant properties.

**ASBESTOS FIBER:** Fiber material made from white asbestos, a naturally occurring fibrous crystalline magnesium silicate. It possesses good heat resistant properties and is used in the composition of paper, boards, and felts requiring this characteristic.

**ASBESTOS LUMBER:** *See* ASBESTOS CEMENT BOARD.

**ASBESTOS MILL BOARD:** A thin, fire-resistant asbestos board used to make fireproof lining in building construction, stoves, ranges, furnaces, and other places where thin sheet materials are needed for protection against fire.

**ASBESTOS PAPER:** A paper sheet made from furnish that contains mostly asbestos fibers together with sizing, fillers, and other additives so that it can be used to make very thin insulating materials where heat and fire protection is required, such as in building construction.

**ASBESTOS ROLL BOARD:** A type of paperboard made primarily of asbestos fibers together with small quantities of appropriate sizes, fillers, and additives in a variety of thicknesses so that it can be wound up into rolls and used for heat and fire protection purposes in building construction.

**ASBESTOS ROOFING FELT:** Asbestos felt saturated with asphalt or coal tar, made up into rolls, and used as roofing material in building construction.

**ASBESTOS SLATERS FELT:** An asphalt-saturated asbestos felt made into rolls and used in building construction to underline slate shingle and tile-type roofs.

**ASBESTOS WALL BOARD:** A flexible building-construction type of board containing high percentages of asbestos fiber. It is manufactured as a relatively thick (1/4 to 1 in.) pressed sheet (sometimes laminated) used for finishing of building structures, both inside and outside.

**ASBESTOS WATERPROOFING FELT:** A thick, strong asbestos felt saturated with asphalt and used in the construction industry for long-lasting waterproofing.

**ASH:** The inorganic residue of complete combustion of paper industry cellulosic materials, such as wood, pulp, and paper, and used as a measure of some of its constituents. *See also* ASH CONTENT.

**ASH CONTENT:** The percentage of ash resulting from the complete combustion of a specific sample weight of cellulosic material, such as wood, pulp, and paper, in which all carbon, combustibles, and volatile compounds are removed. It is an indication of the amount of mineral salts and inorganic foreign matter in wood and pulp. It is also an indication of the filler, coating, pigmentation, and chemical additive content in a paper sheet.

**ASH FREE:** A descriptive term applied to pulp, paper, board, and other cellulosic materials prepared and treated in such a way that essentially no ash is produced when all combustible and volatile compounds are removed by complete combustion.

**ASH HANDLING SYSTEM:** The conveying, storage, and humidifaction of bottom and fly ash on a furnace.

**ASHLESS:** A term referring to grades of paper that are practically ashfree, such as *analytical filter paper*, and which produce practically no ash on combustion.

**ASHLESS FILTER PAPER:** Filter paper used in laboratories for analytical work that produces no ash on complete combustion, indicating that no fillers or additives have been used in its manufacture. *See* ANALYTICAL FILTER PAPERS.

**ASPEN:** The American hardwood most widely used for groundwood manufacture.

**ASPHALT:** A mixture of hydrocarbons prepared from coal tars, made up into the form of emulsions, and used in the paper industry to manufacture roofing and other types of waterproof papers or felts.

**ASPHALT DISPERSION:** The process of freeing asphalt from wastepaper, breaking it up, and scattering it evenly throughout the fiber mass so that the paperboard sheet made from it will contain a minimum of unsightly specks.

**ASPHALT EMULSION:** A sizing emulsion consisting of asphalt, clay, and resins used in the manufacture of some types of unbleached kraft paper and paperboard with low brightness. It is also used to provide a waterproof layer in a two-ply cylinder board when applied between plies and fused by heat when passed over dryers.

**ASPHALT FELT:** An asphalt-saturated, soft, porous paper used in building construction as roof and siding sheathing material made from old cotton rags, woolen rags, and paper stock. *See also* ASPHALT SLATERS FELT.

**ASPHALT LAMINATED PAPER:** A multi-ply paper made with layers of asphalt, used in the construction and container industries or where a high-strength, water-resistant paper is required.

**ASPHALT PAPERS:** Various asphalt-containing papers (impregnated, laminated, or coated) such as: duplex waterproof paper, which consists of two sheets of paper made to adhere by a layer of asphalt; house sheathing paper, which is saturated with more than its own weight of asphalt then coated on both sides with asphalt and finally dusted with talc; and roofing paper, generally a soft, porous, rag paper saturated and coated with asphalt and embedded with crushed slate or other grit.

**ASPHALT ROOFING:** A thick sulfate paper sheet saturated with asphalt. It has waterproof characteristics, and is especially suited for use on roofs.

**ASPHALT SHEATHING PAPER:** A heavily asphalt-saturated paper used where waterproofness is required and usually made from sulfate pulp and waste fibers.

**ASPHALTED BOARD:** A paperboard made of wood pulp and wastepaper, saturated with asphalt or joined together with asphalt to make up a multi-ply sheet, and used in building construction or automobile panels.

**ASPHALTING BOARD:** A paperboard used for the base when making asphalt or asphalted board used by the construction and automobile industries for wrappings and panels.

**ASPHALTING PAPER:** A paper made from sulfate pulp and used to manufacture asphalt papers.

**ASPHALT SLATERS FELT:** Roofing paper made by saturating a sulfate paper sheet and used as sheathing paper under roof shingles, sidings, and in construction of built-up roofs. *See also* ASPHALT FELT.

**ATLAS PAPER:** Type of paper treated to impart high wear, abrasion, and water resistance as well as improved printing qualities. It is usually coated and of a heavy basis weight to minimize show-through so it can be used in atlases, maps, and similar publications.

**ATMOSPHERE:** The ambient air outside and surrounding the manufacturing process, machinery, and various plant facilities of a pulp and paper mill.

**ATMOSPHERIC CURL:** The tendency of paper to take up a curl when standing in the atmosphere due to the difference of one side taking up and releasing moisture from the other side as the humidity changes.

**ATMOSPHERIC PRESSURE:** The absolute pressure exerted by the atmosphere at the point and time of measurement. Also known as *barometric pressure.*

**ATOMIZER:** Equipment used to produce a fine spray of material so it can be directed to a designated area, such as the spraying of atomized water on the surface of paper on the calender of the paper machine to improve finish. Spraying of atomized steam adjusts the caliper or thickness of the paper.

**ATTAPULGITE CLAY:** A coating clay made up of complex hydrated aluminosilicate with clay minerals and 5–15 percent magnesium. The pigment is absorptive, and it is sometimes used in specialty coatings for reproduction papers.

**ATTEMPERATOR:** A device to reduce steam temperature by mixing it with lower temperature steam or water, or by passing it through tubes immersed in the water of a boiler drum, generally by means of automatic control. Same as *desuperheater.*

**ATTRITION MILLS:** Grinding or shredding machines used to split or break chips into splinters for the manufacture of structural boards. They are comprised of two opposite discs equipped with detachable hard iron plate segments having various surface patterns of ribs and dams. One plate may be held stationary while the other rotates (designated a single disc mill), or both plates rotate in opposite directions (designated a double disc mill). The chips, when introduced between the rotating discs, are split or broken into splinters or matchstick-like particles.

**AUTO ADHESION:** The ability of pressure-sensitive, adhesive-coated paper tape to stick to surfaces on contact without the use of any pressure except that provided by the weight of the tape itself. Also referred to as *finger tack, initial adhesive, quick stick, tacky,* and *web grab.*

**AUTOCHROME:** The first successful tricolor screen plate for producing full color transparencies for a single exposure of objects.

**AUTOCHROME PRINTING PAPER:** A paper made from chemical pulp and coated for use on multi-color printing.

**AUTOHYDROLYSIS:** The treatment of wood with water or steam at high temperatures as commercially practiced in the first stage of cooking when manufacturing *dissolving pulp.*

**AUTOGRAPHIC REGISTER PAPER:** A bond-type printing paper with high tensile strength to tolerate perforation and folding. It is primarily used in the manufacture of printed business forms.

**AUTO/MANUAL CONTROL:** A control station (or selector station) that permits an operator to select either automatic or manual operation or equipment at a location remote from the actual physical hardware, e.g., the operator may manually move a knob in the control room to change the position of a damper in an air duct a long distance away. The operator may place the control station in automatic mode and permit the controller to position the damper as required to maintain proper control.

**AUTOMATIC CONTROLLER:** A device or combination of devices that measures the value of a variable quantity, load, or condition, and operates so as to correct or limit deviation of the controlled variable from a preselected reference.

**AUTOMATIC CONTROL SYSTEM:** An operable arrangement of one or more automatic controllers along with their associated equipment connected in loops with one or more processes.

**AUTOMATIC GUIDE ROLL:** A selected roll on a roll-driven machine that automatically adjusts its angle to align the web being driven so that it does not ride from side to side, preventing possible runoff or damage. It is used specifically to guide felts and wires (forming fabrics) on paper machines. Also called *automatic wire guide roll.*

**AUTOMATIC STARTER:** A motor starter designed to automatically control the acceleration of a motor.

**AUTOMATIC SWITCH:** A remote control device for controlling signal circuits.

**AUTOMATIC TRIP:** The automatic opening (other than manually) of a circuit breaker due to changes in current, voltage, or other electrical conditions.

**AUTOMATIC WIRE GUIDE ROLL:** A selected fourdrinier wet end wire (forming fabric) drive roll that automatically adjusts its angle to maintain alignment of the wire to minimize side-to-side movement and prevent possible damage or runoff.

**AUTOMATIC WOOD CHARGING:** The feeding of logs to certain types of grinders, such as magazine and chain grinders, automatically from conveyors running above the top of a row of grinder magazines, as contrasted to manually loading logs in other types of log grinders.

**AUTOMATION:** The planned employment of integrated procedures, machines, and/or systems to maximize the productivity of invested capital. The act or method of making the pulp and paper manufacturing process or system perform with a minimum of operator intervention or supervision. The common word designating the state of being automatic where machinery is applied to perform and control all operations in the mill automatically and continuously.

**AUTOMOBILE BOARD:** Types of plain or specially coated paperboard used in making various panels for use in the automobile industry. It is made from chemical pulp and/or recycled fiber and is waterproofed by the addition of asphalt and resins. This type of board is highly durable and resistant to warpage under extreme temperatures and moisture conditions.

**AUTOMOBILE PANEL BOARD:** *See* AUTOMOBILE BOARDS.

**AUTOMOBILE SEAT COVER:** A fabric made from yarn that consists of twisted narrow slit paper produced from bleached and/or unbleached sulfate pulp. The fabric is used extensively in the automobile industry for trunk linings and rugs. Also referred to as *automobile trunk lining and twisting paper.*

**AUTOMOBILE TIRE ROLL PAPER:** *See* TIRE WRAPPING PAPER.

**AUTOMOBILE TIRE WRAP:** *See* TIRE WRAPPING PAPER.

**AUTOMOBILE TRUNK LINING:** *See* AUTOMOBILE SEAT COVER.

**AUTO SLICE:** A type of slice on a paper machine, provided with an automatic operator, that can be adjusted automatically as well as manually from a remotely generated signal. Sometimes a handwheel is also incorporated so that the slice can be hand-operated in case of a malfunction of the automatic operator.

**AUTOTYPE PAPER:** A type of bond writing or printing paper made from cotton pulp, sometimes in combination with chemical wood pulp having good surface and manifolding properties suitable for making up multi-copy forms.

**AUXILIARY MACHINERY:** Manufacturing equipment in pulp and paper mills on which secondary operations are performed, such as coaters, super calenders, cutters, and trimmers.

**AVAILABLE CHLORINE:** A term used in rating chlorinated lime, hypochlorites, chlorine dioxide, and other chlorine derived chemicals (usually used in water treatment and pulp bleaching operations) as to their total oxidizing power. *See also* ACTIVE CHLORINE.

**AVAILABLE SULFUR DIOXIDE:** *See* FREE SULFUR DIOXIDE.

**AWR:** *See* ALKALI-TO-WOOD-RATIO.

**AXIAL FLOW PUMP:** A type of pump where the flow of material enters along the center axis of the impeller and is discharged from the outer peripheral of vanes. Also referred to as a *centrifugal pump.*

**AXIAL FLOW TURBINE:** A turbine with blades located perpendicular to the rotor shaft and the steam flows parallel to the rotor.

**AZO COMPOUNDS:** Chemical compounds from which synthetic organic pigment dyes are synthesized.

**AZO DYES:** Synthetic organic pigment dyes normally employed in coloring or tinting aqueous mineral coatings. Sometimes referred to as *azoic dyestuffs.*

**AZOIC DYESTUFFS:** *See* AZO DYES.

**AZURE LAID WRITING PAPER:** Types of light blue writing paper made with a laid mark finish.

**AZURE WOVE WRITING PAPER:** Types of light blue writing paper made with a wove finish.

# B

**B:** A reference notation applied to groundwood papers made from unbleached wood pulp. It is also used to indicate finish roughness of drawing paper, where B is rough, B.B. is twice as rough, etc.

**B.B. NOTE PAPER:** (1) A class of paper with good writing qualities, used for writing, bonds, ledgers, and books. (2) Used as an abbreviation for black-bordered note paper.

**B-FLUTE:** The geometric configuration formed by one of the undulations of corrugated medium in corrugated board. It measures 3/32-inch high and runs 50 to 52 per linear foot. *See also* A-FLUTE, C-FLUTE, and E-FLUTE as designations for corrugations of other dimensions and counts per linear foot.

**BABBITT BEARING:** Bearing on pulp and papermaking machinery that is lined with babbitt, an alloy containing tin, copper, and antimony.

**BABY DRYER:** A small diameter drying cylinder located in the dry end section of a paper machine.

**BABY PRESSES:** Low-pressure presses on a cylinder board machine. They are located before the normal press section. Also referred to as *primary presses.*

**BACK COOK:** The result of cooking wood by the sulfite process, which produces darkened pulp due to overheating or raising to maximum temperature too fast. Also called *burnt cook.*

**BACK END:** That portion of a paper machine following the dryer section where the winders, rewinding reels, slitters, and calenders are located.

**BACK FALL:** The section behind the roll in a beater over which the roll discharges the stock after it passes beneath it.

**BACK FLOW:** Any reverse flow of liquid in a pipeline or stream such as occurs in backwashing filters and screens.

**BACK LINER:** The back side liner in a multi-ply paperboard. It is made of lower grade stock than its opposite, top liner.

**BACK LINING PAPER:** Paper used for lining the backs of books.

**BACK LINING:** Paper that is glued to the backbone of a sewed book to bind the signature and keep the backbone of the book free from the backbone of the cover. *See also* BOGUS BACK LINING.

**BACK MARK:** Ridges on loft-dried sheets of paper caused when they are suspended on poles or lines during the air drying process. Also referred to as *pole mark* or *stick mark.*

**BACK PRESSURE EVAPORATOR:** A type of evaporator used to concentrate pulp mill wash liquors and operate with all effects or units under pressure, with low-pressure steam leaving the last effect. It gives best economy in cases where there is a steady use for low-pressure steam, and is commonly used to concentrate bisulfite liquors.

**BACK PRESSURE:** The pressure at any point in a flow system that results from conditions downstream from that particular point. In the paper industry, it is generally considered as the exhaust pressure in any of the unit processes.

**BACK WATER:** Water containing small amounts of bleach, which has been removed from bleached pulp. It is also used to denote filtrates and water around the paper machine. Back water contains one percent or less of fiber. Also called *white water.*

**BACK-PRESSURE TURBINE:** A type of steam turbine found in paper mill power-houses that takes steam at a relatively high pressure and temperature and then exhausts it at the pressure required for the mill's process needs.

**BACK-WATER:** Water returned from the paper machine wire (forming section) pit and saveall trays for re-use.

**BACK:** (1) The lower quality ply in multi-ply boards made up of sides of different stocks. (2) The drive side of a paper machine.

**BACKER LINER:** *See* BACKING PAPER.

**BACKER ROLL:** A large diameter roll in a paper coater that rides against the sheet and opposite the side at which the coating is metered (either by another roll or a metallic blade).

**BACKING AWAY FROM FOUNTAIN:** In the printing industry, an ink that does not flow up to the fountain roller is said to "back away," causing too little and uneven inking. This is due to ink lacking tack and flow.

**BACKING BOARD:** A thick paperboard placed in the backs of mirrors, pictures, etc., or used for partitions in furniture. It is generally made of paper stock on a cylinder machine and is rigid to prevent warping. It may be lined or colored on one side to give the desired finish.

**BACKING PAPER:** A brown-colored paper used to add strength to the flong or mold of alternated sheets and on a stereotyping mold. It also refers to plain wallpaper which is used as an underlayer on the surface before the final printed paper is applied. Also referred to as *backer liner* and *backing sheet.*

**BACKING ROLL MARK:** A paper coating defect caused by the buildup of coating in areas along the machine direction where low spots occur in the coating back-up roll.

**BACKING ROLL SPOTS:** Paper defects caused by an excess of coating which show up as spots in the machine direction of the sheet and are due to a depressed area in the coating backing roll. Sometimes referred to as *butterflies* or *moustache.*

**BACKING ROLL:** The large diameter coating roll or cylinder used on the one-sided application roll-type coater and inverted trailing blade coater.

**BACKING SHEET:** *See* BACKING PAPER.

**BACKING WIRE:** Heavy woven mesh screen stretched over the winding wire of the cylinder mold of a cylinder machine.

**BACKING:** (1) A paperboard used in automobile body construction, usually as panels to be covered with fabrics. *See* AUTOMOBILE BOARD. (2) With reference to pressure-sensitive paper tape, it is a relatively thin material to which the adhesive is applied.

**BACKOFF WATER:** Pressurized water used to move the pressure foot of a pulp grinder back to the position where the magazine can be loaded with logs.

**BACKSIDE:** Paper machine side which is opposite the operating side. A paper machine drive is normally located on the backside.

**BACKSTAND:** A frame to hold rolls of paper on a roll-cutting machine used in the finishing operations of a paper mill. The number of rolls of paper mounted on a stand varies, but it will typically accommodate from 1 to 50 rolls. Also referred to as *roll stand.*

**BACKTENDER:** A member of the paper mill machine crew responsible for the operation of equipment such as winders, rewinders, reels, slitters, and calenders located at the back end of the paper machine.

**BACKUP ROLL:** The roll on a roll-type coater that acts as a backing for the paper web at the point of coating application and runs at web speed at all times. It is usually made of rubber.

**BACKWASHING:** The operation of cleaning a rapid sand or mechanical filter by reversing the flow of water or liquid that is being filtered.

**BACON PAPER:** A single-ply or laminated oil- and grease-resistant wrapping, usually made from glassine paper, used to wrap bacon or other fatty-type meats. Sometimes made from greaseproof and vegetable parchment paper. Also referred to as *bacon wrapper.*

**BACON WRAPPER:** *See* BACON PAPER.

**BACTERIA:** Microorganisms, especially present in the aqueous phase of paper coatings, whose excess activity and growth cause deterioration or spoilage of paper coating formations and ingredients.

**BACTERIAL COUNT:** A laboratory test procedure consisting of a numerical evaluation of the amount of bacterial organisms in a specific sample weight of paper, board, or any material used in its manufacture, which grows under the conducted test conditions.

**BACTERICIDES:** Chemical substances added to paper pulp slurries, coating mixtures, water, and other papermaking materials to inhibit the growth of undesirable bacteria.

**BAD BLEACH:** A deinking mill term referring to a bad printed wastepaper cook in which the ink is not acted upon and the sizing is not thoroughly saponified.

**BAD CORE START:** A general reference to imperfections in a roll of paper resulting from undesirable conditions such as loose winding, door edge alignment, corrugations, and others occurring during the beginning of the wind-up operation and located near the core. Also called *poor core start.*

**BAFFLE BOARD:** Wooden-type dams and plates used in the headbox of a paper and board machine to prevent flocculation and eddy currents. It promotes good mixing and velocity control.

**BAFFLE PLATE:** A flat wooden, metallic, or plastic board used to deflect the flow pattern of gases, liquids, and solid granular material.

**BAFFLES:** Plates placed perpendicular to the flow to change direction of a flowing medium. Also, a system of dams and plates in a paper machine headbox. *See* BAFFLE BOARD.

**BAG LINER:** Protective inner liner layer of a bag, generally made from appropriate types of paper, film, or foils. It is used to prevent penetration of objectionable materials such as moisture, oil, and in certain cases, flavors or odors. Asphalted, creped, kraft papers are used as liners to protect against dust, dampness, and prevent leakage in burlap-type bags.

**BAG MACHINE:** A machine that converts a sheet of paper into bags by automatically cutting, pasting, and printing.

**BAG PAPER:** Any paper made to be used in the manufacture of bags.

**BAG SEALING MACHINES:** Equipment used in a bag mill to seal ends of paper bags. It is also used in automatic bag milling operations to seal the opened end of bags after filling.

**BAG:** Containers of various sizes and shapes made from paper, plastic, etc., that tend to "dimension" flexibly as they are filled with objects. Duplex bags have two layers of paper, and paper bags with three or more layers are called *multiwalled*.

**BAGASSE CUTTER:** Equipment used to shred and cut crushed stalks of sugarcane after the sugar has been extracted.

**BAGASSE FIBER:** The material that makes up pulp, produced by cooking bagasse in chemicals that remove the intrafiber-cementing substance.

**BAGASSE PAPER:** Paper made from the pulp of the residual of sugarcane after the sugar has been extracted from it.

**BAGASSE PULP:** Pulp made from the residual of sugarcane after the sugar has been extracted from it.

**BAGASSE:** Sugarcane residual from the manufacturing of sugar consisting of crushed stalk from which paper pulp can be made.

**BAGGAGE TAGS:** A high-strength tab made of manila or colored board, usually attached to luggage for identification purposes.

**BAGGINESS:** A defective characteristic of a paper sheet having fullness in the center due to high moisture content. Sometimes referred to as *baggy*. *See also* BAGGY PAPER.

**BAGGING MACHINE:** Equipment used in bag mills that automatically converts paper into bags. Also, a machine that automatically fills and seals bags.

**BAGGING:** The forming of the characteristic of fullness of a paper sheet, or *bagginess*. It will cause crackling and puckering if the coated sheet is not dried slowly.

**BAGGY END:** Paper having a fullness on one of the sides of the sheet. *See also* BAGGY PAPER.

**BAGGY PAPER:** A defective web with nonuniform draw, sometimes caused by a localized high-moisture content. The web width does not uniformly support web tension. There are tight and slack sections across the width of the sheet. If severe enough, it results in web tension difficulty in subsequent operations. Sometimes referred to as *baggy end, slack areas, slack center,* and *slack end,* denoting the position of the defect across the face of the roll.

**BAGGY ROLL:** A defective roll of paper made with a wet draw having slack areas, sometimes caused by high-moisture content, not under as much tension as adjacent areas. Also called *wrinkled roll* and, incorrectly, *crushed roll.*

**BAGGY:** *See* BAGGINESS.

**BAGMAKING MACHINE:** Equipment used in a paper-converting plant to automatically convert paper into various types of bags for consumers.

**BAKELITE PAPER:** A hard sheet made by saturating soft paper with a resin that is then polymerized to a Bakelite-like material by drying, curing, and pressing under high temperature and pressure. Sometimes referred to as *hard paper.*

**BAKERS' WRAP:** A type of lightweight paper made from fully bleached pulp. It is used primarily for wrapping bakery goods and typically has high brightness, excellent finish, and good strength.

**BAKERS' YEAST:** A yeast made from sulfite-spent liquor mixed with molasses.

**BAKING:** To use heat to dry surface coating materials on a sheet.

**BAKING CUP PAPER:** Paper made primarily from sulfite pulp especially for use in making liners for baking utensils.

**BALANCE SHEET:** The resin-saturated layer of paper in a laminated sheet which minimizes warping by maintaining uniform stresses throughout the multi-ply composite.

**BALANCED DRAFT:** An operating condition in a furnace where the pressure is controlled slightly below atmospheric. It is usually created by proper adjustment of forced draft and indirect draft fans.

**BALD CYPRESS:** A species of softwood used to make pulp.

**BALE PULPER:** A mechanical disintegrator designed to handle entire bales of paper and paper sheets and reduce them to a fiber slurry.

**BALE:** (1) A large, closely pressed, square- or rectangular-shaped package of rags, wastepaper, pulp, and paper of standard sizes. Bale dimensions and weights vary with the type of contents, grades of paper packaged, and type of press used. (2) The balemaking operation.

**BALED PULP:** Pulp that has been dried and put in bundles weighing 400 to 500 pounds and measuring approximately 30 by 30 by 13 inches.

**BALING PAPER:** Heavy, reinforced, asphalted, or other type of paper made especially to be used as a protective covering for bales.

**BALING PRESS:** A type of pulp compaction machine used to press flash-dried fluff pulp into cakes, which are combined and repressed into bales, or they may be originally compressed into bales. Also, a machine used to compress pulp, rags, wastepaper, and other materials into bales.

**BALING WIRE:** Metallic binding used to secure unbleached sulfite pulp sheet and other wrappings around baled pulp.

**BALING:** The process of making square- or rectangular-shaped bales of rags, wastepaper, pulp, and paper with the use of a mechanical press.

**BALL BEARING:** Ball-type antifriction bearings employed in various locations in the pulp and paper mill–notably on parts such as tables, rolls, and in cylinder molds to provide low resistance to turning.

**BALL MILL REFINER:** A type of refining machine used to grind pulp screenings. The grinding mechanism consists of a body half-filled with flint pebbles.

**BALL MILL:** Equipment used for pulverizing materials such as coal and ground calcium carbonate in an early step of the grinding process. It consists of a slowly revolving steel shell filled with metal balls which grind by impact of balls against the material being pulverized.

**BALL VALVE:** A type of valve commonly used on pulp stock lines, having a plug that is a rotary-motion operated sphere with ports of different shapes, the most common being the V-port, eccentric, and straight-through port.

**BALLOON PAPER:** A well-sized, nonporous, pinhole-free paper made especially to be used for toy hot-air balloons.

**BALSAM FIR:** A softwood species of wood used to make groundwood pulp with good bulk and softness qualities.

**BALSAM POPLAR:** A hardwood species of wood used to produce a chemical pulp for high-grade book and magazine paper.

**BALSTON'S PAPER:** Drawing paper resembling bristol board. It is sometimes put together in multi-layers to make up a pasteboard. Once referred to as *London paper* and *Whalman paper.*

**BAMBOO PAPER:** Paper made primarily from pulp stock derived from the cooking and treatment of bamboo.

**BAMBOO:** A type of large, tall grass that grows naturally, mostly in tropical areas, that is used as a source of pulp and papermaking fiber. It resembles and has some of the same physical properties as straw.

**BAND CONVEYOR:** A type of conveyor using a continuous rubber belt. *See* BELT CONVEYOR.

**BAND MILL:** A mill that prepares logs for chipping by the use of band-type saws.

**BAND SAW:** A continuous, vertical, belt-type saw used in the cut-up mill during the preparation of wood pulp. It removes thick slabs from pulpwood logs.

**BAND STOCK:** A paper made of chemical or mechanical pulp used for banding bolts of cloth or some cloth products. It has good strength characteristics and printability.

**BANDED STOCK:** Paper that is packaged so that the ends of the sheets are exposed.

**BANK NOTE PAPER:** A thin, tough, bond paper made from cotton fiber and/or highly bleached chemical pulp used to make bank notes and various legal documents. Also called *currency paper.*

**BANK PAPER:** A high-strength, durable grade of writing paper used for bank notes and other commercial documents, made primarily from cotton fiber, bleached chemical pulp, or a combination of these fibers.

**BANK STOCK:** High-quality cotton fiber pulp that is used to make *bank note paper.*

**BANK:** An electrical term referring to an assembly of fixed contacts engaged by the moving members of a selector. Banks are usually multiple.

**BANQUET-TABLE COVER PAPER:** Plain, embossed, or printed paper used as a table covering.

**BAR ANGLE:** (1) The angle at which bars of a beater roll are aligned as they pass bars in the beater bedplate. (2) The angle of bars in respect to each other in the opposing members of a stock refiner.

**BAR CODE:** A computer-scannable identification symbol on various products. In the pulp and paper industry, information contained in the proportionally spaced, wide and narrow vertical and horizontal bars of the symbol is fed to process computers and/or stored in order processing systems.

**BAR MARKS:** Undesirable marks made on a coated sheet when carried through air dryers on conveyor bars, if the bars move at the same speed as the web. Bar marks are eliminated when the speed of the bars is faster than that of the web.

**BAR PRESSURE:** The pressure between bars in the beater roll and bedplate as related to the applied load and to the effective average areas of the bars in mutual contact.

**BAR ROUNDING:** The gradual wearing and corrosion occurring on the leading edge of bars on the surface of stock refiner plates as they beat and grind pulp.

**BAR SCREEN:** A mechanical cleaning device used at the inlet of a pulp and paper mill open water supply (lake or river) to remove debris from the water before being pumped into the water treatment area.

**BAR WIDTH:** The distance across the face of the bars in a beater or stock refiner.

**BAR-FIBER ACTION:** The mechanical action imparted on fiber as it passes between opposing sets of bars in a beater or stock refiner.

**BARBER DRYING:** A paper-drying process where the sheet is alternately carried through a medium of hot air over two tiers of driven carrier rolls arranged one above the other.

**BARBER'S HEADREST PAPER:** A type of absorbent, towel-like paper or machine-glazed book paper made up into rolls and used on headrests of most of the older style barber chairs.

**BARITE:** Naturally occurring barium sulfate from which a purified and crystallized form (*blanc fixe*) or paper filler is produced. Also referred to as *baryta* or *baryte*.

**BARIUM CARBONATE:** A paper coating pigment chemical compound ($BaCO_3$) manufactured by chemical processing or derived from witherite, a natural mineral.

**BARIUM SULFATE ($BaSO_4$):** A natural or synthetically prepared mineral filler used in papermaking. Also referred to as *blanc fixe, fast white, pearl white,* or *permanent white*.

**BARIUM SULFATE:** A paper filler and coating pigment chemical compound ($BaSO_4$) derived from *barytes*, a naturally occurring mineral, or prepared by chemical processing. Also referred to as *blanc fixe, fast white, pearl white,* and *permanent white*.

**BARK BOILER:** A furnace designed especially to burn bark as a fuel. It is usually equipped with Dutch ovens.

**BARK BURNING:** The use of bark as fuel for a power boiler, alone or in combination with pulverized coal, oil, or gas, and stoker-fired in the main boiler house.

**BARK DISPOSAL:** The use bark is put to after removal from logs, such as burning in a power boiler or use as landfill.

**BARK DRYER:** Drying equipment designed to dry enough pressed bark (about 40 percent dry) to burn in a bark boiler without the addition of other types of fuel, such as coal, oil, or gas.

**BARK FLUME:** A V-shaped, constructed waterway used to transport bark removed from wood in the wood preparation area of a pulp mill to the bark preparation plant situated close to the boiler house.

**BARK PEELING MACHINE:** Equipment used primarily in the woodlands to remove bark from logs by stripping or peeling.

**BARK POCKETS:** The occurrence of bark embedded in wood. If not removed in the normal barking procedure, this bark shows up as impurities in pulp.

**BARK PRESS:** A machine used to apply mechanical compression to bark removed from wood during its reclamation for burning. It achieves a dryness of at least 40 percent in the bark.

**BARK RECLAIMING:** Putting bark removed from wood to some useful purpose rather than treating it as a waste material. The most common use is to burn it in a power boiler to produce heat, energy, and steam.

**BARK SPECKS:** Paper defects caused by dark specs in the sheet as a result of bark fragments carrying over from the pulping operation.

**BARK:** The rind covering of stems, branches, and roots of trees and plants. Technically, all tissues of woody plants which are outside the cambium layer.

**BARKED WOOD:** Logs which have been through the bark removal stage of the wood preparation process of a pulp mill.

**BARKER:** Equipment used to remove bark from logs prior to further preparation and use in the manufacture of pulp.

**BARKER-CUTTER HEAD:** A wood cleaning machine that removes bark from slabwood with the use of concave knives set in the face of a rotating, concave roll. Also known as a *planer head barker.*

**BARKING DRUM:** A large, horizontally inclined, rotating metal drum equipped with internal steel bars into which pulpwood logs are conveyed to remove bark by mechanical abrasion as they tumble against each other and the inner surface of the drum. Also referred to as DRUM BARKER.

**BARKING LOSSES:** The difference in weight of a pulpwood log before and after the barking operation.

**BARKING MACHINE:** Equipment used to remove bark from logs mechanically by peeling or cutting.

**BARKING PLANT:** That area or the buildings of a wood preparation process where bark is removed from the surface of pulpwood logs.

**BARKING:** The removal of bark from pulpwood logs during wood preparation operations in the woodlands, or at a mill site prior to the chipping and screening process. This is normally accomplished by mechanical means, but chemicals are sometimes used to remove bark or to augment mechanical bark removal. *See also* DEBARKING.

**BAROMETRIC CONDENSER:** A process unit used on black liquor evaporators to create a vacuum by means of a steam-jet air ejector or a dry vacuum pump.

**BAROMETRIC LEG:** A large pipe connected to the center of the cylindrical drum on vacuum filters and dropped some distance below into a filtrate tank. This then will create a vacuum for the filter by the falling filtrate, thus aiding sheet formation on the wire cloth-covered vacuum filter drum.

**BAROMETRIC PRESSURE:** Pressure exerted by the earth's atmosphere. At sea level, it is 14.7 psia or 29.9 inches of mercury absolute, the value decreasing with increasing altitude. Also referred to as *atmospheric pressure.*

**BARREL LINER:** A vegetable parchment or a specially wax-treated, moisture-proof paper used as the inner lining for barrels to provide leakproof retention and maintain freshness of contents. It is also used as a barrier to odor, dust, and other objectionable materials.

**BARRIER MATERIAL:** Types of paper and single- and multi-ply paperboards made especially for the purpose of preventing the penetration of moisture, vapor, water, gases, oils, and other objectionable liquids.

**BARRIER PAPER:** Paper that is specially treated, coated, or laminated to provide resistance to the passage of moisture, vapor, gases, oils, water, and other liquids. Also called *barrier sheet.*

**BARRIER SHEET:** *See* BARRIER PAPER.

**BARRIER:** A term used in the paper industry that refers to specially prepared products that provide resistance to the passage of moisture, vapor, gases, water, oil, and other liquids.

**BARRING AMPLITUDE:** The mean difference between the maxima and minima of machine direction cycle variations in caliper of a finished sheet of paper over a one-second run of a length of paper on a paper machine.

**BARRING FREQUENCY:** The number of peaks of machine direction cyclic variations in caliper of a finished sheet of paper in 1 second, measured in cycles per second.

**BARRING INTENSITY:** A description of the frequency (or spacing) and amplitude of the machine direction cyclic variations in caliper of a finished sheet of paper.

**BARRING:** The cyclic variations in caliper as measured in the machine direction of a finished sheet of paper.

**BARS:** Paper sheet imperfections appearing as uniform parallel undulations running along the machine direction. Also called *buckles, chain marks, rope marks, ridges,* and sometimes referred to as *corrugation marks.* It also refers to metal members fitted on the surface of beater rolls and bedplates that mechanically treat the pulp passing between them during the beating operation.

**BARTEL TRUCK:** Equipment for transporting paper in rolls using spindle-in-core at one end of the roll and clamps to hold the side of the roll.

**BARYTA COATING:** A type of "off-machine" coating process usually used to apply the base coating on photographic-type papers.

**BARYTA PAPER:** A paper coated with barium sulfate to give a smooth, low-gloss surface. It is used chiefly as a base for photographic emulsions and therefore contains no materials injurious to them.

**BARYTA:** Crude barium sulfate ore from which paper filler (*blanc fixe*) is derived. Also called *barite* or *baryte.*

**BARYTE:** Natural mineral ore from which paper coating and filler, barium sulfate ($BaSO_4$), is derived. Also referred to as *barite, baryta,* and *heavy spar.*

**BASE ALUM:** $Al_2(SO_4)_3 \cdot 14H_2O$, $Al_2(SO_4)_3 \cdot 18H_2O$, or a mixture of these hydrates. A papermaking chemical commonly used for precipitating the rosin size into pulp fibers to impart water-resistant properties to paper.

**BASE BOARD/PAPER:** Board or paper made specially to be coated, impregnated, chemically conditioned, or otherwise treated in a particular way so that they can be used for specific purposes. Also referred to as *base stock* and *body stock.*

**BASE SPEED:** The lowest speed obtainable at rated load and voltage and at normal operating temperature, of adjustable-speed motors used in pulp and paper mills.

**BASE STOCK:** Paper or board to be further treated in various ways. *See* BASE BOARD/PAPER. Also called *body stock.*

**BASE:** In the sulfite pulping process, it refers to the cation that combines with the sulfurous acid. The most common are calcium (usually obtained in the form of limestone), magnesium (usually obtained in the form of hydroxide slurry), sodium (obtained as a concentrated solution of caustic soda or soda ash), and ammonia (obtained as anhydrous liquid ammonia).

**BASEWAD PAPER:** Paper used in the manufacture of shotgun cartridges. It is located in the base, and holds and seals the paper tube to the brass wall.

**BASIC COLORS:** High-brightness, low-fastness paper colors obtained by the use of basic dyes, and found in newsprint and coarse grades of paper.

**BASIC DYES:** A bright, high-tinting class of aniline dyes normally having low lightfastness. The color base is made water-soluble by conversion to a salt of the base, such as chloride. They are used in the manufacture of coarse, low-grade types of paper such as newsprint.

**BASIC DYESTUFFS:** Generally hydrochloric acid salts of color bases, which have great affinity for unbleached paper fibers and are used to obtain, at low cost, deep and brilliant shades even without the addition of size or alum.

**BASIC SIZE:** The trade custom for size of sheet and numbers of sheets in a ream. It is used for calculating the basis weight. A few of the specifications are as follows: 25 by 38 inches, 500 sheets for Bible, book, gummed, and offset papers; 24 by 36 inches, 500 sheets for glassine, news, hanging, poster, tissue, waxing, and wrapping papers; 20 by 26 inches, 500 sheets for bond, ledger, manifold, mimeograph, railroad manila, and writing papers. *See also* BOOK BASIS.

**BASIS WEIGHT SCALES:** A balance-type weighing scale used in the paper machine room to determine the basis weight of the grade of paper being manufactured at that time.

**BASIS WEIGHT:** The weight, in pounds, of a ream of paper, usually consisting of 480, 500, or 1,000 sheets of a specified size as stated in the definition of the basic size for the specified grade. The standard ream size varies with different grades of paper according to trade practices. As a general reference, sometimes the weights are converted and expressed as the weight of an arbitrary standard ream of 500 sheets, 24 by 36 inches.

**BASKET LINER:** Type of paper and paperboard making up liners for baskets used in packing and transporting fruits and vegetables. The water-resistant, high-strength paper is usually made from chemical pulps, but vegetable parchments have also been used. The cut and fitted paperboard is made from recycled stock.

**BASSWOOD:** A species of tree which is a source of wood for hardwood pulps.

**BAST FIBER:** Fibers located in the inner bark (phloem) layer of trees and in the outer portions (cortex and pericycle) of other fibrous, woody plants such as hemp, jute, flax, etc.

**BAST PAPER:** Paper made from pulp consisting of *bast fiber.*

**BATCH BEATER:** Mechanical pulp-treating equipment that operates on a predetermined amount of material at one time, as opposed to a continuous operation.

**BATCH BLENDING:** Mixing predetermined amounts of stocks, additives, dyes, and chemicals to produce paper of specified qualities in a stepwise operation, as opposed to a continuous operation.

**BATCH CHLORINATOR:** A vessel used to mix specified amounts of chlorine and pulp so that the chlorination process can be carried out to completion before processing another specific quantity.

**BATCH CYCLING CHEST:** A storage chest in a batch pulping system into which batch pulpers are dumped. The batch of reduced pulp is held and cycled to refiners until refined to the desired degree. Two cycling chests are normally used, thus the second is discharging and filling while the first is cycling through the refiners.

**BATCH DIGESTER:** A cooking vessel, usually pressurized, in which predetermined, specific amounts of wood and cooking liquors are heated so that the wood conversion to pulp is completed and removed before the cycle repeats, as opposed to a *continuous digester.*

**BATCH PROCESS:** A unit process in the manufacture of pulp and paper that is carried out to completion on specified quantities of process materials before repeating the cycle.

**BATCH PROCESSING:** Carrying out a unit process in pulp and papermaking to completion in a discontinuous type of operation.

**BATCH PULPER:** A mechanical paper and pulp sheet disintegrator that reduces predetermined specific amounts of paper and pulp sheets to fiber slurries in one complete, discontinuous operation.

**BATCH REFINER:** Machine used to mechanically treat pulp fiber by putting a predetermined specific amount through a complete refining cycle before starting another cycle.

**BATCH SYSTEM:** A pulp and paper manufacturing unit process consisting of a series of operating units which processes predetermined specific amounts of materials and carries the process to completion before starting another cycle.

**BATCH-CONTINUOUS SYSTEM:** In stock preparation, an arrangement of batch pulpers and continuous refiners and jordans.

**BATCH:** A specified amount of material required to complete one operation in the pulp and papermaking process. Also, an operation in the pulp and papermaking process that is carried to completion on this specific amount before proceeding to the next specified quantity.

**BATH RATIO:** *See* LIQUOR-TO-WOOD RATIO.

**BATHROOM TISSUE:** A grade of lightweight absorbent paper made up into convenient-sized packaged sheets and perforated rolls for use in bathrooms and restrooms.

**BATT-ON-MESH FELTS:** The base material to which the batt is interlocked in a particular type of needled felt.

**BATT:** Material which is interlocked to the base material in the manufacture of certain types of needled paper machine felts.

**BATTERY BOARD:** A strong, metal-free paperboard made of chemical pulp and/or reclaimed stock, and used in the manufacture of some types of dry storage batteries.

**BATTERY PAPER:** A strong, alkaline, moisture-absorbent paper inserted between certain types of storage battery plates. It is made with moisture-bearing paste and used to take up any excess moisture.

**BAUER McNETT CLASSIFIER:** A common piece of test equipment used to determine fiber length distribution in a pulp stock slurry, particularly in mechanical pulp manufacture. Also referred to as *Bauer McNett fractionator.*

**BAUME HYDROMETER:** A float fitted with a scale graduated in units called *Baume.* Used to measure specific gravity (but referred to as *density*) of pulp and paper process liquids and slurries such as black, white, and green liquors, and lime and clay slurries.

**BAUME TEST:** A test to determine the density of pulp and paper liquids and slurries using a float-type hydrometer to arrive at the specific gravity *Baume* value.

**BAUME:** Units of specific gravity measurements made with a float-type hydrometer, and related to the percent solids existing in liquids and slurries that are tested. Used interchangeably with *density* in the pulp and paper industry.

**BCMP:** *See* BISULFITE CHEMIMECHANICAL PULP.

**BCTMP:** *See* BLEACHED CHEMITHERMOMECHANICAL PULP.

**B.D.:** *See* BONE DRY.

**BEADING:** The tight-rolling formation of the edges of paper and paperboard in order to reinforce it and to provide a rigid, protective border, as on drinking cups, paper pails, and other specialty containers.

**BEAM:** A wood or metal cylinder in a loom used in the textile industry on which the warp is wound.

**BEAMING BOARD/PAPER:** A high-strength, heavy-duty, water-finished paper or paperboard used as a wrapping material and made from chemical pulps. It also has been used by the textile industry for winding silk strands on the *beam* just before the weaving operation.

**BEARING HOUSING:** The metal protective covering over a machine bearing, which also serves as an anchor to a firm base.

**BEARING:** A machinery part in which a journal, gudgeon, pivot, or pin turns or rotates.

**BEAT:** To apply the mechanical action of crushing, cutting, and brushing fiber-to-fiber in a beater, in order to impart to the fibers the physical properties that will determine the character of the end product.

**BEATABILITY:** The degree of beating pulp must undergo to impart certain physical properties to it.

**BEATEN:** Describes pulp stock that has been mechanically "worked" in a beater and is ready for refining.

**BEATER ADDITIVE:** Dyes, additives, chemicals, etc. added to pulp in the beater to impart certain colors or properties to the final sheet which will be formed from it.

**BEATER ADDITIVES:** Those materials such as fillers, sizing agents, resins, dyes, slimicides, and anti-foam agents added to the stock furnish during the papermaking process.

**BEATER ADHESIVE:** Additive mixed with pulp stock in the beater to enhance strength properties of the final sheet which will be formed from it.

**BEATER BARS:** Metal bars fitted on the surface of the beater roll just above the bedplate, between which the pulp mass circulates and is mechanically treated.

**BEATER CHEST:** A storage chest located under a beater into which the beaten pulp is dumped after mechanical action is completed. It acts as storage for the refiners.

**BEATER COLORED:** Colored paper made from furnish to which dyes and pigments were added in the beater before sheet formation.

**BEATER DRIVES:** Mechanical means of rotating the roll in a beater, which can be either steam turbine- or electric motor-driven.

**BEATER DYEING:** The process of adding dyes and pigments to the stock in a beater to achieve the desired color of the sheet to be formed from it.

**BEATER FURNISH:** The materials, such as pulps, dyes, and additives, that are put into a beater for mixing and mechanical treatment.

**BEATER KNIVES:** Metal blades located on the peripheral surface of the beater roll. They ride over the bedplate and provide a mechanical action to the pulp slurry in the beater vat when the roll is turned.

**BEATER LINING:** The inner surface of the beater tub, sometimes made of brick.

**BEATER LOADING:** The addition of fillers to the total pulp stock furnish in the beater before the sheet of paper is formed from it. Also, the additions of pulp, dyes, chemicals, etc., to make the total furnish in the beater. Also called *beater filling.*

**BEATER MAN:** Any employee in the beater room who works with the beater. It usually refers to helpers who report to the *beater operator.*

**BEATER OPERATOR:** Person in the beater room in charge of the operation of the beater. Sometimes called *beater engineer.*

**BEATER PADDLE:** A flat wooden tool designed and used for cleaning beater tubs and feeding stock to the beater roll.

**BEATER PLATES:** The metallic plates in the floor of a beater, directly beneath the roll. Both the floor and the roll have bars in their surfaces between which fibers are trapped and subjected to crushing, cutting, and brushing action associated with beating. Also called *bedplates.*

**BEATER ROLL BARS:** Metal bars mounted on the surface of a beater roll. They ride on the bedplate between which the pulp stock slurry passes to be mechanically treated.

**BEATER ROLL:** A driven cylindrical roll located in the beater tub that carries metal fly bars on its surface to ride on the bedplate. The pulp stock slurry passes between the roll and the bedplate to be mechanically treated.

**BEATER ROOM:** That portion of the paper mill where the mixing, beating, and refining of pulp stock takes place prior to forming into paper and paperboard. Modern mills refer to this area as the *stock preparation area.*

**BEATER SIZING:** The addition of sizing ingredients such as alum and rosin to the pulp stock furnish in the beater before it is formed into a sheet of paper.

**BEATER STICK:** A stick usually made of wood and used by paper cutter and trimmer operators to maintain a straight pile of paper by tapping the sides with it.

**BEATER TACKLE:** That part of the beater that has to do with the drive, roll, and its appurtenances, excluding the beater tub or vat.

**BEATER TUB:** The oblong compartment of a beater which holds the ingredients to be mixed and worked on by the roll and bedplate contained within the beater. Also called the *beater vat.*

**BEATER VAT:** *See* BEATER TUB.

**BEATER WASHER:** A beater type washer used to wash rags before they are used in making paper.

**BEATER-SIZED:** Pulp stock and paper made from it whose furnish was made up by the addition of sizing ingredients in the beater before being sent to the paper machine for sheet formation.

**BEATER:** A large, longitudinally partitioned, oval tub used to mix and mechanically "work" pulp with other ingredients to make paper. It also contains a large cylindrical roll extending from one side of the partition (midfeather) and the opposite wall of the tub. The roll is fitted with steel bars (knives) on the surface under which the pulp mixture circulates between it and the *bedplate* below it, where the mechanical action occurs.

**BEATING DEGREE:** The extent to which a stock furnish must remain in a beater under mechanical action to attain the desired characteristics of the sheet.

**BEATING EFFECTS:** The specific properties and characteristics attained by the sheet as a result of the mechanical action the pulp stock has undergone in the beater.

**BEATING ENGINE:** A pulp beater consisting of an elongated tub constructed of a suitable material, such as cast iron, concrete, fabricated steel, or reinforced tile, and fitted with a roll with cast steel heads and used to mechanically mix and treat pulp stock. Sometimes referred to as a *hollander.*

**BEATING RATE:** The speed at which specified physical characteristics, such as freeness, bursting strength, tensile strength, tear, etc., are achieved in the mechanical treatment of pulp in a beater.

**BEATING STONE:** A stone with a smooth surface used as a base for beating papers and books.

**BEATING TIME:** The time that pulp is mechanically treated in a test beater to achieve specified physical characteristics of freeness, bursting strength, tensile strength, tear, etc.

**BEATING:** The process of mixing pulps together with various materials, such as dyes and additives, in a vat-like device called a *beater.* By means of the mechanical actions of crushing, cutting, and brushing the fibers, beating will impart properties that determine the characteristic of the ultimate paper or board product. Beating is the oldest of the stock preparation processes.

**BEAVERBOARD:** Any rigid-type wallboards. It once was a trade name for a laminated type of building wood pulp board.

**BED KNIFE:** Metal plates located on the surface of the bedplate. The beater roll knives roll over the bedplate surface and apply mechanical action on the pulp in the beater vat when the roll is turned.

**BED LOAD:** The amount of pressure that must be applied to the beater roll so that a specific desired contact pressure is obtained between the roll and the bedplate of a beater.

**BED-FIRED FURNACE:** A type of incinerator in which solid fuel is used and burned in a pile formed by feeding the fuel to the floor of the unit called a *grate.*

**BEDPLATE:** *See* BEATER PLATES.

**BEDSTEAD-WRAPPING PAPER:** A heavy kraft wrapping paper used to protect newly manufactured beds during shipment and storage.

**BEECH PULP:** Pulp made from the beech species of hardwood.

**BEECH:** A species of hardwood used to make paper pulp.

**BEEHIVE COOLER:** A type of cooling unit used to cool hot sulfur dioxide gas ($SO_2$) coming from sulfur burners. It is used in the manufacture of bisulfite cooking acid, and consists of lead pipe wound on a conical wooden form, with water being applied in a shower at the top of the cone and flowing over the pipes.

**BEER FILTER PAPER:** A special type of filter paper made from long-fibered pulp and used in the centrifugal clarification of beer.

**BEER MAT BOARD:** A heavy, absorbent type of single- or multi-layered paperboard with good printing surface and little distortion when wet and dried again. It is used to make protective mats and coasters for serving beer and other beverages. Also referred to as *coaster board.*

**BEER'S LAW:** Light transmittance of a stable solution is an exponential function of the product of the concentration and the length of the path in the solution.

**BELLMER BLEACHER:** Specially shaped concrete and tile-lined tub with a worm or screw propeller mounted at one end, used to mix bleach solutions and pulp at 5 to 7 consistency in a single-stage, batch-type pulp bleaching process. Sometimes referred to as a *bellmer chest.*

**BELLMER CHEST:** *See* BELLMER BLEACHER.

**BELLS:** Blisters occurring between two piles of board made on a cylinder machine due to lack of ply adhesion. Also called *blows* and *bubbles.*

**BELT CONVEYOR:** A means of transporting materials such as chips, coal, etc., in a pulp and paper mill. It consists of a continuous belt which rides on rollers and is motor-driven. Sometimes called a *band conveyor* when the belt is made of rubber.

**BELT DRIVE:** A means of transmitting power and mechanical rotary motion from a source such as an electric motor, turbine, etc., to a piece of machinery by use of a flexible belt. A common means used to drive paper machine sections.

**BELT FILTER:** A rotary, cylinder-type vacuum filter for causticized white cooking liquor using a drum-driven belt screen to separate solids from the liquid.

**BELT FILTER PRESS:** Continuous wire-type dewatering unit for use in process and waste applications in the paper industry.

**BELT GRINDER:** A grinder used in grinding paper mill rolls. It consists of a rubber-covered driven contact wheel and take-up or stretcher wheel. The rubber-covered contact wheel is serrated on a helix of approximately 25 degrees.

**BELT PRESS:** A twin-wire pulp press that consists of a gravity drainage zone, a low pressure roll, and a high pressure belt zone with hydraulically tensioned flat belts.

**BELT PRESSES:** Type of presses often used for thickening and settling paper mill clarifier sludge by the use of flat driven wire screens.

**BELT SHIFTER:** A mechanism used on a belt drive to move the belt to various-sized driving rolls to change the speed or stop the rolls.

**BEND QUALITY:** The characteristic of paperboard to withstand many bends before fatigue or damage occurs at the crease. Also called *bendability.*

**BEND TEST:** Test used to evaluate the bending quality of both coated and uncoated paperboard from 0.016 to 0.40 inches (0.4 to 1.0 mm) in thickness. It is made on fourdrinier or multi-ply cylinder machines and reported as "bending number."

**BEND:** (1) A mechanical distortion of paper such as would result from being tightly wrapped around a small core. (2) A crease in a sheet of paper.

**BENDABILITY:** *See* BEND QUALITY.

**BENDER:** The bendability or foldability of boxboard as measured by the angle which it can be bent to, along a previously made score line, before the top liner shows signs of failure.

**BENDING CHIP:** A type of bendable boxboard made from reclaimed, used paper stock and used to make folding containers.

**BENDING FATIGUE TESTER:** A mechanical testing device used to determine the number of times a sample of paperboard can be bent before the top liner is ruptured.

**BENDING MACHINE:** Mechanical equipment that automatically scores and bends boxboard in the process of box manufacturing.

**BENDING STRENGTH:** That physical property of paperboard that allows it to be bent and folded. It is important in the manufacture of boxes and other paperboard containers.

**BENDING:** The process of folding boxboard along a previously made score, as in the making of boxes.

**BENTONITE:** A natural, highly absorptive mineral with high colloidal properties consisting primarily of clay and commonly used in deinking of reclaimed newsprint and as a pitch removal and paper filling agent.

**BETA METER:** An instrument that uses beta ray-emitting, radioactive isotopes to make the primary measurement of a process variable. It is commonly used to measure the weight of the sheet of paper (*basis weight*) as it is being made on the paper machine. It is also used to measure density and level of liquids, solids, and other papermaking materials.

**BETA RAY GAUGE:** An instrument used to measure the weight of a sheet of paper or other material with the use of beta ray-emitting, radioactive isotopes. *See* BETA METER.

**BETA RAYS:** Fast-moving particles of negative electricity emitted from radioactive isotopes such as krypton gas and strontium metal.

**BETA-CELLULOSE:** That portion of the cellulose component of pulp that is alkali-soluble and can be reformed by further acid treatment. Most of it is removed during the manufacturing process of chemical pulps with the use of alkali-type chemicals.

**BEVEL:** A flat area made on the edges of the sides of a pulpwood grindstone face. Also known as *pulpstone bevel* or *edge of stone*.

**BEVELED GEAR DRIVE:** Any mechanical-driving mechanism whose power is transmitted through wheel engagements that have surface teeth arranged at an angle to the face of the wheels.

**BEVELED GEAR:** A type of engaging wheel in any papermaking machinery where the teeth run at an angle rather than straight across the face of the wheel.

**BEVELING TOOL:** An angle-mounted tool used for producing a flat area on the edges of a pulpwood grindstone.

**BEWOID MILL:** Equipment used in the *bewoid process*.

**BEWOID PROCESS:** The preparation of rosin in dispersed form using a small amount of casein dissolved in alkali as a dispersing agent.

**BEWOID SIZE:** A dispersed form of rosin paper size prepared by the *bewoid process*.

**BIBLE PAPER:** An opaque, lightweight printing paper made primarily from fibrillated linen or hemp fiber stock and filled with titanium dioxide. This type of paper was developed in England under the name of Oxford India paper and is used extensively where low bulk is an essential requirement, such as in Bibles, insurance rate books, encyclopedias, etc.

**BIBULOUS PAPER:** A highly absorbent, soft, loosely felted paper made of unsized pulp. *See* ABSORBENT PAPERS.

**BICARBONATES:** Chemicals containing the radical ($-HCO_3$) and used as buffering agents, especially in sulfite-type cooks.

**BIFFAR SCREEN:** A centrifugal-type screen used to remove dirt and other suspended impurities in pulp stock.

**BILL PAPER:** A high-strength, durable type of paper made from cotton fiber and/or bleached chemical pulp. It has good printing and writing surface qualities and is used for commercial papers and documents such as bills, statements, etc.

**BILL POSTER BLANKING PAPER:** A strong, plain, uncoated paper made of chemical pulp well-sized to impart water-resistant and wet-strength properties. It is used as a covering on billboards and remains flat after paste is applied. It has one rough surface to accept the paste and possesses good lightfastness. Also referred to as *blanking paper*.

**BILL STRAPS:** Plain, printed, and colored strips of paper made from kraft, bond, manila, or ledger paper, gummed at one end for fastening into rings and used to bind paper currency into packages.

**BILLBOARD PAPER:** A lightfast, weather-resistant paper made from bleached chemical pulp. It is specifically made for use on billboards exposed to the outside environment.

**BILLHEAD PAPER:** *See* BILL PAPER.

**BILLING MACHINE PAPER:** A high-strength, durable type of paper made from cotton fiber and/or bleached chemical pulp. It has good printing and writing surface qualities and is made up in roll or sheet form of proper size to fit various types of billing machines.

**BILLY STICK:** A wooden stick sometimes used by a backtender to evaluate the operation of a paper machine in a mill. The roll of paper is hit with the stick to obtain an indication of the hardness by the sound produced.

**BIN:** Storage container for chips in a pulp mill with sloped or conical bottom. Main storage is accomplished in large stationary bins, while individual digesters are gravity-fed with chips from bins located above them. Some mills feed digesters from main storage by means of a belt conveyor.

**BINDER'S BOARD:** A gray-colored, glazed board frequently used in the binding of hardcover books. It is usually a single-ply board with such physical characteristics as smoothness, uniformity, high density, stiffness, and strength, and has no offensive odors. Also referred to as *book binding board* and *bookboard*.

**BINDER'S WASTE:** Clean, white wastepaper consisting of book-cutting machine trimmings obtained from book binderies.

**BINDER:** An agent used to bond various materials, such as coatings and laminates, to paper. It may also be a substance that promotes fiber bonding in the sheet, or between sheet plies. *See* BINDING AGENT.

**BINDING AGENT:** A chemical or material used to promote the adherence of fiber to fiber, additives to fibers, and coating to papers. *See also* BINDER.

**BINDING PAPER:** A one-sided, water-activated, adhesive-coated paper made from chemical pulp that is cut into narrow rolls to make binding tape for sealing bundles, cartons, and other packages.

**BINDING POWER:** The ability of materials used in the paper industry to cause substances such as fibers, coatings, and laminates to adhere.

**BINDING TAPE:** Narrow paper strips cut from a one-sided, water-activated, adhesive-coated roll of paper made from chemical pulp and used for sealing bundles, cartons, and other packages. *See also* BINDING.

**BINDING:** (1) White and colored sulfite and kraft paper, coated on one side with water-activated adhesive, cut into roll strips of various widths, and used to seal packages, bundles, and cartons. Also called *binding tape*. (2) In the printing industry, the process of putting the page of a book together between two hardcovers made of processed binder's board and attaching it to the paperboard backbone.

**BINOCULAR VISION:** The act of viewing paper with both eyes in the evaluation of its appearance.

**BIOCHEMICAL OXYGEN DEMAND (BOD):** The amount of dissolved oxygen in parts per million (ppm) utilized in the aerobic biochemical oxidation of the decomposable organic matter in mill wastewater. It is usually measured under specified time and temperature (5 days at 20°C). *See also* BIOLOGICAL OXYGEN DEMAND, and CHEMICAL OXYGEN DEMAND (COD).

**BIOCIDES:** Chemical substances used to prevent microbiological development in pulp and paper process materials.

**BIODEGRADABLE:** Able to decompose as a result of the action of microorganisms. Materials are considered to be particularly biodegradable if they decompose relatively fast; most pulp and paper articles are in this category.

**BIOLOGICAL EFFLUENT TREATMENT:** Process in which living microorganisms are mixed with incoming waste to a paper mill waste treatment plant, and use the biologically degradable organics in waste as foodstuffs or an energy source, thus effectively removing them from applied wastewater.

**BIOLOGICAL OXYGEN DEMAND:** The amount of dissolved oxygen consumed in 5 days by biological processes breaking down organic matter in mill effluent. Sometimes referred to as BOD or BOD5. *See also* BIOCHEMICAL OXYGEN DEMAND (BOD) and CHEMICAL OXYGEN DEMAND (COD).

**BIOLOGICAL TEST:** Laboratory test made on a mill effluent sample to determine its *biological oxygen demand.*

**BIOMASS:** Total amount of forest wood (all growth) used as fuel for power boilers to produce steam to generate energy, heat, and electricity.

**BIOMECHANICAL PULP:** A type of mechanical pulp produced from chips previously treated with a fungi to increase strength properties and decrease energy consumption in subsequent refining operations.

**BIOPULPING:** The use of certain bacteria and other microorganisms to break down the lignin bond in wood (delignification), contributing, in effect, to the production of "pulp."

**BIOTECHNOLOGY:** A general reference to the use of biological microorganisms, enzymes, etc., in the "pulping" of wood (delignification) or the brightening of pulp (bleaching).

**BIRCH PULP:** A hardwood pulp made from the cooking of birch wood chips.

**BIRCH:** A species of hardwood used to make paper pulp.

**BIRD EYES:** Spots in coatings in a finished coated sheet of paper. Also called *fish eyes.*

**BIREFRINGENCE:** A phenomenon associated with the starch granule wherein it produces dark cross patterns by refraction when exposed to polymerized light.

**BISULFITE CHEMIMECHANICAL PULP (BCMP):** Pulp made by pretreating chips with bisulfite cooking liquor with no free sulfur dioxide prior to subjecting it to two stages of atmospheric refining for final mechanical separation of the fibers.

**BISULFITE COOK:** *See* BISULFITE PROCESS.

**BISULFITE LIQUOR:** The cooking chemical with which bisulfite pulp is produced. *See also* BISULFITE PROCESS.

**BISULFITE PROCESS:** The process in which the cooking liquor used contains a predominance of bisulfite ion in the 2 to 6 pH range with little or no true free $SO_2$. The reduction is accomplished at elevated temperature and pressure conditions. Calcium-base liquors were first used, but sodium and magnesium are more common today.

**BISULFITE PULP:** Pulp made by the bisulfite cooking process using bisulfite cooking liquor.

**BISULFITE PULPING:** The process of making pulp from wood chips using bisulfite cooking liquor. *See* BISULFITE PROCESS.

**BISULFITES:** Cooking liquors whose chemical makeup contains the bisulfite radical derived from the sulfur dioxide used in its preparation. The most common bisulfites are calcium, ammonia, magnesium, and sodium bisulfite. *See also* BISULFITE PROCESS.

**BITE:** The surface grain of paper, especially drawing paper, caused by paper machine wire (forming fabric) wove marks on the underside, the felt fabric impressions on the top side, and the slight surface depressions existing in the sheet fibers. The roughness is an indication of the ability of the paper to accept crayon and pencil applications. Also called *tooth*.

**BITUMEN COATING MACHINE:** Equipment used to apply a layer of asphalt-containing coating on one or both sides of paper and board.

**BITUMEN PASTING MACHINE:** Equipment used to apply a layer of asphalt-containing coating between two sheets of paper or board.

**BITUMEN:** Base material used for making asphalt-containing emulsions that internally size dark-colored stocks used for making paper and boards. It is also used for making asphalt-containing coatings.

**BITUMINOUS EMULSION:** Asphalt-containing emulsions that internally size dark-colored stocks used for making paper and boards such as boxboard, structural board, and water-resistant specialties where color is not a factor.

**BITUMINOUS PAPER/BOARD:** Paper or paperboard made from stocks that have been internally sized with an asphalt-type emulsion. *See also* BITUMINOUS EMULSION.

**BLACK ALBUM PAPER:** A high-strength, bendable, jet-black paper made from a mixture of chemical and mechanical pulps, with a very rough finish on both sides. It is used especially for photographic albums.

**BLACK ASH:** The combustion product discharged from a rotary-type recovery furnace resulting from the burning of concentrated black liquor. It contains sodium compounds and large amounts of unburned carbon.

**BLACK BODY LOCUS:** In the appearance evaluation of paper, it is the curve connecting points in the chromaticity diagram representing black bodies at various temperatures. *See also* BLACK BODY.

**BLACK BODY:** In the appearance evaluation of paper, it refers to a thermal radiator of uniform temperature whose radiant existence in all parts of the spectrum is the maximum obtainable from any thermal radiator at the same temperature. (The designation "black body" is appropriate because such a body will absorb all incident radiant flux.) It does not in any way indicate the color of the body. *See also* ABSORPTIVITY.

**BLACK LINE PAPER:** Paper used to make copies or reproductions of drawings and documents. It is processed with chemicals so that the design lines appear black on a white background when exposed to light.

**BLACK LIQUOR BURNING:** The incineration of concentrated black liquor in a specially designed furnace in order to recover spent cooking chemicals and to produce steam and power.

**BLACK LIQUOR DENSITY:** The concentration of organic matter in black liquor which is important in the support of combustion in the recovery furnace.

**BLACK LIQUOR DISPERSER:** A device used in a pulp mill to diffuse agglomerated solids and saltcake in the black liquor just ahead of the liquor firing guns on a sulfate chemical recovery furnace.

**BLACK LIQUOR DROPLETS:** Dispersed spray particles of a black liquor stream leaving the liquor feed guns or burners in a sulfate chemical recovery furnace.

**BLACK LIQUOR EVAPORATION:** The process of concentrating black liquor from pulp washing systems to such a density that, when sprayed into a recovery furnace, it will ignite and continue to burn because of the organic matter it contains.

**BLACK LIQUOR EVAPORATORS:** Multiple-effect combination of steam pressure and vacuum vessels in which black liquor is concentrated. They are arranged in such a way as to minimize the amount of steam used to carry on the process of water evaporation.

**BLACK LIQUOR FEED:** Concentrated black liquor which is pumped into the spray nozzles of a black liquor recovery furnace.

**BLACK LIQUOR GASIFICATION:** The slow oxidation of the organic carbon in spray particles of the black liquor stream from the liquor guns or burners in a sulfate chemical recovery furnace by $CO_2$, $H_20$, and $0_2$ in the furnace gases.

**BLACK LIQUOR HANDLING SYSTEM:** Pumps, pipelines, and holding tanks used during the transportation of black liquor from the brownstock washers, through the evaporators, and to the recovery furnace.

**BLACK LIQUOR OXIDATION (BLO):** The transformation of unstable sodium sulfide ($Na_2S$) to the more stable sodium thiosulfate ($Na_2S_2O3$) in black liquor by the use of air or oxygen ($O_2$) so that objectionable hydrogen sulfide ($H_2S$) and organic sulfur compound formation in its evaporation and combustion are minimized, thus preventing their loss from the sulfate recovery process. The treated liquid is called *oxidized black liquor.*

**BLACK LIQUOR PYROLYSIS:** A process occurring to the black liquor spray particles from sulfate chemical recovery furnace feed guns or burners. The process involves chemical changes in solid sulfur salts in the form of sulfur being released to the gas phase as hydrogen sulfide ($H_2S$) due to the intense heat encountered. Some gasification of organic carbon also occurs, with the major gasifying agents being $CO_2$, $H_20$, and $0_2$ in the furnace gases.

**BLACK LIQUOR RECOVERY BOILER:** A boiler designed especially to recover heat by burning concentrated black liquor (from the cooking of wood by the sulfate process) and to use the heat for steam generation.

**BLACK LIQUOR RECOVERY FURNACE:** A furnace or combustion chamber especially designed to recover desirable chemicals from burning concentrated spent black liquor from the cooking of wood by the sulfate process.

**BLACK LIQUOR SOAP:** The resulting product of the fatty acids and resin acids present in pinewood which is released during the pulp-forming cook, and is dissolved in the liquor. The black liquor soap (sodium salts)

**BLACK LIQUOR:** Liquor from the digester to the point of its incineration in the recovery furnace of a sulfate chemical recovery process. It contains dissolved organic wood substances and residual active alkali compounds from the cook.

**BLACK NEEDLE PAPER:** A black antitarnish-type paper used specifically for making needle wrappings.

**BLACK PHOTO PAPER:** A nonporous paper with a rough finish on one side and a smooth finish on the other side containing no ingredients to affect photographic emulsion and used to package sensitized photographic supplies. Sometimes referred to as *black positive paper.*

**BLACK POPLAR:** A hardwood species of trees from which wood pulp is manufactured.

**BLACK POSITIVE PAPER:** *See* BLACK PHOTO PAPER.

**BLACK PRINT PAPER:** *See* BLACK LINE PAPER.

**BLACK SPRUCE:** A softwood tree from which wood pulp is made.

**BLACK STOCK SCREENING:** The screening of kraft pulp in the black liquor prior to washing.

**BLACK STOCK:** Kraft pulp in the black liquor after cooking and prior to washing.

**BLACK WATERPROOF PAPER:** An asphalt-impregnated paper used as insulation for roofs, walls, and floors in building construction. It is made from jute fiber or sulfate pulp.

**BLACK WRAPPING PAPER:** A high-strength, black-colored, lightproof paper used for packaging and decorative purposes. It is usually made from jute fiber and/or sulfate or sulfite pulp.

**BLACK:** In the appearance evaluation of paper, it refers to the "color" name applied to opaque objects that are highly absorbing throughout the visible spectrum.

**BLACKENED PAPER:** Defective, transparent paper caused by crushing in the calender such that fibers in the sheet are pressed so closely together that the number of light reflecting surfaces decreases appreciably.

**BLACKENING:** A lowering of opacity of paper as the result of poor formation, too much moisture, and crushing caused by too much pressure used during the calendering operation. It produces a darkening effect on the sheet. *See* BLACKENED PAPER.

**BLACKS:** When used as a general reference to a class of papers, it pertains to high-strength, black-colored paper put to a wide range of special uses, such as lightproof wrapping in photography, covers, packaging, etc.

**BLADE BLEEDS:** Type of debris that forms on the dry side of a coating blade when high solids coating and high coating speeds are used. Other types of debris formed are called *stalagmites, spits, whiskers,* and *weeps.*

**BLADE COATER:** A paper coating machine that uses a flexible metallic blade to spread coating material on the web of paper after it has been applied by a dip roll. Also referred to as *flexible blade coater* or a *trailing blade coater.*

**BLADE COATING:** The coating of paper and paperboard by using a flexible blade to spread the coating material after it has been applied by a dip roll. The blade is set at an angle to the coated sheet surface as it passes over a backing roll.

**BLADE CUT:** A paper roll defect referring to a blade scratch that is deep enough to sever the web. *See also* BLADE SCRATCH.

**BLADE MARKS:** Paper coating defect caused during the coating operation with a blade type coater, producing areas which are more transparent than the rest of the sheet. Also referred to as *air knife mark* and *blade streaks.*

**BLADE SCRATCH PICK:** A 1/10-inch (2.4-mm) band of lifted portions on a sheet of paper caused by an excess of coating in areas of a blade scratch in the machine direction.

**BLADE SCRATCH:** Paper coating defect caused during the coating operation with a blade coater and appearing as a thin crack running in the machine direction and having less transparency than the rest of the surface coating. When depth of scratch causes severance of sheet it is referred to as a *blade cut. See also* AIR KNIFE MARK.

**BLADE STREAKS:** *See* AIR KNIFE MARK.

**BLADE WRINKLE:** A blade coating defect which is an irregular line, essentially in the machine direction, where the sheet has been folded over on itself when passing under the blade.

**BLADE:** A thin, flexible, metallic plate used on coaters to spread coating material on paper and used to scrape any adhering materials off the surface of rolls on paper machines and other papermaking machinery such as calenders and pulp washers.

**BLANC FIXE:** A barium sulfate paper filler chemically produced by the action of a sulfate on a barium salt. Also called *fast white, pearl white,* or *permanent white.*

**BLANK BOARD:** A heavy double-lined paperboard used for making showcards and posters.

**BLANK NEWS:** Recycled wastepaper stock made up of trimmings of white unprinted newsprint and used to make up furnish for paperboard.

**BLANK-BOOK PAPER:** Refers to a variety of paper types used to make blank books. They include such grades as writing, bonds, manila, etc., which are selected for the type of book being produced.

**BLANKET MARK:** A mark in the coated surface of paper caused by the continuous blanket (perforated) that runs over the suction box in a conversion coater.

**BLANKET ORDER:** An order for paper that carries no definite specifications as to quantity or date of delivery.

**BLANKET SMASH:** A permanent flaw caused in a paper machine press blanket by foreign materials or extemporaneous extra thick areas of paper coming through the process.

**BLANKING PAPER:** A strong, uncoated paper made of chemical pulp used as a covering on billboards. *See* BILL POSTER BLANKING PAPER.

**BLANKS:** Types of paperboards with good stiffness and printing quality used for making posters, merchandising displays, and other types of signs. They are made from chemical pulp, mechanical pulp, deinked pulp, and clean broke. Blanks are also referred to as *coated blanks* when coated and *plain mill blanks* when uncoated.

**BLASTING PAPER:** A wax-or oil-saturated, heavyweight paper made from chemical pulp and hemp fibers. It is usually highly finished with good water-repellent properties and is used primarily for packaging explosive materials. It is also used by the mining industry as a lining for drilled blasting holes.

**BLEACH BOILER:** Large cylindrical or spherical rotary drums used to cook rags in the manufacture of rag pulp. More commonly called a *rotary digester.*

**BLEACH DEMAND:** The amount of bleaching chemical required to achieve a certain level of brightness and strength in the bleaching of pulp, carried out under specified conditions. It is a measure of how easy or hard it is to bleach the pulp in a bleaching process. *See also* BLEACH REQUIREMENT.

**BLEACH LIQUOR:** A term once commonly used to refer to calcium hypochlorite [$Ca(OCl)_2$] chemical solution in the bleaching of pulp, but now also refers to solutions of other bleaching agents such as sodium hypochlorite (NaOCl), chlorine dioxide ($ClO_2$), etc.

**BLEACH MIXER:** A piece of process machinery used to mix bleaching chemicals, usually hypochlorite, with pulp during the bleaching process.

**BLEACH PLANT:** That portion of a pulp mill where the bleaching process is performed. It usually adjoins the brownstock washing operation but is typically contained in a separate building. Occasionally, this area is referred to as *bleachery* or *bleaching plant.* It also refers to the area where hypochlorite bleach solutions are prepared.

**BLEACH PLANT OPERATOR:** The person in charge of bleach plant operations, also referred to as *bleacher man.*

**BLEACH REQUIREMENT:** The amount of bleaching chemical required to achieve a certain level of brightness in the pulp bleaching process. It is usually stated in terms of chlorine equivalent. *See also* BLEACH DEMAND and CHLORINE REQUIREMENT.

**BLEACH RESISTANT PAPER:** Paper that has been colored with the type of dyes and pigments which will withstand the action of bleaching chemicals and materials used to remove images imprinted on it with a minimum of discoloring.

**BLEACH SCALE:** A paper defect resulting from carryover of bleach residues into the paper mill and appearing as small brownish-white brittle areas in the sheet.

**BLEACH SLUDGE:** The settlings or dregs left after chlorine is bubbled and mixed thoroughly with limewater to produce bleach liquor (which is decanted from the top of the reaction tank).

**BLEACH SOLUTION:** A chemical solution used to purify and whiten pulp. It is usually of the oxidizing or reducing type such as sodium or calcium hypochlorite and chlorine dioxide.

**BLEACH TOWER:** A tall, cylindrical retention chest where pulp, mixed with the bleaching agent, is retained the required time for the bleaching action to be completed in a continuous system of pulp bleaching. An upflow-type is used when bleaching low consistency pulp, and a downflow-type is used when bleaching medium and higher consistency pulp. Also referred to as *bleaching tower.*

**BLEACH TUB:** A vessel used to mix pulp and chemical bleaching agents and retain them for a specified period of time for the bleaching action to be completed in a batch system of bleaching pulp.

**BLEACH WASHER:** A filter (washer) located after a bleach tower in the bleaching sequence of pulp where the pulp is washed free of the residual bleaching agent and the products of the bleaching action.

**BLEACH:** (1) A chemical used to purify and whiten pulp. It is usually of the oxidizing or reducing type, such as chlorine-based solutions, oxygen, hydrogen peroxide, ozone, and similar chemicals. (2) The process of purifying and whitening pulp by chemically treating it to alter the coloring matter and to impart a higher brightness to the pulp.

**BLEACHABILITY:** Capacity of a pulp to be bleached to a given whiteness, approximately and indirectly related to lignin content. Common tests used to determine bleachability are *permanganate number (kappa or k number)* and *roe chlorination number.* They indicate the amount of bleach that will be required in the bleaching process and serve as a guide and control for the cooking process.

**BLEACHABLE PULP:** Pulp which has the capability of being purified or whitened by chemical treatment to remove or change coloring matter so that it will achieve a higher brightness characteristic.

**BLEACHED BROKE:** Paper trimmings or damaged paper resulting from breaks on the paper machine and finishing operations or anywhere in the papermaking process where bleached pulp has been utilized.

**BLEACHED CHEMITHERMOMECHANICAL PULP (BCTMP):** A high brightness mechanical pulp (generally sold as market pulp) produced by the *chemithemomechanical pulping* process and bleached to 90-plus G.E. Brightness. It has characteristics similar to some bleached kraft pulp grades, but with a much higher yield and generally lower capital costs for production capacity.

**BLEACHED CORRUGATING:** A sheet made from bleached chemical pulp and used to make a corrugating medium for displays used in merchandising and special wrapping material.

**BLEACHED LINED CHIPBOARD:** A multi-ply paperboard made with a recycled paper stock base and lined with a bleached white layer having good bending properties necessary in making boxes and cartons.

**BLEACHED PULP:** Pulp that has been purified or whitened by chemical treatment to alter coloring matter and has taken on a higher brightness characteristic.

**BLEACHED SEMICHEMICAL MECHANICAL PULP (BSCMP):** A *semichemical mechanical pulp (SCMP)* that has been bleached to relatively high brightness for use in higher value printing and writing papers. Similar to, but not the same as, *bleached chemithermomechanical pulp (BCTMP).*

**BLEACHED SPECIMEN:** A specimen whose light absorption has been decreased by chemical or radiant means.

**BLEACHER MAN:** *See* BLEACH PLANT OPERATOR.

**BLEACHER:** A vessel used to mix pulp and chemical bleaching agent and retain them for a specified period of time for complete bleaching action in a one-step process of bleaching pulp.

**BLEACHERY:** *See* BLEACH PLANT.

**BLEACHING AGENT:** A variety of chemicals used in the bleaching of wood pulp such as chlorine ($Cl_2$), sodium hypochlorite ($NaOCl$), calcium hypochlorite [$Ca(OCl)_2$], chlorine dioxide ($ClO_2$), hydrogen peroxide ($H_2O_2$), sodium chlorite ($NaClO_2$), oxygen ($O_2$) and others. Also referred to as *bleaching chemical.*

**BLEACHING CHEMICAL:** *See* BLEACHING AGENT.

**BLEACHING EFFLUENT:** Liquid discharge from a pulp bleaching process which contains chemicals and other contaminants considered a source of pollution if allowed to flow into natural water resources.

**BLEACHING LIQUOR:** Reference to any chemical solution used in bleaching pulp such as calcium hypochlorite, chlorine dioxide, hydrogen peroxide, etc.

**BLEACHING LOSS:** The amount of pulp removed by bleaching chemicals and lost in the washing phases during the bleaching process.

**BLEACHING PLANT:** *See* BLEACH PLANT.

**BLEACHING POWDER:** Bleaching chemical, calcium hypochlorite [$Ca(OCl)_2$], in powder form.

**BLEACHING PROCESS:** *See* BLEACHING.

**BLEACHING SEQUENCE:** The order in which bleaching chemicals are added. It determines the number and arrangement of stages used in a continuous system of bleaching pulp.

**BLEACHING STAGE:** A unit process operation in the bleach plant in which one of the bleaching chemicals is added in the sequence of a continuous system of bleaching pulp.

**BLEACHING TIME:** The retention time or the time interval that it takes the pulp being bleached to pass through the entire bleaching process or a bleaching stage in the sequence of a continuous system of bleaching pulp.

**BLEACHING YIELD:** The percentage of pulp remaining after reaction with bleaching chemicals and washing during the bleaching process, as compared with the original amount of unbleached pulp.

**BLEACHING:** The process of purifying and whitening pulp by chemical treatment to remove or change existing coloring material so that the pulp takes on a higher brightness characteristic. It is usually carried out in a single stage or a sequence of several stages. Chlorine, peroxides, calcium hypochlorite, carbon dioxide, and, lately, oxygen and ozone are most generally used to bleach chemical pulps. For groundwood pulp, sulfur dioxide and sodium peroxide are used, as well as reducing agents.

**BLEED TRIM:** Reference to a type of waste-cutting resulting from book binder trimming of white printed book paper.

**BLEED:** The discoloration of dyed pulp and paper due to the removal of color by liquids, thereby making it susceptible to staining other wet materials it contacts.

**BLEEDING FASTNESS:** The resistance of a dye to staining when the sheet comes in contact with water or other solvents. Such dyes are used in paper made for automobile seatcovers and packaging for greasy foods.

**BLEEDING RESISTANCE:** A characteristic of colored paper or board that exhibits minimum discoloration when it comes in contact with liquids and does not stain wet materials it touches.

**BLEMISH:** An undesirable mark on the surface or coating of paper and paperboard.

**BLEND POINT:** A phase of a blending operation in the papermaking process where the combining solutions and suspensions form a continuously uniform material.

**BLEND:** A continuous uniform material made by combining solutions or suspensions such as stock slurry made from a variety of pulps, additives, dyes, and chemicals.

**BLENDER:** A piece of machinery or system used to combine solutions or suspensions to form a continuously uniform material.

**BLENDING:** (1) Combining solutions or suspensions to form a continuously uniform material. (2) The mixing of a variety of types of pulp, additives, dyes, and chemicals to make up a stock slurry satisfactory to produce a particular grade of paper with specific characteristics.

**BLIND DRILLED ROLL:** A suction roll on a yankee tissue machine that dewaters the wet web and transfers it from the felt to the yankee dryer surface.

**BLINDING:** An undesirable condition created in pressure screens when the screening medium becomes dirty or plugged, requiring flushing or cleaning to restore satisfactory operation.

**BLISTER CUT:** Damage to a paper sheet occurring when a fold, caused by a blister entering the calender stack, results in a small cut. Sometimes called a *blow cut.*

**BLISTER PACK:** A paper sheet defect occurring when a fold, caused by a blister entering the calender stack, results in a dense crease mark in the web. Also called a *blow pack.*

**BLISTER PICK:** A paper defect appearing as a round or oval-shaped area and caused by partial separation of the sheet in layers.

**BLISTER PLY SEPARATION:** A defect in multi-ply paperboard caused by the separation of layers at a location where a blister is formed due to poor adhesion.

**BLISTER:** Paper and paperboard defect caused either during the drying process when, due to poor felts, air becomes entrapped under the sheet or when the paper is dried too fast. With multi-ply, cylinder-type paperboards, the defect is the result of poor binding between layers. Also referred to as *blow* and *bubble. See also* AIR BELL.

**BLISTERING:** The developing of defects in paper by drying too suddenly on drying cylinders or by air getting between the felt and the sheet. It also occurs in cylinder boards and is caused by plies pulling apart when insufficient adhesion exists.

**BLO:** *See* BLACK LIQUOR OXIDATION.

**BLOCK BIN:** A storage place for pulpwood.

**BLOCK VALVES:** Hand valves located in proper positions with reference to automatic or relief valves so they can be closed for the performance of maintenance on the automatic or relief valves.

**BLOCK:** A term sometimes used to refer to a piece of pulpwood before grinding. It also refers to an abrasive pulpstone segment used to dull another pulpstone.

**BLOCKING RESISTANCE:** Ability of paper rolls and sheets to withstand temperature, pressure, and humidity without coating materials sticking together or adhering to calender and supercalender rolls.

**BLOCKING:** Contiguous paper and paperboard sheets in rolls and stacks sticking together because of a combination of pressure, temperature, and humidity conditions, and faulty coating. It is also a condition that occurs in calenders and supercalenders when paper adheres to the rolls and is called *calender roll blocking.*

**BLOOD RESISTANCE:** That quality of paper or paperboard that prevents its absorption of blood. It is a very important property when used for wrapping and packaging certain fresh meats and meat products in the food industry.

**BLOOD SERUM PENETRATION TEST:** A laboratory test used to determine the degree of blood resistance of paper.

**BLOODPROOF PAPER:** A high-strength paper having maximum blood resistance. It is typically sized with wax emulsion or other antiabsorption-imparting materials and used extensively by the food industry to wrap, package, and dispense fresh meat and meat products.

**BLOOM:** In the appearance evaluation of paper, it refers to the scattering of light in directions near specular direction as determined by a deposit on a specimen. Bloom can be removed by rubbing or polishing.

**BLOTCH:** Spotty or mottled appearance of a paper or of the ink film printed on that surface.

**BLOTTING BOARD:** A type of highly absorbent, thick, heavyweight blotting paper.

**BLOTTING PAPER:** A low-strength, porous, unsized paper with high liquid absorbency properties made from chemical and mechanical pulps, rag fibers, cotton linter fibers, or a combination thereof. It is generally used to absorb ink from written materials and other liquids where required. Sometimes referred to as *blottings*.

**BLOTTINGS:** A class of unsized papers having high liquid absorbency with a good surface appearance suitable for printing. Blottings are made from chemical and mechanical pulps mixed with rag and cotton linter fibers. Traditionally they are used to take up ink from written material but they are also used whenever the absorption of excess liquid is required. Also referred to as *blotting paper*.

**BLOW BACK:** An operation used in a kraft batch pulping process to clean relief screens during a cook so that they will pass relief gases freely. It consists of closing the relief valve and opening a valve on a steam line, introducing steam between the closed relief valve and the screen, and thus blowing off any accumulated fiber back into the digester and cleaning the screen.

**BLOW BOXES:** Devices used in air drying of paper to provide support and a guide plane for the sheet. They are designed to keep the sheet from blowing away as drying air velocity is increased.

**BLOW CUT:** *See* BLISTER CUT.

**BLOW GAS:** Sulfur dioxide ($SO_2$) gas that is released during the blow of a digester. It is usually recovered and used to fortify freshly made cooking liquor in the sulfite pulping process.

**BLOW HEAT RECOVERY:** A process of recovering heat from the steam released while discharging a digester into a blow tank. It essentially consists of a combination of heat exchangers, cooling towers, pumps, and storage tanks.

**BLOW HEAT:** The heat in the steam released during the discharge of a digester into a blow tank after the cooking process is completed.

**BLOW LINE REFINING:** Any mechanical action taking place or applied to the cooked fiber immediately after the digester and before the blow tank in the blow line.

**BLOW LIQUOR:** The spent cooking liquor that drains from the blow pit after a completed digester cook is discharged into it. It is a common practice in the older type pulp mills.

**BLOW OFF:** (1) The operation of suddenly releasing pressure from a steam-generating boiler when it reaches a prescribed value. This is done automatically by a quick-opening pressure relief valve. (2) The release of cooked pulp from sulfite batch-type digesters by the use of internal pressure.

**BLOW OUT:** The action of placing a small amount of dyestuff on the point of a knife or spatula and gently blowing onto a piece of wet pulp or filter paper to check the dye being used.

**BLOW PACK:** *See* BLISTER PACK.

**BLOW PIT WASH:** A method of washing cooked pulp in the blow pit by filling it with water after the cooking liquor is drawn off, permitting the water to drain off, then repeating this operation several times. The wash is finished by the use of high-pressure hoses.

**BLOW PIT:** A large, round or elliptical tank, usually built of sapwood-free southern pine or Douglas fir, into which sulfite digesters are discharged for storage and/or washing. It may also be a square or rectangular chamber of wood or of concrete lined with wood, acid-resistant brick, or tile. Also called *wash tank*.

**BLOW POINT:** The time to blow a digester, which is predetermined to occur before or during the cooking phase.

**BLOW PRESSURE:** In the batch digester pulping process, it is the digester pressure at which the blow valve is opened to empty the digester.

**BLOW ROLL:** A roll located between the fourdrinier part of a paper machine and the felt press part under which the paper web passes as it travels from the suction couch roll to the lead felt roll.

**BLOW TANK:** A storage tank in which cooked pulp is stored after it is blown tangentially from a digester into a separator, with pulp dropping into the tank, and the steam and gases escaping from the top vent to a steam condenser.

**BLOW TESTS:** Determining the time to blow digesters (blow point) by withdrawing samples at intervals during the cook.

**BLOW UNIT:** A mechanical device by which a continuous digester discharges its contents into the blow line.

**BLOW VALVE:** A valve at the outlet of a digester which is opened to allow the contents to be discharged through the blow line into the blow tank. It is closed during the cooking cycle on batch-type digesters.

**BLOW WRINKLE:** A paper defect appearing as a wrinkle occurring on the sheet when a fold caused by a blister enters the calender stack and passes between the nip.

**BLOW ZONE:** That section of a downflow, vertical continuous digester below the point where hot liquor is extracted and replaced by cool liquor.

**BLOW-THROUGH DRYERS:** A type of paper machine dryer system where the flow of steam and condensate through the individual dryer sections depends upon a drop in steam pressure between them. An ejector is usually used for evacuation of noncondensibles.

**BLOW-THROUGH STEAM:** The steam used to create a differential and flow between paper mill dryer sections to evacuate condensate formed during the paper drying process.

**BLOW:** (1) The discharge of the contents of a digester into a blow tank by the sudden release of built-up pressure during a cook. It also refers to the discharge of the oxygen reactor into the blow tank in the process of bleaching pulp with oxygen. (2) In a paperboard mill, it designates blisters occurring between two layers of a multi-ply board made on a cylinder machine due to the lack of adhesion. *See also* AIR BUBBLES, BELLS, and BUBBLES.

**BLOWBACK VALVE:** A valve in the blowback steam line on a kraft batch digester which must be opened when the relief line is closed to clean the relief screen so that it will freely pass relief gases from the digester to the relief line *See also* BLOW BACK.

**BLOWDOWN:** (1) Drains from the power and recovery boiler and/or waterwall headers to remove spent precipitated feedwater treatment chemicals. They may be periodic or continuous. (2) In the batch sulfite process, the relief of gas and steam to lower the digester pressure from hold to the blow pressure in order to recover excess $SO_2$ at the end of the cook. Also called *gasdown, pressure blowdown,* or *side relief.*

**BLOWER:** Mechanical equipment such as an enclosed fan designed to move large volumes of air in power and recovery boiler burner and draft operations. It is also used in drying processes in the manufacture of pulp and paper.

**BLOWING OF OXYGEN BLEACHING STAGE:** The process of discharging the oxygen reactor into the blow tank using the vessel pressure in the operation of bleaching pulp with oxygen.

**BLOWING TIME:** Duration of time in the batch digester cooking process, from the opening of the blow valve until the digester is empty.

**BLOWING:** (1) A phenomenon caused by an accumulation of air between a sheet of paper and the felt ahead of the press nip on a paper machine. (2) The process of discharging the contents of a digester into a blow tank by the release of built-up pressure. (3) The process of discharging the oxygen reactor into the blow tank when bleaching pulp with oxygen.

**BLOWLINE:** A pipeline connecting a number of digesters to a common or a number of blow tanks. It is a means of conveying the pulp coming through the blow valve, when discharging after the cook is completed, to the blow tank.

**BLUE CARPET PAPER/BOARD:** A paper or paperboard with a heavily textured surface (honeycombed or bubbled) used as carpet lining or as wrapping/packing for delicate items. The cushioned texture is acquired by special forming rolls in the primary manufacturing stages of the product, which is made primarily of recycled fibers.

**BLUE GLASS METHOD:** A procedure of examining a slurry of pulp and water using a blue glass to look for imperfections such as flocs and bundles.

**BLUE GLASS TEST:** A standardized test procedure for examining a slurry of pulp and water with the use of blue glass to look for imperfections such as flocs and bundles.

**BLUE GLASS:** A piece of dark blue glass on which a thin film of pulp slurry and water are placed for examination.

**BLUE HAZE:** Emissions of volatile wood resins from veneer driers or from growing forests on warm days.

**BLUE PLASTERBOARD:** A thick, heavy paper made from reclaimed wastepaper pulp on a cylinder-type machine and used primarily in building construction for lining walls. It is characterized by its light blue color.

**BLUE ROSIN SHEATHING PAPER:** A blue, well-sized, hard-finished sheathing paper made of wastepapers on a cylinder machine. It is used to protect against wind, dust, moisture, and water in locations such as under sidings, between floors, and in linings of walls and linings of boxcars.

**BLUE SPOTS:** Blue-colored specks in a sheet of paper resulting from undispersed particles of color pigment, undissolved dyes, color reactions with rosin, pitch particles, or other elements of the furnish that give rise to colored froth or foam accumulating in back of the paper machine slice and dropping onto the wire (forming zone) when dried.

**BLUE TRACING PAPER:** A blue-colored, translucent, greaseproof sheet or a bond or manifold sheet chemically treated or oiled to increase transparency. It is used for tracing original drawings, figures, patterns, curves, etc. It is usually made from cotton fibers and/or chemical pulps.

**BLUE WHITE PAPER:** A white paper with a bluish cast to make it look whiter.

**BLUEPRINT BASE PAPER:** A highly-sized and well-formed paper used for making blueprint paper. It has excellent surface characteristics and good wet strength with uniform surface absorbency. This paper contains no chemicals to react with the sensitized coating and will not shrink or warp when moistened.

**BLUEPRINT PAPER:** A paper produced from blueprint base stock by treatment with chemicals so that when it is exposed to light and subsequently developed (e.g., with ammonia) those areas will turn blue. It is used to make copies or reproductions of drawings and documents.

**BLUSHING:** The appearance of a white and grayish cast during the drying of lacquer film.

**BOARD BENDING MACHINE:** Mechanical equipment used to automatically score and bend paperboard during the box and container manufacturing operations.

**BOARD BOXES:** Cartons and containers made from paperboard.

**BOARD CONVERTING MACHINE:** Mechanical equipment designed to automatically convert paperboard into usable items such as boxes and cartons.

**BOARD FOOT:** A 1 foot by 1 foot by 1 inch cubic measurement usually associated with wood.

**BOARD LINER:** The exterior layers of a multi-ply, built-up paperboard. The term also refers to an automatic pasting machine that applies a paper lining to board before it is made into boxes.

**BOARD LINING:** A sized-type paper made from a mixture of mechanical and chemical pulps. It is used as a lining or for covering low-quality paperboard.

**BOARD MACHINE:** A machine used to manufacture paperboard grades. The most common ones are the cylinder and fourdrinier types.

**BOARD MILL:** That portion of a paperboard mill where the actual board is formed, dried, and prepared for further use.

**BOARD SHEARS:** A manually operated shear knife used to cut small lots of paperboard.

**BOARD:** *See* PAPERBOARD.

**BOB-TAIL TRUCK:** A type of truck commonly used to gather and transport pulpwood. It has a short bed and self-loading equipment.

**BOBBIN FIBER:** A grade of vulcanized fiber used to manufacture textile bobbin heads. They are soft enough to punch but firm enough to resist edge denting in use, with the capability of being machined to a very smooth finish.

**BOD5:** *See* BIOLOGICAL OXYGEN DEMAND.

**BOD:** *See* BIOCHEMICAL OXYGEN DEMAND.

**BODY BOARD:** A paperboard used in automobile body construction, mostly as panels. *See* AUTOMOBILE BOARDS.

**BODY DAMAGE:** A paper roll defect on the body of a roll consisting of gouges and cuts. It is observable when the roll wrapper has been removed and, often, before.

**BODY PAPER:** Any type of paper intended for additional processing such as coating or finishing, prior to conversion into user products.

**BODY SOFTNESS:** *See* SOFTNESS.

**BODY STOCK:** *See* BASE BOARD/PAPER.

**BODY WRAP:** Paper and/or board specially prepared to be used for moisture-resistant paper roll wrapping. It is used to protect the roll from damage and to help provide against distortion of the paper due to gain or loss of moisture from the surrounding atmosphere. It usually includes body wrap, heads, and bands. Often referred to as a *wrapper.*

**BOGUS BACK LINING:** An inferior strength kraft paper used as part of the backbone of sewed books to separate the backbone from the backbone of the cover, thus binding the signature. Also referred to as *bogus paper* or *bogus board* when it is of lower quality than the paper normally used. See BOOK BACK LINER PAPER.

**BOGUS BOARD:** *See* BOGUS BACK LINING.

**BOGUS BRISTOL:** An inferior-type of single- or multi-ply bristol board usually made from waste newsprint, trimmings, shavings, and other materials, on a cylinder-type machine to imitate the higher quality product.

**BOGUS CORRUGATING MEDIUM:** An inferior-type corrugating medium made from recycled paper stock.

**BOGUS DRAWING PAPER:** A lower quality type of drawing paper made from reclaimed paper stock, but with sufficient surface bite and texture to make it suitable for working with crayons, pencils, or watercolors.

**BOGUS DUPLEX:** A low-quality, smooth, glazed, calendered paper, dyed on one side.

**BOGUS KRAFT:** A brown-colored paper made of mechanical pulp, unbleached sulfite, or wastepapers, which is colored brown so it can be used as a substitute for kraft paper where strength and durability are not prime factors.

**BOGUS LINING PAPER:** A sized paper used for lining board and boxes. It is made from recycled stock and is of lower quality than conventional lining paper, which is made from chemical and mechanical pulps.

**BOGUS MANILA:** A type of paper made from recycled paper stock and colored to imitate, and used in place of, higher quality manila paper which is usually made from more expensive chemical pulp.

**BOGUS MILL WRAPS:** Heavyweight paper made of screenings, jute, rope, sulfate, and wastepaper used for wrapping paper mill runs, large rolls, or paper put into bundles.

**BOGUS PAPER:** *See* BOGUS BACK LINING.

**BOGUS PASTING PAPER:** A class of inferior-type paper made specifically to replace higher quality paper and suitable for adhesion to other materials, such as boards and other paper products.

**BOGUS SATURATING PAPER:** A lower quality saturating paper used in place of regular higher quality material as a base for impregnation with asphalt, tar, and similar substances.

**BOGUS SCREENINGS:** A class of inferior papers made from trimmings, cuttings, recycled wastepaper, and screenings. It is used as a substitute for the original paper from which they have emanated.

**BOGUS TAG:** A type of tag board made from old newsprint and other wastepapers, lined to provide suitable surfaces for specific use where the higher quality conventional product is not required.

**BOGUS WRAPPING:** An inferior-type wrapping paper made primarily from recycled wastepaper stock with characteristics of, and used in place of, higher-quality product, especially where there is no requirement for higher strength.

**BOGUS:** Refers to a product that has been made from wastepaper or other inferior materials to imitate higher quality grades. The term is applied to many different types of paper and board falling into this category.

**BOILER BLACKOUT:** The loss of chemical reduction in the smelt bed of a sulfate recovery furnace due to the lack of sufficient energy being produced from the combustion reaction to sustain it. This is caused primarily by a decrease in black liquor feed density below a prescribed minimum for the particular burner nozzle design, usually accompanied by a loss of flame.

**BOILER BLOWDOWN:** Periodic or continuous drains from the drum and/or waterwall headers to remove spent precipitated feedwater treatment chemicals from the unit.

**BOILER DRUM:** A feedwater storage vessel located at the upper end of the evaporator section of a boiler. At the last stage of feedwater heating, by direct contact with steam from the evaporator section, the subcooled feedwater is heated to saturation. Mechanical separation of vapor and liquid takes place here.

**BOILER EFFICIENCY:** The ratio of the heat absorbed in the boiler to the heat supplied by the fuel. It can be calculated by taking the total heat value in BTUs produced in the steam in the boiler and dividing by the total heat value in BTUs in the fuel inputs, and is expressed in percentage.

**BOILER FEED PUMP:** The pump which supplies high-pressure feedwater to the drum of the power or recovery boiler. The pump is usually driven by a turbine or a motor.

**BOILER FEEDWATER:** Treated water that is pumped under pressure to the boiler section of a steam generator consisting of two steel drums connected by steel tubes, with the level closely controlled in the upper drum.

**BOILER HORSEPOWER:** Heat energy required by a boiler to evaporate water from 212°F into dry saturated steam at the rate of 34.5 lbs per hour.

**BOILER HOUSE:** *See* POWERHOUSE.

**BOILER TUBES:** Steel tubes that make up the evaporator section of a steam generator. They are arranged so that they connect two steel drums on upper and lower ends or are along the walls of a recovery furnace. Hot gases heat the tubes on the way to the stack and evaporate the water they contain into steam.

**BOILER WATER TREATMENT:** The chemical conditioning of boiler feedwater to soften it and make it less corrosive to the boiler tubes.

**BOILER:** Broad or general term for a steam-generating unit. It is referred to as an industrial boiler when primarily used to generate steam for process requirements such as in a pulp and paper mill, or as a recovery boiler when used in the chemical recovery cycle of a pulp mill.

**BOILING POINT (BP):** The temperature at which water is converted into saturated steam at a specified pressure in the evaporation section of a boiler.

**BOLE:** The trunk of a tree.

**BOLT:** (1) A short, round section of a log. (2) A roll of wallpaper of specified length.

**BOND CIRCULAR PAPER:** A lightly-sized, vellum-finished writing or printing paper having good folding characteristics. It was once used primarily for publicizing bond issues, but is now used for making other types of circulars and additional purposes suitable for this type of paper.

**BOND STRENGTH:** The intralayer binding force in a multi-ply paperboard or laminate. The term also refers to the degree of adherence of coating and film on a sheet surface, and to the interfiber binding force within a sheet.

**BOND:** Name applied to a class of writing and printing paper made from bleached chemical wood pulps and cotton fibers. It has good surface, strength, and formation characteristics. It was first used for the printing of bonds, stock certificates, and legal documents, but now also used for business letterheads, forms, and other printing purposes.

**BONDED AREA:** Term used to designate the external surface area of the fiber in a sheet of paper involved in the interfiber bonding.

**BONDING:** The sticking together of papermaking fibers, which provides toughness and strength to the paper

**BONDING AGENT:** A material added to stock that will increase the strength of paper made from it by increasing the cohesion of the fibers.

**BONDING STRENGTH:** The intralayer binding force in a multi-ply paperboard or laminate. It also refers to the degree of adherence of coating and film on a sheet, and to the interfiber binding force within a sheet that provides resistance to lifting or picking of its fibers on the surface and splitting of the paper when used.

**BONE DRY (B.D.):** (1) A descriptive term for the moisture-free conditions of pulp and paper. *See* OVEN DRY (O.D.). (2) Refers to air containing absolutely no vapor.

**BONE DRY WEIGHT:** Moisture-free weight of materials involved in the pulp and paper manufacturing process.

**BONE FIBER:** A type of fiber made from chemically gelled cotton or alpha-wood cellulose fibers. It is used for general and electrical purposes, but has characteristics of higher stiffness, hardness, and density than other types within this classification.

**BONE GLUE:** A protein derived from animal bones, skins, sinews, and tendons. It is used as a surface-sizing agent for high-quality rag and rag content papers such as bank note papers. *See also* ANIMAL GLUE.

**BOOK BACK LINER PAPER:** A good-strength paper made from sulfate chemical pulp. It is used in bookmaking to bind the signatures of sewn books by gluing it to the book backbone. This keeps the book backbone away from the cover backbone.

**BOOK BACKING PAPER:** Similar to or the same as *book back liner paper.*

**BOOK BASIS:** The weight of the trade custom ream of book paper consisting of 500 sheets, 25 by 38 inches in size, expressed as 25 x 38-500. *See* BASIC SIZE.

**BOOK BINDING BOARD:** *See* BINDER'S BOARD.

**BOOK BULK:** The thickness of a given number of pages of a book as contributed by the caliper of the paper used to make the book.

**BOOK COVER PAPER:** A type of coated or uncoated paper made from long-fiber chemical pulp and cotton fiber. It is used for printed title and protective fold-over covers for hardcover books. The term also refers to a heavyweight sulfate paper used as a protective covering for school, library, and other books. Also called *book wrapping paper.*

**BOOK END PAPER:** A good fold-strength paper made from chemical pulp and/or cotton fibers. It has good pasting qualities and is used to cover the inside of book covers.

**BOOK FASHION INSPECTION:** A paper mill finishing room paper inspection and sorting procedure where a few inches (thickness) of a paper pile is checked at a time. One-half the area of one side of the sheet is inspected in a fan-like fashion and rolled back to its original position. Paper flags are inserted when defects are noted, and these sheets are removed when the accepted sheets are transferred to a new pile.

**BOOK JACKET PAPER:** *See* BOOK COVER PAPER.

**BOOK LINING PAPER:** *See* BOOK END PAPER. Also hard-sized for use as a chip and strawboard liner.

**BOOK MATCH BOARD:** A stiff, absorbent, cylinder machine paperboard made from recycled wastepaper stock with a lining of mechanical pulp. It is used to make stems for book matches. Sometimes called *match stem stock.*

**BOOK MILLS:** Converting plants that manufacture tablets, notebooks, memo pads, and other types of writing, recording, and bookkeeping ledgers from paper and paperboard shipped in from paper mills.

**BOOK PAPER PREMIUMS:** High-cost papers whose most important qualities are high opacity and low caliper. These papers also have very high longevity properties and are used in such products as Bibles and encyclopedias.

**BOOK PAPER:** Types of paper with similar physical characteristics and properties that make it suitable for printing and other uses in the graphic arts industry. Also, papers used for making tablets, envelopes, gumming, and other types of converted paper products. *See also* BOOK MILLS.

**BOOK SHAVINGS:** Reclaimed book paper stock consisting of unused, unprinted trimmings and other waste from bookmaking operations. Sometimes called *soft white shavings.*

**BOOK STOCK:** Book paper made from fiber reclaimed by removing ink, coloring material, and filler from wastepaper. It is mixed with fresh fiber, depending on the quality required for its intended use.

**BOOK WRAPPING PAPER:** *See* BOOK COVER PAPER.

**BOOKBINDER'S PAPER:** A general reference to papers used in binding books. It has good fold-tear strength and is usually made from chemical pulp and/or cotton fibers.

**BOOKBOARD:** A general reference to paperboards used in the book-binding industry to make front and back covers of hardbound books. *See* BINDER'S BOARD.

**BOOKKEEPING-MACHINE PAPER:** Grades of paper made with appropriate characteristics, especially to be cut into sheets and rolls for use in bookkeeping machines.

**BOOKLET COVER:** A reference to coated or uncoated plain, decorated, or embossed paper and board used to bind small books.

**BOOM:** A pocket formed by floating logs, with end touching end, during the process of hauling logs from a river into a log haul-up, ahead of a log slasher.

**BOOSTER:** A generator inserted in series in a circuit to change its voltage. A booster may be driven by an electric motor (in which case it is termed a *motor booster*).

**BOOT BOARD:** Fiberboards made of wood pulp (mainly recycled fiber), leather, and other natural and/or synthetic fibers (sometimes a combination of such materials), and used for counter, heel, innersole, shank, and other parts in the manufacture of boots.

**BORATE:** A term used in reference to sodium borate ($Na_2B_4O_7$) from which borax is prepared. It is used with casein in the preparation of coating for paper and paperboard.

**BORAX:** A sodium-borate chemical compound used in dissolving casein when used in coating paper and paperboard.

**BORED ROLLS:** Paper machine rolls that have been drilled with passages through which fluid heating or cooling medium is passed to control the surface temperature.

**BOROHYDRIDES:** Bleaching chemicals containing sodium borohydride ($NaBH_4$) used alone and with chlorine dioxide ($ClO_2$) to obtain very high brightness pulps.

**BOSS MACHINE TENDER:** The person on the paper machine crew who is first in command and responsible for the overall operation of the paper machine.

**BOSS:** A localized projection on the external surface of a pipeline, fitting, tank, or process vessel provided for the attachment of vents, drains, measurement connections, or other accessories.

**BOTTLE CAP BOARD:** A heavily-sized paperboard, usually saturated with wax, and cut into round disks for use in bottle stoppers.

**BOTTLE LABELING PAPER:** Specially coated and sized paper made primarily from chemical pulp. It has a good printing surface and other characteristics to make it suitable for use as labels on all types of glass bottle containers.

**BOTTLE PACKING BOARDS:** An inferior grade of wrapping paper (indented, corrugated, and strawboard) used to provide packaging protection in the shipment of bottles.

**BOTTLE WRAPPING PAPER:** Types of paper used to make wrappers for bottles to provide protection from breakage, exposure to light, and tampering.

**BOTTLE WRAPS:** Types of bottle-packing boards and wrapping papers used by the bottle industry to provide protection against breakage and light exposure of the bottle contents.

**BOTTLE-UP:** The temporary stoppage of the operation of a continuous digester in such a way that the pressure is sealed in the system.

**BOTTOM COUCH ROLL:** A hollow-drilled roll located under the fourdrinier wire (forming fabric) at the press section end of the paper machine. It contains a suction box to aid in the removal of waste from the sheet, and is usually the main drive roll for the entire fourdrinier section.

**BOTTOM FELT PRESS:** The press through which the bottom felt and paper sheet are run in the press section of a paper machine.

**BOTTOM FELT:** The felts in a multi-felt press section of a paper machine that are located on the wire (forming fabric) or bottom side of the paper and serve as a carrier of the sheet through the press.

**BOTTOM PRESS ROLL:** A rubber-covered felt press roll which is located at the bottommost location of the press and contacts it with applied pressure. Normally it contains suction boxes to improve water removal capacity.

**BOTTOM PRESS:** Refers to the press through which the bottom felt and paper sheet pass in the press section of a paper machine.

**BOTTOM SIDE:** The wire (forming fabric) side of a sheet of paper formed on a fourdrinier wet end of a paper machine.

**BOUGUER'S LAW:** In the appearance evaluation of paper, it states that equal layers will absorb equal fractions of radiant flux entering them.

**BOUND WATER:** The form in which hydroscopic water is present in a sheet of paper, such as mechanically and chemically bound to the fibers. Mechanically bound water exists within the fiber and fiber network spaces and adheres due to surface tension forces. Chemically bound water exists as a hydration bond with cellulose molecules in the fiber or other substances in the paper.

**BOW WAVE:** A type of finish that produces a spreading two-toned, ripple-like appearance. It is caused by varying the thickness of a sheet, with thicker areas appearing lighter than the thinner areas.

**BOWED BAR:** *See* SPREADER.

**BOWED ROLLS:** Types of spreader rolls on a paper machine, especially designed and constructed with a bend to prevent wrinkle formation in a sheet. On some rolls, this bend is adjusted mechanically or by applying internal hydraulic pressure.

**BOWL:** Calender and supercalender rolls. Also refers to paper used in the construction of supercalender rolls.

**BOWL GLAZING:** Applying a high surface finish on paper during the calendering operation. This is accomplished by passing the sheet through chilled metal rolls turning at different peripheral speeds.

**BOWL PAPER:** A soft paper made from rope, jute, and rag stock, used in making up embossing machine and supercalender rolls. *See* CALENDER ROLL PAPER.

**BOX CLIPPINGS:** Reclaimed stock made up of trimmings and cuttings from paperboard and used in boxmaking operations.**BOX COVER PAPER:** A plain or coated paper made from chemical pulp mixed with groundwood and rag pulps, generally colored and embossed, used to cover paper boxes.

**BOX ENAMEL PAPER:** A one-side coated, highly glazed paper used for covering boxes.

**BOX LAYER BOARD:** Paperboard made from reclaimed paper stock and used to make separators in shipping containers and packages.

**BOX LINER:** A class of waterproof parchment and papers used by the food industry to line boxes, crates, and other food containers to retain product freshness. Also, a grade of paper made from mechanical and chemical pulp used to cover chipboard to be made into less expensive boxes.

**BOX MACHINE:** Equipment used to automatically cut, score, bend, and paste paperboard together to form box containers.

**BOX STAY TAPE:** Rolls of narrow-width, high-strength paper, usually gummed on one side, made from sulfate pulp and rope fiber, and used to reinforce the edges and corners of boxes and shipping cartons. Sometimes it is cloth-lined to add more strength.

**BOX:** A container made by cutting, scoring, folding, and pasting paperboard together.

**BOXBOARD CONTAINER:** A container constructed from plain, lined, or clay-coated paperboard. It is made from wood pulp or paper stocks, or any combination of these.

**BOXBOARD:** Any plain, lined, or clay coated paperboard made from chemical and mechanical pulps mixed with reclaimed paper stock, and used to produce boxes and other containers.

**BOXED WRITINGS:** Paper that has been cut to a specified size to fit prescribed boxes for merchandising and other uses.

**BOYLE'S LAW:** The volume of a given mass of gas varies inversely to the pressure, provided the temperature remains constant.

**BP:** *See* BOILING POINT.

**BRACKET TRIMMER:** Paper cutting machine equipped with automatic spacers to trim paper. *See* CUT SIZE.

**BRAILLE PRINTING PAPER:** A paper made to be embossed with a series of dot patterns used by the blind in touch-reading. The sheet must have a smooth surface so the dots will be pronounced, uniform caliper so the dots are all the same height, and the right amount of stiffness so the dots will become compressed. Usually the paper is embossed or "printed" wet.

**BRAKE:** The mechanism on unwind and windup stands that will slow down the speeds of each as the size of the windup roll increases and the unwind roll decreases. The term also refers to other machine speed control mechanisms anywhere in the pulp and papermaking operation.

**BRAKING:** The process of controlling speeds of pulp and papermaking machines by retarding the motion or speed of the moving member.

**BRASH PAPERS:** Paper made from pulp stock consisting of brittle types of fiber and additives resulting in sheets that are brittle. *Also see* BRASH PULP.

**BRASH PULP:** Pulp stock made up of fibers that have been treated to make them brittle.

**BREAD BAG PAPER:** A machine-glazed or finished paper made for converting to bags to be used for packaging bread. It is made from both bleached or unbleached chemical and mechanical pulp.

**BREAD LABEL:** A lightweight paper, coated on one side, and made specially to be used to print brand names which are attached to the bread wrappers by bakeries.

**BREAD WRAPPER:** A thin, waxed paper, generally made opaque for printing by titanium dioxide, and used to wrap bread.

**BREAK DETECTORS:** Mechanical and optical devices located at strategic positions on a paper machine in order to sense and alert the operator of any disruption of the paper web (break) during the machine operation.

**BREAK IMPULSE:** An impulse in the mill electrical system due to a temporary interruption of current.

**BREAK:** (1) Denotes a tear completely across the sheet of paper or paperboard that occurs while the paper machine is running or during some subsequent conversion operation. (2) When referenced to an electrical circuit device in the mill, it is the minimum distance between the stationary and movable contacts when these contacts are in the open position.

**BREAKDOWN MILL:** A facility that breaks down large logs into smaller ones before they are cut up into pulpwood logs.

**BREAKER BARS:** Metal members mounted on the surface of the bedplate and roll of a beater. They serve as a pulper to break up dried sheets of pulp.

**BREAKER BEATER:** A beater used as a pulper to break up dried sheets of pulp.

**BREAKER ROLL:** A roll that rides on a sheet of paper over the suction couch on the fourdrinier part of the paper machine when an open draw transfer is used. Sometimes referred to as a *lump breaker roll* or a *presser roll.*

**BREAKER STACK:** Calenders located between coating operations, when more than one coat is applied, to smooth the paper following the base coat application. Also used in some dryer arrangements.

**BREAKER TRAP:** A device used to perform a coarse fiberiation on unbleached pulp, especially semichemical pulp, before refining and washing.

**BREAKER:** An electrical current interrupter.

**BREAKING DOWN:** *See* SLABBING DOWN.

**BREAKING LENGTH:** The calculated limiting length of a strip of paper or board of any uniform width beyond which, if such a strip were suspended by one end, it could break by its own weight. Its value is calculated from the tensile strength and the basis weight of the sheet in expressing paper and pulp testing results. *See also* TENSILE STRENGTH.

**BREAKING STRENGTH:** Resistance of paper to break or tear under tension as determined by applying a weight at one end of a strip of paper of standard dimensions.

**BREAKING:** (1) Process of reducing dry mill waste, old papers, dried pulps, and lapped pulps to pulp form, with enough suspension water to make it suitable for beating and/or refining and sufficiently free from bunches or sheets. (2) The cracking of the adhesive layer on gummed paper so that it will not bend when fed to the printing press.

**BREAST ROLL:** A large metal or fiberglass-covered roll located so as to support the fourdrinier wire (forming fabric) at the headbox end of a paper machine where the stock inlet admits the stock to the wire. It is usually driven by the wire which runs over it.

**BRICK:** A brick such as an ordinary building brick or refractory brick, or an abrasive block (pulpstone segment) used to dull a pulpstone.

**BRICKING:** Dulling the surface of a pulpwood grindstone with a brick or abrasive block pressed against its face while revolving.

**BRIDGE TREE:** A heavy cast iron or steel bar separating and holding the side frames of a pulpwood grinder to which holding bolts are attached.

**BRIDGING:** The arching of chips, pulp stock, etc., in a storage bin, usually located over the discharge opening or outlet device at the bottom. This impedes the movement of material, creating a discharging problem. It can also occur in other locations in storage bins. *See* ARCHING.

**BRIGHT ENAMELS:** A class of highly polished paper that has been coated on one side, finished on calenders or brush polishers, and used for making labels.

**BRIGHTENER:** Fluorescent dye or pigment that absorbs ultraviolet radiation and re-emits blue light, thereby effecting a white appearance to a sheet of paper when added to the furnish stock. Also called *whitener.*

**BRIGHTNESS CEILING:** The brightness value level of pulp which cannot be exceeded regardless of additional amounts of bleaching chemicals used or number of stages and combinations used.

**BRIGHTNESS DROP:** The percentage decrease in brightness due to the aging of pulp, paper, and paperboard.

**BRIGHTNESS METER:** Any of several instruments used to measure luminosity (brightness) of pulp and paper during bleaching and sheet formation stages. Also, meters used to measure recovery furnace luminosity.

**BRIGHTNESS REVERSION:** The loss of brightness of bleached pulp, paper, and paperboard over a period of time. Also known as *color reversion* and *yellowing.*

**BRIGHTNESS TEST:** Laboratory procedures used to determine the degree of whiteness of pulp, paper, and paperboards, and reported in values as specified by the type of equipment and method used.

**BRIGHTNESS:** A measure of the degree of reflectivity of a sheet of pulp or paper for blue light measured under specified standard conditions. Same as *whiteness. See* G.E. BRIGHTNESS and ELREPHO BRIGHTNESS.

**BRIMSTONE:** Sulfur used in the preparation of cooking acid in the sulfite process.

**BRISTLE MARKS:** Indentations in the surface of coated paper caused by bristles that have fallen off coating brushes and become stuck to the calender roll surface.

**BRISTOL BOARD:** A good-quality type of cardboard once commonly made from plied-together, high-rag content papers (called *bristols*) with a caliper thickness of more than 0.006 inch. Now, it is more commonly made on a fourdrinier or cylinder machine.

**BRISTOL MARK:** A type of paper defect in brush-coated papers consisting of a small dent caused by coating brush bristles coming off and sticking to calender rolls or any other surface on which it will impress its image.

**BRISTOLS:** Types of heavy grade papers possessing higher than average quality characteristics. They are rigid enough to be used in the manufacture of various announcements, invitations, postcards and other similarly used products. Originally made in Bristol, England, from reclaimed rags.

**BRITISH THERMAL UNIT (BTU):** A unit of energy used in the measurement of heat. It is equivalent to $1/180$ of the quantity of heat required to raise the temperature of 1 pound of liquid water from 32° to 212°F at standard atmospheric pressure. (A slightly less accurate definition is the quantity of heat required to raise the temperature of 1 pound of liquid water 1°F.

**BRITTLENESS:** Refers to the extent that paperboard will crack when bent, or to the degree that it can be bent before cracking.

**BROAD FOLD:** In the folding operation of paper finishing, it refers to the fold made with the grain of the sheet running in the shorter dimension. The term is also used in bookmaking to describe the type of fold made on a printed page so that it is wider than the book width.

**BROADLEAF TREE:** A deciduous (hardwood) tree having characteristically broad, flat leaves, as distinguished from coniferous (softwood) trees having long, needle-like leaves.

**BROADSIDE:** A piece of paper printed on one side.

**BROCADE PAPER:** A type of decorative covering paper whose characteristics are raised and marble-like surface designs created by an embossing procedure.

**BROKE BEATER:** A beater equipped with steam jets adjacent to the back fall area and used for repulping wet strength broke from paper machines.

**BROKE BOX:** A large wood container on wheels which the broke boy uses to haul broke from the source to the broke pit.

**BROKE BOY:** A term used to refer to the paper mill employee who hauls broke from the machine to the broke pit. Sometimes called *reel boy, broke hustler,* and *plugger.*

**BROKE BUNDLE:** A gathering of a quantity of broke which is sometimes tied up into large size bales to facilitate transportation.

**BROKE HANDLING:** A combination of pneumatic and mechanical conveying equipment which carries paper trimmings and paper damaged as a result of breaks on the paper machine to the broke pulper.

**BROKE HUSTLER:** *See* BROKE BOY.

**BROKE PIT:** A hole in the floor of a paper mill building, usually located over broke pulpers or broke conveyors to pulpers, into which broke from the process is dumped.

**BROKE PULPER:** A piece of machinery which disintegrates paper trimmings and paper damaged as a result of breaks on the paper machine and finishing operations, or anywhere in the process of paper manufacturing, and mixes the trimmings and damaged paper with water to form a stock slurry to be returned to the papermaking operation.

**BROKE:** Paper trimmings or damaged paper as a result of breaks on the paper machine and in the finishing operations, or anywhere in the process of paper manufacturing. It is usually returned for reprocessing.

**BROKEN CARTON:** Refers to any specified quantity of paper and paperboard which is less than the standard carton quantity established for it.

**BROKEN CASE:** Reference made to any specified quantity of paper and paperboard which is less than the standard case quantity established for it.

**BROKEN EDGE:** Defect in the outer edges of paper rolls and sheets which have been caused by mechanical malfunctioning of process equipment or through careless handling in preparation for and during transportation from one place to another.

**BROKEN PACKAGE:** Package of paper from which sheets have been removed so that it will meet the number specified in an order.

**BROKEN REAM:** Reference made to any specified quantity of paper which is less than the standard ream established for the grade under consideration.

**BROKEN:** An older paper mill reference to paper trimmings or paper damaged as a result of breaks in the paper machine and finishing operations (broke) which might be reprocessed or considered as waste.

**BROMIDE PHOTOGRAPHIC PAPER:** A silver bromide emulsion-coated paper base used in photography as a printing paper for making enlargements.

**BRONZE CREPE PAPER:** A type of crepe paper that has a bronze surface finish achieved by coating it with copper, gold, or silver metallic-based substances.

**BRONZE EMBOSSING PAPER:** A bronze-finished paper used by an embosser to imprint designs and patterns.

**BRONZE FLECK:** A type of paper defect with a characteristic crystal-line bronze speck appearance caused by metallic particles picked up during processing. Also called *bronze specks.*

**BRONZE PAPER:** Paper that has been coated with a special finely divided metallic powder-base coating material to impart a bronze-like appearance to the surface.

**BRONZE SPECKS:** *See* BRONZE FLECK.

**BRONZE:** A metallic, copper-base alloy with tin and sometimes other metals, once used extensively in the paper industry to make paper machine wire, rolls, and other items in the papermaking process.

**BRONZY COLOR:** In the appearance evaluation of paper, it refers to the metallic-like color observed by viewing the surface in the general direction of specular reflection, producing a hue which is usually quite different from that observed in other directions.

**BROOMED LOG ENDS:** Log ends frayed during the barking operation which will produce undesirable shives on grinding to produce groundwood.

**BROWN GROUNDWOOD:** Type of groundwood pulp in which logs are steamed in digesters prior to grinding. The steam is used to soften the wood but also discolors it so that the resulting pulp becomes dark brown. Also called *brown pulp.*

**BROWN PAPER:** Grades of paper made from unbleached pulp stock such as corrugating medium, linerboard, and kraft paper.

**BROWN PRINT PAPER:** A cotton fiber and chemical pulp paper once used extensively in reproduction work (e.g., making blueprints). It has now been largely replaced by other types of reproduction papers.

**BROWN PULP:** *See* BROWN GROUNDWOOD.

**BROWN WRAPPING PAPER:** A high-strength, brown-colored paper used for general wrapping purposes and most commonly made from unbleached kraft paper.

**BROWN:** In the appearance evaluation of paper, it refers to the color name for perception of dark yellow or dark orange.

**BROWNSTOCK WASHER:** Pulp mill equipment that continuously removes washable spent cooking liquors and intercellular materials dissolved by the cooking liquors during the cooking process. There are various designs, with the most common being those operations based on the principles of vacuum and diffusion.

**BROWNSTOCK WASHING:** A pulp mill process consisting of the removal of washable spent cooking liquor and intercellular material dissolved by the cooking liquors during the cooking process.

**BROWNSTOCK:** Brown-colored pulp from the cooking process and up to the bleaching process in a chemical pulp mill.

**BRUISING:** The mechanical action on the fibers during refining that causes the loosening of thread-like elements from the fiber inner wall. *See* FIBRILLATION.

**BRUSH CALENDER:** A paper calender with brush rolls usually located immediately following a paper machine calender in order to achieve higher surface gloss and smoothness.

**BRUSH COATER:** Paper coating equipment that applies coating material to the surface of the paper by the use of a revolving cylindrical brush, and uses oscillating flat brushes to smooth it out.

**BRUSH COATING:** A method of coating paper using various types of brushes.

**BRUSH DAMPENER:** A machine used to moisten paper, usually of the glassine type, before it is calendered by employing rotating brushes.

**BRUSH DUSTER:** A sheet-cleaning device used on a cylinder paperboard machine. It consists of brushes enclosed in a suction box extending across the sheet. It is used to remove undesirable felt hairs that may become embedded in the sheet during manufacture.

**BRUSH ENAMEL PAPER:** A specially coated, smooth paper with a highly brilliant surface obtained by polishing the sheet with brushes before it is calendered.

**BRUSH FINISH:** The high finish applied to coated paper by running it under stiff brushes before calendering.

**BRUSH GLAZING:** The process of obtaining a high glaze finish on a sheet of paper by the use of cylindrical brushes.

**BRUSH MARK:** Coated paper surface defect caused by faulty or misaligned brushes during operational adjustments made when brush coaters are used to apply coating material to paper.

**BRUSH POLISHING MACHINE:** A machine used to obtain a high gloss on paper by the use of cylindrical brushes polishing the surface of the paper.

**BRUSH POLISHING:** Applying a high gloss to the surface of specially coated paper by the rubbing action of a revolving cylindrical brush turning at a higher peripheral speed than the paper.

**BRUSH ROLLER:** The cylindrical drum on some types of brush polishing machines that carries the brushes.

**BRUSH-FINISH COATING:** A paper coating procedure in which paper, after the coating material is applied, is given a highly polished finish. This is accomplished by the rubbing action of a number of revolving brushes fitted on a revolving drum, over which the sheet runs.

**BRUSH:** (1) The cylindrical and flat brush elements of a paper coater which applies and smooths the coating material on the paper surface. (2) The carbon contacts made to the rotor of electric motors and generators.

**BRUSHING OUT:** The process of producing thread-like elements extending from the inner and outer wall of fibers during the refining operation of pulp. This aids formation of a sheet on the paper machine by providing increased fiber-to-fiber bonding. Also called *fibrillation.*

**BRUSHING:** (1) The mechanical action on the fibers during refining that causes the loosening of thread-like elements from the fiber outer wall. *See* FIBRILLATION. (2) The act of applying a wire brush to a revolving pulpstone surface to clean the stone. It is used sometimes in lieu of a burr to extend the burring cycle.

**BSCMP:** *See* BLEACHED SEMICHEMICAL MECHANICAL PULP.

**BTU:** *See* BRITISH THERMAL UNIT.

**BUBBLE COATING:** A paper coating procedure using a technique in which the brightness and opacity of a coating film containing little or no pigments is increased by the introduction and dispersion of microbubbles.

**BUBBLES:** Blisters occurring between two plies of a board made on a cylinder machine due to a lack of ply adhesion. Also called *air bubbles, bells,* and *blows.*

**BUBBLING:** A technique of introducing microbubbles into dry coating to increase its brightness and opacity.

**BUCKET CONVEYOR:** A means of transporting dry material from one place to another in a pulp and paper mill. It consists of buckets attached to single or multiple chains, ropes, or belts.

**BUCKET ELEVATOR:** Elevating conveyor consisting of buckets attached to single or multiple chains, ropes, or belts. It is used to transport dry material in a pulp and paper mill.

**BUCKLE:** A paper roll defect caused by variable thickness in the paper sheet resulting in an undulating surface on the outside of the roll. Also referred to as *bars, corrugation marks, rope marks,* and *tire tracks.*

**BUFFER LIQUOR:** A chemical solution added to the cooking liquor in a neutral sulfite semichemical (NSSC) cook in order to neutralize the organic acids formed in the initial stages. It also helps regulate the pH. The most common buffer liquors are sodium bicarbonate ($NaHCO_3$), sodium carbonate ($Na_2CO_3$), and kraft green liquor ($Na_2CO_3$ and $Na_2S$).

**BUFFER STORAGE:** The use of huge pulp storage chests between the pulp mill and the paper mill in order not to disturb the operation of the pulp mill by changes in pulp consumption and, in the same way, to eliminate disturbances in paper production originating from the pulp mill. It is also used as a means of equalizing the pulp quality.

**BUFFERING AGENT:** Chemical added in the sulfite pulping process to regulate the pH of the cooking liquor, e.g., $Na_2CO_3$ added to $Na_2SO_3$ or aqueous ammonia added to $(NH_4)_2SO_3$.

**BUFFERING RATIO:** Ratio of the equivalents of buffering agent to equivalents of sulfite in the sulfite pulping process.

**BUFFERS:** (1) Various devices or materials used in pulp and paper mills to reduce shock due to physical contact. (2) Chemicals used in the pulp and papermaking process capable of neutralizing both acids and bases in solution and thereby maintaining the original hydrogen ion concentration (pH).

**BUFFING PAPER:** A type of paper coated with flint grains (to make it abrasive) and used to rub a smooth, velvety finish on materials such as leather, etc.

**BUILDING A REEL:** The process of winding up the continuous sheet of paper coming off the dry end of a paper machine into a large roll of paper called a *reel.*

**BUILDING BOARD:** General reference to all types of paperboards, such as wall, plaster, panel, sheathing, and insulating board, used in the construction of buildings.

**BUILDING PAPER:** Types of paper produced from strong fiber stocks, such as kraft, rag, and selected pulp screenings, processed and treated for use in the building trade as sheathing, roofing, and vapor barriers.

**BUILT-IN STRESS OR STRAIN:** A physical property of paper attributed to the type of stress or strain developed as a result of tension caused by shrinkage during the drying process.

**BULGE RESISTANCE:** A property of felts that resists the formation of slack places. *See* BULGE.

**BULGE:** A slack piece in the felt, usually about 2-feet wide and 3-feet long, caused by that part of the felt being held during a jam in the dryer, or by allowing a thick, bulky wad of wet broke to run around the circumference of a dryer or dryer felt roll.

**BULGING:** The process of producing a slack place in a felt. *See* BULGE.

**BULK DENSITY:** Bone dry weight of wood used in a pulp mill cooking process. It is expressed as pounds per cubic feet in North America. Also referred to as *packing factor.*

**BULK FACTOR:** (1) A value of sheet bulk obtained by dividing bulk thickness by air-dried basis weight, and is the same as the apparent specific volume, i.e., the inverse or reciprocal of specific density. (2) Wood used in a pulp mill cooking process.

**BULK HANDLING:** The transporting, treating, and processing of materials in large masses and volumes.

**BULK INDEX:** The value of bulk obtained by dividing the caliper of the sheet by the air-dried basis weight.

**BULK MODULUS:** A characteristic of a material which is related to the ratio of pressure to the fractional change in volume produced by the applied pressure.

**BULK NUMBER:** The caliper of eight sheets of paper divided by eight, the result divided by the basis weight, and this result multiplied by ten.

**BULK STORAGE:** Storing of materials in large masses and volumes.

**BULK THICKNESS:** A physical dimension for a sheet of paper in a pile, calculated by dividing the total thickness of the pile by the number of sheets it contains.

**BULK:** The compactness property of a sheet in relation to its weight (whose value decreases as the compactness increases). It is measured as the thickness of a pile of a specified number of sheets under a specified pressure.

**BULKHEAD:** An upright partition in storage chests, bins, and tanks, separating them in a number of compartments. It also refers to entrances to these storage areas.

**BULKING BOARD:** A paperboard made to have less weight per sheet or thickness by applying less calender pressure than the same paperboard calender using normal pressures.

**BULKING BOOK PAPER:** A paper made (with a minimum of fillers) out of pulp grades especially chosen and mixed to impart maximum bulk and finished to specifications and caliper required for its use as book paper.

**BULKING DUMMY:** A number of standard sheets used as a bulk gauge for sheets to be made into printing, book paper, etc.

**BULKING NUMBER:** A bulk measurement made by taking the caliper of a specified number of sheets of paper and dividing that number by its basis weight.

**BULKING PAPER:** *See* BULKING BOOK PAPER.

**BULKING PRESSURE:** The pressure used in a test to determine the bulk of paper.

**BULKING THICKNESS:** The thickness of paper determined by applying a known specified pressure to a fixed number of sheets. *See* BULK.

**BULKY:** A descriptive term for paper and paperboard made with a maximum of bulk.

**BULL SCREEN RAKERS:** Devices used on pulpwood grinders to facilitate the discharge of rejects from the bull screen.

**BULL SCREENS:** Coarse screens with large perforations onto which pulp from a grinder flows by gravity at one to three percent consistency. They are used to remove knots, slabs, large slivers, and other large objects before further screening. These may be vibrating flat screens or large, horizontal, rotating, perforated cylinders.

**BULL'S EYE:** A roll of paper made from one continuous sheet in which no splices were required.

**BULLNOSE:** The indented portion of a kraft recovery boiler located just below the superheater.

**BUNCH PLATEN FINISH:** Additional smoothness and uniformity added to the surfaces of paper by subjecting small stacks of it to mechanical pressure between the plates of a platen press.

**BUNCHED PLATED:** A special finish applied to a small stack of paper by passing it through the plates of a platen press and applying mechanical pressure in such a way that both sides of the sheets are alike.

**BUNDLE:** (1) Fifty pounds of board consisting of a unit made up of a number of sheets as determined by size and thickness. (2) A package of paper and paperboard made up usually for shipping purposes.

**BUNDLING:** The making up of packages of paper and paperboard, generally for shipping and transporting purposes.

**BUOYANCY:** The property of a fiber and other paper making materials that tends to make it rise or float when submerged in water or other fluid.

**BURLAP FINISH:** A type of surface finish that has the characteristics of burlap cloth, effected on various kinds of paper by applying mechanical pressure to a pile of sheets alternately layered with burlap, covering calender rolls with burlap, or embossing it with burlap texture etched rolls on an embossing machine.

**BURLAP-LINED PAPER:** A special high-strength, water-resistant type of wrapping paper commonly made from sulfate chemical pulp lined with asphalted burlap cloth.

**BURLAP:** A coarse, woven fabric, usually made of jute.

**BURN OUT:** The phenomenon of a sheet of paper losing color or brightness during the drying process.

**BURNER GAS:** Sulfur dioxide ($SO_2$) gas produced in a sulfur burner during the preparation of sulfite cooking acids.

**BURNER:** (1) Rotary- or spray-type production equipment used to burn molten sulfur to produce sulfur dioxide ($SO_2$) gas which is used in the preparation of sulfite cooking acids. Also called *sulfur burners.* (2) A device for the introduction of auxiliary fuel and air into a furnace at the required velocities, turbulence, and concentration to establish and maintain proper ignition and stable combustion of fuel within the furnace.

**BURNING OF CHIPS:** The result of improper cooking of wood in digesters due to cooking chemicals imbalance, increasing temperature too fast, and overheating. It causes the chips to char before proper separation of the fibers, producing darkened pulp and incompletely cooked chips with burned centers.

**BURNING:** A method of determining if a sheet is a solid or pasted Bristol by igniting a corner. If it is a pasted sheet, it will char and layers will part and peel free. The number of plies can then be counted.

**BURNISHED FINISH:** A high-gloss paper and paperboard surface finish produced by polishing coated paper on calenders, burnishers, and glazing machines.

**BURNISHER:** A machine equipped with rubbing stones used to polish a hard, high-gloss finish on paper and paperboards.

**BURNISHING:** The polishing of paper and paperboard to produce a hard, high-gloss surface finish.

**BURNS:** Damage to a stone face, specifically an area on a pulpstone face in which embedded charred wood or other burned or charred material occurs, usually due to lack of shower water. It may also be caused by the presence of a material difficult to grind, such as stone, glass, etc., which raises the temperature in the area beyond the charring point.

**BURNT CENTERS:** Damaged, uncooked centers of wood when cooking liquor fails to penetrate through it during the cooking process, leaving that portion to be damaged by high temperature.

**BURNT COOK:** *See* BLACK COOK.

**BURNT PAPER:** Defective paper sheet which has lost its color and become very fragile.

**BURNT PULP:** Darkened pulp resulting from cooking wood in a digester under chemically unbalanced conditions, with a too-fast temperature rise and other improper procedures that will cause discoloration. Also referred to as *black cook.*

**BURR HOLDER:** A pulpwood grinder device made up of bearings and a spindle on which a sharpening tool is mounted.

**BURR IMPRESSION:** The design pattern or mark made by a pulpwood grindstone sharpening tool on the surface of the stone. Also known as *burr pattern.*

**BURR LEAD:** The slope angle of the teeth on the surface of a spiral-type pulpwood grindstone sharpening tool. Sometimes referred to as the *burr tooth angle.*

**BURR NUMBER:** A method of grading burrs on a pulpwood grindstone based on the number of points they have per square inch of surface.

**BURR PATTERN:** *See* BURR IMPRESSION.

**BURR TOOTH ANGLE:** The angle formed by the axis and the tooth of a pulpwood grindstone sharpening tool. Similar to *burr lead.*

**BURR:** A special tool used to sharpen pulpwood grindstones. Also called a *bush roll* or *stone dresser.*

**BURRING:** (1) Resurfacing pulpstone with a surfacing tool called a burr. Also known as *dressing* or *jigging.* (2) A metallic paper machine wire problem which occurs when wear causes metal to feather out, resulting in decreased drainage. Normal causes are heavy loading on the wire, excessive vacuum on the suction boxes, or misaligned forming boards or deflectors.

**BURST FACTOR:** The strength of a sheet, determined by dividing its bursting strength by its basis weight.

**BURST RATIO:** The bursting strength of a paper or board in relation to its basis weight under specified conditions.

**BURST STRENGTH:** The maximum uniformly distributed pressure applied at right angles to its surface that a test piece of paper will stand before it breaks under the conditions defined in the standard test method. Sometimes called *Mullen* or *pop strength.*

**BURST TEST:** A rupture test made on a testing machine that determines the bursting strength of paper and board. Also called *pop test.*

**BURST TESTER:** A machine on which paper and board are tested for burst strength. Among the more common ones are the Cady, Mullen, and Ashcroft. Also called *bursting strength tester.*

**BURST:** (1) The pressure in pounds required to rupture one square inch of paper under stipulated testing procedures. The most common procedure to break the paper is by expanding a flexible diaphragm against a securely held circular area of paper. The Mullen Tester is an example. *See also* BURST STRENGTH. (2) Refers to a defect in paper characterized by a rupture in the web not extending to the edges.

**BURSTING STRENGTH TESTER:** An instrument used to determine the resistance of paper to rupture by the application of pressure or force to one side.

**BURSTING STRENGTH:** The resistance of paper to rupture when pressure is applied to a side by a specified instrument. It is usually determined on a Mullen tester and expressed in pounds per square inch (kilopascals per square meter). Sometimes referred to as *burst, Mullen,* and *pop strength.*

**BURSTING:** The rupturing of paper and board when pressure is applied to one of its sides. It is usually done on a specified instrument under specified conditions to determine its bursting strength.

**BUS:** A main electrical conductor which carries a heavy current for a number of users, such as machines.

**BUSH ROLL:** *See* BURR.

**BUSHING:** A detachable metal liner which forms the inner diameter of a grinder flange and comes between the grinder shaft and flange.

**BUSINESS COMMUNICATION PAPER:** Paper used in making commercial stationery such as checks, forms, copier and printing paper.

**BUTTER CARTON:** A white coated manila-backed paper board especially made to be waxed after printing for making containers for butter.

**BUTTERFLY VALVE:** A type of shutoff or regulating valve, distinguished by having a cylindrical body with a disk or vane on a shaft installed perpendicular to the axis of the cylinder. It is used in pulp and paper mills on low-pressure, large-size process lines, and can be hand, remote, or automatically operated.

**BUTTON BOARD:** A paperboard made from reclaimed wastepaper stock and used to make the inside base for inexpensive buttons.

**BUTTON CARDBOARD:** A bristol-type, lightweight board used for displaying and merchandising buttons.

**BUTTON SPECKS:** A defect appearing as tiny, scattered, light-colored spots on the surface of paper made with a rag content. It is caused by small pieces of buttons inadvertently carried over in the papermaking process.

**BUZZER:** A machine used to remove bark from lateral sections of trees before chipping.

**BY-PASS:** The routing or deflecting of liquid or any other flowing material from the main process stream around a part, such as a valve, meter, pump, etc., and back into the main stream.

**BY-PRODUCT:** Something produced during the pulp and papermaking process in addition to the principal products, such as tall oil, alcohol, carbon black, etc.

# C

**C-FLUTE:** The undulating geometric formation in corrugated board that is 5/32 inch high and runs 39 to 42 per linear foot. *See also* A-FLUTE, B-FLUTE, and E-FLUTE as designations for corrugations of other dimensions and counts per linear foot.

**C1S:** A notation designating paper or paperboard that is *coated one-side*, i.e., on one side only.

**C2S:** A notation designating paper or paperboard that is *coated two-sides*, i.e., on both sides.

**CABLE MARKING PAPER:** A lightweight, colored, lightfast paper made from sulfate pulp which is slit into narrow strips and used as a means of providing identification of ropes and cables.

**CABLE PAPER:** A strong, high dielectric paper with a low power factor, commonly made on a cylinder-type paper machine from sulfate pulp or manila rope fiber, and properly treated with insulating liquids so that it can be used as insulation on electrical wires and cables. *See also* ANTACID MANILA PAPER.

**CADMIUM YELLOWS:** Papermaking pigments varying from pure to slightly greenish yellow, and are modifications of cadmium sulfide (CdS).

**CADOXEN:** A common name for a solution made up of the proper proportions of ethylenediamine and cadmium oxide in water; once used as a solvent for cellulose.

**CAKE:** The dewatered pulp mat as it comes off a filter or press. Same as FILTER CAKE.

**CAKE BOARD:** A rigid manila-lined board or a waxed, corrugated board used by bakeries beneath and between layers of cakes prior to decorating, packaging, and shipping.

**CAKE PAN LINER PAPER:** Coated vegetable parchment or paper made from chemical pulp and used by bakeries to line pans, trays, and other baking utensils.

**CALCINATING:** The process of burning limestone or limemud ($CaCO_3$) to quicklime (CaO).

**CALCIUM BISULFITE [$Ca(HSO_3)_2$]:** The primary chemical ingredient in the calcium-base liquor used to cook chips by the sulfite process.

**CALCIUM CARBONATE ($CaCO_3$):** (1) A white pigment commonly used in the paper industry as a paper coating material and filler. It is a compound that can be chemically prepared but also occurs in a natural form. (2) A naturally occurring mineral, called limestone, usually combined with magnesium carbonate ($MgCO_3$), and used primarily in the pulping industry to produce lime (CaO).

**CALCIUM HYPOCHLORITE [$Ca(OCl)_2$]:** A common pulp bleaching chemical that has largely been replaced by sodium hypochlorite ($NaOCl_2$)

**CALCIUM OXIDE (CaO):** A chemical obtained by burning limestone ($CaCO_3$), also called lime and used in the causticizing area of a pulp mill, and mixed with green liquor to convert sodium carbonate ($Na_2CO_3$) into sodium hydroxide (NaOH). The calcium carbonate precipitate (lime mud) can then be burned in a kiln to produce more lime for reuse.

**CALCIUM SULFATE:** A white pigment, chemically prepared in the hydrite form ($CaSO_4 \cdot 2H_2O$), but also found in nature in both the anhydrite ($CaSO_4$) and the hydrate form sometimes called *crown filler.*

**CALCIUM SULFITE ($CaSO_3$):** (1) A white pigment used in papermaking as a paper coating material and filler. It is chemically prepared from the reaction of calcium chloride with sulfurous acid. (2) A paper filler used primarily in sized paper, but has also been used in unsized paper where bright, opaque, and high ink receptivity properties are desired. It is usually made by passing hot sulfur dioxide gas from a sulfur burner into a mechanically produced spray of milk of lime.

**CALENDAR BACK BOARD:** A type of stiff, lined paperboard made from chemical pulp mixed with recycled fiber and used as a mounting baseboard for various kinds of printed calendars.

**CALENDAR PAD BOND PAPER:** Various kinds of bond papers suitable for making up pad-type calendars.

**CALENDAR PAPER:** Any type of paper from which printed calendars are made.

**CALENDAR STOCK:** Lightfast types of paper and paperboard from which printed calendars are made.

**CALENDER:** A piece of paper processing equipment located at the dry end of the paper machine. It consists of a set of horizontal, metallic, cylindrical rolls, vertically stacked one upon the other, through which the paper sheet is passed for smoothing, leveling, and gloss improvement. Sometimes referred to as *calender stock.*

**CALENDER BLACKENED:** A reference made to paper whose color has been darkened because of calendering it while it is still wet or with too much pressure. This crushes the fibers and disorients the formation, resulting in not only a darkened color but also a reduction in opacity and surface characteristics. Also called *calender crushed.*

**CALENDER BROKE:** Paper damaged in some way while being calendered and discarded as waste or recycled back to the papermaking process.

**CALENDER COLORED:** Color-dyed paper and paperboard in which the dyeing medium has been added to the sheet during the calendering process. Also called *calender dyed.*

**CALENDER COOLING:** The operation of applying ventilating air to the calender rolls to eliminate excessive heat produced by friction between the high spots.

**CALENDER CRUMBS:** Tiny particles of paper that are created by and accumulate during the calendering process.

**CALENDER CRUSH FINISH:** A paper surface finish obtained by moistening the sheet while it is being calendered.

**CALENDER CRUSHED:** *See* CALENDER BLACKENED.

**CALENDER CUT:** A mark or cut occurring in the paper sheet as a result of wrinkles in the web while being calendered.

**CALENDER DYED:** *See* CALENDER COLORED.

**CALENDER FINISHED:** Reference made to paper and paperboard that has been passed through a calender to improve its surface characteristics by the application of pressure and friction, along with the addition of moisture or other surface-conditioning materials during the calendering operation.

**CALENDER MARKS:** Paper defects caused by faulty calender roll surfaces or by extraneous matter that has entered the nips and attached to their surfaces. *See also* CALENDER SCAB, CALENDER SCALE, and CALENDER SPOT.

**CALENDER PICK:** Type of paper defect produced during the calendering operation when areas of the paper surface stick to the roll as scales or as a lifted band. This occurs especially if the calender has to be stopped for some reason.

**CALENDER ROLL BLOCKING:** The adherence of paper to the calender due to improper temperature, pressure, humidity, or coating conditions. *See* BLOCKING.

**CALENDER ROLL PAPER:** An unsized, soft-type paper made especially for the manufacture of supercalender and embossing machine rolls. It is made from rope, jute, or rag fiber. Sometimes referred to as *bowl paper.*

**CALENDER ROLLS:** Cylindrical, metallic, or compressed cotton and paper rolls that are positioned one on top of the other in calenders and supercalenders.

**CALENDER SCAB:** A paper surface defect that is caused by extraneous material building up and sticking to the calender roll surface sufficiently to imprint itself on the paper while being calendered. *See also* CALENDER MARKS, CALENDER SCALE, and CALENDER SPOT.

**CALENDER SCALE:** A type of paper defect caused by extraneous material that collects on calender roll surfaces and then becomes actually embedded in the paper during the calendering process.

**CALENDER SECTION:** That portion of the paper machine where the calender is located.

**CALENDER SIZING:** The process of treating the surface of paper and paperboard with materials such as starches, polyvinyl alcohol, adhesives, and other sizing solutions at the calender stage of manufacture.

**CALENDER SPOT:** Transparent mark on paper made during calendering by extraneous material, such as fragments of paper, sticking to the surface of a calender roll. *See also* CALENDER MARKS, CALENDER SCAB, and CALENDER SCALE.

**CALENDER STACK:** *See* CALENDER.

**CALENDER-STACK CRUMBS:** A condition created at the calender when flecks of paper are produced by crushing action of rolls on small, thicker areas of the sheet.

**CALENDER STAINED:** Paper that has been surface-colored by the application of a staining solution to the sheet during the process of calendering it.

**CALENDER STREAK:** A type of paper defect resulting from improper pressing and drying of the paper web before calendering. It appears as relatively high gloss bands extending in the machine direction.

**CALENDER VELLUM FINISH:** A smooth, vellum-like paper surface finish achieved by the use of special surface-prepared rolls in calenders.

**CALENDER WRINKLE:** A paper roll defect appearing as a small unopened fold in paper caused by a blister entering the nip of rolls on the calender stack.

**CALENDERED PAPER:** Any paper that has been passed through a calender, after being dried on the paper machine, to further improve its surface levelness and smoothness while glazing it to improve its gloss.

**CALENDERING:** The process of leveling, smoothing, and the applying of gloss to the surface by running it through a calender.

**CALENDERMAN:** A common paper mill title used to refer to the calender operator.

**CALF BOARD:** A heavy paper embossed to resemble leather and used in the manufacture of books.

**CALF PAPER:** A bookmaking type of paper made with a surface that looks like leather.

**CALIBRATION:** (1) Ascertaining, usually by comparison with a standard, the locations at which scale/chart graduations should be placed to correspond to a series of values of the quantity which the instrument is to measure, receive, or transmit. (2) To adjust the output of a device to bring it to a desired value, within a specified tolerance, for a particular value of the input, or to ascertain the error in the output of a device by checking it against a standard.

**CALICO PAPER:** Type of paper with a surface resembling calico; used primarily for decorating purposes.

**CALIPER:** (1) The thickness of a sheet of paper or paperboard, measured under certain specifically stated conditions, expressed in units of thousandths of an inch. The units are called "mils" when referring to paper, and "points" when referring to paperboard. Also known as *thickness.* (2) The instrument for and the measurement of the diameter of logs and trees. It consists of a graduated beam at a right angle to a fixed arm and a movable arm. Also called a *micrometer.*

**CALIPER SHEAR BURST:** A paper roll defect referring to an irregular separation or rupture of the web, generally occurring between an area of relatively high and low caliper, and normally extending some distance in the machine direction.

**CALORIE:** A unit of thermal energy defined as the amount of energy required to raise 1 gram of water from 14.5°C to 15.5°C, and also referred to as the 15°C gram calorie. A kilogram calorie is the amount of energy required to raise a kilogram of water the same amount.

**CALORIFIC VALUE:** In the analysis of fuel, it is the number of BTUs that can be obtained from the combustion of a specified quantity, such as in 1 pound of coal.

**CALORIMETER:** An optical test instrument or device for determining and specifying colors, sometimes by comparison with a standard color set.

**CAMBIUM:** A thin formative layer of cells between the bark and wood of trees and other woody plants. It gives rise to new cells and is responsible for the growth in width.

**CAMBRIC:** Thin, fine linen used to emboss the surface of paper to resemble the surface of linen or cotton cloth.

**CAMBRIC FINISHED PAPER:** Paper whose surface has been plated or embossed to resemble fine linen or cotton cloth.

**CAMPAIGN BRISTOL:** A specialty type of inexpensive, light-colored bristol used for advertising postcards and political campaign posters.

**CAN:** (1) One of the drying cylinders in a paper machine dryer section. (2) Drum, canister, or container that can be made of paperboard or fiberboard, sometimes in combination with other materials.

**CAN BOARD:** A class of well-sized leakproof paperboards made from bleached chemical pulps. It is sometimes lined with protective plastics and foils, and used to make spiral wound cans and similar-type containers for the food industry.

**CAN STOCK:** Paperboard base used to make board for food containers. *See* CAN BOARD.

**CANADIAN PULP & PAPER ASSOCIATION:** Headquarters: Montreal, Que., (514) 866-6621.

**CANADIAN STANDARD FREENESS (CSF):** A standardized *freeness* test in which a sample of stock is placed in an instrument chamber having a slotted screen at the bottom. The amount of water that drains from the stock during a specific period of time is carefully measured and calculated to determine a numerical *Canadian standard freeness (Csf)* value, with high numbers indicating maximum drainability. *See also* FREENESS.

**CANAL:** (1) A moat running under pulpwood grinders through which pulp and water from grinders flow to the bull screen. (2) Open space in the structure of coniferous wood that contains the resins and pitch.

**CANARY WRITING PAPER:** A type of pen- and ink-sized writing paper used in making tablets, advertising flyers, business forms, etc. It is usually made from unbleached chemical pulp and mechanical pulp in a canary color.

**CANARY YELLOWS:** A variety of yellow pigments used in papermaking that are modifications of lead chromate ($PbCrO_4$). Also called *chrome* and *chrome yellows.*

**CANDLE PAPER:** A well-sized, colored, lightweight paper made from chemical and mechanical pulps with little or no strength characteristics. It is used extensively in the manufacture of pyrotechnic materials.

**CANDLEPOWER:** The unit of luminous intensity of a light source.

**CANDY BOX DIVIDER BOARD:** An even-surfaced, chocolate-colored, manila-lined paperboard used to make separators between layers and pieces of candy in chocolate boxes.

**CANDY CUP PAPER:** A very dense, lightweight paper (glassine) used to make "cups" for holding pieces of boxed candy.

**CANDY SLAB PAPER:** A vegetable parchment paper treated with a release agent so that it can be used in candymaking as a liner that will not absorb fats, sugars, oils, and moisture, and will not stick to candy products.

**CANDY TWISTING TISSUE:** A high-strength, lightweight, soft, waxed paper made from chemical pulp and used for twist-wrapping certain types of candy and photographic film.

**CANDY WRAPPING PAPER:** High-gloss, glazed, or waxed papers used to make printed wrappers for candy products. Glassine and lacquer coated paper are also used.

**CANISTER PAPER:** A hard sized paper usually made from bleached chemical pulp especially for use in making cylinder containers.

**CANT ANGLE:** A measure of the relative position between the cutting shear blade and knife ring it rides against on winder and rewinder slitters in paper mills. Also referred to as *toe-in angle.*

**CANT HOOK:** A leverage-type tool equipped with a hook-like element and used to move and handle pulpwood logs.

**CANTING:** Reducing large logs to smaller sizes by cutting them longitudinally on a bandsaw prior to grinding or chipping in a pulp mill.

**CANTS:** Another name for pulpwood logs.

**CANVAS:** A common paper mill reference to a type of cotton-base paper machine dryer fabric used to maintain intimate contact between the sheet and drying cylinder. It helps control sheet shrinkage and minimize cockling and edge curl in the sheet. Also called *canvas duck* or *canvas asbestos duck.*

**CANVAS PAPER:** A paper whose surface has been embossed to give it a canvas cloth appearance.

**CAP MAKER'S STRAWBOARD:** A heavyweight type of paperboard used in making the inner base for cap visors. It is usually made from cooked straw and/or recycled used paper stocks on a cylinder machine.

**CAPACITANCE:** The ratio of the electrical charge on a condenser to the potential difference between the terminals of the condenser.

**CAPACITOR PAPER:** A chemically pure, uniform, very thin paper with good dielectric properties made from sulfate pulp or hemp fiber. It is used to make electrolytic absorbent separators between foils in the manufacture of condensers and capacitors. Also called *condenser paper* or *condenser tissue.*

**CAPACITY:** (1) The maximum output or production capability of a piece of process machinery, unit process, or entire mill in the pulp and paper industry. It is usually stated in units of pounds or tons per day. (2) The quantity of material a process unit can handle. For instance, a pump capacity is the number of gallons per minute of water it will pump under a specified set of conditions.

**CAPILLARY RISE:** A measure of the height a liquid will rise into a strip of paper or board when suspended vertically with its lower end immersed in the liquid.

**CAPTIVE PULP:** Pulp used by a paper mill that is made in a pulp mill on the same site and integral to it. May also refer to pulp (market pulp, in particular) made by one mill and used by another belonging to the same company. Could also apply to a "customer" who is part owner (or some other contractual arrangement) in a joint-venture market pulp mill.

**CAR CARD:** A smooth rigid heavy blank board coated on one side and used for car signs and showcards.

**CAR LINER:** A coarse type of single or asphalt bound, laminated, heavy paper used as a freight car liner to protect shipping contents. It is usually made from sulfate pulp, screenings, or recycled wastepaper stock.

**CAR LINING PAPER:** Wrapping-type paper used as a liner to cover bottom and walls of box cars for the prevention of leakage of bulk material during shipment.

**CAR SIGN BOARD:** A type of clay-coated, supercalendered, multi-ply paperboard made on a cylinder machine with the outer layers made from chemical pulp and the inner layer made from recycled wastepaper. It is used for making advertising posters usually found in buses, railroad cars, subways, and street cars.

**CARBOHYDRATES:** Any of a group of neutral compounds comprised of carbon, hydrogen, and oxygen, e.g., sugars, starches, and celluloses.

**CARBON BLACK:** A pigment used in the manufacture of gray and black papers.

**CARBON COATING:** A coating mixture used in the manufacture of duplicating paper that is made up of a mixture of carbon black in a wax or some other oil-soluble vehicle.

**CARBON DIOXIDE ($CO_2$):** A gas found in lime kiln flue exhaust, the amount of which is related to the operating efficiency of the kiln in converting calcium carbonate ($CaCO_3$) to lime ($CaO$). It is also found in other plant effluents.

**CARBON MONOXIDE ($CO$):** A constituent of power and recovery boiler flue gases, the amount of which is related to combustion efficiency in the boiler. It is a very toxic gas.

**CARBON PAPER:** A lightweight paper coated with a mixture of carbon black or similar coloring material and wax or some other oil-soluble vehicle; used for making pencil, pen, and typing machine copies.

**CARBON SPOTS:** A type of paper defect caused by small carbon- or coal-like particles inadvertently contained in the sheet. Similar to and also called *cinder specks* and *coal specks*.

**CARBON TISSUE:** Thin paper made with a hardened carbon containing dichromated gelatin, especially to be used for transfer work in marking plates for rotogravure printing.

**CARBONATE PAPER:** An unsized, supercalendered paper made from chemical pulps and containing high percentages of calcium carbonate filler. It is used as a general printing paper requiring good color and opacity.

**CARBONATES:** Types of fillers, such as calcium carbonate, that will react with acidic chemicals or will produce alkaline conditions in water solutions. Also used in coating formulations. *See also* ALKALINE PAPERMAKING.

**CARBONIZING PAPER:** A lightweight, uncoated paper made from unbleached chemical and/or mechanical pulps. It is surface-coated with a carbon solvent or wax so that it will take up carbon inks and release them under pressure, thereby duplicating the inked areas being printed.

**CARBONLESS PAPER:** A type of copying paper that is treated or coated so it can be used without the need of a carbon coating or interleaved carbon paper. *See also* NO CARBON REQUIRED (NCR) and ACTION PAPER.

**CARBORUNDUM PAPER:** Class of abrasive-type papers that have been coated on one side with coating materials made from carbon and silicon. *See also* ABRASIVE PAPERS.

**CARDBOARD:** Stiff, relatively thick (greater than 0.006 inch) single or multi-ply paperboards. It does not define a specific grade or product, and generally is not used in the paper industry.

**CARDBOARD FINISH:** A type of surface finish applied on paper by using paperboard on the presses during the process of plating.

**CARDBOARD LININGS:** Types of papers used in the manufacture of lined chip and news boards to achieve improved surface appearance.

**CARDBOARD MIDDLES:** Inferior grade of boards used as the inner layer of multi-ply paperboards, and made from mechanical and recycled waste pulp.

**CARLOAD:** The quantity of paper a mill ships in one freight car that must exceed a freight classification zone minimum weight to qualify for carload freight rate.

**CARPET BROWN:** *See* CARPET LINING.

**CARPET LINING:** A thick, soft, spongy paper made from recycled paper and rag stock for use as a cushioning base for rugs and carpets. Sometimes called *carpet brown.*

**CARPET LINING BOARD:** A soft-finished paperboard from which carpet and rug cushioning pads are made.

**CARRYING ROLL:** The rolls on a paper machine that support the wet end wire (forming fabric), wet end and dry end felts, and are not driven.

**CARRYOVER:** (1) Water droplets that come over with the steam from boilers due to poor separation in the drum because of improper drum level. (2) The cooking liquor that comes over with the relief gases during a sulfate cook in a batch-type digester. Also known as *pullover.*

**CARTON:** (1) A unit weight of paper used in shipping determined by the grade. (2) A box, usually manufactured from corrugated board, such as folding and rigid setup boxes. (3) A box made of solid fiberboard used for shipping.

**CARTON LINING PAPER:** A heavyweight kraft paper, waxed, asphalted, glassine, and any other type of paper used as an inner liner for cartons to protect the enclosed product(s).

**CARTON SEALING PAPER:** A lacquer or wax-coated paper that can be sealed by heat to make outer protective wrappings for food cartons.

**CARTRIDGE PAPER:** A multi-ply type of paper, usually made on a cylinder machine from recycled paper stock or chemical pulp mixed with rag or flax fiber. It is used to make shotgun shell tubes.

**CASCADE EVAPORATOR:** A type of evaporator used to concentrate sulfate black liquor before burning in a recovery furnace. It consists of a series of rotating vertical disks or tubular elements partially immersed in pre-evaporated liquor and partially subjected to contact with the hot combustion gases from the furnace.

**CASCADE SYSTEM:** Piping flash and blow through steam as on a paper machine such that it flows from one dryer section to another downstream in a series configuration in order to conserve energy.

**CASE:** A shipping box made from corrugated board or solid fiberboard.

**CASE LINING PAPER:** A heavyweight kraft paper, waxed, asphalted, glassine, or other type of paper used as an inner liner for shipping boxes to protect the contents.

**CASE LOT:** A shipping weight unit of flat paper or paperboard, consisting of a container of packaged sheets. The number of sheets and weight of the package depends on the grade of paper.

**CASE MAKING BOARD:** *See* CONTAINER BOARD.

**CASEIN:** A paper sizing and adhesive coating material produced by coagulating the protein of skimmed milk with a suitable acid, such as hydrochloric, sulfuric, or lactic acid.

**CASH REGISTER PAPER:** Paper made from chemical and/or mechanical pulp, generally but not always white, that is slit into narrow strips and wound into rolls to record cash register transactions.

**CASING:** (1) The metallic outer container enclosing the pulpstone of a pulpwood grinder. (2) The outer shell of a pump that encloses the impeller. Also called the *housing*.

**CASKET PAPER:** Colored paper made from chemical and mechanical pulps and used as an underliner for cloth used in caskets.

**CAST COATING:** The process of producing a very high-gloss finish on coated paper by pressing it against a highly polished surface (usually a steam-heated, chrome-plated drum) while it is setting up so that it will take on a finish similar to the one on the drum surface. *See also* CAST-COATED PAPER.

**CAST IRON:** Molded iron used to make the frames of many pulp and papermaking machines.

**CAST-COATED PAPER:** A very high-gloss, coated paper and paperboard whose surface characteristics are produced by allowing the applied coating to harden while in contact with the surface of a steam-heated, highly polished, chrome-plated drum. *See also* CAST COATING.

**CATALOG PAPER:** A class of fairly good strength, lightweight, opaque papers made from deinked stock, mechanical pulp, and chemical pulp. It is used for making mail order catalogs and telephone directories to reduce their mailing weight.

**CATALYST:** A substance or agent that is added to a process to either initiate or increase the rate of chemical reaction without entering into or chemically affecting the reactants or end product.

**CATHODE:** The negative pole in an electrolytic cell.

**CATHODE RAY TUBE (CRT):** A vacuum tube in which cathode rays, usually in the form of a slender beam, are projected on a fluorescent screen. It is used to produce video displays on remote viewers and computer system operator interfaces in the mill.

**CATIONIC STARCH:** A type of starch that has been given a positive charge in order to increase its adhesion to the fibers and retention in the paper in which it is used.

**CATIONICS TECHNOLOGY:** The science and practice of using cationic fillers and coating formulation ingredients to improve retention, coating adhesion, etc.

**CATWALK:** Bridging passageways between or along process equipment or areas of a mill.

**CAUSTIC:** A term used to refer to corrosive alkaline chemicals and substances used in pulp and paper manufacturing.

**CAUSTIC EXTRACTION:** A stage in the pulp bleaching sequence (E) that normally follows the chlorination stages to remove alkali-soluble, chlorinated lignins. It is now also being used in some non-chlorine sequences.

**CAUSTIC EXTRACTED K. NO.:** A measure of the bleachability of pulp tested immediately after the caustic extraction stage in a pulp bleaching process.

**CAUSTIC SODA:** Sodium hydroxide (NaOH), a chemical used as a raw material and found extensively in pulp and paper processing materials, especially in the soda and sulfate process.

**CAUSTICITY:** The NaOH content divided by the active alkali content (all expressed as $Na_2O$) of alkali cooking liquor and expressed as a percent.

For sulfate cooking liquor:

$$\left(\frac{Na_2S}{NaOH + Na_2S + Na_2CO_3} \times 100\right)\%$$

For soda cooking liquor:

$$\left(\frac{NaOH}{NaOH + Na_2CO_3} \times 100\right)\%$$

**CAUSTICIZING:** Converting green liquor to white liquor by the use of slaked lime [$Ca(OH)_2$] that reacts with the sodium carbonate ($Na_2CO_3$) in the green liquor to form active sodium hydroxide (NaOH) in the white liquor. Also called *recausticizing.*

**CAUSTICIZING EFFICIENCY:** A percentage found by dividing the NaOH by NaOH and $Na_2CO_3$ in the white liquor from the causticizing process (both chemicals being expressed as $Na_2O$). A correction is made for NaOH content of the original green liquor in order to represent only the NaOH produced in the actual causticizing reaction:

$$\left(\frac{NaOH_{(WL)} - NaOH_{(GL)}}{NaOH_{(WL)} - NaOH_{(GL)} + Na_2CO_{3(WL)}} \times 100\right)\%$$

Where WL = White Liquor

GL = Green Liquor

**CAVITATION:** The alternate formation and collapse of vapor bubbles inside control valves and pump casings.

**CD:** *See* CROSS DIRECTION.

**CEDAR WALLPAPER:** A wood grain surface pattern-type of wallpaper used for lining closets and some clothes boxes. It is made primarily from cedarwood pulp mixed with a little groundwood pulp.

**CEDAR WRAPPING PAPER:** A protective type wrapping paper that has been treated with cedar oil.

**CELL:** One of the minute units or elements that form plant life.

**CELLOPHANE:** A thin, transparent, moistureproof sheet made from chemically regenerated cellulose used extensively as a packaging wrapper.

**CELLULOID:** A trademark for a thin, high-strength, flammable, transparent plastic material made from cellulose nitrate and camphor.

**CELLULOSE:** The chief substance in the cell walls of plants used in pulp manufacturing. It is the fibrous substance that remains after the nonfibrous portions, such as lignin and some carbohydrates, are removed during the cooking and bleaching operation of a pulp mill.

**CELLULOSE ACETATE:** Ester of cellulose and acetic acid used in making up acetate fiber for textiles.

**CELLULOSE FIBER SHOEBOARD:** A special type of paperboard made on a wet machine from stock that is made up primarily of cellulose fibers, and used in the manufacture of shoe parts.

**CELLULOSE LACQUER:** Solution made up of various cellulose compounds in a suitable organic solvent with other materials so that it can be applied as a coating, such as on paper, that will leave a protective, high gloss, film-like finish when the solvent evaporates.

**CELLULOSE NAPKIN PAPER:** A facial-type tissue paper made from a special mixture of bleached chemical and mechanical pulp with sufficient wet-strength properties so it can be used to make multi-ply table napkins.

**CELLULOSE WADDING:** *See* WADDING.

**CELLULOSE XANTHATE:** The product of reaction between carbon disulfite and alkali cellulose. It is a viscous, orange-colored mass from which a sugar can be made.

**CEMENT SACK PAPER:** A high-strength, pliable-type of paper made from sulfate chemical pulp. It is used in the manufacture of multiwall bags for such products as cement and lime.

**CENTER STOCK:** Mechanical and recycled wastepaper pulp that is used to make the inner layers of a multi-ply paperboard.

**CENTRIFINER:** Machine used to remove dirt from diluted pulp slurry by use of centrifugal force to separate it.

**CENTRIFUGAL PUMP:** A type of pump commonly used to move liquids and stock slurries through a pipeline. Its principle of operation depends on the centrifugal force to develop a pressure rise on the flowing medium proportional to the speed and diameter of its internal spinning impeller. *See also* AXIAL FLOW PUMP.

**CENTRIFUGAL SCREENS:** A type of screening equipment that separates the finer fibers from mechanical pulp stock by centrifugal action of vanes throwing the pulp against a perforated plate.

**CENTRIFUGAL SEPARATORS:** Types of stock cleaners used in a paper mill to remove heavy dirt particles by separating and removing them from the fiber using centrifugal force, the most common being the rotating and pressure drop types.

**CERIN:** A wax-like substance found in the bark of certain trees.

**CFM:** Abbreviation for cubic feet per minute.

**CHAD:** Small disk resulting from the punching of a paper sheet.

**CHAIN GRINDER:** A continuous type of pulpwood grinder in which the logs are fed against the top of the grinderstone by a chain system. The speed of the chain can be changed to vary the pressure between the wood and the stone.

**CHAIN MARKS:** Chain-like calendered paper surface defect showing up as alternating dark and light streaks due to unequal strains in processing caused by poor formation of the sheet. Sometimes called *bars, buckles, corrugation marks, rope marks, ridges, tire tracks,* and *laidlines.*

**CHAIR SEAT BOARD:** A type of rigid board made from chemical and/or mechanical pulp under high compression. It is used in furniture manufacture to make seats for chairs and has other similar uses.

**CHALK:** (1) A natural form of calcium carbonate ($CaCO_3$) used to make paper filler and coating material. Sometimes called *Paris white.* (2) Limestone that is used to produce lime in lime kilns for recausticizing green liquor. *See* CALCIUM OXIDE.

**CHALK PAPER:** Any suitable lightweight paper used for wrapping chalk and crayon products.

**CHALKING:** (1) The breaking away of pigment from the paper surface during finishing and converting operations. (2) The rubbing off of the pigment of ink from printed paper surfaces.

**CHALKY APPEARANCE:** A reference to a low-luster, coated paper surface related to the type of coating material used.

**CHANNELING:** An undesirable movement of part of the material mass in a reactor tower or storage tank at a greater speed than the remainder of the mass moving through the vessel. Channeling disrupts homogeneity and uniform distribution as well as retention time and reaction time in such operations as towers in bleach plants, pulp storage chests, chip bins, continuous digester, etc.

**CHARCOAL BAG PAPER;** A heavyweight, high-strength kraft paper used to make bags or sacks for packaging and transporting charcoal and coal products. Also called *fuel sack paper.*

**CHARCOAL PAPER:** Paper that has been physically decomposed by heat during a malfunctioning drying process.

**CHARGE:** (1) To fill a batch-type digester with chips and cooking liquor. (2) The chips and cooking liquor in the digester.

**CHARGING FLOOR:** The floor level in a digester house of a pulp mill where chips are fed into the digester and usually where the digester is operated from.

**CHART PAPER:** A well-sized, high-strength, durable, bond-type paper with good printing qualities, made from cotton fiber and chemical pulp. It is used for making flat circular and strip roll instrument recording paper, maps, and charts.

**CHARRED PAPER:** Any paper that has lost its physical properties due to heat.

**CHECK PAPER:** A relatively good-strength paper with good writing qualities, made of cotton fiber and chemical pulp. It is dyed and chemically treated so that alterations to recorded information cannot be made without a marked effect to the surface. *See* ANTIFALSIFICATION PAPER.

**CHECK VALVE:** A valve that automatically stops backflow when the fluid flow in a pipeline reverses.

**CHEESE WRAPPING PAPER:** Reference to any type of paper used by the food industry to wrap cheese. Depending on the specific product, it can include waxed papers, vegetable parchment, paper laminates, and glassine.

**CHELATE:** *See* CHELATING AGENT.

**CHELATING AGENT:** (1) Any chemical used to stabilize other chemical solutions, such as using magnesium sulfate to stabilize hydrogen peroxide bleaching liquor. (2) Chemicals used in the pulp bleaching process to control the brightness restricting and reversion effects of iron salts and other heavy metals in the pulp. (3) Organic chemical compounds that combine with iron and contribute to corrosion of steel digesters during alkaline pulping. Also called a *sequestering agent* or *chelate.*

**CHEMITHERMOMECHANICAL PULP (CTMP):** Pulp made by the thermomechanical process, in which the wood chips are pretreated with a chemical, usually sodium sulfite, either prior to or during presteaming, as an aid to subsequent mechanical processing in refiners. Sometimes referred to as *thermo-chemimechanical pulp (TCMP).*

**CHEMI-WASHER:** A continuous, horizontal (fourdrinier-type) multi-stage, counter-current, enclosed-type brown stock washer.

**CHEMICAL CELLULOSE:** A highly chemically purified wood or cotton pulp usually produced by a special bleaching process to produce cellulose that can be used to manufacture chemical derivatives for other products such as nitrates, acetates, etc.

**CHEMICAL CHARGE:** (1) The amount of chemicals put into a batch-type digester with the chips to make one cook. (2) In a single-stage batch bleaching process, the quantity of chemicals added and usually consumed in bleaching one batch of pulp.

**CHEMICAL CONSUMPTION:** The chemical used by the sulfate pulping process. It is determined by the pounds of salt cake ($Na_2SO_4$) added per day divided by the tons of air-dry pulp produced per day. In the soda pulping process, it is the percent found by dividing the quantity of sodium carbonate added to the new liquor by the total alkali in the cooking liquor fed to the digesters.

**CHEMICAL DEBARKING:** The peeling of bark from wood after the tree has been treated with appropriate chemicals that loosen the bark from the wood after a period of time.

**CHEMICAL FIBER PAPER:** Paper that is made only from chemical pulp stocks.

**CHEMICAL FILTER PAPER:** A type of high-strength, unsized, porous paper used in chemical filter presses and similar equipment to remove suspended materials from process liquids. It is usually made from chemical pulp and/or cotton fiber.

**CHEMICAL FOG:** Chemical entrainment in recovery boiler flue gases that are not removed by mechanical devices such as scrubbers, and exist in an extremely fine state of division.

**CHEMICAL LOSS:** The percentage of total chemical not recovered in the recovery process of a sulfate pulp mill. It is determined by dividing the quantity of sodium sulfate added to the new liquor by the total alkali in the cooking liquor fed to the digesters.

**CHEMICAL LOSS IN EVAPORATORS AND FURNACES:** Percentage of total chemical lost in the evaporation and burning process of a sulfate recovery cycle. It is determined by dividing the total alkali to the evaporator plus sodium salts in the new chemical minus the total alkali in the green liquor, by the total alkali to the digesters.

**CHEMICAL LOSS IN PULP WASHING:** The percentage of total chemical lost in the sulfate pulp washing process after cooking. It is determined by dividing the total alkali to the digester minus the total alkali to the evaporator, by the total alkali in the cooking liquor fed to the digesters.

**CHEMICAL LOSS IN RECAUSTICIZING AND MUD WASHING:** Percentage of total chemical lost in the sulfate liquor recausticizing and mud washing process of a pulp mill recovery cycle. It is determined by dividing the total alkali in the green liquor minus the total alkali in the white liquor, by the total alkali in the cooking liquor to the digesters.

**CHEMICAL OXYGEN DEMAND (COD):** A measure of the amount of oxygen required to oxidize organic and oxidizable inorganic compounds in water. The COD test is used to determine the degree of pollution in mill effluents. *See also* BIOCHEMICAL OXYGEN DEMAND (BOD) and BIOLOGICAL OXYGEN DEMAND (BOD).

**CHEMICAL PAPER DEFECTS:** A class of paper defects that are the result of faults caused by the chemical constituents of a sheet as contrasted to the mechanical defects caused by its structure formation faults.

**CHEMICAL PULP:** The mass of fibers resulting from the reduction of wood or other fibrous raw material into its component parts during the cooking phases with various chemical liquors, in such processes as sulfate, sulfite, soda, NSSC, etc.

**CHEMICAL RECOVERY:** The recovery of chemicals in sulfate cooking liquor after it is used to cook wood in the digester (spent liquor). It is expressed as a percentage determined by dividing the total alkali to the digesters minus the sodium sulfate added to liquor, by the total alkali in the cooking liquor going to the digester after correcting for any change in liquor inventory.

**CHEMICAL-TO-WOOD RATIO:** Ratio of the weight of chemicals charged to the weight of oven dry wood charged to a digester.

**CHEMICALLY BOUND WATER:** Water in a sheet of paper that has entered into a hydration bond with the cellulose molecules or other substances in the paper.

**CHEMIGROUNDWOOD:** A process of producing groundwood pulp in which whole barked logs are given a pressurized heat treatment in sodium sulfite and sodium carbonate solution to soften the wood prior to grinding.

**CHEMIMECHANICAL PULP (CMP):** Pulp made by pretreatment of chips with chemicals at a temperature usually below 100°C, followed by atmospheric refining.

**CHEMIMECHANICAL PULPING:** Wood grinding or chip refining process employing a mild chemical treatment to improve strength or increase production.

**CHEST:** A storage tank in the paper industry used especially for stock, pulp, water, white water, etc.

**CHESTNUT BOARD:** A class of brown paperboards made with pulp produced from southern hardwoods using the semichemical cooking process.

**CHEVIOT:** A granite paper surface effect produced by adding dark-colored fibers to a lighter-colored furnish.

**CHEVIOT BOX COVER:** A paper made from mechanical and chemical wood pulps with a small amount of darker colored fibers added. It is used primarily to cover boxes.

**CHIEF CHEMIST:** A common title referring to the person in charge of the pulp and paper mill testing laboratory.

**CHIEF ENGINEER:** A common title for the head of the local plant engineering department who is responsible for such activities as new construction, installation of machinery, and repair and maintenance work.

**CHINA BOARD:** A cylinder, machine-coated, multi-ply paperboard made primarily of recycled newsprint stock in a variety of colors to be converted into tags, tickets, and similar items.

**CHINA CLAY:** A superior type of naturally occurring clay comprised mainly of aluminosilicate mineral ($Al_2O_2 \cdot 2SiO_2 \cdot 2H_2O$) known as *kaolinite* or *kaolin*, and commonly used in conjunction with other similar materials as a paper coating ingredient or sheet filler.

**CHIP ANALYSIS:** A wood chip quality control test used in pulp mills that includes the separation of chips by size, and classifying and weighing each size fraction.

**CHIP BIN:** Storage chamber for wood chips located directly over batch digesters from which they are filled.

**CHIP CHARGE:** The quantity of wood chips that are fed into a batch-type digester to make up one cook.

**CHIP CRUSHER:** A hammer or pin mill where oversize fractions of chips are returned from the chip screens to be reduced in size.

**CHIP DIGGER:** A unit that is driven into a chip-filled freight car and sucks the chips up into a suction unloading system.

**CHIP FEEDER:** A positive displacement-type device designed to feed chips at essentially atmospheric pressure into a pressurized processing unit, such as a presteaming vessel or digester, without loss of pressure. These units are usually rotary-type valves or enclosed screw-type conveying devices.

**CHIP FEEDER PLUG SCREEN COMPRESSION RATIO:** The ratio of a specific volume of chips at the inlet to the specific volume of chips at the discharge of a screw type chip feeder.

**CHIP FILLING:** The process of loading a batch-type digester from a chip bin above the digester, or from a conveyor transporting wood chips from a silo to the floor above the digester top.

**CHIP METER:** A volumetric chip feeding device whose action can be monitored and correlated to the quantity of chips passing through it.

**CHIP PACKING:** A method of removing air from wood chips as they are being charged into a digester, usually by the use of a steaming device, in order to achieve greater penetration of cooking liquor into the chips.

**CHIP PILE:** Chips that are stored outside in a mound type of structure usually located near the pulp mill so that chips can be conveniently conveyed from it to the digester storage.

**CHIP RECLAIM:** A hopper used in a pulp mill to accept chips from storage so that they can be fed to an air blower for conveying to the digester.

**CHIP SCREENMAN:** The pulp mill employee who looks after the screening of pulpwood chips.

**CHIP SCREENS:** Rotating, gyratory, vibrating, or inclined stationary screening devices used to separate oversized chips, slivers, and sawdust from the accepted chips sent to the digester for cooking. Oversized chips are returned to crushers for reprocessing, with sawdust usually being sent to the boiler house for burning.

**CHIP SILOS:** Large concrete or tile chip storage areas usually sized so that they will hold enough chips for about 8 hours of pulp mill operation.

**CHIP STORAGE:** An area or means for accumulating chips from screening operations in a pulp mill so they may be temporarily stored between chip preparation and digesters, such as silos, bins, and outdoor piles.

**CHIP TESTER:** (1) The person responsible for testing chips for density, moisture, dirt, etc., in a pulp mill. (2) Any testing equipment used to test chips for such things as density, moisture, dirt, etc.

**CHIPBOARD:** An inferior quality, low density, solid, or lined paperboard made primarily from recycled wastepaper stock, and used in low-strength applications.

**CHIPPER:** A piece of equipment in the woodyard/pulp mill area used to "chip" whole logs. It consists of an enclosed, rapidly revolving disk fitted with surface-mounted knives against which the logs are dropped in an endwise direction in such a manner that they are reduced to chips, diagonally to the grain.

**CHIPPER KNIVES:** The actual knife (cutting/slicing) blade components of a *chipper*. Some chipper units use disposable knives.

**CHIPS:** Pieces of wood (approximately 1 inch square and 1/8 inch thick) resulting from the cutting of pulpwood logs in chippers in the wood preparation area of a pulp mill prior to conversion into pulp in the digester(s). Sometimes called *wood chips*.

**CHIPS AND SAW:** A pulp mill expression used to refer to a mixture of chips obtained directly from pulpwood and extraneous chips obtained from wood rejects supplied by sawing operations in lumber mills and other sources.

**CHLORALKALI:** Reference made to chlorine and alkali when manufactured in a process as coproducts.

**CHLORINATED ORGANICS:** A group of *organochlorine* compounds, such as DDT, *polychlorinated biphenyls (PCBs)*, *dioxin*, *furans*, and *chloroform*, that tends to remain in the environment for long periods of time and to accumulate inside the tissues of living organisms, interfering with biological processes. Some of these compounds are produced during the bleaching of pulp with chlorine and chlorine containing compounds, and eventually are discharged with bleach plant effluents into receiving waters. Also referred to as *organochlorine*.

**CHLORINATION:** (1) The mixing and reacting of chlorine water or gas with pulp in the bleaching operation. (2) The application of chlorine to mill water supply and sewage for disinfection or oxidation of undesirable compounds.

**CHLORINATION STAGE:** The step in a multi-stage bleaching process ("C" stage) where chlorine water or gas is mixed, allowed to react, and then washed as an initial operation in a complete pulp bleaching system.

**CHLORINATOR:** A device for adding a chlorine-containing gas or liquid to mill wastewater. Sometimes the term is also used to refer to the chlorine mixer in the bleach plant.

**CHLORINE:** A greenish-yellow, poisonous, gaseous chemical element ($Cl_2$) used in bleaching pulp and water purification in a pulp and paper mill.

**CHLORINE CONSUMPTION:** Actual amount of chlorine consumed to bleach pulp, expressed as pounds of chlorine used per air dry ton of pulp bleached, or a percentage on the same basis. It may also be expressed on a bone dry basis.

**CHLORINE DIOXIDE SOLUTION:** A very unstable water solution of chlorine dioxide gas ($ClO_2$) produced in the chemical preparation area of a pulp mill. It is used in the pulp bleaching process.

**CHLORINE DIOXIDE STAGE:** The step or steps in a multi-stage bleaching process (the "D"-stages) where chlorine dioxide solution is mixed with pulp, allowed to react, and then washed as one of the operations making up a complete pulp bleaching system.

**CHLORINE EVAPORATOR:** A specially constructed, thermostatically controlled vessel using hot water or steam to vaporize liquid chlorine transferred from tank cars to a pulp mill bleach plant. This vaporized product is used in the chlorination stage of a bleaching process, as well as to make up hypochlorite bleaching liquor. Also called *chlorine vaporizer.*

**CHLORINE INSTITUTE INC.:** Headquarters: Washington, D.C., (202) 775-2790.

**CHLORINE MIXER:** A mixing device used in the bleach plant to mix chlorine water or gas with unbleached pulp.

**CHLORINE NUMBER:** A measure of chlorine consumed by a sample of pulp (100 grams, 15 minutes, 25°C) and used to determine the bleach required for a quantity of unbleached pulp. *See also* BLEACH DEMAND.

**CHLORINE REQUIREMENT:** The amount of elemental chlorine ($Cl_2$) required to achieve a specified final brightness level of pulp in the bleaching process. It is supplied in the form of elemental chlorine and/or bleaching agents such as hypochlorites, chlorine dioxide, etc.

**CHLORINE VAPORIZER:** *See* CHLORINE EVAPORATOR.

**CHLOROFORM:** A *chlorinated organic* ($CHCl_3$) sometimes produced during the bleaching of pulps with chlorine and/or chlorine containing compounds.

**CHLOROLIGNIN:** The product of reaction between lignin in unbleached pulp and chlorine in the chlorination stage of a multi-stage pulp bleaching system.

**CHLOROPHYLL:** The green coloring matter in leaves that assists in the combining of water and carbon dioxide to form carbohydrates by a natural process called photosynthesis. Wood substance is then produced from the carbohydrates.

**CHLOROTOXINS:** A general reference to toxins such as *dioxins, furans,* and other *chlorinated organics* produced during the bleaching of pulp with chlorine and/or chlorine containing compounds.

**CHOCOLATE BOX BOARD:** Paperboard, colored brown on one or both sides, made especially for use in producing setup and folding boxes for packaging chocolate candy.

**CHOKER:** A noose of steel cable fastened around a log to pull it to a landing.

**CHROMA:** The intensity of the color of paper and paper products related to its depth or weakness. Also called CHROMATICITY.

**CHROMATIC ABERRATION:** An image imperfection caused by the light of different wavelengths focusing in different planes.

**CHROMATIC PAPER:** (1) A type of paper used to cover boxes and cartons. It has a mottled-effect surface produced by the use of different colored fibers to make up the sheet. (2) Sometimes used as a general, descriptive reference to paper with color.

**CHROMATICITY:** *See* **CHROMA.**

**CHROME YELLOWS:** *See* CANARY YELLOWS.

**CHROMES:** *See* CANARY YELLOWS.

**CHUCK:** A device that produces a means of holding a paper roll in winding, rewinding, or unwinding machines. It usually consists of a metal expandable core inserted in the ends of a roll or roll core to ensure concentric running on the spindle.

**CIGARBAND PAPER:** Paper possessing fairly good strength, suitable for embossing, and having a printable surface. It is used for making cigar bands.

**CIGARBOX BOARD:** A rigid, pasted chipboard used to make boxes for packaging cigars.

**CIGARETTE CARTON BOARD:** A coated, single-ply paperboard made from sulfate chemical pulp with a good printable surface and used to make cartons for cigarette packages.

**CIGARETTE PAPER:** An unsized, high-strength, uniform-type of tissue paper made from flax pulp. It is used to make tobacco wrappers in the commercial manufacture of cigarettes, and to make tobacco wrappers for self-made cigarettes. *See also* COMBUSTIBLE PAPER.

**CIGARETTE TIP PAPER:** Paper usually made from bleached chemical pulp and flax. It has a smooth, soft texture and is used to make cigarette mouthpieces.

**CINDER SPECKS:** A type of paper defect caused by small cinder-like particles that are inadvertently contained in the sheet. Similar to *carbon spots.*

**CIRCUIT BREAKER:** An electrical protective device that automatically opens a circuit under overload or short-circuit conditions. It can also be used for making, breaking, or changing connections in an electrical circuit under other abnormal conditions.

**CIRCULATION:** (1) In steam-generating boilers, it refers to the continuous movement of water down the downcomers, up the generating tubes, and through he drum. (2) In batch digesters, it refers to the movement of cooking liquor during the pulping process, which can be natural or forced.

**CLAD DIGESTER:** A digester to which the inner surface is covered with a layer of metal to make up for that which has been worn away. *Cladding* is the metal coating applied to the inner surface of a digester, or the process of applying the coating.

**CLAFLIN REFINER:** A short, wide-angled refiner used to process rejects and very brash pulp.

**CLAMP MARKS:** Surface disfigurations occurring during the paper trimming operation in the finishing room of a mill. It is caused by the paper clamping mechanism on guillotine trimmers.

**CLAMP TRUCK:** *See* GRAB TRUCK.

**CLARIFICATION:** (1) The removal of turbidity and suspended solids by settling in mill wastewater treatment. (2) In the causticizing plant in a pulp mill, it refers to the settling out of suspended materials from green and white liquors.

**CLARIFIERS:** Storage tanks in which suspended solids are allowed to settle and be removed from green and white liquors in the causticizing areas of a pulp mill from mill wastewater.

**CLASSIFIER:** (1) A device for separating pulp fiber in liquid suspension according to fiber length and particle size by depositing the fractions on known mesh wire cloth for final drying to determine bone dry weight. (2) A device on a coal pulverizer outlet that separates oversize particles and recycles them back to the pulverizer.

**CLAY:** A naturally occurring, earthy, fine-grain material comprised of a group of crystalline clay minerals with a natural basic structure of aluminosilicates whose hydrous chemical form is $2H_2O \cdot Al_20_3 \cdot 2Si0_3$. It is commonly used in the paper industry to make up paper filling and coating materials. Clays are sometimes altered by further refining, heat treatment, etc., to enhance or extend their end uses, e.g., calcined clay and delaminated clay. *See also* CHINA CLAY and KAOLIN.

**CLAY LUMP:** A paper defect occurring as a lump caused by clay agglomeration in the sheet filler and coating.

**CLAY STARCH SYSTEM:** A process in which starch and clay are prepared and mixed to make paper coating material.

**CLAY-COATED BLANKS:** A paperboard from which printing board for calendars, cards, and other advertising mediums are made. It is coated with clay coating material or has been lined with clay-coated paper.

**CLAY-COATED BOXBOARD:** A paperboard from which boxes are made. It is coated with clay coating materials to enhance surface qualities such as smoothness, whiteness, brightness, and printability.

**CLAY-FILLED PAPER:** Paper made with clay being the primary component of the filling material used.

**CLEAN BLOW:** When a batch digester is discharged into a blow tank with no uncooked chips or pulp remaining in the bottom or in the blow line from the digester to the blow tank.

**CLEAN ID FAN:** An induced draft fan on a pulp mill recovery furnace located ahead of the precipitator.

**CLEANLINESS:** The absence of dirt or spots on paper.

**CLEANSING TISSUE:** A good-strength, soft, absorbent, sanitary, lightweight paper made from bleached chemical and mechanical pulp. It is used to make up toweling and wiping materials for both domestic and industrial purposes.

**CLEARANCE:** (1) With reference to piston operators, it is the short distance at each end of a cylinder after the stroke cycle of the piston is completed. (One half of the total length of the cylinder minus the stroke of the piston.) (2) The distance between the surfaces of the plates in a refiner through which pulp is passed for mechanical treatment. (3) The distance between any moving and stationary parts of machinery.

**CLEARCUT:** A tree harvesting technique used in even-aged management to speed regeneration of valuable species with all the timber removed when the site is harvested.

**CLEARWELL:** A reservoir containing mill supply water which has been filtered or purified before going into standpipes or distribution systems.

**CLIPPER SEAM:** A method of joining the ends of paper machine canvas-type dryer fabric to form an endless belt.

**CLOSE FORMATION:** A term used to denote the fiber orientation in a web structure that produces a uniform density distribution throughout a sheet of paper.

**CLOSED AIR HOOD:** A canopy, located over the dryer section of a paper machine, with side panels that extend to the paper machine room floor or into the basement. It is used to confine and control vapor-laden air rising from the dryers so that the heat can be recovered by using it to heat the air entering the machine room.

**CLOSED CYCLE BLEACHING:** A pulp bleaching process in which all normally polluting effluents are recovered.

**CLOSED CYCLE MILL:** A pulp mill concept in which all liquid effluents are recovered, thereby eliminating water pollution by the mill. Also called *effluent-free mill.*

**CLOSED LOOP:** A family of automatic control units linked together with the process in such a way as to form a chain so that the effects of the control action are constantly measured. If the process deviates beyond the desired limits, the control units react to bring it back into line.

**CLOSED TRANSFER:** A means of transferring the wet sheet web from the wire (forming) section to the wet press of a paper machine, and includes the suction and yankee pickup.

**CLOSURE:** *See* CLOSED CYCLE MILL.

**CLOTH BOARD:** A stiff, nonwarping paperboard made with a type of surface that will accept cloth material pasted on it.

**CLOTH FINISH:** A paper sheet finish that resembles the woven surface of cloth. It is obtained by pressing cloth materials such as burlap, linen, etc., on the paper sheet.

**CLOTH PAPER:** A high-strength, durable paper made with the intention of lining it with cloth material. *See* CLOTH-LINED PAPER.

**CLOTH-CENTER PAPER:** A special type, multi-layered paper made up of a variety of cloth inner layers.

**CLOTH-LINED BOARD:** Paperboard that has been covered with pasted-on cloth to improve surface appearance.

**CLOTH-LINED PAPER:** A high-strength, durable, multi-layered paper sheet made with a lining of cloth material. It is used to make tickets, maps, special envelopes, etc.

**CLOTHING:** A general reference to press and dryer felts on the dry end of a paper machine. It can also include wires (forming fabrics) on the wet end former(s).

**CLOUD FINISH:** A surface effect produced by dripping white pulp slurry on a colored paper web being formed on the paper machine wet end wire (forming fabric).

**CLOUDY:** A reference to the fiber orientation in a web structure as a result of uneven or irregular distribution throughout a sheet of paper. Also called *wild formation.*

**CLOUDY FORMATION:** The evenly distributed light and dark areas in a sheet of paper as seen when viewed by transmitted light.

**CLOVERLEAF PRESS:** A type of paper machine wet end press that has three rolls, with two suction rolls located 45° below the horizontal centerline of the center roll.

**CLUTCH:** A coupling device used to connect or disconnect a driving and a driven part of a machine, such as parts or sections of a paper machine.

**CMP:** *See* CHEMIMECHANICAL PULP.

**COAGULATION:** (1) A process whereby finely divided particles are made to combine to produce a particle capable of settling. It is commonly done in the clarification of the mill water supply by chemical means. (2) The clumping of particles in order to settle out impurities in mill supply and wastewaters by the use of lime or alum.

**COAL FIRED:** A term referring to various types of power boilers that use coal as a fuel.

**COAL SPECKS:** *See* CARBON SPOTS.

**COAL TAR DYES:** Aniline organic paper coloring materials obtained from coal tar.

**COARSE PAPERS:** A variety of medium-weight papers made from bleached and unbleached, chemical and mechanical pulps, and suited especially for commercial and industrial uses.

**COARSENESS:** A property of pulp related to the refining, fiber perimeter, and wall thickness characteristics as it affects the weight per unit length of the fiber.

**COASTER BOARD:** A rigid, nonwarping, absorbent type of heavyweight paperboard made from chemical pulp, mechanical pulp, and/or recycled paper stock. It is used to make small protective pads placed under drinking glasses. Also called *beer mat board.*

**COATED:** (1) Various types of papers and paperboards that contain a layer of coating material such as clay, various pigments, and/or other special substances in combination with adhesive(s) of some type. Also, paper and paperboard treated with a surface layer of varnish or lacquer. These products are "coated" to improve their surface characteristics, e.g., opacity, smoothness, color, printing quality, etc.

**COATED ART PAPER:** A book paper that has been especially coated for use in halftone printing.

**COATED BLANKS:** A class of coated single-ply or multi-ply rigid paperboards with good printing surfaces. These boards are made from chemical pulp, deinked pulp, and clean broke, and are used for making posters, merchandising displays, and other signs.

**COATED GLASSINE:** A glassine-type paper coated with a water and water vapor-resistant transparent film applied by the evaporation of a proper organic solvent or by passing through a molten bath.

**COATED ONE-SIDE (C1S):** Reference to paper that is coated on one side only. Also expressed as *one-side coated.*

**COATED PAPER:** A term applied to any type of paper whose surface has been treated in such a way as to apply a coating in order to enhance its finish characteristics. *See* COATED.

**COATED SECONDS:** Slightly defective coated sheets of paper and paperboard that are separated from good sheets and sold as inferior.

**COATED TWO-SIDES (C2S):** Reference to paper that is coated on both sides. Also referred to as *two-sides coated.*

**COATED WEIGHT:** Refers to the amount, thickness, etc., of the layers of mixed clays and adhesives etc. that are applied to the surface of paper or paperboard to create a new surface.

**COATING:** (1) The process of treating a sheet of paper or paperboard so that a pigmented coating material layer or a clear film is applied to its surface. (2) Refers to the coating material or film substance before coating. (3) The coating layer that is formed on the paper and paperboard sheet.

**COATING BASE:** A term referring to any type of paper or paperboard before it is coated. Also called *coating raw stock.*

**COATING CLAY:** The types of finer size clays with higher whiteness used to make up coating materials for paper and paperboard surfaces. *See also* CHINA CLAY, CLAY, and KAOLIN.

**COATING COLOR:** A coloring mixture made up of desired pigment and dyes with necessary adhesives and other modifiers and mixed in a carrier liquid. It is used to coat the surfaces of paper and paperboard. Sometimes referred to simply as *color.*

**COATING GRADE CLAY:** The type of clay used to make up coating material. It has such properties as size, viscosity, color, whiteness, etc., as determined by the grade of paper to which it is to be applied. *See also* CHINA CLAY, CLAY, COATING CLAY, and KAOLIN.

**COATING IMMOBILIZATION:** The technology of preparing a paper coating that has flow characteristics until applied to the sheet, at which time it no longer flows and does not strike further into the base sheet. This is accomplished by methods that increase solids on application and reduce solids after they are applied. *See also* STRUCTURED COATINGS.

**COATING MOTTLE:** Gloss variations on the surface of a coated sheet of paper that has been calendered.

**COATING KITCHEN:** An area or room in a paper mill located in the stock preparation area where paper coating and coloring materials are stored, prepared, and mixed in proper proportions. Sometimes referred to as *color kitchen, color room,* or *mixing room.*

**COATING LUMP:** A random spot or speck on the paper sheet surface that is higher in basis weight than the surrounding area. It is caused by broke in the coating or dried particles of coating in the coating mixture.

**COATING MACHINE:** A machine used in a paper mill to apply coating slurry to a base paper, employing a variety of methods including brushing, rolling, air jets, etc.

**COATING PAN:** A container in which a reservoir of coating material is maintained on a coater. The coating is picked up from the pan for applying to the paper surface.

**COATING PICK:** A type of paper surface defect caused by inadvertent lifting of coating particles or patches of coating.

**COATING PIGMENTS:** Paper coating materials of a mineral nature such as clay, ground and precipitated calcium carbonate, titanium dioxide, talc, zinc oxide, satin white, etc.

**COATING PLANT:** A part of the paper mill or a separate area where the paper coating operation is carried out as a separate operation from papermaking.

**COATING RAW STOCK:** *See* COATING BASE.

**COATING SKIPS:** Large, irregular shapes caused by nonuniform or uncoated areas of paper.

**COATING SPLASH:** An irregular-shaped paper surface defect that is higher in basis weight than the surrounding area. It is caused by wet coating splattering on the sheet during or after the original coating application.

**COATING STREAK:** A paper surface defect appearing as narrow light and heavy strips that are more or less transparent than the surrounding area due to buildup of material behind the coater blade, or a bent or damaged blade.

**COBB TEST:** *See* WATER ABSORPTION.

**COCK:** A form of valve having a tapered plug with a hole that is rotated to provide a passageway for fluid.

**COCKLE:** A paper defect appearing as a wrinkle caused by nonuniform shrinking due to uneven drying or sheet formation.

**COCKLE FINISH:** A paper surface rippling effect produced by controlled air drying in which uneven shrinkage occurs under minimum sheet tension conditions during the manufacture of various types of writing papers.

**COCKLING:** The process of producing a ripple-like defect or intentionally creating this type of paper surface by controlled uneven drying of the sheet under minimum tension conditions.

**COD:** *See* CHEMICAL OXYGEN DEMAND.

**COEFFICIENT OF EXPANSION:** The increase in unit length, area, or volume for a 1° rise in temperature.

**COEFFICIENT OF FRICTION:** A characteristic of linerboard determined by measuring the sliding resistance of sheets made from the material when they are pressed tightly together.

**COGENERATION:** Generation of power in an industrial power plant to produce both steam and electricity for in-plant use, as well as for sale to outside utility companies.

**COHESION:** The shearing stress between surfaces, such as in a duplex sheet of paper, and fibers in a pulp slurry.

**COHESIVENESS:** A refining term referring to the intrinsic attraction of fiber surfaces, which equals the shearing stress divided by the relative bonded area.

**COIL:** A small roll of paper made by slitting a larger roll to the desired width and rewinding in a continuous roll for a particular use on business machines.

**COIN WRAPPER:** A strong kraft paper used for the packaging of coins.

**COLD BLOW:** The introduction of cool wash liquor to the bottom zone of a Kamyr continuous digester to cool the pulp going to the discharge devices. This eliminates fiber degradation if blown at higher digester temperature. Also, an experimental method of emptying a batch type digester with compressed air or steam pressure from an accumulator tank after the free liquor is drained from it into the accumulator tank.

**COLD CAUSTIC:** Sodium hydroxide (NaOH) solution used to cook hardwood chips at ambient temperature as part of one of the semichemical pulping processes called "cold soda pulping."

**COLD CAUSTIC EXTRACTION:** A stage of the pulp bleaching process in the preparation of high alpha-cellulose dissolving pulps in which the high consistency pulp is steeped in over 10 percent caustic soda to dissolve out the hemicellulose, other short-chained celluloses, and resin.

**COLD CAUSTIC PULP:** A semichemical-type pulp produced by cooking hardwood chips with sodium hydroxide (NaOH) solution at ambient temperature.

**COLD COLORS:** Paper colors in the blue, green, and violet range that produce a cool psychological effect as opposed to those in the red, orange, and yellow range known as *warm colors*.

**COLD FLOW:** (1) A shifting of components or materials bonded together due to conditions that exceed the resistance limits of the adhesive used. (2) The deformation of a material or substance due to excessive exposure conditions. Also termed *creep*.

**COLD GRINDING:** The stone grinding of wood logs at a controlled low temperature by the use of large amounts of water, usually below 50°C. Sometimes referred to as *color grinding*.

**COLD PRESSED FINISH:** A very rough surface finish applied to a heavyweight, surface-sized paper made from chemical pulp and cotton fibers. This paper is used for artwork, diplomas, charts, etc.

**COLD PRESSING:** Forming multi-layered paper and board by the use of a proper adhesive and applying mechanical pressure without the use of heat.

**COLD SEAL COATING:** A type of cohesive coating applied to the surface of paper and paperboard which will adhere to the fiber surface and nothing else without the use of heat.

**COLD SETTING ADHESIVE:** (1) An adhesive that cures at temperatures below approximately 68°F (20°C) after addition of a catalyst or curing agent. (2) An adhesive that sets at or below room temperature without a catalyst or curing agent.

**COLD STEEP BLEACHING:** A method of bleaching a high consistency pulp sheet formed on a feltless wet machine, or a vacuum thickener equipped with press rolls so that it is self-supporting. The bleach liquor is applied to both surfaces of the sheet by spraying it onto the rolls for even spreading.

**COLD WAXED BOARD:** Paperboard that is made moisture-resistant by the application of hot paraffin wax and immediately cooled by ice water so that it solidifies primarily on the surface with little penetration into the sheet.

**COLD WORKING PRESSURE:** The maximum pressure rating of a pipeline, valve, fitting, etc., that is coincident with the ambient temperature or otherwise understood to be in the range of from -20°F to +100°F (-29°C to +38°C).

**COLLOID:** Uncrystalline material resembling jelly or glue. Starch used in papermaking is a typical example.

**COLLOTYPE PAPER:** A good-quality printing paper sufficiently durable to withstand excess moisture from collotype printing.

**COLOPHONY:** Another term for *rosin*, a residue from the distillation of gum from resinous southern pine trees. It is used for internal sizing of paper and paperboard.

**COLOR:** (1) The dyes and pigmented mixtures used to treat paper or paperboard and the visual effect it produces in the product and on its surface. *See* COATING COLOR. (2) The act of accomplishing a visual effect on paper and paperboard by the use of dyes and pigmented mixtures. (3) A characteristic of paper and board sheet surfaces consisting primarily of *chromaticity* and *brightness*..

**COLOR EDGE:** The presence of colored coating on the uncoated side of a one-side color coated paper. It is usually trimmed off before winding.

**COLOR FASTNESS:** The resistance of colored paper and dyestuffs to color deterioration under prolonged exposure to heat, light, and other normal ambient conditions or uses.

**COLOR GRINDING:** *See* COLD GRINDING.

**COLOR KITCHEN:** *See* COATING KITCHEN.

**COLOR LAKE:** An artificial coloring pigment consisting of an inorganic base, such as clay, barium sulfate, or similar material, on which dyes are deposited. Also known as *lake* or *lake pigment*.

**COLOR LUMP:** A color-coated paper surface defect consisting of an agglomeration of the coating material in the form of a lump.

**COLOR MAN:** The paper mill employee responsible for the care, preparation, and mixing of pigments and dyes.

**COLOR MATCHING:** Procedure of adjusting a quantity of dyes and other chemicals or additives being mixed with paper stock so that in the final sheet and standard sheet visually apparent differences are minimized under specified illuminating and viewing conditions.

**COLOR REVERSION:** Loss of brightness during or after the bleaching process due to the development of unsatisfactory operating conditions or to aging over a period of time. *See also* BRIGHTNESS REVERSION, REVERSION, and YELLOWING.

**COLOR ROOM:** *See* COATING KITCHEN.

**COLOR SPOTS:** Unwanted multi-colored specks appearing on the sheet surface caused by dyes collecting in various surface areas of the headbox of a paper machine as the result of drying, and subsequently falling off onto the sheet or onto the wire (forming fabric).

**COLOR STRENGTH:** The intensity of color of a pigment used in dyes.

**COLOR VALUE:** Closeness of a color to white or black or degree of lightness or darkness.

**COLORED BROKE:** Dye-containing papers from trimmings, paper machine broke, and various papers and paper cuttings from finishing operations.

**COLORIMETRY:** Science of color measurement.

**COLORING:** The physical and colloidal process of using a dyestuff to impart color to paper or paperboard. Also termed *dyeing.*

**COLORING PIGMENT:** A hard, small-particled organic or inorganic coloring substance used to make up color coating materials. It remains suspended in the coating mixture carrier. *See also* COATING COLOR and COLOR.

**COLUMN STRENGTH:** Load-carrying property of paperboard, determined by the force that an unsupported section will withstand before bending, when the force is applied parallel to its plane.

**COMBED EDGES:** Edges of paper having a marbled appearance produced by combing the color used to dye it.

**COMBINATION BOARD:** A multi-layered, cylinder-made paperboard in which the outer layers are made from different pulp stocks than the inner layers. They may also be of different colors.

**COMBINATION BOILER:** Power boiler designed to use oil, coal, or gas, or any combination of these as fuel.

**COMBINATION CUTTERS:** Types of paper cutters that perform more than one cutting operation in the finishing room, such as trimming and cutting paper to various user sizes.

**COMBINATION FUEL:** Power furnace fuel that consists of more than one different and individual combustible material being fed into and fired in a furnace at the same time, such as bark in combination with oil and gas in a pulp and paper mill power plant.

**COMBINED BOARD:** A multi-layered board made by uniting a number of boards with the use of proper adhesives.

**COMBINED SULFUR DIOXIDE:** (1) That portion of sulfur dioxide ($SO_2$) in sulfite cooking liquor that is combined with the cation base to form the principal cooking chemical, such as with calcium to form calcium bisulfite, with magnesium to form magnesium bisulfite, with sodium to form sodium bisulfite, or with ammonia to form ammonia bisulfite. (2) Sulfur dioxide ($SO_2$) in sulfite cooking acid that is combined with the base as a monosulfite (nonavailable $SO_2$). It is determined by the difference between the total $SO_2$ and free $SO_2$, and expressed as a percentage. Also known as *nonavailable sulfur dioxide.*

**COMBUSTIBLE PAPER:** A type of paper that is treated so that it will burn at a desired rate under prescribed conditions. *Cigarette paper* is an example.

**COMBUSTION CHAMBER:** Usually a large steel brick lined chamber located after the sulfur burner in an acid plant of a sulfite pulp mill in which gases are mixed and more completely converted to $SO_2$.

**COMBUSTION CONTROL:** The automatic control of burning fuel in power and chemical recovery furnaces so that it can be carried out to its highest efficiency.

**COMMERCIAL DRY:** An expression used to refer to the moisture content of a product as pounds of water per pound of solid as it leaves a drying process. Also called *commercial dry basis.*

**COMMERCIAL DRY BASIS:** *See* COMMERCIAL DRY.

**COMMERCIAL MATCH:** A mill term describing an acceptable paper that is slightly off-spec from a match sample used for that particular run.

**COMMERCIAL TISSUES:** Types of tissue paper made for commercial use, such as protective wrapping and packing for shipping materials.

**COMMERCIAL WRITINGS:** All types of writing papers, such as bond, tablet, ledger, etc., used throughout the business world for a variety of purposes.

**COMMUNICATION PAPERS:** All types of papers used in the communication field that have been converted into convenient form for use with communications equipment (e.g., teletype machines), such as rolls, tapes, special cut-size, etc.

**COMPACTION BAFFLE FILTER:** A pulp washer utilizing a pressurized drum filter that achieves high capacity by using high feed consistency and operating speeds, resulting in a smaller machine than an equivalent conventional vacuum or pressure filter.

**COMPOSITE CAN AND TUBE INSTITUTE:** Headquarters: Alexandria, VA, (703) 549-2233.

**COMPOSITE FUEL:** A fuel for power furnaces made up of a blend of combustible solids, such as coal, waste materials, etc., in a liquid carrier such as oil.

**COMPLETE TREE CHIPPING:** *See* WHOLE TREE CHIPPING.

**COMPRESSIBILITY:** The amount that sheet caliper will decrease under a specified load or use condition. In printing, it is referred to as *cushion.*

**COMPRESSING:** A refining action that densifies the fiber in the pulp.

**COMPRESSION:** A characteristic of paper board sheets determined by the resistance to force applied to the edge of representative samples.

**COMPRESSION STRENGTH:** The physical property of paper and paperboard that is resistant to compression loads.

**COMPRESSION WOOD:** Pulpwood with lower cellulose and higher lignin content than normal, usually originating from the lower sides of branches and trunks of softwood trees that contain larger, less dense summerwood growth rings.

**COMPRESSOR:** A mechanical device for increasing the pressure of air or gas.

**COMPUTER:** (1) A device capable of accepting information, performing prescribed operations on this information, and providing the results of these operations. Its major elements usually include memory, control, arithmetic, logic, and input and output facilities. (2) An automatic electronic machine for performing calculations used in pulping and papermaking to monitor, control, and optimize manufacturing operations.

**CONCORA CRUSH:** A specific test for edge compression resistance of *linerboard* and *corrugating medium*, in which strips of either grade are crushed edgewise. The force necessary to attain edge crush is carefully measured to determine standardized test values. *See also* RING CRUSH, EDGE CRUSH TEST, RULE 41.

**CONDENSATE:** The liquid formed from the lowering of temperature of a saturated vapor, such as water formed from the saturated steam being cooled in paper machine dryers and condensing-type turbines.

**CONDENSATE REMOVAL SYSTEM:** A system of process equipment, syphons, rotary joints, headers, condensate separators, condensers, and pumps designed and assembled to remove condensate from within paper machine dryer rolls, separate it from noncondensible vapors, and condense any remaining steam so that it can be pumped to and used in the power and recovery boilers.

**CONDENSER PAPER:** *See* CAPACITY PAPER.

**CONDENSER TISSUE:** *See* CAPACITY PAPER.

**CONDENSING RATE:** Pounds per hour steam condensed–as in paper machine dryer drums, dryer sections, or entire machines, etc.

**CONDENSING TURBINE:** A turbine in which steam from a boiler is passed through to expand to a low pressure exhaust and then discharged to a condenser in which it changes into water as it cools in preparation for returning to the boiler.

**CONDITIONED PAPER:** Paper that has been prepared and treated so that the moisture is uniformly distributed throughout the sheet and has reached equilibrium with the surrounding air. Also called *screened paper* or *seasoned paper.*

**CONDITIONER:** A machine used to bring paper into equilibrium with atmospheric humidity by increasing or decreasing its moisture content.

**CONDITIONING:** Providing the most suitable controlled atmospheric conditions under which paper and paperboard are tested, stored, or used. It is accomplished by controlling the temperature and humidity of the air.

**CONDUCTION:** (1) The passage of heat through or along a mass or between two bodies in direct contact. (2) The passage of electric current in a device such as a wire.

**CONDUCTIVE HEAT TRANSFER:** The movement of heat from one portion of a mass to another portion of the same mass or between separate contacting masses without particle displacement.

**CONDUIT:** Various size metal pipes used to protect and carry wire around the mill.

**CONE PAPER:** Unsized paperboard with a smooth finish used to make up conical cylinders on which yarn is wound and shipped.

**CONE PRESS:** A cone shaped press used to thicken rejects from some paper mills.

**CONICAL REFINER:** A type of mechanical pulp refiner consisting of a conical plug whose surface is equipped with individual bar and wood spacers, and is fitted inside a similarly equipped shell between which the pulp slurry is moved as the motor-driven plug rotates. It applies a refining action on the fibers and is commonly known as the *jordan refiner.*

**CONIFEROUS:** Species of cone-bearing softwood trees, such as firs and pines, commonly used to make sulfite and sulfate chemical pulps.

**CONIFERS:** Cone-bearing types of trees used as a source of pulpwood. Also referred to as *gymnosperms.*

**CONSISTENCY:** (1) A measure of the fibrous material in pulp solutions, e.g., pulp and water, or stock (pulps and additives) and water. It is expressed as a percentage of this material in the solution, in terms of *bone dry (B.D.), oven dry (O.D.),* or *air dry (A.D.)* weight. (2) That property of adhesives or other coating material related to viscosity, plasticity, etc., that makes it resistant to deformation or flow.

**CONSISTENCY PENCIL:** A graduated, round, pointed, metallic rod once used to make quick consistency determinations on high density pulp by dropping it into a sample from a fixed height and reading the depth of its penetration from the graduations.

**CONSISTENCY REGULATOR:** A control unit designed to detect the consistency of pulp stock in the flow line or sample bypass, and used to activate a water dilution valve to bring the pulp stock to a desired consistency level.

**CONSTANT HUMIDITY ROOM:** A room whose temperature and humidity is closely maintained to a specific level, and in which paper is conditioned and tested to determine whether it meets certain specifications.

**CONSTRUCTION PAPER:** A heavy type of paper used for watercolor and crayon artwork made primarily from groundwood pulp and in a variety of colors.

**CONTACT ADHESIVE:** An adhesive, apparently dry to the touch, that will adhere to itself instantaneously upon contact.

**CONTACT DRYER:** A type of dryer where the greatest amount of heat required for evaporation is supplied by conduction from a heated cylinder type drying can.

**CONTAINERBOARD:** Any type of single or multi-ply, solid or corrugated boards used to make up boxes and other containers for shipping materials, and the type of paperboard used to make them up. Also called *casemaking board.*

**CONTINUOUS:** (1) Pulp and papermaking processes and process equipment that operate to completion in an uninterrupted series of actions or events. (2) An entire sheet of paper in a roll.

**CONTINUOUS DIGESTER:** A wood-cooking vessel in which chips are reduced to their fiber components in suitable chemicals under controlled temperature and pressure in a continuous operation.

**CONTINUOUS LAID DANDY:** A dandy roll which has been made with the laid surface wires wound in such a manner that they run parallel to the machine direction.

**CONTINUOUS PULPING PROCESSES:** Any pulping process in which the fibrous raw material and cooking chemicals move through the successive processing phases in a continuous fashion.

**CONTINUOUS RATING:** An electrical term referring to a machine rated for continuous service and able to operate continuously at its rated output without exceeding any of the specified limitations, such as permissible temperature.

**CONTRAST RATIO:** An opacity measurement related to paper and calculated from the ratio of reflectances of a sample when backed up by a black body and white body, respectively.

**CONTROLLED BURNING:** Use of fire to destroy logging debris, control disease, reduce fire hazard, and enhance growth of pines in newly planted areas by eliminating competitive broad-leafed species, which are much more susceptible to fire.

**CONTROL POINT:** The average value of the controlled variable being maintained by a control system under steady conditions.

**CONTROL SET POINT:** The value of a controlled variable of a control system which is the desired value to be maintained.

**CONTROL SYSTEM:** A group of integrated control elements that act together to produce corrective action based on signals provided by the process measurement element.

**CONTROL VALVE:** Types of process control valves that are remotely operated, either from the output of an automatic controller or from a manual control station.

**CONTROLLED VARIABLE:** A process quantity or condition that is measured and controlled by an automatic controller.

**CONTROLLER:** The device in a control system that dictates the occurrence of events in that system.

**CONVECTION:** The conveying of heat from a heat source by contact with a moving fluid or gas that has a cooler temperature than the heat source.

**CONVECTIVE HEAT TRANSFER:** The movement of heat from one point to another point in a gas or liquid by physical internal mixing and motion.

**CONVERSION COATING:** The process of applying coating material to a sheet of paper or paperboard after it is made. It is accomplished in an off-machine location and is sometimes referred to as *off-machine coating*.

**CONVERTER:** One who makes user products such as envelopes, bags, containers, tapes, etc., from finished paper and paperboard. This is sometimes carried out at the paper mill site. *See also* CONVERTING.

**CONVERTING:** The operation of treating, modifying, or otherwise manipulating the finished paper and paperboard so that it can be made into end-user products such as special coating, waxing, printing, and gumming, and envelope, bag, and container manufacturing.

**CONVERTING MILL:** A name sometimes applied to a paper or paperboard mill that does not produce the pulp it uses on-site.

**CONVERTING PAPER:** Paper made especially to undergo further treatment, modification, and manipulation so that it can be converted to end-user products such as envelopes, bags, tapes, etc.

**CONVEYOR:** A mechanical apparatus, usually consisting of an endless moving belt, a chain of receptacles, or a series of rollers, used to transport pulp and papermaking materials in and around a mill.

**COOK:** (1) The process of reacting fiber-containing materials, such as wood, rag, straw, bagasse, etc., with suitable chemicals, usually under high temperature and pressure, to reduce them into their component parts so that the acceptable fibers can be separated and made into pulp for further use in making paper. (2) A mill name for the *digester operator* or *digester man.*

**COOKING CYCLE:** The cooking operation in a batch-type digester from the start of one cook to the start of the next in the same digester.

**COOKING LIQUOR:** Chemical solution added to digesters to reduce chips into their fiber components by dissolving the cementing material, usually lignin, thereby forming pulp.

**COOKING TIME:** The time lapse between the start of the steaming to the start of the blow of one cook in a batch-type digester.

**COOLING CYLINDER:** A water cooled drum normally located after the dryers on a paper machine and used to cool the paper, thereby restoring some of the moisture in the paper.

**COOLING ROLL:** Hollow, metallic, cylindrical roll located in the dryer section of a paper machine and operated at a lower temperature by the use of cooling water in order to restore some moisture to the sheet. Also called *sweat roll.*

**COPPER NUMBER:** Weight in grams of copper reduced to the cuprous state by 100 grams of pulp. It is determined by a laboratory procedure under standardized conditions and is a measure of the degree of degradation of the pulp.

**COPPICE:** Sprouts which grow from hardwood stumps after a tree has been cut.

**COPYING TISSUE:** A high-strength, soft, lightly sized type of tissue paper made for manifold bill copying and letterpress applications.

**CORD:** (1) A pulpwood volume measurement comprised of a pile measuring 4 feet by 4 feet by 8 feet, equaling 128 cubic feet (3.62 cubic meters). (2) Yarn or twine made from twisted or spun, narrow width paper strips that can be woven or knitted into fabric.

**CORE:** A metal, wood, or convolute-wound fiber tube used to wind paper or paperboard into rolls for transporting and shipping purposes. Sometimes referred to as a *shaft.*

**CORE BOARD:** A well-sized chipboard made to be cut into strips for spiral or involuted windings in making cores for winding paper and textile material.

**CORE CURL:** The tendency of reel or core rolled paper to become set to its contour, causing a curl in the sheet. Also called *roll set.*

**CORE PLUGS:** Plugs used to prevent core crushing by fitting them into the ends. They are usually made from wood and metal, but plastic and compressed paper have also been used. Sometimes known as *bungs.*

**CORE SLIPPAGE:** *See* SLIPPED ROLL.

**CORE STOCK:** A type of paperboard made from chemical pulps mixed with screenings and used to make cores for paper rolls.

**CORE WASTE:** Any paper that inadvertently remains on the core of a roll after the sheet has been rewound from it as a result of improper splicing, faulty brake tension on the winder, uneven tension on the sheet, etc.

**COREWOOD:** That structural portion of a pulpwood tree formed during early years of its life and located in the central regions of the lower sections of the tree. Also located in the top sections of older trees. Also called *juvenile wood.*

**CORK PAPER:** A cork-like sheet of paper made in a variety of ways depending on its end use. Ground cork is mixed with chemical pulps and proper adhesives and glycerin additives to make a heavyweight sheet for gasketing material. Laminates of thin cork and paper sheets are used to make some types of paper "tips." An example of a product using cork is protective packing material made from heavy manila paper coated with a mixture of ground cork and adhesive.

**CORN STARCH:** A material used to make pigment binder, adhesive, and sizing material used in the paper industry. It is obtained from corn granules.

**CORN STRAW:** Corn stacks harvested for use in making certain types of heat and sound resistant paper board.

**CORONA TREATMENT:** An electrical power treatment used in extrusion coating to increase tension of the substrate for better adhesion and wetability.

**CORRESPONDENCE PAPER:** A high-quality paper made from chemical pulps and/or cotton fiber. It is especially finished by embossing, watermarking, mottling, etc., to achieve an attractive surface effect, and has good pen and ink qualities for use in making all types of correspondence paper items.

**CORROSION:** The gradual wearing away or the effect upon equipment caused by exposure to process and environmental conditions.

**CORRUGATED BOARD:** (1) A pasted, single- or double-faced, multi-layered type of board in which the bottom or middle layer is fluted. (2) The fluted paperboard after it has gone through the corrugating operation and before it is pasted to the flat facing board sheets. It is used as a protecting, separating, or support medium in packaging fragile items.

**CORRUGATING:** The process of producing parallel, wave-like shapes (flutes) on the corrugated medium during the manufacture of corrugated board on special equipment. Sometimes called *fluting. See also* CORRUGATING MACHINE.

**CORRUGATING MACHINE:** A piece of mechanical process equipment consisting of two alternately grooved, metallic rolls geared to operate so that the ridges run within the grooves in such a manner to impart parallel, wave-like shapes (flutes) to a sheet passed through it. *See also* A-FLUTE, B-FLUTE, C-FLUTE, E-FLUTE, and FLUTE.

**CORRUGATING MEDIUM:** Paperboard that is made from chemical and semichemical pulps sometimes mixed with straw or recycled paper stock that is to be converted to a corrugated board by passing it through a corrugating machine. The corrugated board is primarily used as the center segment of corrugated shipping containers, being faced with *linerboard* on both sides.

**CORRUGATION:** The parallel, wave-like configuration impressed on a sheet when pressed through special equipment. *See also* CORRUGATING MACHINE.

**CORRUGATION MARK:** Parallel, wave-like paper markings primarily caused by ambient temperature and humidity conditions around wound rolls and uneven caliper. Sometimes called *buckle, rope marks,* or *tire tracks.*

**COTTON FIBER:** Hair-like fibers removed from the cottonseed and used to make a high-quality pulp for paper and dissolving purposes. The fiber may also be obtained and processed in the form of new and used cotton cloth.

**COTTON FIBER CONTENT:** A reference to paper made from a stock furnish containing cotton fibers.

**COTTON LINTERS:** The shorter length cotton fibers used to make a higher quality pulp for dissolving purposes in the manufacture of cellulose derivatives.

**COTTON WIPE TEST:** A superficial method of approximating the roughness of a surface over which pulp stock will flow. It consists of passing a wad of cotton over the surface and observing the number of fibers that adhere as an indication of the possible fiber hang-up when pulp stock runs over it. The test is usually done on stock handling equipment, the inside surface of pipelines, and certain flow-through type meters.

**COTTONING:** The formation of feathery filaments of adhesive coating that forms between the coating and coating machine parts.

**COUCH FELT:** (1) A cylindrical-shaped woolen blanket used to cover the couch roll on a paper machine. (2) In making paper by hand, it refers to a piece of wool blanket to raise the sheet from the paper mold.

**COUCH JACKET:** A cylinder-shaped wool felt once used as a couch roll cover.

**COUCH MARKS:** Paper defects appearing as the pattern of the couch roll holes. They show up when transmitted or direct light strikes the paper. In older felt-covered couch rolls, the marks would resemble the jacket surface. Sometimes called *shadow marks.*

**COUCH PIT:** That portion of the white water pit beneath the wet end of a fourdrinier paper machine that is located below the couch roll end of the wire (forming section). It collects water draining from that section so that it may be returned to the white water system.

**COUCH ROLLS:** Rolls that are involved both in the removal of water after the web is formed on the paper machine and picking up the web so that it can be transferred to the wet press section. The couch roll under the wire (forming section) on a fourdrinier machine also serves as the main drive as well as a dewatering device. *See* BOTTOM COUCH ROLL. The couch roll on the cylinder machine serves as the main pickup felt drive as well as a water removal device.

**COUCHING:** Pickup of the sheet formed on the cylinder wire by a felt as the web passes through the nip made by the couch roll contacting the felt, web, and cylinder face.

**COUNT:** The quantity of sheets of a specific size, weight, and caliper, contained in a standard unit package established for the respective kind and grade of paper or paperboard. Also refers to the number of cross machine direction strands per inch on paper machine forming fabrics. *See also* MESH.

**COUNTER:** A device that continuously registers single occurrences and from which a total number of occurrences can be extracted for a given period of time. Also, the person or instrument that counts sheets of paper in a paper mill.

**COUNTER BOARD:** A heavily calendered paper board made primarily from pulp fibers especially for use in the manufacture of shoe counters.

**COUNTER ROLL:** (1) A roll of standard newsprint paper made outside of the regular runs and used for other purposes, such as for wrapping. (2) A smaller sized roll of paper fitted with wooden end plugs and used in counter-top wrapping paper dispensers in retail stores.

**COUNTERCURRENT COOKING:** The addition of part or all of the white liquor into the second and third zones of a Kamyr continuous digester so that liquor and chips move in a countercurrent manner against each other below the second zone.

**COUNTERCURRENT WASHING:** (1) A method of washing pulp by running the wash water countercurrent to the flow of pulp through the process. Examples include countercurrent intra-stage washing in a multi-stage bleaching process (to minimize effluent) and the countercurrent flow of wash water to pulp flow on vacuum-type brownstock washers (to minimize water use and maximize black liquor recovery). (2) The washing of pulp within a Kamyr continuous digester (before blowing) in which the wash water flows countercurrent to the pulp flow in the process.

**COVER PAPER:** Any paper normally used as a protective covering for such items as books, pamphlets, magazines, catalogs, boxes, etc. They are usually made from chemical and mechanical pulp, and sometimes are mixed with rag fibers.

**COVER-TO-COVER:** The lapse of time from the securing of the digester cover at the start of one cook to the securing of the digester cover at the start of the next cook in the same batch-type digester in a pulp mill.

**COVERAGE:** A papermaking adhesive term referring to the ability of the adhesive to spread uniformly over the surfaces of paper and paperboard sheets.

**CPPA:** *See* CANADIAN PULP & PAPER ASSOCIATION.

**CRACK:** Defects occurring in folds of paper because of coating and/or fiber separation and appearing as very small sheet and surface fissures. Also referred to as *crack at fold*.

**CRACK AT FOLD:** *See* CRACK.

**CRACKED EDGE:** A paper web and fourdrinier wire (forming fabric) defect occurring as small breaks on the edges, with very little migration toward the center.

**CRACKING:** A coated paper defect showing up as surface fissures (during printing, folding, or other converting operations) which cause the coating layer to separate. *See also* CRACK.

**CRACKLE:** A stiffness-related property of a sheet which causes it to produce noise when shaken. It is caused by extended refining of the fiber, and is known as *tinny* when the noise is a ringing sound.

**CRATERS:** A paper coating defect caused by entrained air bubbles bursting after the coating is applied to the sheet surface. They resemble volcanic craters under magnification.

**CRAWLING:** Defect of a finishing composition in which the adhesion to the surface is not sufficient to prevent the finish from pulling together while still wet, leaving uncoated areas.

**CRAZING:** Defects in paper coating, appearing as short, random, intersecting cracks in the surface of the coating.

**CREAM WHITE PAPER:** A paper which has a cream white paper appearance.

**CREASE:** (1) A physical deformation of the structural property of a sheet of paper along the line of a fold made into it. (2) To put an indentation on the surface of a board or heavy paper to make it easier to fold. Also called *score.*

**CREASE RETENTION:** The ability of paper to retain a fold.

**CREASEABILITY:** Physical property of paper that resists the breaking away of the surface coating along the crease on folding.

**CREASING STRENGTH:** The strength of paper after being folded under specified load conditions.

**CREST:** The top of a dam, dike, spillway, or weir to which fluid must rise before passing over the structure.

**CREEP:** The distortion of a material or substance structure due to being subjected to excessive conditions beyond its resistance limitations. Also referred to as *cold flow.*

**CREPE:** A simulated crinkly finish effect generally produced on a sheet of tissue or lightweight paper by crowding it on a roll with the aid of a doctor blade on the paper machine (*machine crepe*) or during converting.

**CREPE PAPER:** Low basis weight paper made from sulfite, sulfate, and/ or mechanical pulp, and given a simulated crinkly finish by crowding the web sheet over a roll by means of a doctor blade.

**CREPE RATIO:** A means of expressing the degree of crepe that exists in a sheet. It indicates the extent to which it will stretch or can be pulled out before breaking and is usually expressed in length of extension or in percentage.

**CREPE WRINKLE:** An accordion-like paper defect resulting from paper slippage occurring in poorly built-up rolls and excessive caliper variation of the sheet.

**CREPED KRAFT PAPER:** Wrapping or liner paper made from sulfate chemical pulp which has been creped to impart various degrees of stretch.

**CREW:** A team of operating personnel in various areas and activities throughout a pulp and paper mill.

**CRILL:** Very small pieces of fiber that occur on the surface of the pulp and which are rubbed off during the refining process.

**CRIMP:** (1) The rolling of the edge of a sheet of paper or paperboard. It can be intentional or a defect. (2) A fold made in a sheet in such a way that the grain is broken, forcing it to retain a flat position.

**CRINKLE:** (1) A simulated finish effect similar to *crepe*, except that it is produced on heavier papers. (2) A property of paper that allows it to be stretched in both with- and cross-grain directions.

**CRITICAL ANGLE:** Maximum angle, measured from the perpendicular, at which refraction takes place when light rays pass from medium of higher index of refraction to one of lower index. It is also the minimum angle at which total internal reflection takes place under those conditions. The phenomenon is used in a method of measuring black liquor density.

**CRITICAL PRESSURE:** The saturation pressure corresponding to the critical temperature. For water it is 3,206 *psia*.

**CRITICAL SOLIDS CONTENT:** A condition of pulp fibers that is coincidental with the critical percentage of solids at which no free water remains.

**CRITICAL SPEED:** The speed at which rotating shafts or tubes will tend to vibrate excessively in a transverse direction. Also referred to as *whirling speed* or *whipping speed*.

**CRITICAL TEMPERATURE:** The temperature above which a substance remains a compressible vapor. For water it is 705°F.

**CROCK:** The action of scuffing or abrading the surface dye or pigment from a sheet of paper.

**CROB FIBERS:** Pulp and papermaking fibers derived from nonwoody plant species. They are only considered suitable when harvested on a seasonal basis (usually only one growth period per year). Sometimes referred to as annual crop fibers.

**CROSS CUTTER:** A type of paper-cutting machine designed to cut rolls of paper into sheets of specified lengths, suitable for the intended end use.

**CROSS DIRECTION (CD):** A term referring to the side-to-side direction of a paper machine, or to the papersheet made on it, as opposed to along the machine or machine direction which runs from head to exit end. Sometimes expressed as *across machine direction. See also* CROSS GRAIN.

**CROSS FLOW DISTRIBUTOR:** A type of paper machine headbox stock inlet in which the delivering of stock is made at the sides of the machine, perpendicular to the machine direction.

**CROSS GRAIN:** Side-to-side direction of the paper sheet opposite to the fiber alignment (grain of the sheet) tendency during the sheet forming phase. Sometimes referred to as *across the grain. See also* CROSS DIRECTION.

**CROSS HEAD:** The part of a pulpwood grinding stone sharpening tool lathe that holds the assembly and that moves across the face of the pulpstone on the lathe rods.

**CROSS LAMINATION:** A multi-layered laminate term denoting that the grains of the various lengths run alternately in a perpendicular direction to each other.

**CROSS LINKING:** A means of improving wet strength of a paper sheet by modifying the fiber surfaces.

**CROSS-MACHINE TENSION BURST:** A paper roll defect referring to an irregular separation or rupture of the web. While the defect is normally in the machine direction, the fiber separation is in the cross-machine direction. Normally, it is not visible at the edge of the roll.

**CROSS RECOVERY:** The practice of recovering chemicals from a sulfite or neutral sulfite semichemical process spent liquor together with sulfate process spent liquor in a sulfate recovery system when the two processes are associated with each other.

**CROSS SECTION:** A section of a body at right angles to its length, such as a section cut at right angles to the grain in wood.

**CROWN:** The surface curvature shape or diameter profile of paper machine rolls necessary to compensate for deflection, in order to obtain a uniform nip pressure distribution during operation.

**CROWN AMOUNT:** The difference in diameter between center and ends of the crown face of paper machine rolls. It can be positive for convex shape or negative for concave shape, and is usually expressed in thousandths of inches.

**CROWN FACE:** The axial distance over which the crown of paper machine rolls is measured.

**CROWN FACE CENTERLINE:** The centerline of the crown face of the paper machine roll which is normally the centerline of the machine, except on the suction press roll where it is centered on the drilled portion of the shell.

**CROWN FILLER:** A naturally occurring calcium sulfate base filler used when a high white color is desired in high-grade papers. Sometimes referred to as *gypsum* or *pearl filler. See also* CALCIUM SULFATE.

**CROW'S FEET:** *See* RIDGE.

**CRT:** *See* CATHODE RAY TUBE.

**CRUDE TALL OIL (CTO):** Tall oil before it is put through the fractional distillation process. Also referred to TALL OIL and TALL OIL ROSIN.

**CRUISE:** To estimate the volume and quality of a timber stand by visual examination.

**CRUMBS:** Disintegrated caustic soda-treated wet pulp to be used for making viscose rayon.

**CRUSH:** A rippled paper surface defect that is caused by the use of too high pressure and/or too high moisture at the paper machine or in the press section.

**CRUSH FINISH:** A paper surface finish resembling a coarse linen.

**CRUSHED:** A term used to describe paper that has a surface defect known as *crush*.

**CRUSHED CORE:** A paper roll defect in which the core of the roll has collapsed.

**CRUSHED FINISH:** A paper surface effect produced by applying high pressure at the dandy roll while the sheet is wet, causing a lumpy formation. This effect can also be purposely imparted onto the sheet after it is made by passing it through an off-line embossing machine.

**CRUSHED NEWS:** Old newspapers that have been bundled without regard to the quality of the paper or to whether they had been circulated or unsold.

**CRUSH RESISTANCE:** The physical property of paper and paper board that is resistant to compression loads. *See also* CONCORA, RULE 41.

**CRUSHED ROLL:** Defective roll of paper caused by stacking rolls on end so that the lower rolls fail in an axial direction, or rolls which have been inadvertently deformed by dropping or other accidental occurrences.

**CRUSHER:** A machine for disintegrating chips into specified weights and thicknesses directly after passing through the chipper in the wood preparation area of a pulp mill.

**CRUSHING:** A refining action producing permanent fiber deformation.

**CRUSHING STRENGTH:** *See* CRUSH RESISTANCE.

**CRYSTALLINE REGIONS:** When referring to cellulose fiber it pertains to areas in which the component molecules are found to be in a regular arrangement usually determined by X-ray diffraction techniques.

**CSF:** *See* CANADIAN STANDARD FREENESS.

**CTMP:** *See* CHEMITHERMOMECHANICAL PULP

**CTO:** *See* CRUDE TALL OIL.

**CULLER:** One who measures and classifies wood logs in the woodlands or on arrival at the woodyard of a pulp mill.

**CULLS:** Paper that does not meet manufacturing specifications and is sold as a lower quality product or returned to the papermaking process for recycling.

**CUNIT:** A type of pulpwood measurement that is equivalent to a volume containing 100 solid cubic feet of unpeeled wood.

**CUP PAPER:** Paper made of bleached pulp that is hard sized, waxed, and is to be used for converting to drinking cups.

**CUPRAMMONIUM HYDROXIDE:** A standard paper mill laboratory solution of ammonium hydroxide and cupric hydroxide used to dissolve cellulose in pulp samples to determine its viscosity as a measurement of the quality of the pulp.

**CUPRAMMONIUM VISCOSITY:** The viscosity measurement of dissolved cellulose in cuprammonium hydroxide solution under stated test conditions. It indicates the molecular weight and structure and the quality of the pulp from which it was derived.

**CUPRIETHYLENEDIAMINE HYDROXIDE:** A standard paper mill laboratory solution of cupric hydroxide and ethylenediamine used to dissolve cellulose in pulp samples to determine its viscosity as a measure of the quality of the pulp.

**CUPRIETHYLENEDIAMINE VISCOSITY:** The viscosity measurement of dissolved cellulose in cupriethylenediamine solution under stated test conditions. It indicates its molecular weight and structure and the quality of the pulp from which it was derived.

**CURL:** A paper or paperboard deformation caused by nonuniform distribution of strains and stresses throughout the sheet as a result of uneven internal moisture and conditioning.

**CURL TEST:** A method of testing the sizing of paper, based on the fact that when paper is floated on water, the edges curl upward, and the axis of the curl will parallel the grain of paper.

**CURLING:** A refining action which permanently reduces the radius of curvature of a fiber.

**CURRENCY PAPER:** A strong, durable type of writing paper made from silk, cotton, and bleached chemical pulps. Its primary use is in making monetary bills, bank notes, bonds, and other commercial documents. Similar to and sometimes called *bank note paper.*

**CURTAIN COATER:** A paper coating system which consists of dropping an extruded laminar film of coating material on the paper or paperboard sheet traveling on a conveyor.

**CURVED BAR:** *See* SPREADER.

**CUSHION:** A printer's term for *compressibility,* which is the amount that the sheet will reduce in thickness during the printing process.

**CUTLERY PAPER:** Thin, tarnish-resistant paper made to be used to wrap cutlery and silverware. *Also see* ANTI-TARNISH PAPER.

**CUTS:** Defects in sheets of paper that occur as small ruptures that are not so extensive as to render the sheet in two. *See also* BLADECUT, BLISTER CUT, CALENDER CUT, FIBER CUT, and HAIR CUT.

**CUT SIZE:** Fine paper cut to specific end-use dimensions (16 inches by 21 inches or less) on a paper trimmer, usually of the guillotine or rotary type.

**CUT TO REGISTER:** A watermarker paper-cutting operation that produces sheets in which the watermark design appears at the same location on each sheet.

**CUTIN:** A wax-like substance found in plant cell walls.

**CUTOFF SQUIRTS:** Fine jets of water directed onto the paper machine wire (forming fabric) with enough force and so located, one on each side, to cut the wet web to correct width. Sometimes one squirt is made adjustable so that it can be gradually fed across the wire to form a tail or lead strip on the wet sheet to facilitate feeding it to the wet presses and through the dryers.

**CUTTER:** A type of paper-cutting machine designed to cut rolls of paper into sheets of specified lengths suitable for the intended end use.

**CUTTER BROKE:** Waste and trim paper produced by the paper-cutting operation and recycled to the papermaking process.

**CUTTER DUST:** Tiny, coated paper fiber fragments that break off the edge of the sheet during the cutting operations and cause problems if carried through to printing.

**CUTTER SET:** A finishing room term referring to matched rolls of paper that will produce the same number of sheets when cut simultaneously on a cutter.

**CUTTING:** A refining action causing a physical separation of a fiber across the axis into two or more pieces. Also termed *shortening.*

**CUTWATER:** The point at which water flowing around the perimeter of a pump casing is forced to travel towards the pump discharge.

**CYCLE:** As an electrical term, it represents one complete set of positive and negative values of an alternating current.

**CYCLOIDAL VACUUM PUMP:** A type of vacuum pump used on a paper machine vacuum system that is constructed with two lobe-type impellers rotating in opposite directions, entrapping and expelling a volume of air directly proportional to the pump velocity.

**CYCLONE CLEANER:** A pulp stock cleaner in which pressurized slurry is tangentially injected into a truncated, cone-shaped vessel to produce a rapid spinning action of the suspension, causing a centrifugal force that moves the heavier particles to the outside wall where they drop down to the bottom and are removed. The inner vortex column with the acceptable fiber moves to the top where it is removed for further processing. Also called a *vortex cleaner.*

**CYCLONE EVAPORATOR:** A type of evaporator used to concentrate sulfate black liquor before burning in a recovery furnace by spraying partially evaporated liquor through tangential nozzles into a cyclone, with hot gases being introduced at the bottom.

**CYLINDER:** (1) A mechanical part of a pulpwood grinder, located over the log charger, which contains a piston. The piston is fed by high-pressure water and, in turn, forces the pressure foot against the wood so that it is pushed against the pulpstone. (2) *Also see* DRYER DRUM.

**CYLINDER BOARD:** Paperboard made on a cylinder-type machine. *See also* CYLINDER MACHINE.

**CYLINDER DRIED:** Paper and paperboard that has been dried by running it over steam-heated cylindrical metal rolls. Also referred to as *machine dried paper.*

**CYLINDER HOLE:** A light sheet defect made on a cylinder-type paper machine, appearing as a hole, caused by pitch and dirt sticking to the face of the wet end wire.

**CYLINDER MACHINE:** A type of papermaking machine primarily used to make paperboards. The forming cylinders are constructed or covered with wire so that, as they turn within a tank or vat particularly filled with a stock solution, fibers are picked up to form a web on the surface, with water draining through and passing out at the ends. The wet sheet is then transferred off the cylinder onto a felt for possible combining with the other sheets (multiple cylinders on same machine) and subsequent pressing and drying. Sometimes called a *vat machine.*

**CYLINDER MOLD:** The wire-covered portion of a cylinder machine that rotates while partially immersed in the diluted stock in the vat from which the fibers are picked up to form a web on its surface.

**CYLINDER MOLD EXHAUST:** An exhaust unit used on some cylinder-type paper machines to maintain suction in the cylinder roll, thus achieving better formation at the higher speeds of operation.

**CYLINDER OPERATOR:** A device that converts fluid power into linear mechanical force and motion. It consists of such movable elements as a piston and piston rod, plunger, or ram operating within a cylindrical bore.

**CYLINDER PRESSURE:** The pounds per square inch of water pressure applied to a cylinder, or the water pressure times the cross-sectional area of the cylinder, giving a result of total pounds applied.

**CYMENE:** *See* SPRUCE TURPENTINE.

# D

**DAF:** *See* DISSOLVED AIR FLOTATION.

**DAM:** (1) An adjustable baffle board extending across the full width of a pulpwood grinder over which pulp flows to the grinder ditch. It is raised or lowered to control pulpstone immersion. (2) A barrier constructed across a water stream (e.g., river) to confine the flow and create a depth or heal that could be used to drive electric generators, providing power for pulp and paper mills.

**DAMP STREAK:** Darkened machine-direction stripe in the paper sheet resulting from too much moisture when passing through calenders, super calenders, and other high pressure operations. It causes decreased opacity.

**DAMPER:** (1) A manual or motor-driven, plate-like valving device usually used in air-ducted lime kilns, furnaces, etc., to control the air flow through them. (2) Water or steam spray used on a sheet of paper as it enters the calender stack to moisten it and thus improve its finish.

**DAMPING ROLL:** A type of roll usually located just ahead of a coater or at the end of a paper machine. It is used to cool the sheet by being operated at a temperature that is low enough to cause moisture to condense on its surface and slow up the drying.

**DAMPING STRETCH:** Dimensional instability caused by wetting or dampening a paper sheet.

**DANCING ROLLS:** *See* SPRING ROLL.

**DANDY BLISTERS:** Paper blemishes that have the appearance of blisters and have been pressed down on the surface of the sheet, causing concentric wrinkles. This condition is brought about by the dandy roll becoming filled up in various spots.

**DANDY CRUSH MARKS:** Paper defects having a curdled appearance, spread over the surface of the sheet. These defects are caused by too liberal use of water on the dandy roll, with the roll wading in water that cannot normally drain through the sheet.

**DANDY MARK:** (1) A paper defect consisting of a small, thin area left on the sheet due to the disturbance or picking up of fibers in the web by the dandy roll. Also known as *dandy pick.* (2) The design in a paper sheet made by impressing from a raised pattern on the dandy roll. It can be seen as translucent when held to a source of light.

**DANDY PICK:** The lifting or picking up of fibers from the wet paper web by the dandy roll, causing thin spots in the sheet. *See* DANDY MARK.

**DANDY ROLL:** A woven wire cloth skeleton cylinder usually located on top of the paper machine wire (forming fabric) and just ahead of the couch roll. Sometimes covered with an arrangement of thin longitudinal wires, crossed at close intervals by heavier circumferentially wound wires used to produce a wave and laid effect on the paper surface. Designed with numbers and figures in the wires to impart water marks on the sheet while wet. *See* WATERMARK DANDY ROLL.

**DANDY ROLL MARKS:** Defects, appearing as streak marks, on the top surface of a paper sheet caused by dirty dandy rolls on paper machines.

**DATA:** Any process or off-line information with particular reference to that which can be processed or produced by a computer or control system.

**DC:** *See* DISTRIBUTED CONTROL.

**D.C.:** *See* DIRECT CURRENT.

**DEAD BAND:** The range of a controller through which a signal or action can be varied without initiating any response.

**DEAD BEATEN:** A term sometimes used to refer to stock that has been heavily refined to a very low freeness. *See also* GREASY.

**DEAD FINISH:** A smooth, low-gloss type of paper surface finish.

**DEAD SPOTS:** Small, low-gloss areas on the surface of glossy finished paper.

**DEAD STEAM:** Steam that has completed work. Sometimes called *exhaust steam.*

**DEAD TIME:** The interval of time between the initiation of a signal or action change in a controller and the start of the resulting response, or any definite delay between two related actions measured in units of time. Also known as *process lag.*

**DEAD WHITE:** The surface color of a white paper that shows no shade variations.

**DEAD ZONE:** A zone within the range of a controller in which no value of output or response exists regardless of the change in a signal or action.

**DEADENING FELT:** A very heavy grade of paper, usually made from old rags and papers, used by the building trade as environmental and sound insulation.

**DEAERATION:** (1) The process of removing free air in stock slurries by the use of a vacuum, usually prior to the headbox of a paper machine, to minimize foaming and slime problems and improve the drainage and sheet formation. (2) The driving off of entrained gases from condensate going to the feedwater pump of a boiler by heating to saturation, using steam.

**DEAERATOR:** (1) A unit where condensate trickles over a large series of baffles and is heated to saturation by the admission of steam, which causes entrained gases to boil off. Also called *direct contact heater* for direct contact heating of the feedwater by steam. (2) A large vessel under vacuum, located just ahead of the wet end of a paper machine, through which the stock slurry is passed to remove entrained air.

**DEALKALIZATION:** The removal of bicarbonates and carbonates from boiler feedwater.

**DEBARKING:** The process of removing bark from logs or from the tree. Although this term most frequently refers to chemical bark removal, it is sometimes used to indicate mechanical bark removal as well. *See also* BARKING.

**DEBONDERS:** Chemicals that prevent the adhesion of fibers in a sheet of paper, resulting in softness.

**DEBRIS:** The result of external fibrils rubbing off the surface, and parts of the wall being removed from the fiber during beating and refining. It determines the characteristics of the produced pulp stock, such as freeness, specific surface, and coarseness.

**DEBUG:** To detect, locate, and remove mistakes from a program or malfunctions from a computer.

**DECALCOMANIA PAPER:** A medium-weight, absorbent-type paper having a smooth finish with good wet strength, and coated with an adhesive. It is usually made from chemical pulp and cotton fibers, and is used for making decals. Also called *decal paper.*

**DECALCOMANIA SOLUTION:** A mixture of starch, gum arabic, and water used as a coating to make decalcomania paper. Also referred to as *decal solution.*

**DECAL PAPER:** *See* DECALOMANIA PAPER.

**DECANTATION:** A unit process utilized in the clarification of green and white liquor in the causticizing operation of a pulp mill. It is accomplished in a multi-compartment settling tank in which solids are allowed to settle and clear liquor is decanted off in each sector.

**DECIDUOUS:** Refers to those species of trees that lose their leaves annually, with some exceptions such as birch and cypress.

**DECKER:** A drum vacuum filter used to thicken pulp slurries for increased fiber storage and to produce white water for reuse in the process.

**DECKER MAN:** The paper mill employee who operates and is in charge of the drum vacuum filter (*decker*) used for removing excess water from pulp.

**DECKLE:** Fixed rubber strips or moving straps located on the top of and inside of the side edges of the paper machine wet end wire (forming fabric), and also found in deckers and pulp washers. It is used to prevent the spill of fibers over the sides and determines the width of the web of paper that can be run on a particular machine.

**DECKLE EDGE:** (1) The rough edge of a sheet of paper formed by the flow of the stock against the deckle or accomplished by means of a jet of air or water. It is also a descriptive term applied to paper showing such an edge. (2) Paperboards that have been made without trimming off the feathery, full-width side edges.

**DECKLE POSITION:** A number assigned to a roll in a set cut from a reel of paper, counting from the front side of the paper machine.

**DECKLE PRODUCTION:** The weight in pounds or tons of material passing over the paper machine wet end and the moisture content obtained at the dry end at the time of manufacture, over a period of time numerically equal to the sum of trimmed production and trim or shavings.

**DECKLE STAINED:** Enhanced appearance of deckle edge paper and paperboard by coloring the side edges.

**DECKLE STRAPS:** Endless, rubber-based belts that lie on the side edges of the paper machine wet end wire (forming fabric) and move along with it to retain the fiber slurry and prevent spillover. *See* DECKLE.

**DECOMPOSITION:** Reduction of the net energy level and change in chemical composition of organic matter because of the actions of aerobic or anaerobic microorganisms.

**DECORATED:** A term applied to certain types of paper and paperboard such as blotting, building, cover, and sheathing paper and cardboards, whose appearance has been enhanced by embossing, printing with a characteristic design, lining, or some other process.

**DEED PAPER:** A highly durable, surface-sized paper made primarily of cotton fibers and selected bleached chemical pulps. It is used for long-term record preservation documents such as deeds.

**DEFIBERIZE:** To mechanically reduce fibrous materials such as chips, pulp sheets, paper sheets, etc., into their fiber components. It is accomplished by use of machinery such as refiners and pulpers.

**DEFIBRATED PULP:** Pulp made by mechanically reducing chips into their fiber components at elevated temperatures, usually after presteaming in a preceding chamber.

**DEFLAKER:** A mechanical fiber slurry lump breaker, consisting of a mixer or agitator operated at high speed so that there is a complete separation of undefiberized bundles.

**DEFLECTION:** The bending of paperboard when supported at the ends with a specific force applied to the middle section.

**DEFLOCCULATION:** The chemical or mechanical breaking up of agglomerated substances in slurries in such a way that the particles produced will stay in suspension by repelling each other.

**DEFLOUR:** Screening pulp slurries to remove the short fibers and small fiber particles.

**DEFOAMER:** (1) Material such as kerosene, fatty acids, fatty alcohols, and other hydrocarbons added to an aqueous solution to prevent the accumulation of air bubbles by bursting them and releasing the air before they can form foam. (2) A mineral such as bentonite or diatomaceous earth added to pulp stock furnish to prevent the formation of foam that may cause *foam marks* in the paper sheet.

**DEGRADATION:** The deterioration of quality of a fibrous material such as pulp, paper, and paperboard. It is the result of the breakdown of the cellulose polymer chain structure due to hydrolysis or oxidation usually associated with time.

**DEGREASING:** The process of removing greases and oils from mill sewage, waste, and sludge.

**DEGREE DAY:** A unit used in heating and air-conditioning to indicate a deviation of 1°F in the mean temperature of a day from the accepted standard of 65°F.

**DEGREE OF POLYMERIZATION (DP):** The average number of base units per molecule in the polymer chain structure of cellulose. It is indicative of the quality of the fiber makeup of pulp.

**DEHUMIDIFICATION:** The reduction of the water vapor content of ambient air.

**DEHYDRATION:** The evaporation of moisture from a sheet of paper and paperboard due to high temperature and/or low humidity ambient conditions.

**DEINKED:** A reference term used with any type of pulp and/or paper made from printed wastepaper. Most of the ink, filler, and other extraneous materials are removed by a combination of mechanical disintegration and chemical treatment.

**DEINKED NEWSPRINT:** Pulp or newsprint paper made from used newsprint (from which the ink was removed) and mixed with wood pulp.

**DEINKED PAPER STOCK PULP:** Pulp made from deinked printed paper using a combination of mechanical disintegration and chemical treatment. It is mixed with other wood pulps to make additional printing paper.

**DEINKING:** The processing of printed and other used, reclaimed wastepaper by mechanical disintegration, chemical treatment, washing, and bleaching in order to remove ink and other undesirable materials so that it can be reused as a source of papermaking fiber.

**DEINKING AGENTS:** Materials added to a slurry of printed wastepapers that will aid the ink-removing process.

**DEINKING MILL:** An industrial plant concerned with the removal of ink and other objectionable nonfibrous materials from wastepaper. It is usually associated with other pulp and paper mills. Also referred to as deinking plant.

**DEIONIZATION:** The removal of ionized solids, especially from boiler feedwater.

**DEKATHERM:** A purchase unit of fuel gas, on a heat content basis, equivalent to 1 million Btu.

**DELAMINATION:** The separation of pasted paper layers of a laminated sheet due to adhesive breakdown.

**DELICATESSEN PAPER:** Dry-waxed, grease resistant paper used as an inner wrap for greasy food and meat so that there is no penetration to the outer wrapping.

**DELIGNIFICATION:** Separation of the lignin components from the cellulose and carbohydrate materials of wood and woody materials by chemical treatment and washing, such as the cooking of chips in a digester (usually elevated temperature and pressure steam conditions) and subsequent bleaching of the cooked pulp.

**DELIVERY TABLE:** The table portion of a paper-cutting machine on which the sheets are automatically stacked after cutting. Also called a *lay boy*.

**DELTA G:** The apparent difference in gloss between ink printed on a sheet (type and solid/halftone photos and graphics) and the gloss of the sheet itself. *See* PRINT SNAP.

**DELTA P:** The difference in pressure at two different locations in a pressurized system.

**DEMAND FACTOR:** The ratio of the maximum demand of an electrical system, or part of the system, to the total connected load of the system, or part of the system, under consideration.

**DEMINERALIZATION:** Removal of mineral contaminants from water, such as that used for feedwater to a boiler.

**DEMY:** The size of a sheet of writing paper; in the U.S. it measures 16 by 21 inches.

**DENDROLOGY:** The science of identifying and classifying tree species.

**DENNISON WAX:** Specially made hard resin, non-oily wax sticks, numbered according to their adhesive power, and commonly used in paper-testing laboratories to measure paper surface strength. The testing is accomplished by softening the sticks with heat and applying them to the surface to determine the highest number that does not disturb the surface. *See* PICK RESISTANCE.

**DENSIFICATION:** The compaction of fibers caused by pressure applied to high caliper portions of a sheet of paper as it passes through the calender stack on a paper machine.

**DENSITY:** (1) Weight per unit volume, expressed in units of pounds per cubic foot (lb/cu. ft.) pounds per gallon (lb/gal) or grams per milliliter (g/ml). **(2)** In reference to paper and paperboard sheets, it is the ratio of the weight to the volume, with high-density paper and paperboard having a high weight-to-volume ratio. Sometimes referred to as *finish* in paperboards. *See also* BAUME, APPARENT DENSITY, and APPARENT SPECIFIC VOLUME.

**DENSOMETER:** A paper laboratory tester used to measure the porosity of paper as the time required for a standard volume of air to penetrate through a standard area of paper.

**DEODORANTS:** Pleasant smelling substances added to pulp mill gaseous emissions to mask unpleasant odors.

**DEPITHING:** The separation of the pith from the fiber so that bagasse can be used as a source of pulp for papermaking.

**DEPOSITABLE PITCH:** Pitch in pulps, refined virgin fibers, refined secondary fibers, machine furnishes, and white water that has a tendency to settle and collect on papermaking equipment under certain conditions. *See also* STICKIES.

**DEPOSITS:** A very broad term used in the pulp and paper industry that encompasses aggregations of all types of undesirable accumulations anywhere in the process such as slime, pitch, scale, foam, dirt, size, alumina gel, etc.

**DESICCANT:** A chemical agent that may be used to remove moisture from air or gases.

**DESIGN PAPER:** Paper with design printing usually used as wrapping paper or for making decorative boxes and bags. *See* DESIGN PRINTING.

**DESIGN PRINTING:** Surface coloring and achieving a printed effect at the same time by immersing a sheet of paper in color solution and subsequently passing it through designed-marked rollers so that the design stands out heavily colored or printed against a lighter but same-colored background. *See also* DESIGN PAPER.

**DESORPTION:** The drying out or giving up of moisture or water as with wood, pulp, and paper.

**DESUPERHEATER:** A unit used to reduce the temperature of superheated steam from a boiler by the injection of atomized water or a heat exchanger located in the bottom half of the drum, through which steam is bypassed. Same as *attemperator.*

**DETAIL DRAWING PAPER:** Paper made with a grainy surface usually in buff or manila color and used to make working drawings.

**DETAIL PAPER:** A thick, high quality, hard-sized, semi-translucent paper.

**DETERGENT:** A synthetic washing agent, sometimes used to clean paper machine felts, that acts like soap by lowering the surface tension of water and holding dirt in suspension.

**DEVELOPING PAPER:** Paper used in photography that has been coated with light-sensitive emulsion. The emulsion produces visible images on subsequent chemical treatment.

**DEVIATION:** Departure from a desired or expected value or pattern.

**DEVIATION ALARM:** An alarm caused by a variable departing from its desired value by a specified amount.

**DEVICE:** Any apparatus or piece of equipment that performs a specified function.

**DEW POINT:** The temperature at which air or gas becomes saturated, or the temperature to which a mixture must be cooled under constant pressure and composition to reach saturation.

**DEWATER:** (1) The tendency of solids in a slurry to aggregate and cause the draining of water from standing or flowing sludge or pulp slurry in a pipeline, sometimes to the point where the remaining solids become thick enough to make removal difficult or to obstruct free flow through the line or a restriction such as a valve. (2) The process by which some of the water is removed from the pulp stock, increasing the consistency.

**DEWAXED:** A term used to refer to waxed paper from which wax was removed to reclaim the base.

**DEXTRIN:** An adhesive material used for gumming papers, tapes, or envelopes made by hydrolysis of starch with acids, enzymes, and/or heat.

**DIAGONAL SPLICE:** The joining of two ends of paper in making up continuous rolls; joint cuts are made at an angle to make them run better through a printing press or other processing equipment.

**DIAMOND BURR:** A pulp grinderstone sharpening tool (burr), having a pattern of straight grooves and crossing circular grooves resulting in rows of pyramidal points.

**DIAPHANIC PAPER:** A well-formed transparent thin sheet of paper made to achieve a stained glass effect by lithographic printing, treating it with adhesive, and applying it to a glass surface.

**DIAPHRAGM PAPER:** A type of filtering paper with an asbestos base.

**DIAPHRAGM SCREEN:** A type of screen once used extensively in the pulp and paper industry to clean stock. It was also called a flat screen because it consisted of perforated screen plates laying in the same horizontal, slightly inclined plane to aid stock flow over them.

**DIATOMACEOUS EARTH:** A material used as a filtering medium in mill water and waste treatment processes. It is also used in the amorphous form of its silica derivative in making up paper coating and filler materials.

**DIATOMACEOUS SILICA:** The amorphous form of diatomaceous earth, used primarily in the preparation of paper coatings and fillers. It is naturally produced by aquatic diatoms.

**DIAZOTYPE PAPER:** A chemically pure type of paper with good strength properties, made from chemical pulps mixed with cotton fiber and coated with light-sensitive diazo chemicals for use in copying machines.

**DIBBLE:** A spade-like tool used in pulpwood tree farming to make and prepare holes for planting seedlings.

**DICOTYLEDONS:** Trees grown from and bearing seeds with two lobes, such as walnut.

**DIE:** Various devices for imparting a desired shape, form, or finish to a material, or for impressing an object or material.

**DIE CUT:** A term used in reference to paper and paperboard products cut by a metallic die to specified dimensions or form.

**DIE EMBOSSED:** Paper or paperboard in which a design has been impressed by the use of metallic dies on a mechanical press.

**DIE WIPING PAPER:** Paper used for wiping the surface of printing plates in the printing industry.

**DIELECTRIC:** Material that will not conduct direct electric current.

**DIELECTRIC CONSTANT:** Ratio of capacitance produced by a material between the plates of a condenser, to the capacitance produced by vacuum in the same condenser. It is used as an indication of the nonconductance of direct electric current in paper used for electrical insulation and in condensers.

**DIELECTRIC PAPER:** A very thin, electrically nonconductive type of paper used for electrical insulation in condensers and other devices to prevent transmission of electrical current. It is specially made to be free of metallic and other electrically conductive materials.

**DIELECTRIC STRENGTH:** The voltage difference level across a sheet of paper or paperboard at which a spark will jump through under specified conditions.

**DIFFERENTIAL CONTROL:** Actuated from differential controller, can be tied in to dryer section supply or blowdown.

**DIFFERENTIAL PRESSURE:** (1) The difference between pressures existing at two points in a system which permits mass flow between those points, such as in condensate removal from paper machine dryers. (2) The basic principle used in making many process fluid flow measurements throughout the mill.

**DIFFERENTIAL WINDER:** A type of paper winder used in a paper mill. Its gears are designed so that it will allow the simultaneous winding of different size rolls with the same peripheral speed.

**DIFFUSE POROUS:** Referring to wood from trees whose pores do not differ abruptly in size and are more evenly distributed throughout both spring and summer wood.

**DIFFUSER:** A large tank with a false bottom screen plate constructed for the drainage of cooking liquor from pulp. It is also used to wash the pulp with fresh water.

**DIFFUSION BLEACHING:** *See* DISPLACEMENT BLEACHING.

**DIFFUSION WASHER:** A pulp washer used to wash brownstock and interstage pulp during bleaching, and is also being used in some oxygen delignification systems. It utilizes the principle of diffusion or displacement of liquors with wash water to remove the liquor.

**DIGESTED SLUDGE:** Solids removed from mill waste effluent and allowed to digest under aerobic or anaerobic conditions to drive off volatile contents.

**DIGESTER:** (1) A pressure vessel used to chemically treat chips and other cellulosic fibrous materials such as straw, bagasse, rags, etc., under elevated temperature and pressure in order to separate fibers from each other. It produces pulp. (2) In a waste treatment plant, it is a closed tank that decreases the volume of solids and stabilizes raw sludge by bacterial action.

**DIGESTER BLOW:** *See* BLOW.

**DIGESTER BLOW VALVE:** A handwheel or motor-driven valve located at the bottom end of a batch digester that is opened at the end of each cook. The cooked chips are discharged through it into the blow line and blow tank by virtue of digester pressure.

**DIGESTER CAP:** *See* DIGESTER COVER.

**DIGESTER CAPPING VALVE:** A large, automatically operated, rotary or gate-type valve, located over the top opening of a batch digester head through which the chips are loaded. It is then closed tightly to contain the pressure during the cooking operation.

**DIGESTER COVER:** The metal-flanged cover over the top of a batch digester head which is opened or taken off while the chips are loaded into the digester. It is then bolted down to seal in pressure during the cooking operation. Also referred to as the *digester cap* and *digester lid*.

**DIGESTER HEAD:** The structure over the top opening of a batch-type digester through which the chips are loaded into the digester.

**DIGESTER HOUSE:** The building or section where the digester is located or housed in a pulp mill.

**DIGESTER LID:** *See* DIGESTER COVER.

**DIGESTER LIQUOR:** Cooking liquor as it exists in the digester during the cooking operation.

**DIGESTER MAN:** *See* COOK.

**DIGESTER NECK:** The narrow, cylindrical section of a batch digester, located in the upper feed section just below the head and above the main body section.

**DIGESTER OPERATOR:** The person in charge of the digester operations. Commonly referred to in the pulp mill as the *cook*.

**DIGESTER RELIEF:** Gases (and sometimes liquor) withdrawn from the top of the digester and sent to a by-product or reclaiming system, depending on the type of digester.

**DIGESTER ROOM:** The area in the digester house where the operations of the digester are conducted. With batch-type digesters, it is located on the top or charging floor of the digester house.

**DIGESTER SIDE RELIEF:** Digester liquor withdrawn by a side relief from the upper section of a sulfite batch digester in the earlier part of the cooking operation, and returned to storage (accumulator) to conserve sulfur dioxide and heat.

**DIGESTER TEST:** A test made on a sulfite batch digester liquor for total, free, and combined sulfur dioxide during the cooking process.

**DIGESTING:** (1) The process of treating cellulosic fibrous raw materials with chemicals, usually at elevated temperatures and pressures, for the purpose of separating the fibers and removing impurities to produce pulp suitable for papermaking. (2) In waste treatment, it is the biochemical decomposition of organic matter.

**DIGITAL:** Pertaining to quantized signals or data that may be expressed in the form of digits.

**DIGITAL COMPUTER:** A *digital* based computer that performs arithmetic and logic computations on discrete data.

**DIKE:** An artificial structure designed to prevent a body of liquid from inundating an area of land.

**DILATANT:** Types of slurries whose viscosity and apparent volume increases or decreases with the applied rate of shear, sometimes existing with high solid content coating materials such as clay, which has a tendency to increase and affects its pumpability.

**DILUTION:** The act of decreasing the density or consistency of a solution or slurry by the addition of a dilutent (e.g., the adding of water to a pulp slurry), usually by use of a consistency controller or regulator.

**DILUTION FACTOR:** Pounds of water put into the liquor system per pound of air-dry pulp during the washing operation in a pulp mill.

**DILUTION WASHING:** A pulp-washing procedure that consists of alternate dilution with the washing medium and thickening until clean.

**DIMENSIONAL STABILITY:** A sheet characteristic indicative of its ability to maintain original machine and cross-machine dimensions with time and under variable moisture and relative humidity conditions.

**DIMETHYL SULFIDE [$(CH_3)_2S$]:** One of the chemical mercaptans whose vapor causes a very disagreeable odor in relief and other effluent gases emitted from the sulfate pulp manufacturing process.

**DIMETHYL SULFOXIDE [$(CH_3)_2SO$]:** A chemical by-product of the sulfite pulping process used as an industrial solvent and chemical reactant. It has very high penetrating qualities and has been used medicinally as a locally applied pain depressant under the common name of *DMSO.*

**DIOXIN:** An undesirable toxic chlorinated contaminant generated when chlorine is used in the bleaching of wood pulp. It is considered to be highly carcinogenic, particularly in its most potent form–2, 3, 7, 8-tetrachlorodibenzo-p-dioxin (2, 3, 7, 8-TCDD).

**DIP DYEING:** Surface-coloring of a sheet of paper by passing it through a dye solution either in a bath or through the flooded nips between two applicator rolls.

**DIPPERS:** Pipes installed inside slow speed paper machine dryers, attached to the dryer head, and arranged to remove condensate as it is formed in the dryer.

**DIPPING:** Coloring of paper after it is formed and dried on the machine and cut into narrow rolls. It is usually performed by converters who do not manufacture paper but transform it into colored end-use products. Sometimes called *surface coloring* or *tub coloring*.

**DIRECT ACTING:** The operation of equipment and devices directly proportional to the action applied.

**DIRECT CONTACT HEATER:** A unit used to heat condensate feedwater by direct contact with steam, which also drives out entrained gases. Also called a *deaerator*.

**DIRECT CONTACT EVAPORATOR:** Type of black liquor evaporator used just ahead of the recovery furnace that concentrates the liquor by removing water through direct contact with the hot furnace flue gases.

**DIRECT COUNTERCURRENT WASHING:** A pulp-washing procedure used in a multi-stage bleach plant in which the filtrate from one pulp washer is used in the previous washer and runs countercurrent to the flow of pulp through the bleaching system.

**DIRECT CURRENT (D.C.):** A unidirectional electrical current or nonpulsating current, such as supplied by batteries, or rectified *alternating current (A.C.)*.

**DIRECT DRIVE:** A type of equipment drive in which the motor and the equipment it drives are connected through a coupling without gears or auxiliary apparatus.

**DIRECT DYES:** Aniline-type dyes that are light resistant and have an especially high attraction to cellulose fibers.

**DIRECT FLOW VAT:** The vat of a cylinder-type paper machine in which the flow of stock is in the direction of the cylinder rotation.

**DIRECT PROCESS PAPER:** A fine grade of paper made from bleached chemical pulps mixed with cotton fiber. It is coated with a light-sensitive material so it can be used for copy reproduction with the diazo process. Sometimes called *ozalid*.

**DIRECT STEAM COOKING:** A procedure for cooking chips in a batch-type digester in which the steam for heating is injected directly into the digester.

**DIRECT STEAMED DIGESTER:** A batch-type digester in which chips are cooked by injection of heating steam directly into the digester body.

**DIRECTION:** Paper and paperboard dimensions running parallel to the paper machine (machine direction) or at right angles to it (cross direction).

**DIRECTORY PAPER:** Lightweight grades of printing paper having good strength properties. They are made from a mixture of bleached chemical, mechanical, and reclaimed pulps and used in printing such things as catalogs, pamphlets, and telephone directories. *Telephone directory paper* is a specific grade.

**DIRT:** Undesirable, extraneous, contaminating material in pulp, paper, or paperboard that is visible in transmitted or reflected light.

**DIRT COUNT:** A paper quality check made by passing a sample of paper over a lamp and reporting the number of small pieces of foreign material observed per square foot.

**DIRTY BLOW:** When a batch digester is discharged into a blow tank with uncooked chips and pulp remaining in the bottom of the digester or in the blow line from the digester to the blow tank.

**DIRTY ID FAN:** An induced draft fan on a pulp mill recovery furnace located after the precipitator.

**DISC CHIP SCREEN:** A type of wood chip screen that uses a circular motion and inclined plates with elongated slots to remove a maximum amount of fines.

**DISC FILTER:** *See* DISC SAVE-ALL.

**DISC SAVE-ALL:** A process unit used to remove fibers from white water by filtering it through a series of rotating, circular, flat screening elements. Sometimes called a *disc filter.*

**DISCHARGE HEAD:** The pressure generated by a pump.

**DISHBOARD:** A stiff grade of plain or plastic-coated paperboard, made so that food-serving dishes can be mechanically stamped from it on a die press.

**DISHED:** A condition of a pile of paper that lies in a concave fashion with the sides higher than the center, or the end of a roll of paper that is not flat.

**DISINFECTION:** The killing of large portions (but not necessarily all) of the harmful and objectionable microorganisms in or on a medium by means of chemicals, heat, ultraviolet light, etc., or the use of a chemical additive and other treatment to reduce the number of bacteria, particularly pathogenic organisms in mill supply and wastewater.

**DISK REFINER:** A type of machine used to mechanically treat fibrous material to convert it to a form suitable for making a paper or paperboard sheet. It consists of one or more surface-patterned disks, located in a casing, and operated in rotary opposition to each other. The fiber material is passed between the disks as lateral pressure is applied against them to abrade, cut, or bruise the material as required. *See also* DOUBLE DISK REFINER.

**DISINTEGRATION RESISTANCE:** The ability of pulp to resist breaking up and dispersing when immersed in water.

**DISPERSE VISCOSITY:** A viscosity measurement made on a pulp, usually by means of a falling ball or pipette, after it has been dissolved in a solvent such as cupriethylenediamine.

**DISPERSIBILITY:** The comparative ease and/or completeness with which coating pigments are resolved into their individual discrete particles in a coating color.

**DISPERSING AGENT:** A substance added to a slurry that consists of finely divided, suspended materials to keep the individual particles from agglomerating, and improve uniformity, as in making up certain types of pigment-dyeing solutions for paper coloring.

**DISPLACEMENT BLEACHING:** A type of pulp-bleaching process in which the chemicals are displaced through a pulp mat, rather than being mixed into the pulp. This provides a more rapid bleaching operation. Sometimes referred to as *diffusion bleaching* or *dynamic bleaching.*

**DISPLACEMENT PRESSING:** A relatively new concept in pressing based on the displacement of liquid (water) in the sheet with pressurized air or gas (heated), together with conventional pressing principles, in order to increase sheet dryness levels out of the press section to 65 percent or more.

**DISPLACEMENT RATIO (DR):** In the washing of pulp in filters, it is the ratio between the recovered solids in a volume corresponding to the liquor volume of pulp leaving the washing operation and the solids in a corresponding volume of liquor entering it. It characterizes the efficiency of the washing process.

**DISPLACEMENT WASHING:** Method of washing pulps in which the cleaning is done by displacing liquors or chemicals from a thickened mat with the washing medium.

**DISPLAY BOARD:** A rigid, single- or multi-ply paperboard made from mechanical and chemical pulp to meet the requirements for its use in advertising and merchandising displays.

**DISSOCIATION:** A process whereby a chemical combination breaks up into simpler constituents usually capable of recombining under different conditions.

**DISSOLVED AIR FLOTATION (DAF):** A waste water pretreatment process used by some pulp and paper mills in which air is injected into the waste stream to aid the separation of low-density contaminants such as fibers, oils, grease, etc.

**DISSOLVED OXYGEN (DO):** The amount of oxygen in mill effluent streams, usually expressed in milligrams per liter, parts per million, or percent of saturation, related to the level of contaminants, or the degree of pollution being contributed to the receiving waters.

**DISSOLVED SOLIDS (DS):** The organic and/or inorganic dissolved material in solution, such as black liquors, white water, and other process liquids and liquid wastes. They can be determined by evaporating a filter sample to dryness and calculating the weight of the residual in lb/gal.

**DISSOLVING PULP:** A very pure, fully bleached chemical pulp or cotton linter pulp with a high alpha-cellulose content. It is made specifically for dissolving in suitable solvents for regeneration to form cellophane, rayon, etc., or combining with other chemicals to produce such derivatives as acetates, nitrates, etc. *See also* ALPHA PAPER.

**DISSOLVING TANK:** A storage tank into which the molten chemicals, called smelt, from a recovery furnace drop and are mixed with water or weak liquor to form green liquor.

**DISTILLATION:** Removal of impurities or recoverable vapors from liquids by boiling and condensation, as in turpentine and tall oil production.

**DISTRIBUTED CONTROL (DC):** An advanced series of functional or geographically located, millwide control devices dedicated to performing sequential and process control for a specific area of operation. It communicates over long distances to interconnected, centralized stations.

**DITCH:** A trough or canal, located below a number of pulpwood grinders in a row, that receives the stock discharge from the grinders in a groundwood mill.

**DITHIONITE:** *See* HYDROSULFITE.

**DLK:** *See* DOUBLE LINED KRAFT CORRUGATED CLIPPINGS.

**DMSO:** *See* DIMETHYL SULFOXIDE.

**DO:** *See* DISSOLVED OXYGEN.

**DOCTOR:** A device consisting of a thin blade of material in contact with and extending across the face of a roll, usually to keep the surface clean by a scraping action. Also referred to as *doctor blade* or a *roll doctor.*

**DOCTOR BLADE:** A thin metal or wooden blade mounted on paper machine press rolls, drying cylinders, and calender rolls to scrape the surface clean and smooth.

**DOCTOR BOARD:** A board used on vacuum-type filters (deckers) to deflect pulp from the point of discharge to a receiving hopper or directly into a chest.

**DOCTOR BROKE:** Wet paper that is scraped off of wet press rolls by doctors during a paper break and is returned to the papermaking process.

**DOCTOR DUST:** Undesirable accumulation of fuzzy-like paper dust on roll doctor blades. It has the potential to fall onto the sheet and cause a defect as it passes between the rolls.

**DOCTOR MARKS:** A paper defect caused by press roll ridges made by doctor blades.

**DOCUMENT MANILA PAPER:** An inferior type of thick paper made from a combination of mechanical and chemical pulps. It is used to make up tickets, tags, and similar items.

**DOCUMENT PARCHMENT:** Specially sized paper made from linen and cotton fiber pulp to look like animal parchment. It is used for long-lasting documentation purposes such as diplomas, certificates, official government agreements, and other recordkeeping purposes.

**DODGER:** A small handbill made from newsprint or paperboard with surface characteristics suitable for printing, advertising, merchandising, etc.

**DOG EAR:** A paper-finishing error caused by the corner of a sheet being bent under before trimming, resulting in an oversized corner when the trimmed sheet is unfolded.

**DOG HAIRS:** Fuzzy-like fibers that stick up from the coated surface of a paper sheet.

**DOLOMITE:** A limestone consisting of a calcium and magnesium carbonate base that is used in the preparation of cooking liquor for the sulfite digesting of wood.

**DOPE:** A viscose coating solution of cellulose acetate or nitrate and ethyl or benzyl cellulose in a suitable volatile solvent. It is also used as a lacquering material.

**DOUBLE CALENDERED:** A term referring to coated paper processed through supercalenders two times to achieve desired surface characteristics.

**DOUBLE CAP:** A paper and paperboard size whose dimensions are 17 by 28 inches.

**DOUBLE COATED:** Paper or paperboard sheet with a heavy coating on one surface achieved by applying a thick layer of coating material or two thinner coats of the same or different materials. It is sometimes used when both sides of the sheet are coated.

**DOUBLE COATERS:** Types of paper coating machines used to apply coating material to both sides of paper at the same time.

**DOUBLE COVER PAPER:** A cover-type paper with increased weight achieved by pasting two regular weight sheets together. It is used in bookmaking.

**DOUBLE DECKLE:** Paper machine wet ends that have deckle straps or strips on both sides of the wire (forming fabric).

**DOUBLE DECOMPOSITION:** A chemical reaction in which the metal of a salt changes place with the hydrogen of water, with both substances–salt and water–being decomposed. It is represented by a typical reaction that takes place in a sulfate digester in which the sodium sulfide in solution is partly hydrolyzed in sodium hydroxide and sodium hydrate ($Na_2S + H_20 = Na_2S + HOH \leftrightarrows NaOH + NaSH$).

**DOUBLE DILUTION:** A method of decreasing the consistency of a high density pulp slurry to a desired consistency by the addition of water in two steps, usually by the use of a consistency controller.

**DOUBLE DISK REFINER:** A type of machine used to mechanically treat fibrous materials (chips and pulp), essentially consisting of two metallic surface-patterned disks with one stationary or both rotating in opposite directions on horizontal shafts. Chips or pulp are fed into the center of one of the disks and move outward between the surfaces as they are subjected to mechanical treatment.

**DOUBLE-FACED CORRUGATED BOARD:** A multi-ply, sandwich-type paperboard constructed of two outer layers of heavyweight kraft linerboard glued to an inner layer of fluted board. It is a very common carton-making material.

**DOUBLE-FACED PAPER:** A paper sheet made to have two distinctly different surface appearances, usually in color.

**DOUBLE-FELTED PRESS:** A type of press on a linerboard machine handling two separate felts, one contacting the top surface of the sheet and the other contacting the wire (forming fabric) side or the bottom surface of the sheet.

**DOUBLE FELTING:** (1) The use of two felts on a linerboard machine press. *See* DOUBLE-FELTED PRESS. (2) The addition of a second felt (normally a top felt) to a single-felted press.

**DOUBLE-LINED KRAFT (DLK) CORRUGATED CLIPPINGS:** An important source of fiber, considered to be recycled, derived for cuttings and trimmings at corrugated box conversion plants, specifically boxes made with linerboard facings on both sides of a corrugated center section.

**DOUBLE MANILA-LINED CHIPBOARD:** A multi-ply combination board made up of chipboard that has been lined on both sides with manila layers. It is used in containers and construction.

**DOUBLE REFLECTORS:** Type of supports used on higher speed paper machines at both ingoing and outgoing nips of the table roll. It is commonly used on newsprint-type paper machines.

**DOUBLE SIZED:** A type of paper that has been passed through the sizing solution, dried, and passed through a second sizing solution before final drying.

**DOUBLE THICK COVER:** A very rigid, dual-ply cover paper made by gluing two standard sheets of cover paper together.

**DOUBLE WHITE PATENT-COATED BOARD:** A paperboard used to make boxes and cartons. It is normally made on a cylinder-type machine and lined on both sides with white fiber.

**DOUGHNUT:** A roll of rewound facial tissue paper slit to various widths depending on the size of the finishing machine on which it will be further processed into consumer products.

**DOWNCOMERS:** The large diameter pipelines in a power or recovery boiler that carry water down and out of the drum to distribution headers at the bottom of the boiler to feed the steam-generating tubes.

**DOWNFLOW DIGESTER:** A continuous-type of chip cooking vessel in which the chips are fed into the top, travel downward as they are being cooked, and discharge from the bottom.

**DOWNFLOW TOWER:** A retention tower, usually in a multi-stage bleaching operation, in which the pulp slurry enters in the top and discharges from the bottom.

**DOWNPASS:** A chamber or gas passage placed between two combustion chambers in a furnace to carry the products of combustion downward.

**DOWNSTREAM:** A reference to an event that occurs later than those under current consideration.

**DOWNTIME:** The amount of time a piece of equipment or process is out of service or not operating due to malfunctions, breakdown, repair, and maintenance.

**DP:** *See* DEGREE OF POLYMERIZATION.

**DR:** *See* DISPLACEMENT RATIO.

**DRAFT:** The differential pressure which causes the movement of air, gases, and other vapors through furnaces, kilns, ducts, and up chimneys and flues.

**DRAFTING PAPER:** Paper used for pen or pencil drawing. It is made from various fiber combinations, depending on grade and end use.

**DRAG:** The phenomenon of "pulling" created on the fiber suspension being ejected onto the paper machine wire (forming zone) from the headbox. It is caused by the wire speed being greater than the slice jet velocity of the stock. The opposite condition of "pushing" the suspension is known as *rush.*

**DRAG SPOTS:** Paper defects showing up as thin streaks and lumps caused by fibers gathering and sticking to the slice of a paper machine wet end preventing the local flow of stock. It produces a lump on the paper surface when it breaks away.

**DRAG WEAR:** Paper machine fabric wear caused by contact with stationary machine components or by not moving at the same speed as a contacting moving machine part.

**DRAIN:** (1) The removal of water from a pulp slurry by allowing it to drop away through a screen by the force of gravity. (2) A ditch or sewer for spillover and liquid waste removal.

**DRAINAGE AIDS:** Substances added to pulp stock slurry to increase its capability to lose water by gravity when on a screen or paper machine wet end wire (forming fabric).

**DRAINAGE FACTOR:** The ratio of the amount of time required for water to drain from a quantity of stock slurry to the weight of stock under specified conditions, and expressed as seconds per gram of pulp.

**DRAINAGE RATE:** The rate at which water will drop away from the pulp slurry by gravity as the sheet is formed on the paper machine wet end wire (forming fabric).

**DRAINAGE TIME:** The amount of time it takes for water to drain away from a stock slurry and form a mat on a standard size screen under specified conditions in a laboratory test procedure.

**DRAINER:** Storage cells where washed, bleached rag pulps are placed for a number of days so that water will drain from them.

**DRAINER STOCK:** Pulp slurries from which water has been removed by letting them stand in chests with perforated bottoms so that as the water drains, the fibers remain.

**DRAINING DOWNWARD:** The draining away of the adhesive from a layer of coating into the base sheet of a coated paper during the drying process.

**DRAW:** (1) The dryer intersection tension applied to a paper sheet on the machine to maintain the proper run of the paper between the sections operated at different speeds while it is drying. Applied to prevent breakage of the paper sheet. (2) The displacement of the cut sheet by the thickness of the trimming knife, in the finishing of paper.

**DRAW DOWN TANKS:** Cooking liquor storage tanks located above the pulp mill batch digesters into which white and black liquors are pumped, and from which they are "drawn" or dropped batch-wise in prescribed amounts into the digester. Also called *measuring tanks.*

**DRAW SPEED CONTROL:** Controlling the speed of groups of dryers of a paper machine so that each section runs faster than the previous one to maintain proper sheet tension between sections. It is accomplished while the sheet goes through dimensional changes during drying.

**DRAWING BOARD:** *See* ILLUSTRATION BOARD.

**DRAWING PAPER:** Numerous grades of paper used for pen and pencil artwork. It has fiber compositions dependent on the type of work for which it is intended.

**DREGS:** Solids that settle and comprise the underflow from clarifiers in the causticizing process.

**DRESSING:** The act of applying a sharpening tool to the face of a pulpwood grindstone so that the teeth will be impressed into the grinding surface, forming a pattern on, or sharpening it. Also called *burring* or *jigging.*

**DRIED-IN STRAIN:** The strain property that exists in a paper sheet formation. It is caused by the drying process, which can be dissipated over a period of time or hastened by conditioning under high humidity or dampening with moisture. *See also* DRIED-IN STRESS.

**DRIED-IN STRESS:** That portion of the stress property existing in a paper sheet formation which is the result of drying. *See also* DRIED-IN STRAIN.

**DRIER:** Having a lower moisture content. It is commonly used interchangeably in paper mills with *dryer,* referring to paper machine equipment used to dry paper by evaporation of moisture by the use of steam.

**DRIFT:** The undesirable slow movement in equipment operation over a period of time, below what is considered normal.

**DRILL STICKS:** Tools used to punch holes in paper.

**DRILL TESTING:** Method of making moisture determination of baled or rolled pulp in which samples are taken with the use of a special boring tool.

**DRILLING PATTERN:** The arrangement of holes on the surface of the press suction rolls of a paper machine.

**DRINKING CUP PAPER:** A hard, rosin-sized paper made from long-fibered, bleached chemical pulps, suitable for coating or dipping in wax, and with physical characteristics necessary for converting it into cups to hold liquids for drinking purposes.

**DRINKING STRAW PAPER:** A clean, high-strength paper made from bleached chemical pulp with the necessary physical characteristics suitable for spirally winding it into tubes used for the drinking of liquids.

**DRIP ASH:** Ash in a coal-powered furnace caused by firing coal with a high fusion temperature.

**DRIP PUMP:** A pump that collects condensate from heater exchangers and returns it to the system, usually through a deaerator.

**DRIVE:** Any equipment, electric motor, belts, chains, gears, etc., that are connected to a piece of machinery to make it operate.

**DRIVING:** A means of transporting pulpwood logs by floating them in a stream with the current.

**DROP MARKS:** A paper defect caused by drops of water falling onto the wet web while on the wire (forming fabric) of the wet end of a paper machine.

**DROP OFF:** (1) Paper that is left over after an order of specified widths is cut from a reel off the machine. (2) The falling off of a wet sheet from the underside of the bottom felt on a cylinder machine.

**DROP TEST:** A quick test for determining the coater resistance of paperboard by timing how long it takes for a drop of water to soak into it.

**DROP TESTER:** A type of paperboard box-making plant testing instrument that drops plain or corrugated paperboard shipping containers, under specified conditions, to determine their capabilities to stand up under free fall impact.

**DROPLEG:** A vertical run of pipe, attached to the side of vacuum pulp washers, which develops suction inside the drum due to the wash water falling down through this pipe by the force of gravity. The water normally falls into a seal box or tank.

**DRUBBER:** A special defibering machine used for reducing chips into their fiber components without cutting the fibers or grinding up the foreign matter. It consists of a horizontal shaft carrying metallic impellers spaced along it and having small vanes on each face at the periphery, all located in a metal casing.

**DRUM:** (1) Horizontal, cylindrical pressure vessel, located at the top of the boiler, which receives a steam/water mixture from the evaporating sections of the boiler, and is the point of introduction of feedwater. It serves in the storage of water, separation of steam and water from the evaporator, and feedwater heating. (2) A wire-covered cylinder located in the vat of a vacuum pulp washer onto which the pulp pad is formed.

**DRUM BARKER:** *See* BARKING DRUM.

**DRUM CHIPPER:** A wood log chipper, consisting of a rotating drum with grooving tools to cut transverse grooves of a fixed depth into the wood, and with trailing blades to slice the wood between the grooves.

**DRUM FILTERS:** Pulp washing and stock slurry dewatering types of process units using one wire screen-covered, cylindrical drum, partially immersed in pulp suspension and white water, from which it picks up fiber and forms a sheet as a means of separating the solids from the liquid. It is commonly used in multi-stage brownstock washing and pulp bleaching processes.

**DRUM SAVE-ALL:** A white water fiber and filler reclamation unit located in a paper mill. It consists of a wire-covered drum partially submerged in a vat with sweetener stock and white water, with internally applied vacuum so that a recoverable fiber-filler mat forms on its surface.

**DRUM TEST:** A method of determining the durability of paper and paperboard boxes under shipping conditions by tumbling in a rotating drum.

**DRUMHEAD MANILA PAPER:** A special, high-strength, durable type of paper made from manila hemp fibers for use in making various musical drumheads.

**DRY BARKING:** The removal of bark from logs using a single log friction barker immediately after cutting and before the bark dries.

**DRY BONDING:** The process of applying an adhesive to a web of paper and drying it before being mated to a second web.

**DRY BOTTOM:** Ash removed from furnaces in solid or "drip" state. It is caused by firing coal with a high ash fusion temperature.

**DRY BROKE:** Paper returned for repulping and reprocessing, resulting from a process malfunction after the felt press section of the paper machine through calenders, reels, rewinders, and finishing operations.

**DRY BULB TEMPERATURE:** The temperature of air as determined by a thermometer with a bare bulb or primary element. It is a measure of the "dry" air temperature, as opposed to air under saturated conditions. *See also* WET BULB TEMPERATURE.

**DRY CHIP REFINING:** Mechanically processing chips by feeding them directly into disk refiners using only water from the casing water showers.

**DRY COATING:** A method of applying a dry-coating pigment in a confined fine air suspension to a paper sheet after an adhesive is applied.

**DRY COOK:** A bad wastepaper cook in which the ink is not acted upon and the sizing is not thoroughly saponified, making it impossible to remove these during subsequent washing operations. Also commonly called a "bad bleach" in a deinking mill.

**DRY CUT:** A paper defect showing up as a fairly long split in the machine or cross-machine direction.

**DRY END:** (1) That portion of a paper machine in which the paper sheet moisture is removed by evaporation. It consists of a number of dryer sections or air dryers, depending on the type and size of the machine. (2) Sometimes it also includes all of the paper machine units from the felt presses through the rewinders, such as size presses, coaters, calenders, reels, and slitters.

**DRY FINISH BUTCHERS:** Paper made from mechanical and chemical pulps, highly sized and calender-finished without moisture, to make it blood-resistant so it can be used for wrapping fresh meat.

**DRY FINISHED PAPER:** A wrapping-type paper with a rough unglazed surface finish.

**DRY FINISHING:** Smoothing the surface of a paper or paperboard sheet by the application of pressure or friction in a calender without the use of steam or moisture.

**DRY FOIL:** A foil with surfaces that are essentially free of oil residue sometimes caused by the rolling process during manufacture, and it is water-wettable (hydrophilic).

**DRY FORMING:** Producing a sheet of paper from fibers without the use of water, usually air.

**DRY INDICATOR TEST:** A test to determine the sizing or water resistance of paper by floating a sample on water, sprinkling a mixture of dye and sugar on the upper surface, and timing it for a color change due to the penetration of the water.

**DRY LAID:** *See* AIR LAID.

**DRY LAID FORMING:** *See* AIR LAID.

**DRY LAKE:** Ground, artificial pigment made by depositing a dye on a mineral base, such as clay, and drying it. It is used to make up color pastes and color paper.

**DRY LIMESTONE PROCESS:** Removal of oxides of sulfur from gaseous mill effluents by passing them through crushed limestone.

**DRY LINE:** The line at which the free, glossy, water surface disappears on the forming wire (fabric) of a fourdrinier paper machine.

**DRY NIP:** The line of contact between rolls that is normally dry, such as in the calender section of a paper machine.

**DRY PRINTINGS:** Types of paper treated so that they have a high-surface affinity for the reception of inks, allowing them to be used with rapid drying inks.

**DRY PROOFING PAPER:** An inferior type of quick-drying printing paper used in the printing industry to check the type setup for mistakes.

**DRY PULP:** Cut sheets of wood pulp that are dried on paper machine dryers.

**DRY RUB:** A test made on a pigment coated paper and board, consisting of rubbing the surface of the sample against black velour paper under controlled conditions and comparing the amount of coating picked up on the black paper with reference standards. Also called *dusting.*

**DRY RUB RESISTANCE:** A paper and paperboard surface property that withstands the disruptive effects of extended rubbing or scuffing.

**DRY SIZE:** A water-soluble, completely saponified type of rosin size, prepared by spray drying or drum drying, that can be added directly to pulp stocks or pre-dissolved to improve and control distribution in the stock.

**DRY STEAM:** Steam that contains no moisture and may be either saturated or superheated.

**DRY STRENGTH AGENTS:** Substances added to the pulp stock slurry that will tend to increase the strength of the sheet made from it.

**DRY WAXED PAPER:** Applying wax to a paper sheet with the use of roll-type machines so that most of the wax is forced inside the sheet.

**DRY WRINKLES:** Paper defects caused by malfunctions or maladjustments of equipment at the dry end of a paper machine. *See* WRINKLES.

**DRYER:** That portion of a paper machine where water is removed from wet paper by evaporation through such methods as passing it over rotating, steam-heated, cylindrical, metal drums or running it through a hot air stream. Also erroneously spelled *drier.*

**DRYER CAN:** *See* DRYER DRUM.

**DRYER CYLINDER:** *See* DRYER DRUM.

**DRYER DEFLECTOR:** Special air doctors located at critical points along paper machine dryer felts to reduce undesirable air flows induced by the movement of the fabric blanket.

**DRYER DRUM:** One of the steam-heated, cylindrical, metal vessels that make up the dryer section of the paper machine over which wet paper sheets are passed and dried by evaporation. A common paper mill term is *dryer can* or *cylinder.*

**DRYER FELT:** (1) The felt blanket on a paper machine dryer that keeps the paper sheet in contact with the surface of the dryer drums as it passes through the dry end. (2) A continuous belt made of cloth material and used in the dryer section of a paper machine to press and maintain positive contact of the sheet against the surface of the dryer cylinder. Also incorrectly spelled *drier felt. See* FELT.

**DRYER FELT MARKS:** Paper or paperboard defects appearing as thin spots caused by particles of foreign materials on the felt and the impression of the press felt on the sheet surface.

**DRYER GROUP:** *See* DRYER SECTION.

**DRYER HEADS:** The circular end sections of cylindrical dryers in the dry end section of a paper machine that are normally attached to the drums by bolting.

**DRYER HOOD:** A canopy located over the dryer section of a paper machine used to confine and control vapor-laden air rising from it. This allows heat to be recovered and used to heat air entering the machine room. Normally, there are two types, "open" and "closed," that are also called *air hoods*.

**DRYER PICK MARKS:** Paper defects caused by paper machine dryer rolls picking up small particles of coating or paper from the sheet as it passes over the drying surface.

**DRYER POCKET:** The area between the dryer rolls in a multi-drum dryer section on paper machines.

**DRYER RATIO:** The percentage of drums using flash steam from separators to primary steam drums in a dryer section of a paper machine.

**DRYER SCREEN:** Porous dryer felts used in conjunction with hot air ducts to improve drying efficiencies on paper machines.

**DRYER SECTION:** A grouping of a number of gear-driven, cylindrical metal drums in series that are steam-heated to dry wet paper passed over it. The dry end of a paper machine may contain several of these sections. Sometimes referred to as DRYER GROUP.

**DRYER SPEAR POLE:** *See* SPEAR POLE.

**DRYER SPEARS:** *See* SPEAR POLE.

**DRYER STEAM MAKE-UP:** Amount of steam fed to a given dryer section to maintain proper pressure differential with other sections.

**DRYER SYPHONS:** Drop pipes built in dryer drums of a paper machine such that they remove condensate and blow-through steam. They are oriented to the plane, perpendicular to the long axis of the drum, and may be rotary, in which case the tip rotates with the drum. They may also be stationary, in which case the tip remains fixed relative to dryer surface rotation.

**DRYER SYPHON SHOE:** Part of the dryer internal structure connected to the extreme end of a dryer syphon pipe. Also called DRYER SYPHON TIP.

**DRYER SYPHON TIP:** *See* DRYER SYPHON SHOE.

**DRYING:** The function in pulp, paper, and paperboard making during which moisture is removed from the wet pulp, paper, and paperboard sheet to bring it down to the desired moisture content.

**DRYNESS:** Refers to moisture content of pulp, paper and paperboard. *See* MOISTURE CONTENT, AIR DRY, BONE DRY, and OVEN DRY.

**DRYING CRACKS:** Coated paper surface defects appearing as tiny fissures caused by uneven or too-fast drying of the coating on the paper sheet.

**DRYING PAPER:** A lightweight, thin, absorbent-type paper extensively used in the printing industry to place between freshly printed material so that it will absorb excess ink.

**DS:** *See* DISSOLVED SOLIDS.

**DUAL PRESS:** Suction press section of a paper machine, usually consisting of three rolls set horizontally to each other. The first roll is a rubber-covered suction press, the second a hard-surfaced roll, and the third a rubber-covered roll.

**DUB:** The end relief ground on rolls. It may be in the form of a taper, step, or both, which is specified depending on the application.

**DUB AMOUNT:** The measurement of the radius of the *dub* of a roll, in thousandths of an inch "on a side."

**DUCTILITY:** The degree to which a foil or sheet can be drawn or elongated without rupture.

**DULL COATED PAPER:** Dull, flat, surface-finished paper made from a high freeness pulp stock and coated on both sides with calcium carbonate-containing materials. It is used where low-gloss surfaced papers are required, as in some book printing papers.

**DULL FINISH:** Any smooth paper and paperboard surface finish with very low gloss or luster.

**DULLING:** The act of applying a brick or abrasive block to the face of a revolving pulpwood grinderstone to dull the grinding surface in order to lower the freeness of the produced pulp.

**DUMBELLS:** Wet-lay nonwoven paper machine defect caused by paired clumps of fiber connected by one or more long fibers.

**DUMMY:** A pattern or set of pages cut and assembled to help in the planning of a proposed printed piece.

**DUMP:** A land site for disposal of solid mill waste material.

**DUMPERS:** Hydraulically operated elevating platforms on which chip trucks, trailers, and freight cars are tilted to unload them into a receiving hopper, feeding onto a conveyor belt.

**DUMPING:** (1) Unloading pulp stock from a beater after treatment by dropping it into a beater chest located below it. (2) Unloading of some types of batch digesters by dropping their contents into a pit or tank.

**DUNNAGE:** Cargo-protecting and stowing materials used in transportation, including various types of braces, props, and heavy polyethylene-coated inflated kraft paper bags.

**DUNNAGE PAPER:** A heavyweight paper or paperboard used as a cargo protective liner, especially in freight cars. It is usually made from sulfate pulp, screenings, and wastepaper.

**DUOTONE FINISH:** A colored paper surface effect resulting from the use of uneven pressure during finishing. It causes variations in brightness or depth of color.

**DUPLEX:** (1) A reference to papers and paperboards with each side having a different color, finish, or surface texture, which is produced on the paper machine or by pasting. (2) A general term referring to multi-ply paper and paperboard, and to bags made up of two separate sheets of paper.

**DUPLEX ASPHALT PAPER:** Paper made by pasting together two sheets, usually kraft, with asphalt or some other bituminous base material, making it highly water resistant.

**DUPLEX BLOTTING PAPER:** A dual-ply type of blotting paper made by combining two different colored plies on a cylinder machine or pasting the two sheets together after drying.

**DUPLEX BOARD.** A combination-type paperboard made from two different stocks of the same or different colors on either a cylinder or fourdrinier machine, equipped with two separate stock approach and headbox systems.

**DUPLEX BRISTOL:** A heavyweight printing paper that is colored different on the sides.

**DUPLEX COATED BRISTOL:** A heavyweight paper used in advertising covers, postcards, flyers, folders, etc., made from bleached or chemical pulp stock, and coated with a dark color on one side and a light color on the other side.

**DUPLEX COATER:** A type of coating machine that applies coating material to both sides of a sheet of paper at the same time.

**DUPLEX COLORS:** Coating color mixtures that are different and are used to color each side of a sheet of paper at the same time.

**DUPLEX ENAMEL BOOK:** A book type of paper used in advertising and having a dual surface finish of different color hues.

**DUPLEX FINISH:** Two-sided surface finishes of different color and characteristics.

**DUPLEX LITHO:** A two-sided paper with a rough surface on one side and a smooth finish on the other surface. As its name implies, it is used primarily for lithography.

**DUPLEX PAPER:** A general term for all types of two-sided, surface-finished papers with different colors, textures, and other surface characteristics on each side.

**DUPLEX SANDWICH PAPER:** Paper consisting of two pasted sheets.

**DUPLEX SUPER:** Paper that has been supercalendered in such a way that one surface has a glassier finish than the other.

**DUPLEX TEXTURE:** A dual texture finish produced to be different on each surface of paper or paperboard sheets by multi-ply formation on the machine, or by lining both sides with dissimilar stocks.

**DUPLEX WRAPPERS:** Paper used for protective coverings of such items as rolls and packages of paper, clothing, and other materials. It is usually made from a heavyweight, water-finished grade of paper with dull surface finish.

**DUPLICATING PAPER:** A highly finished paper made from bleached chemical pulps and used as masters and copy sheets in the reproduction of multiple copies on various types of duplicating machines.

**DURABILITY:** A characteristic of paper and paperboard relating to its ability to stand up and retain its original properties under constant use over extended periods of time.

**DUST:** Any fine grain, particulate matter capable of being suspended in air, or very fine, loose material located on the sheet surface.

**DUSTER:** A machine used in a rag pulp mill consisting of a rotating, perforated, cylindrical drum in which old rags and papers are tumbled to remove dust and other undesirable foreign materials. Sometimes called *rag duster.*

**DUSTING:** (1) The undesirable separation and collection of minute quantities of surface coating materials, additives, fillers, and fibers from the sheet in the finishing operation, or during its use in printing and conversion into user items. (2) A test made on pigment-coated paper and paperboard to determine the quality. Also called *dry rub.*

**DUTCH OVEN:** A special furnace, usually located in the boiler house, used to burn wood and bark after the water has been pressed out of it.

**DYE:** (1) A natural or synthetic, organic or inorganic substance used to make up materials to impart a color to pulp slurries or the paper or paperboard sheet in papermaking, or to make up coating material to color their surfaces. The name is used interchangeably with the common paper mill term, *dyestuff.* (2) The act of coloring (or changing the color of) any material (stock, paper, etc.) by bringing it into contact with another material (dye) of a different color in such a manner that the resulting color will be more or less permanent.

**DYEING:** *See* COLORING and DYE.

**DYESTUFF:** *See* DYE.

**DYNAMIC BLEACHING:** *See* DISPLACEMENT BLEACHING.

**DYNAMO:** An electrical generator.

# E

**E-FLUTE:** Geometric configuration formed by one of the undulations of the corrugated medium in corrugated board used in folding cartons. It measures approximately 1/16-inch high and runs 88 to 100 per linear foot. *See also* A-FLUTE, B-FLUTE, and C-FLUTE, as designations for corrugations of other dimensions and counts per linear foot.

**EARLYWOOD:** *See* SPRINGWOOD.

**EASILY WETTED:** *See* HYDROPHYLIC.

**EASY BEATING PULP:** Pulp that reaches its maximum bursting strength in a short time during the beating process and shows only a slight drop in tear and burst after it reaches its maximum.

**EASY BLEACHED PULP:** Pulp that has not been completely bleached.

**EASY BLEACHING:** Pulp made by cooking under conditions that increase the activity of the digesting process more than normal to produce pulp requiring less chemicals to bleach to the desired brightness. Sometimes known as *soft pulp.*

**ECONOMIZER:** A heat exchanger used to preheat boiler feedwater by recovering heat from the flue gas. It is usually a continuous tube design and is generally located in the flue gas flow between the superheater and air heater.

**EDGE:** The outer extremity of a sheet or web of paper.

**EDGE CRUSH TEST:** A test used to evaluate the compressive strength of containerboards used in the manufacture of corrugated shipping boxes. *See* RULE 41 and RING CRUSH.

**EDGE FAILURE:** A common wet press felt problem caused by improper roll crown and roll width, along with improper guiding.

**EDGE OF STONE:** A mechanical pulp mill reference to the flat area made on the edges of the sides of the pulpwood grindstone face. Also called *pulpstone bevel* or *bevel.*

**EDGE PROTECTORS:** Types of hard, heavy paperboards used to make paper roll end protectors that prevent damage to the roll during packing and shipping.

**EDGE SEALING:** The soldering of frayed synthetic forming fabric edges on the paper machine.

**EDGE TEAR RESISTANCE:** The ability of a paper sheet to withstand forceful pulling apart at the edges.

**EDGEWISE COMPRESSION STRENGTH:** The ability of corrugated paperboard to withstand crushing pressure applied in the direction of the fluting.

**EDUCTOR:** A device used in the sulfite pulping process for reducing the back pressure on the high pressure relief header, and mixing and absorbing the digester relief liquor and gases with a flowing stream of raw acid before entering the accumulator. Sometimes called an *injector.*

**EF:** *See* ENGLISH FINISH.

**EFFECT:** One unit of a multiple vessel-type evaporator train.

**EFFECTIVE ALKALI:** The NaOH + $\frac{1}{2}Na_2S$ content in sulfate cooking liquors. It is expressed as $Na_2O$, which is an indication of the actual amount of alkali available for or remaining during sulfate pulping reactions. Usually expressed in grams per liter but can also be expressed in other equivalent units.

**EFFECTIVE HEAT:** The heat supplied to a pulp and paper mill building that is dependent on the difference between the temperature of the entering air to be heated and the air already inside the building.

**EFFECTIVITY:** A term sometimes used in relation to alkaline pulping cooking liquor. It refers to a percentage value determined by dividing the effective alkali by the total alkali, with both expressed as $Na_2O$ grams per liter, multiplied by 100.

**EFFICIENCY:** The amount of useful work done in relation to the potential capacity of the complete process or any piece of machinery or equipment in a mill.

**EFFLUENT:** Pulp or paper mill discharge that tends to pollute the environment, particularly water streams, lakes, and other bodies of water.

**EFFLUENT CLOSURE:** *See* CLOSED CYCLE MILL.

**EFFLUENT FREE MILL:** *See* CLOSED CYCLE MILL.

**EFFLUX RATIO:** The headbox slice jet velocity divided by the wire speed on the wet end section of a paper machine. Also referred to as *rush/ drag ratio* and *push/drag ratio.*

**EGG CARTON BOARD:** A rigid, water-resistant paperboard with good folding properties. It is used for making cartons for eggs.

**EGG CASE FILLER BOARD:** A paperboard used to make interlocking, die-cut strips that are assembled to form small compartments to separate eggs in a layer packed in a case.

**EGG CASE FLATS BOARD:** A paperboard made from waste newsprint and mechanical pulp stock. It is made into indented or molded sheets and used between the layers of eggs in a case.

**EGGSHELL FINISH:** A rough paper surface finish made up of randomly aligned, small, round hills and valleys similar to an eggshell surface. It is produced by the use of special-textured felts in the press section.

**EJECTOR PUMPS:** Types of pumps that have no moving parts and depend on the movement of a motive fluid in the form of a liquid, gas, or vapor flowing through the pump at a high velocity, thereby imparting a motion to the other fluid to be moved.

**ELASTIC LIMIT:** The smallest amount of stress applied to a sheet of paper that will produce a permanent disorientation and alteration of its structure as related to the time rate at which it is applied.

**ELASTIC SOLIDS:** Materials such as paper that deform instantaneously under applied load to a degree linearly proportional to the load, and return to their original dimensions when the load is removed.

**ELASTICITY:** The ability of a material to change shape in response to forces applied to it and return to the original shape when the deforming forces are removed.

**ELASTOMERS:** (1) Latex-containing materials used as additives in papermaking to impart improved dimensional stability, edge tear, folding endurance, burst, internal tears, tensile strength, elongation, and wet strength to paper products. (2) The natural and synthetic rubbers that are sometimes compounded and used as pressure-sensitive paper-coating material.

**ELBOW:** A pipe fitting used to change the direction of a pipe in varying degrees. Also known as an *ell.*

**ELDER:** A broadleaf hardwood tree species used as a source of pulpwood.

**ELECTION BRISTOL:** A heavyweight printing paper used as a base stock to make up cards and posters for election campaigns and other advertising purposes.

**ELECTRIC BOARD:** A dense, hard paperboard used in the electrical industry for insulation on switch blocks or other electrical equipment.

**ELECTRIC DRIVE:** A drive arrangement for modern papermaking machinery sections and/or process equipment that utilizes electric motors.

**ELECTRICAL CABLE PAPER:** A clean, high-strength, dielectric paper made from sulfate chemical pulp and manila hemp fibers. It is used to make strips that are treated so they can be used as insulation on electrical wires and cables.

**ELECTRICAL CONDUCTIVITY:** The ability of a substance (specifically a sheet) to pass electric currents through it. The term is generally used in reference to treated electrical papers used in condensers and insulation having very low conductive properties.

**ELECTRICAL INSULATING PAPER:** Grades of paper with high dielectric properties used for electrical insulation purposes.

**ELECTRICAL PRESSBOARD:** A calender-finished, high-density paperboard having good dielectric properties. It is made from sulfate chemical pulp and cotton rag fiber and is used extensively in the manufacture of electrical apparatuses such as transformers.

**ELECTRICAL SHOP:** A separate room or area in a pulp and paper mill where repairs on electrical equipment are carried out.

**ELECTRODIALYSIS:** A process that uses electrical current and an arrangement of permeable membranes to separate soluble minerals from water or effluents.

**ELECTROLYSIS:** The process of decomposing chemical solutions by passing an electric current through them.

**ELECTROLYTES:** Substances whose molecules divide into positive and negative particles, called ions, when dissolved in water. They form solutions that will conduct electricity.

**ELECTROLYTIC BATH:** A pool of electrolytic solution used for electroplating metals by the decomposition of the solution with an electric current.

**ELECTROLYTIC CAPACITOR PAPER:** A highly dielectric paper made from sulfate chemical pulp, cotton, and hemp fibers. It is used between foils of insulating liquid-filled condensers and as spacers between foils of electrolytic condensers.

**ELECTROMAGNETIC FLOWMETER:** A flow sensing device commonly used to measure pulp slurry flow. It operates on the principle that a conductive fluid moving through a magnetic field will generate an electromotive force (voltage) that is proportional to its velocity, and which can be detected by a pair of electrodes in a pipeline. This voltage signal can be converted to a rate of flow measurement.

**ELECTROSTATIC COPYING PAPER:** A smooth-finished, stable, medium-weight type of bond paper made from chemical pulps. It is generally treated with a zinc oxide coating material and used on dry-type office copying machines.

**ELECTROSTATIC PRECIPITATOR:** Process equipment used to remove suspended materials from gaseous process emissions by imposing an electrical charge on the particles so that they would be attracted by oppositely charged electrodes.

**ELEMENT:** A component of a process, a piece of process equipment, or a device.

**ELEMENTAL CHLORINE:** *See* CHLORINE.

**ELEVATING TRUCK:** Types of vehicles used in paper mills to pile up rolls, bales of paper, and other materials.

**ELEVATOR:** A transporting device used to hoist, lift, or carry materials from one floor to another in a mill.

**ELL:** *See* ELBOW.

**ELM:** A broadleaf hardwood tree species used as a source of pulpwood.

**ELMENDORF TEST:** A very common test used in paper mill laboratories to determine the tear-resistant property of paper. The test measures the force required to split apart a 2-½-inch wide by 3-inch long specimen held between two clamps after an initial cut has been made with a sharp knife. Sometimes called simply *tear test.*

**ELONGATION:** A physical property of a paper sheet that will allow it to experience a certain degree of stretching.

**ELUTRIATION:** A process of sludge conditioning in which certain constituents are removed by successive flushing with fresh water or plant effluent, thereby reducing the demand for conditioning chemicals.

**ELREPHO BRIGHTNESS:** The measure of paper or pulp *whiteness* or *brightness* using a Zeiss Elrepho test instrument. It is the reflectance value relative to a standard in the blue region of visible light, expressed as a percentage of the standard's brightness. This test, used in most parts of the world other than the U.S., is described in TAPPI standard T525. It is conducted by diffusely illuminating a 34-mm circular specimen at a light wavelength and a viewing perspective identical to that used with the *G.E. brightness* test. Because the specimen is diffusely illuminated, its values are generally not affected by surface roughness, making it particularly applicable to pulp. The measure is sometimes referred to as "diffuse brightness." *See also* G.E. BRIGHTNESS, which is the brightness measure generally used in the U.S. There is no direct relationship between the two brightness measures, although *Elrepho brightness* tends to be 0.5% to 1.0% higher on the average.

**EMBOSSED:** (1) A type of paper finish obtained by mechanically impressing a design on the sheet with engraved metallic rolls or plates. (2) Types of papers and paperboards decorated in the above manner.

**EMBOSSER:** A piece of equipment used to mechanically impress a design on a sheet of paper by the use of raised or depressed engraved metallic rolls or plates riding over matching metal or compressible rolls or plates between which the paper is passed.

**EMCC:** *See* EXTENDED MODIFIED CONTINUOUS COOKING.

**EMERY:** An abrasive material, usually ground-up corundum, used to make a coating formulation for abrasive-type papers.

**EMERY PAPER:** A heavyweight sheet made from sulfate chemical pulp. It is coated with a mixture of pulverized emery stone (corundum) in the manufacture of grinding and polished abrasive papers.

**EMISSION FACTOR:** The ratio of the amount of pollutant being discharged from a process to the amount of product manufactured.

**EMISSION STANDARD:** The total amount of a pollutant as established by a governing body statute that is legally permitted to be discharged from a particular source.

**EMISSIVITY:** The radiant energy emitted by the surface of a material at a given temperature, divided by the radiant energy emitted by a standard material (blackbody) at the same temperature.

**EMPIRE STATE PAPER RESEARCH INSTITUTE (ESPRI):** Headquarters: State Univ. of New York, College of Environmental Science & Technology, Syracuse, NY 13210.

**EMULSIFIED FUEL:** An emulsified mixture of fuel oil and water.

**EMULSIFIER:** A surface-active agent for promoting the formation and stabilization of an emulsion. Same as *emulsifying agent* and *surfactant.*

**EMULSIFYING:** Mixing two immiscible liquids in such a way that one of them will disperse into the other in tiny droplets, such as water in oil.

**EMULSIFYING AGENT:** A material that enhances and stabilizes the ability of one liquid to disperse itself into another. Also called *emulsifier* or *surfactant.*

**EMULSION:** A liquid dispersed in an immiscible liquid in droplets larger than colloidal size, such as water in fuel oil and wax in paper-coating materials.

**ENAMEL:** Clay coating on coated paper.

**ENAMELED:** A common general reference to any paper and paperboard whose surface has been treated with a coating mixture.

**ENAMELED BLOTTING PAPER:** A multi-ply blotting-type paper lined on one side with a one-sided coated paper.

**ENCASED KNOT:** That portion of a branch of a tree which becomes embedded in the bole after the branch dies. Also called *loose knot*, it does not cook well in digesters during the pulping process.

**END BANDS:** Paper roll end shipping protectors made from heavy, hard types of paperboard.

**END CAP:** *See* ROLL HEAD.

**END CAPPING:** *See* ROLL HEADING.

**END DAMAGE:** Roll defects that consist of gouges, cuts, and core indentations at either end.

**END DUMP FREIGHT CAR:** Freight cars designed for transporting and unloading chips with an end door opening hinged at the top that swings open to discharge chips when tilted on a hydraulically operated elevating platform.

**END LEAF PAPER:** A good-strength, heavyweight paper used to make end leaves or sheets for binding books to their covers. It is made from chemical pulps.

**END LEAVES:** Heavyweight sheets of paper with good tear and folding properties. They are used to bind books to covers in bookmaking. Also called *end sheets*.

**END PAPER:** *See* END LEAF PAPER and END LEAVES.

**END PRODUCTS:** Finished and converted paper and paperboard made into items ready to be utilized by the user.

**END SHEETS:** *See* END LEAVES.

**END THRUST:** The tendency of a pump impeller to move toward, and sometimes rub against, the casing due to the pressure across it.

**ENDOTHERMIC:** Type of process that absorbs or requires the addition of heat to maintain itself.

**ENERGY:** The capacity to do work.

**ENGINE SIZING:** A common paper mill term often used when it was common to size paper by adding the sizing material to the beater. The meaning has been expanded to include any size addition to the stock furnish before the formation of the sheet on the paper machine.

**ENGINEERING UNITS:** Units of measurement as applied to the process variable, such as psi, °F, feet, gpm, etc.

**ENGLISH FINISH (EF):** A type of medium finish applied to a sheet of paper that is smoother than that coming off the dryers or calenders, but not as smooth as the finish coming off the supercalenders.

**ENGRAVING:** Any printing plate produced by etching or cutting.

**ENGRAVING BRISTOL:** A resilient, heavyweight printing paper suitable for use in plate engraving.

**ENGRAVING PROOF PAPER:** A heavyweight, high-grade, coated paper with a glossy surface to prepare printed proofs of the engraving plates in the printing industry.

**ENP:** *See* EXTENDED NIP PRESS.

**ENRICHMENT:** The discharge of nitrogen, phosphorous, carbon compounds, or other nutrient contaminants into mill waste receiving waters that greatly increase the growth potential for algae and other aquatic plants.

**ENRICHMENT OF WASTEWATERS:** *See* ENRICHMENT.

**ENTHALPY:** The total amount of energy contained in a unit mass of a substance due to its condition or state. It represents the sum of the intrinsic energy plus the pressure-specific volume of the product.

**ENTRAINED AIR:** Air that is either dissolved or suspended in a liquid or slurry, such as pulp stock, and is commonly caused by vortexing in pumps or other mixing action.

**ENTRAINMENT:** The entrapment and carrying over of material in a stream. It is sometimes undesirable, such as air in black liquor or stock slurries causing foam and droplets of cooking liquor and fiber in the relief gases of a sulfate digester.

**ENTRAINMENT SEPARATORS:** Components of a black liquor multiple-effect evaporator that remove entrained liquor from the vapor. They consist of a centrifugal-type catch-all mounted inside or outside of the vapor head on each effect, as well as in the upper portion of the heating element.

**ENTROPY:** A property describing (mathematically) that portion of heat added to a process which cannot be converted into useful work; it is a measure of unavailability. The most theoretically efficient process is the one carried out at constant entropy, or isentropically.

**ENVELOPE MANILA:** A manila-colored paper made from chemical and mechanical pulps on a fourdrinier paper machine. It is generally converted into envelopes.

**ENVELOPE PAPER:** Any types of paper used in making envelopes.

**ENVELOPE-LINER TISSUE:** A lightweight tissue paper made especially to be used as a decorative lining for envelopes, increasing their opacity.

**ENVIRONMENT:** The sum of all external conditions and influences affecting the life, development, and, ultimately, the survival of an organism or various groups of organisms.

**ENVIRONMENTAL PROTECTION AGENCY (EPA):** Headquarters: Washington, D.C., (202) 260-2090.

**ENZONE:** A non-chlorine, environmentally benign pulp bleaching process that combines enzyme treatment with oxygen, ozone, and alkaline hydrogen peroxide bleaching chemicals. It has been shown to bleach softwood and hardwood kraft pulps to high brightness with competitive viscosity and good strength properties. The trademarked process was developed principally by Emerging Technology Transfer Inc. of Raleigh, N.C.

**ENZYME:** A substance that activates or accelerates biological processes and is used in liquid, powdered, pellet, or capsule form in a paper mill to convert starch so it can be used in the sizing of paper. Also used in chemical pulping/bleaching stages to boost delignification efficiency and further reduce the need for chlorine and chlorine containing compounds in the brightening of pulps.

**EPA:** *See* ENVIRONMENTAL PROTECTION AGENCY.

**EPILIMNION:** The upper layer in a body of water, being warmer and less dense than the lower water.

**EQUALIZATION:** A type of waste water treatment used by some pulp and paper mills in which the effects of fluctuations of flow, pH, *biological oxygen demand*, and total solids is minimized by proper design on size and configuration of the settling or equalization chamber.

**EQUILIBRIUM:** A condition when all inputs and outputs (supply and demand) have settled down and are in balance.

**EQUILIBRIUM MOISTURE CONTENT:** That state in which a substance will remain unchanged as long as it is exposed to the same atmosphere of constant temperature and humidity.

**EQUILIBRIUM RELATIVE HUMIDITY:** Relative humidity conditions of the atmosphere under which exposed paper will neither lose nor gain moisture when the temperature is constant.

**EQUIVALENT PAPER WEIGHT:** A term used to denote the difference in weight between quantities of the same kind of paper in two different sheet sizes.

**EQUIVALENT SPHERICAL DIAMETER (ESD):** A nominal diameter of nonspherical particles of clay that subside in a slurry at the same rate as a theoretical equivalent spherical clay particle.

**EQUIVALENT STEAM:** A common powerhouse term for *latent heat of evaporation*, which is the heat, in Btu's (970.4), required to convert one pound of water at 212°F (100°C) and atmospheric pressure into saturated steam at the same pressure and temperature.

**ERASABILITY:** The ability of a sheet of paper to resist scuffing and surface disorientation by the removal of writing with an erasure.

**EROSION:** The wearing away of land surface by wind or water, occurring naturally from weather or runoff. It is intensified by clearing the land of trees and other vegetation.

**ESD:** *See* EQUIVALENT SPHERICAL DIAMETER

**ESPARTO:** A short-fibered, coarse, grass-like plant used to make alkaline chemical pulps from which book papers are made.

**ESPARTO PAPER:** A bulky paper with good formation made from a furnish consisting of a high percentage of esparto chemical pulp mixed with chemical wood pulps. It is used extensively in the manufacture of book, writing, and other papers. *See also* ALPHA PRINTING PAPER and ALPHA WRITING PAPER.

**ESPARTO PULP:** A pulp made by cooking *esparto*, usually in an alkaline cooking liquor.

**ESPRI:** *See* EMPIRE STATE PAPER RESEARCH INSTITUTE.

**ESTER NUMBER:** The difference between the saponification number and the acid number in the laboratory analysis of rosin size used in papermaking.

**ESTER PULPING:** The cooking of wood chips in a liquor made up of water, acetic acid, and ethyl acetate.

**ETCH:** (1) To produce an image on a plate in photoengraving by the use of chemical or electrolytic action. (2) In offset lithography, it is the acidified gum solution used to desensitize nonprinting areas of a plate.

**ETCHING PAPER:** An even-surfaced, strong, high-grade type of drawing paper used in artwork.

**ETHANOL:** An alcohol that can be produced by the fermentation of the sugars of sulfite waste liquor.

**EUCALYPTUS:** A large broadleaf tree with short fibers. It is a major source of pulpwood in many parts of the world and has been used extensively in trials at U.S. mills.

**EVAPORATION LOAD:** The evaporation capacity required to dry paper based on tonnage per day or by operating conditions such as speed and width. It is used to determine the number of dryers needed and exhaust requirements.

**EVAPORATION PONDS:** Shallow, artificial ponds where mill effluents are pumped or allowed to flow and permitted to dry, and either removed or buried by more effluent.

**EVAPORATIVE COOLER:** A bare evaporator coil through which cool air passes and over which water is sprayed to accomplish heat transfer in the water through evaporation.

**EVAPORATOR CAPACITY:** The quantity of water an evaporator is capable of removing from a liquor solution. It is directly proportional to the difference in temperature between the steam and the liquor in each effect.

**EVAPORATOR ROOM:** The location where the black liquor evaporators are housed. Evaporators are located outside in most mills where the weather is not hostile.

**EVAPORATORS:** Process equipment in which various liquids, such as spent liquors from the pulp washing process, are concentrated by the heat removal of water so that chemicals may be readily removed from it.

**EVEN AGED:** A pulpwood timber stand consisting of trees primarily of the same age.

**EVENER ROLL:** A perforated roll located just prior to the slice in a paper machine headbox. It is used to stabilize the flow.

**EVENERS:** Devices inserted just prior to the slice in paper machine headboxes to stabilize the flow. They can be racks of parallel plates, bars, or rolls.

**EVERGREENS:** Trees that do not shed their leaves or "needles" on a seasonal basis. These are generally softwoods.

**EXCELSIOR:** Strands of curly, slender pieces of wood made by shredding.

**EXCELSIOR TISSUE:** Strips of paper made by passing lightweight papers through a shredder. It is used to pack delicate materials for handling and transporting.

**EXCELSIOR WRAPPING:** A medium-weight sheet made from chemical pulp. It is used as a covering for straw and shredded wood or paper to make up protective pads used in the shipment of breakable materials.

**EXCESS AIR:** The weight of air introduced to power and recovery furnaces in excess of the theoretical requirement, calculated by subtracting the theoretical from the total air. It is usually expressed as "percent excess air," determined by multiplying the weight of excess air by 100, and dividing by the weight of the theoretical air.

**EXHAUST STEAM:** Steam that has completed work. Also referred to as *dead steam.*

**EXHAUSTER:** (1) A fan on a coal pulverizing outlet used on some boilers to air-convey fuel from the pulverizer to the burners. (2) A type of pump used to produce a vacuum for use on paper machine drainage and dewatering systems. It is made with a series of impellers rotating at a high rate of speed (typically 3,600 rpm).

**EXPANSION:** The increase in the dimensions of a paper or paperboard sheet due to absorption of atmospheric moisture. Also called *hygroexpansivity.*

**EXPANSION WRINKLES:** A roll defect caused by the top layers of paper absorbing moisture from the atmosphere, thereby expanding and causing wrinkles in the sheet.

**EXPERT SYSTEMS:** A branch of *artificial intelligence* that enhances decision-making capabilities of pulp and paper mill computer systems by turning data into decision support tools. These systems use a set of rules to emulate the reasoning process of a person. The rules include "rules of thumb," process behavior, design criteria, constraints, and diagnostic procedures.

**EXPLODED PULP:** Pulp made by subjecting wood chips to high steam pressure and suddenly releasing it in order to separate the fibers by causing the internal cell spaces to explode.

**EXPRESS PAPER:** A heavy, dense type of paper made for use in the manufacture of plastic sheets by impregnating and laminating it with the use of pressure.

**EXPRESS SEAM:** A dryer fabric seam made by using a polyester spiral with a flap folded back and each machine direction yarn woven back into the fabric, effectively producing a zero profile.

**EXTENDED DELIGNIFICATION:** The carrying out of the delignification process further than what is normally done to produce wood pulp in order to obtain stock intended for special purposes, or to improve its subsequent bleachability.

**EXTENDED IMPULSE PRESS:** *See* HIGH IMPULSE PRESS and IMPULSE PRESSING.

**EXTENDED MODIFIED CONTINUOUS COOKING (EMCC):** A method of extending the pulp delignification capacity of a Kamyr continuous digester by applying white liquor in the final high temperature wash stage, comprising, in effect, a third cooking zone in the digester. *See* MODIFIED CONTINUOUS COOKING (MCC).

**EXTENDED NIP PRESS (ENP):** A type of press nip design in which a bottom shoe, conforming to the shape of a top roll, presses a rubber belt riding on it and a pair of felts with the sheet of paper between them, against the rotating top roll. By properly loading the shoe, the nip can be extended from the usual 1 to 1-½ inches to about 8 to 10 inches, thus increasing the residence time of the sheet in the nip. The Extended Nip Press is a product of Beloit Corp.

**EXTENDED SURFACE RADIATORS:** Heating radiators that embody the principle of adding indirect surfaces to direct heating surfaces by means of fins or projections attached to the ordinary surface of coils or radiators. This is done to increase the amount of surface exposed to the air to be heated.

**EXTENDER:** A material added to adhesive coating formulations to increase the bulk in relation to the active or primary components so that the amount of primary binder required per unit area is reduced.

**EXTENSIBLE PAPER:** A high-bursting-strength kraft paper achieved by treating the paper sheet in such a way that it will stretch when put under tension, making it suitable for use in shopping bags and wrappers.

**EXTERNAL SIZING:** Any sizing method where the application of the material is done when the formed sheet is partially dried. Also referred to as TUB or SURFACE SIZING.

**EXTRA STRONG:** Paper or paperboard having greater strength than that normally specified for a specific grade.

**EXTRACTION:** (1) Steam extracted from intermediate stages of a turbine as a heat source for process or feedwater heating. (2) Cooking liquor removed from a continuous digester and sent to the chemical recovery cycle. (3) A stage during the cooking process of wood in which the solubilized lignin and hemicellulose are removed from the wood structure.

**EXTRACTION-CONDENSING TURBINE:** A type of steam turbine found in paper mill power houses that takes in the throttle flow through an intermediary port at the pressure and quantity required for the mill's needs. The remaining steam is passed through the turbine's low pressure section and condensed.

**EXTRACTION STAGE:** That stage in a multi-stage pulp bleaching operation ("E"-stage), usually following a chlorination stage, in which sodium hydroxide (NaOH) is used to remove water insoluble chlorinated lignin and other colored components not removed in an intermediate washing operation. Also referred to as the *caustic stage* or *alkaline extraction stage.*

**EXTRACTIVES:** Woody plant components that are not part of the cell wall structural elements and can be removed with neutral solvents such as ether, alcohol, and water. Also referred to as EXTRANEOUS WOOD COMPONENTS.

**EXTRACTOR:** A piece of process equipment used primarily to thicken recycled chemical fiber stock slurries prior to high consistency bleaching. The thickening is accomplished by removing water when the slurry is passed through a perforated cylindrical screen.

**EXTRANEOUS WOOD COMPONENTS:** Materials that are not part of the cell walls in the structure of wood, and are the basis for some by-products of the cooking process.

**EXTRAS:** The sheets in a ream that exceed the number established as standard for the grade of paper involved.

**EXTRUSION COATING:** Coating of paper or paperboard with heated, solvent-based plastic materials.

**EYES:** A paper defect caused by slime and appearing as spots, which sometimes have small center holes.

# F

**FABRIC:** A material fabricated of any fiber; usually a planar structure comprised of interlacing yarns, fibers, or filaments and produced by weaving, knitting, felting, or other means.

**FABRIC FORMING:** The process of forming a paper web on the wet end of a paper machine equipped with plastic fabric "wires."

**FABRIC PATCHING:** The repair of small holes on synthetic forming fabrics on the machine with a common household iron after drying with a hair dryer.

**FABRIC PRESS:** A type of paper machine wet press that uses a special multiple weave fabric belt sandwiched between the regular felt and the rubber-covered roll, increasing the capacity to receive and remove water from the nip between the rolls.

**FABRIC PURGE:** Water that is squeezed from the sheet of paper and felt into the voids of a fabric on a fabric-type wet press. It is removed from the fabric by the suction action of the fabric leaving the fabric roll of the press. This is accomplished by centrifugal action of the fabric wrapping a fabric carrying roll and by passing the fabric over a slotted suction box.

**FABRIC SHRINK SLEEVE PRESS:** A type of paper machine wet press in which a special multiple weave fabric is shrunk over the rubber-covered roll to increase the capacity to remove water from the nip between the rolls.

**FABRIC SHRINK SLEEVE PURGE:** Water squeezed from the sheet and felt into the voids of the fabric of a fabric shrink sleeve press and removed from the fabric voids by means of an air shower consisting of jets of compressed air.

**FACE STOCK:** A comparatively thin, flexible material to which adhesives are applied and used in making up pressure-sensitive items.

**FACE-WIRE:** A fine mesh wire screen stretched over a heavy, woven, mesh backing wire and used on the cylinder mold of a cylinder-type paper machine.

**FACER:** A paperboard machine having components specifically designed to make corrugating medium.

**FACIAL TISSUE:** A lightweight, soft, absorbent, sanitary paper made from bleached chemical and mechanical pulp. It is usually converted and packaged directly into napkins, toilet tissue, toweling, and other facial-type wiping paper.

**FACING PAPER:** A liner for paper or paperboard made of lightweight paper.

**FADE:** The slow loss or change in color of paper. It is usually caused by exposure to light over a period of time.

**FADING TEST:** A paper mill laboratory test made on a paper sheet to determine its degree of resistance to discoloration.

**FAIL SAFE OPERATION:** A characteristic of a piece of mill equipment, such as valves, switches, and machinery, which will fail to a safe condition when the actuating energy source is interrupted. The safe condition may be in a closed or shutdown position, fully open or operating condition, or remain in a fixed position.

**FAIR PAN:** A special device used to apply carbonate filler in dilute solution directly on the surface of the wet web of paper on the wire (forming section).

**FALLING FILM EVAPORATOR:** A type of long-tube vertical evaporator sometimes used to concentrate spent liquors in pulp mills by having them flow through steam-surrounded tubes to evaporate the water from the liquor. Sometimes called a *long-tube evaporator* or *vertical film-type evaporator.*

**FALSE BOTTOM:** A partition (often perforated) located just above the actual bottom of a tank or chamber. It is used by the paper industry in storage and process vessels when it is desired to drain liquids from a mixture with the possibility of returning the liquids for recirculation through the mixture.

**FALSE PRESSURE:** That pressure built up in the upper part of a batch digester during a cook which is higher than the equivalent saturated steam pressure and is caused by an accumulation of non-condensible gases that prevents temperature rise to proper cooking temperature if not removed or relieved. *See also* NONCONDENSIBLES (NCs).

**FAN:** An apparatus to create air movement and circulation by blowing air in or out of an area.

**FAN DUSTER:** A machine used in the preparation of rags to be used in papermaking in which the rags are beaten to produce dust. The dust is drawn away by means of the fans.

**FAN INLET AREA:** The inside area of the inlet collar on a fan.

**FAN OUT:** A dimensional change in paper during its use, such as processing through various types of printing machines.

**FAN OUTLET AREA:** The inside area of a fan outlet, which is usually the area of the flange on the outlet side.

**FAN PUMP:** A large centrifugal-type pump used to pump and mix pulp stock and large quantities of recirculated dilution water and stock to the headbox of a paper machine.

**FAN SHOWERS:** Types of water showers used in cooling pulpstones on wood pulp grinders.

**FANCY END:** A medium-weight paper with good tearing and folding properties made from bleached chemical pulp. It is used to bind books to their covers.

**FANCY FINISH:** Types of paper sheet finishes that are usually embossed, plated, or decorated. It is most often used with book covers and box liners.

**FANNING:** A manual method of visually inspecting paper on skids by spreading the edges and corners in a fan-like fashion to identify flawed or damaged sheets.

**FARAD:** A unit of electrical capacity. The capacity of a condenser (capacitor) charged to a potential of 1 volt by 1 coulomb of electricity.

**FAST COLOR:** A paper color that undergoes a minimum of fading or change when exposed to chemicals, light, and other hostile-type environments over an extended period of time. Also called *resistant color.*

**FAST WHITE:** *See* BLANC FIXE.

**FASTNESS:** The resistance of paper or dye material to fading or change in color when exposed to hostile-type environments such as light, alkalies, acids, etc., over an extended period of time.

**FATIGUE:** The tendency for paper and paperboard to fail under repetitive applications of strains or loads.

**FATTY ACID:** A by-product obtained from the tall oil removed from black liquor during its evaporation in the sulfate pulping chemical recovery cycle. It is used as a base for drying oils and soap manufacture.

**FBA:** *See* FIBRE BOX ASSOCIATION.

**FD:** *See* FORCED DRAFT.

**FEATHER EDGE:** The border of a sheet of paper and paperboard that is irregular and thin due to the tapering of the thickness from the body towards the edges at the deckles. Also called feather-edged deckle.

**FEATHER-EDGED DECKLE:** *See* FEATHER EDGE.

**FEATHERWEIGHT:** A descriptive term for paper and paperboard that is of very light weight, or of very small weight-to-bulk ratio.

**FECULOSE:** A translucent paper-sizing material made by treating starch with glacial acetic acid.

**FEED RING:** The large ring that surrounds the grinding stone in a ring-type grinder and forces (feeds) the wood against the revolving stone.

**FEEDBACK CONTROL:** A process control scheme in which a measured variable is compared with a desired value to produce a signal used to perform a corrective action that will reduce the magnitude of the difference between them.

**FEEDER:** A device used to supply materials to a process such as pulp or chips to the digester and subsequent bleaching stages.

**FEEDFORWARD CONTROL:** A process control scheme in which the information concerning one or more conditions that can disturb the controlled variable is used to perform corrective actions to minimize the deviations of the controlled variable from a desired value.

**FEEDSTOCK:** Pulp stock slurries entering a unit process during pulp and papermaking operations.

**FEEDWATER:** High-pressure water supply pumped to a boiler drum of a power or recovery furnace.

**FEEDWATER HEATER:** A heat exchanger used for heating the boiler feedwater by using the extraction steam from turbines.

**FEEL:** The evaluation of paper surface finish by the use of the sense of touch.

**FELLER-BUNCHER:** A mechanical, self-propelled pulpwood logging machine that shears off trees at the base and piles them up for removal from the woodland areas.

**FELT:** A woven belt of wool, cotton, and/or synthetic fibers used to transport the sheet of paper between the rolls of the press section (*wet felt*) and against the dryer drum (*dryer felt*) in the dry-end section of a paper machine. Sometimes referred to as *papermaker's felt*.

**FELT AIR SUPPLY UNIT:** A part of the paper mill air supply system used on paper machines to help dry the top and bottom felts and provide a source of warm air around the sides of the dryers to pick up moisture driven from the pockets, thereby increasing felt life.

**FELT CLEANING COMPOUNDS:** Nonionic chemical detergents and chemicals used for pH control that are continuously added to the shower water of the felt cleaning system on a paper machine.

**FELT CLEANING SHOWERS:** A set of showers dispensing detergent, acid, and/or alkaline agents. These showers are used on paper machine wet presses to clean and maintain felt openness. This procedure is followed by water rinse showers.

**FELT CONDITIONER:** A device consisting of one or more conditioning boxes attached to the frame of the paper machine. When passed over the conditioners, the felts are kept clean and open to assure uniform and maximum water removal at the presses.

**FELT DRYER:** A cylindrical, steam-heated dryer in the paper machine dry end that runs in direct contact with the felt and dries any moisture it has picked up.

**FELT FINISH:** Types of paper finishes produced by felts such as: specially woven felts on the wet press; dry, specially marked felts holding paper against the dryer in a yankee-type machine; and by mechanically pressing the felt against the sheet after it is made in a platen.

**FELT GUIDE:** A mechanism that senses the position of a felt on a paper machine and automatically adjusts an offset felt guiding roll to compensate for any weaving motion of the felt, thus preventing the felt from running off the machine.

**FELT GUIDE ROLL:** The adjustable offset roll in a guiding system that compensates for felt weave motion and maintains the felt on the machine.

**FELT HAIRS:** Paper defects caused by stringy pieces of the felt fabric being removed and becoming embedded in the surface of the sheet.

**FELT LIFE:** The elapsed time from the initial installation of a felt on a paper machine to the time it has to be replaced due to wear or malfunction.

**FELT MARK:** Paper defect caused by the worn or dirty surface condition of the wet felt in the press section. It appears as spots or thin areas on the sheet.

**FELT MARKING:** The process of weaving special patterns into a felt so that it will impart a specific design mark on the sheet of paper.

**FELT PAPER:** A heavyweight paper made from rags and used newsprint pulp. It is used in wall and floor construction as a barrier to sound and drafts, and is impregnated with asphalt-base material for use in roofing, shingles, and siding.

**FELT ROLL DOCTOR:** A doctor blade often provided on the first outside felt roll after the paper machine wet press nip. It is used to lift off any part of the sheet that may follow the felt when passing the sheet through the press or at the time of a break.

**FELT ROLLS:** Metallic rolls located in each space between paper machine dryers and used to turn the felt and/or carry it back over on its return run.

**FELT ROOFING PAPER:** A heavyweight paper used as a base sheet. It is impregnated with asphalt material and surfaced with crushed minerals for use in the construction industry for roofing.

**FELT RUN:** The arrangement and direction of the felts on the paper machine.

**FELT SIDE:** The top side of a sheet of paper as it is being formed on the wire (forming fabric) of the wet end of a paper machine, which later comes in contact with the felts during subsequent drying phases. Also known as *top side*.

**FELT STRETCHER:** A device on a paper machine used for taking up the slack in the felt and applying the proper tension.

**FELT STRIPS:** Non-metallic paper machine felt supports usually mounted on a stainless steel base.

**FELT WHIPPER:** Variable-speed, motor-driven beating device located on the outside of the wet felt on paper machines. It is used to forcibly loosen any dirt picked up by the felt, which is then washed off by a strong water shower.

**FELTING:** The process of fiber interweaving during the formation of a paper web.

**FEMALE THREAD:** The internal thread in pipe fittings, valves, machine parts, etc., for making screwed connections.

**FESTOON DRYING:** A paper-drying process in which the paper is dried in a temperature and humidity controlled chamber as the continuous sheet is carried through it while draped over a pole-like carrier, forming short loops.

**FIBER:** An elongated, tapering, thick-walled cellular unit that is the structural component of woody plants. Fibers are separated from each other during the pulping operation and reassembled into the form of a sheet during the papermaking process in the paper mill. Frequently spelled "fibre."

**FIBER ANALYSIS:** Determining the types of fibers in a paper specimen by the use of microscopic examination.

**FIBER AXIS RATIO:** Ratio of fiber width to fiber thickness.

**FIBER BINDING:** The interfiber cohesion between adjacent fibers in a sheet of paper arising from the hydrogen bond developed between hydroxyl groups of the carbohydrates.

**FIBER COARSENESS:** Weight per unit length of fiber.

**FIBER COMPOSITION:** The fiber makeup of a stock slurry or a sheet of paper, expressed in percentages of various fibers. The slurry or sheet is analyzed for its composition by various methods such as staining and microscopic analysis.

**FIBER CONTENT:** A listing of cellulose fibers used in papermaking that come from different sources and are usually identified by microscopic examination or by their reaction to staining agents.

**FIBER CUT:** A paper defect occurring as a small, straight, smooth, randomly located cut in the paper sheet caused by a larger than normal, less compatible fiber or shive passing through high-pressure roll nips.

**FIBER DEBRIS:** Pieces of the fiber wall that may have been separated from the body of a fiber.

**FIBER FLEXIBILITY:** The ability of a fiber to bend or deform.

**FIBER KNOTS:** Bunches of fibers that have not been properly disintegrated and are visible to transparent light in the paper.

**FIBER LENGTH:** The average length of fibers in the mass of pulp or stock furnish.

**FIBER NETWORK:** The orientation of fibers in suspensions relative to one another. This action allows them to exhibit such properties as improved tensile strength, shear strength, elasticity, and extensibility.

**FIBER PICK:** A disturbance in the paper sheet surface that manifests itself in the lifting and complete removal of discrete fibers or fiber bundles.

**FIBER PILING:** *See* FUZZING.

**FIBER PRODUCTION:** The weight of fiber in the product manufactured. It is calculated by deducting the weight of the nonfibrous materials in the product from the total weight of the product produced.

**FIBER RISING:** A surface-change phenomenon in which fibers rise above the paper surface.

**FIBER ROLL:** *See* PAPER ROLL.

**FIBER SATURATION POINT:** The point where all water is evaporated from the cell cavities of wood fiber, but the cell walls are still fully saturated with water.

**FIBER TEAR:** The damage and rupture of paper fibers when two adhered plies are separated.

**FIBERBOARD:** (1) Types of paperboards including container boards and vulcanized fiberboards. They are usually made from a combination of chemical pulp and recycled stock. (2) Defibrated wood formed under heat and pressure and without the use of adhesives.

**FIBERGLASS REINFORCED PLASTIC (FRP):** Various plastic materials in the form of sheeting, piping, and molded products whose strength and other performance characteristics have been reinforced by the inclusion of *fiberglass* strands, bars, mesh, and/or fabrics.

**FIBERGLASS:** Glass fibers used in the preparation of special papers that will not rot, burn, absorb moisture, and are generally inert or insensitive to most outside influences.

**FIBERIZE:** To mechanically reduce chips, dry pulp, dry broke, and used paper into their component fiber elements by violent agitation with impellers or in refiners and pulpers.

**FIBERIZED FORMING:** A method of dry forming fiber batts by shredding roll pulp and air transporting the fiber to a moving wire (fabric). There, suction under the wire deposits the fiber onto the web. It is used to make the absorbent layers for disposable diapers and other personal products.

**FIBRAGE:** A continuous layer of fibers of uniform thickness such as those caught on the leading edge of bars on a beater or refiner as they move through a stock slurry.

**FIBRE BOX ASSOCIATION (FBA):** Headquarters: Rolling Meadows, IL., (708) 364-9600

**FIBRETS:** A highly fibrillated form of cellulose acetate used in the production of wet laid wovens, specialty papers, and coatings.

**FIBRIL:** Fine, thread-like components of the cellulose fiber structure that are loosened and may become partially separated from the wall during mechanical refining. Also referred to as *fibrillae*.

**FIBRILLAE:** *See* FIBRIL.

**FIBRILLATION:** The external disruption of lateral bonds between surface layers of a fiber that results in partial detachment of fibers or small pieces of the outer layers of the fiber and the internal or lateral bonds between adjacent layers within a fiber. Fibrillation occurs during the mechanical refining of pulp slurries. Also referred to as *bruising, brushing, brushing out*.

**FIFTH HAND:** A member of the paper machine operating crew whose job is to assist the third hand in operating the winder and perform duties other than the ones performed by the fourth hand, but be able to replace him or her if necessary.

**FILL:** (1) The widest trimmed sheet of paper or paperboard that is possible to be manufactured on a specific paper machine. (2) A common pulp mill term referring to one complete charge of chips and cooking liquors to a batch-type digester.

**FILL FACTOR:** A term applying to chip feeding devices such as used on continuous digesters. It is related to the area or volume a fixed amount of chips will occupy in the displacement chambers moving the chips.

**FILLED BOARD:** Paperboard made on a multi-cylinder type of machine with the outside liners and body of the sheet made of different stocks.

**FILLED BRISTOL:** A laminated, heavyweight printing paper with outer plies made of different pulp than the center. It is produced on a cylinder type of machine.

**FILLED PAPER:** Any type of paper in which mineral pigments are used as a filler or loading material.

**FILLER:** (1) A substance added to pulp stock to fill the spaces between fibers and enhance the printing properties of the paper made from it. (2) The inner layers of multi-ply paperboards.

**FILLERBOARD:** The center plies or portion of multi-ply combination boards, wallboards, and container boards that may have different characteristics than the outer layers.

**FILLER CLAY:** A white, aluminum silicate, natural mineral added to pulp stock to enhance opacity, brightness, and printing surface smoothness of the paper made from it.

**FILLINGLESS FELTS:** *See* KNUCKLE FREE FELTS.

**FILM COATED:** Paper with a pigmented coating that is lighter than normal. It provides a more uniform and smoother surface.

**FILM PRESS:** A *premetered size press* that allows a high solids content to be applied to the sheet along with sizing materials. It utilizes a wire-wound coating rod or various types of blades to premeter pigmented size onto the size press roll, for transfer to the paper. The traditional size press pond and its associated "splashing" problems at high speeds is thus eliminated. These units include Voith's Speedsizer, Jagengerg's Filmpress, Beloit's Metering Blade Sizepress, Valmet's Symsizer, and BTG's Twin HSM Coater.

**FILTER:** Any kind of device that can be used to remove or partially remove suspended solids from a liquid or gas.

**FILTER AIDS:** A material, such as diatomaceous silica, added to dilute a pulp suspension that is to be filtered in order to prevent immediate blocking of the filtering mediums by the fines and colloidal constituents, as is typically done to white water in a paper mill.

**FILTER CAKE:** *See* CAKE.

**FILTER PAPER:** A chemically pure, good-strength, unsized paper. It has high porosity and is made from chemical and/or cotton fiber pulps. It is used for separating suspended solid materials in liquids and gases in laboratory analysis procedures and for domestic and industrial purposes.

**FILTERING:** The removal of suspended particles in liquids and gases by passing them through a porous material such as paper, cloth, sand, diatomaceous earth, etc.

**FILTRATE:** Pulp washer and filter effluents.

**FILTRATION RATE:** The rate at which liquid is fed through a filter, expressed in gallons per minute per square foot of filter media surface.

**FINE PAPERS:** High-quality writing, printing, and cover-type papers having excellent pen and ink writing surface characteristics.

**FINE SCREENS:** Types of screening equipment used throughout the papermaking process having a fine mesh sieve and used to remove fiber knots, shives, and particles (debris) from low consistency, diluted stock slurries. The particles removed are very much smaller than those removed by regular coarse screens.

**FINENESS:** The particle size and distribution of materials such as clay which impacts on their suitability for making up paper-coating pigments.

**FINES:** Small particles of fiber that are shorter than normal wood pulp fibers. Sometimes called *wood flour.*

**FINGER BARS:** The mounting structure for the steel plate with protrusions (*fingers*) used on pocket-type pulpwood grinders. They are used to catch and hold shives and slivers in the pocket and allow pulp and water to flow through.

**FINGER TACK:** *See* AUTO ADHESION.

**FINGERS:** The protrusions on a horizontal steel plate located in the pocket of a pocket-type pulpwood grinder. They are situated on the downrunning side and close to the stone so that the fingers will catch shives and large slivers and allow the pulp and water to flow between them.

**FINISH:** (1) Surface characteristics of a sheet of paper, such as smoothness, appearance, and gloss, as determined visually. (2) Sometimes used to indicate the density of a sheet of paperboard. *See* APPARENT DENSITY, DENSITY, and YIELD.

**FINISH POINTS:** The thickness of 50 pounds of standard-size sheets of paperboard (usually 25 by 40 inches) expressed in points, mils, or thousandths of an inch.

**FINISH VARIATION:** Differences in surface characteristics of paper, such as evenness, glossiness, texture, etc., occurring particularly during manufacture of a specific run of paper in a paper mill.

**FINISHED PRODUCTION:** The weight, in pounds or tons, of material manufactured over a period of time, at the moisture content obtained at the time of manufacture, and suitable for sale.

**FINISHING:** The processing of paper after the papermaking operations are completed. Included are supercalendering, slitting, rewinding, trimming, sorting, counting, and packaging prior to shipment from the paper mill.

**FINISHING BROKE:** Trimmings, cuttings, and other wastepaper from finishing operations, most of which is recycled and reprocessed again into paper.

**FINISHING PAPER:** A high-strength, pliable paper made from chemical wood pulp, coated with an abrasive material, and used for hand sanding.

**FINISHING ROOM:** The department of a paper mill where the paper is cut, trimmed, sorted, counted, packed, and prepared for shipment.

**FINISHING WASTE:** Trimmings, cuttings, and other scraps of paper produced during the finishing operations in a paper mill which may or may not be recycled and reprocessed again into paper or paperboard.

**FIR:** A coniferous tree used as one of the sources of wood for the manufacture of sulfite pulp.

**FIRE LANE:** Clear space around log piles, chip piles, mill buildings, and equipment suitable for fire fighting operations by personnel and motorized fire apparatus.

**FIRE-RESISTANT PAPER:** Types of paper made from a variety of fibers and chemically treated so that they will not support combustion. Such paper is self-extinguishing when a direct flame is applied and removed.

**FIRE RETARDANT:** A compound that can be added during the manufacture of tissue, kraft papers, and lining papers for building applications, board, and packaging products to overcome both flame and residual glow.

**FIRE TUBES:** The piping section of a boiler surrounded by water and through which the flue gases pass.

**FIREPROOF PAPER:** Special types of papers made from asbestos, ceramics, or glass fibers, and used for fire-protective purposes in heating appliances and other applications to provide protection from fire hazards.

**FIRING HOOD:** A refractory-lined steel plate enclosure around the discharge or firing end of a lime kiln.

**FIRMWARE:** Types of pulp and paper computer process programs that are placed into read-only type memories.

**FIRST DRYER SECTION:** The dryer section of a multi-section dry end of a paper machine into which the wet sheet from the press section is fed.

**FIRST EFFECT:** The first vessel in a multi-effect evaporator train in which live steam is usually introduced into the system and the concentrated black liquor leaves the system.

**FIRST PRESS:** The first set of press rolls through which the wet sheet of paper passes after leaving the wet end section of the paper machine.

**FIRST PRESS ROLL:** The first roll of the press section over which the wet sheet of paper passes after leaving the wet end section.

**FIRST STAGE:** A pulp mill reference to the chlorination stage (C-stage) of a multi-stage pulp bleaching operation, which traditionally has been the first step. Recent technological developments have introduced other chemicals for use in the first step, such as ozone.

**FISH EYES:** Paper sheet defects appearing as glazed, translucent spots, caused by agglomerated fiber bundles, slime, or improperly prepared chemical additive in the stock furnish. Also called *bird eyes*.

**FISH PAPER:** Paperboard specially prepared and treated with chemicals for use as electrical insulation in the manufacture of motors and dynamos.

**FISH WRAPPING:** (1) An odorless, sanitary, waxed-type paper made with wet tensile strength so it can be used in markets for the packaging of fish. (2) Vegetable parchment paper used for fish packaging.

**FITTINGS:** Fixtures and various accessories used to adapt and adjust parts to piping, machines, and other equipment in and around a pulp and paper mill.

**FIXING:** A term used in the coloring of paper that refers to the combining of a dye, which has little affinity for the fiber, with a mordant, which adheres to the fiber, in order to produce the desired color.

**FIXTURE:** (1) In a mechanical pulp mill, the device used to dress or sharpen pulpstones. Also called *lathe*. (2) A device used to adapt machine and piping components to each other.

**FLAG:** A paper or paperboard marker placed at and extending beyond the edge of a roll of paper to identify the location of a defect or splice in the sheet. Also known as a *splice tag.*

**FLAGGING:** The process of inserting a tab of paper or paperboard in a roll so that it extends beyond the edge to identify locations of faults and splices in the sheet.

**FLAKEBOARD:** A wood board formed from knife-cut, flat, flake-like chips and adhesives. Also called *particle board* and *waferboard.*

**FLAKES:** Small pieces of coating material that have broken away from coated paper.

**FLAKING:** The process of small particles of paper-coating material breaking away from the surface of a paper sheet.

**FLAKT DRYER:** (1) A wet pulp sheet dryer designed to remove moisture by blowing hot air over and under the suspended sheet. Manufactured by ABB Flakt. (2) Certain types of similar hot-air dryers used on pulverized fuels such as hog fuel, coal, pellets, and peat.

**FLAME-RESISTANT PAPER:** Types of chemically treated paper that have the property of resisting combustion and will not support a flame. This is accomplished by released gases that combine with the surrounding oxygen, making it unavailable for support of combustion.

**FLAME SAFETY SHUTDOWN SYSTEM:** A system that detects when the burner on a recovery furnace has lost ignition and closes the safety shutdown valves in the auxiliary fuel supply.

**FLAMEPROOF PAPER:** A chemically treated paper sheet that will not support combustion but will char when a flame is applied and will extinguish itself when the flame is removed.

**FLANGE:** (1) Large, metallic, circular plate-like fittings that are installed on a pulp grinder shaft, on either side of the pulpstone, and hold it in position. (2) A type of fitting on equipment, such as ends of pipelines, to provide a means of joining or connecting them.

**FLASH DRYING:** A technique in drying groundwood pulp in which thickened, disintegrated pulp drops into the suction side of a fan. The fan carries combustion gases from an oil burner into a drying tower. The pulp is later separated from the combustion gases and the process repeated until the proper dryness is achieved.

**FLASH MIXING:** The rapid and uniform distribution of a coagulant through the mass of raw water in the mill water treatment process.

**FLASH POINT:** The temperature at which a fluid first gives off sufficient flammable vapor to ignite when approached with a small flame or spark.

**FLASH STEAM:** The steam produced when pressurized hot liquid enters into a lower pressure chamber. Examples are liquids extracted from digesters and condensate removed from paper machine dryers.

**FLASH TANKS:** Vessels into which pressurized hot liquid is introduced at a lower pressure so that steam will be produced and recovered for reuse in the process.

**FLAT-BOTTOM BURR:** A type of pulpwood grindstone sharpening tool on which the roots of the surface teeth are separated to form a flat area.

**FLAT BOX:** Suction boxes equipped with drilled or slotted covers, over which the paper machine wire (forming fabric) slides, so that water and air are pulled from the wet paper sheet being formed on the wire.

**FLAT BUNDLE:** A standard package of single paper sheets made up so that it does not contain any folded ones that may cause creases.

**FLAT CRUSH RESISTANCE:** The ability of corrugated paperboard flutes to withstand a perpendicularly applied load to the surface. It is measured under specified testing conditions.

**FLAT DRYERS:** Conveyor-type dryers used to dry coated papers. Also called *ladder dryers* or *tunnel dryers.*

**FLAT FINISHED:** The surfaces of a sheet that have a minimum of gloss.

**FLAT NEWS:** Newsprint sheets of paper after they have been cut from a roll in the printing of newspapers.

**FLAT PAPER:** Packaged paper of various grades that leaves the finishing operation and is shipped from the paper mill in a flat position having no folds to cause creases.

**FLAT REAM:** Standard ream of paper that is packaged and shipped, and contains no folded sheets.

**FLAT SCREEN:** A general type of screen used for fine screening. It consists of a long rectangular box with three blank sides and an outlet on the fourth side. The outlet is fitted with flat, slotted metal plates over which the pulp suspension is allowed to flow. Screening action is induced by vibrating diaphragms below the plates. The diaphragms produce pulsations in the stock suspension causing acceptable fibers to pass through the slots with showers removing the rejected material at the end of the screen box for subsequent rescreening and final rejection.

**FLAT WRITINGS:** Types of writing paper with a smooth finish and a minimum of glare.

**FLATNESS:** That quality of a sheet of paper that makes it resistant to curling and distortion.

**FLATTENED ROLLS:** Rolls of paper which have been crushed due to accidental dropping or squeezed by sideward pressure being applied to them.

**FLAX:** A fiber that is the main constituent of the flax plant straw used in making cigarette and other similar low ash tissue. It is also used to make linen cloth, which is in turn a source of rags used in making fine writing papers.

**FLAX PAPERBOARD:** A type of paperboard made with a high content of prep from flax straw fiber and used extensively for insulation purposes.

**FLEX RESISTANCE:** The ability of paperboard to withstand a center load with minimum bending while suspended at the ends.

**FLEXIBILITY:** The ability of stiff grades of paper and boards to bend without damage.

**FLEXIBLE BLADE COATER:** *See* BLADE COATER.

**FLEXIBLE DIAPHRAGM CHIP SCREEN:** A type of wood chip screen used to screen out fines in cold climates by feeding chips onto one end of a downward angled flexing screen plate, and passing through slots or discharging off the end.

**FLEXIBLE FIBER:** A plasticizer softened, chemically gelled, cotton cellulose paper used to make up electrically resistant items such as spacers, gaskets, and packings.

**FLEXING RESISTANCE:** The ability of paper and paperboard to resist an applied bending force.

**FLEXOGRAPHIC PRINTING:** *See* ANILINE PRINTING.

**FLEXURAL RIGIDITY:** The ability of paper and paperboard to resist an applied bending force.

**FLIGHT AND DRAG CONVEYOR:** A type of materials-carrying device consisting of one or more parallel chains on which plates or flights are fitted. The plates drag or scrape material such as coal ashes, chips, flaked pulp, etc., along to the point of delivery as the chain is driven over powered sprockets on either end.

**FLIMSY PAPER:** Very thin types of paper such as tissues and onion skins.

**FLINGERS:** Metal attachments located radially and projecting from the outside of the pulpstone positioning flanges. They are used to propel the stock forward and break up accumulations of shives and stock between flanges and the side frame in the cylinder pits.

**FLINT BACKING PAPER:** A well-sized, high-strength paper on which abrasive materials such as sand, aluminum oxide, silicon carbide, and garnet are glued on one surface. It is used for smoothing wood or metals.

**FLINT GLAZE:** A type of surface polish applied to coated paper by the use of smooth stone burnishers. By their rubbing action, these units impart a hard, brilliant finish to the sheet.

**FLINT PAPER:** A heavy, strong, hard-sized rope or kraft paper with abrasive materials such as flint, garnet, aluminum oxide, or silicon carbide glued to one surface. It is used for smoothing wood and metal.

**FLITTING:** A condition in which a paper web flaps between top and bottom rolls as it leaves the nip. It is most likely to happen between applicator rolls during some method of coating.

**FLOATER:** A type of air dryer used for drying coated paper.

**FLOATING DRYER:** The dryer roll immediately following the doctor, used in the creping of tissue paper to impart more crepe to the sheet by running it at a slower speed than the subsequent dryers.

**FLOC:** (1) Small, gelatinous masses formed in a liquid by the addition of coagulants through a biochemical process, or by agglomerations, commonly occurring in water and waste treatment processes. (2) Fiber aggregates in pulp slurries.

**FLOCK:** (1) Very short or pulverized wool, cotton, or rayon fiber used to create patterns on cloth or paper used for wall and other types of coverings. (2) Agglomerated bundles of fibers that become suspended in stock slurries.

**FLOCK PAPER:** Types of wallpaper and other paper coverings with surface designs produced by surface sizing followed by sprinkling with pulverized wool, cotton, or rayon fibers. Collectively referred to as *flock*.

**FLOCKING TISSUE PAPER:** A lightweight paper made with a high finish to minimize fuzzing and used as a protective material in clothing manufacturing.

**FLONG PAPER:** A pulp-like, coated-type paperboard.

**FLOODED NIP COATER:** A type of paper coater where the sheet wraps a backup roll and excess coating is applied by a rotating roll forming a nip. The nip becomes flooded with coating material.

**FLORIST'S TISSUE:** A colored, lightweight, waxed tissue used to wrap plants and flowers.

**FLOTATION:** (1) A method of raising suspended matter to the surface of a liquid by means of various procedures, including aeration, vacuum, the evolution of gas, chemicals, electrolysis, heat, or bacterial decomposition. (2) The process of fiber entanglement and aggregate formation with a tendency to rise or float in low consistency pulp stock slurries.

**FLOTATION DEINKING:** The process of removing ink from used paper by aerating chemically treated pulped paper with vigorous agitation to cause the separated ink particles to adhere to the air bubbles that are formed and rise to the surface as froth.

**FLOTATION TEST:** Floating a small sample of paper on the surface of a liquid, such as water, writing ink, oil, etc., and measuring the time for the liquid to penetrate through. It is a test of the resistance of paper to penetration by that liquid.

**FLOUR:** Small particles of fiber that are shorter than normal wood pulp fiber. *See* FINES.

**FLOUR-SACK PAPER:** A general term used to refer to any paper made for use in the manufacture of bags to package flour. It is made from a mixture of hemp and chemical pulp and has good strength properties.

**FLOW:** The progressive movement of materials in a process system.

**FLOW EVENERS:** Paper machine headbox assemblies, such as explosion-proof chambers, tapered ducts, rotating perforated rolls, vanes, baffles, etc., designed to even out the streaky, turbulent flow and large cross eddies coming from the spreaders of the approach flow system.

**FLOW-ON COATING:** A method of coating paper by applying the coating preparation (suspension of pigment and adhesive) on the wet sheet during its formation on the wire (forming fabric) of the wet end of the paper machine.

**FLOW POINT:** The temperature at which a gelled material like glue adhesive becomes soft enough to deform and start to flow. An important factor in the adhesive coating of paper.

**FLOW RATE:** The volume or mass which passes a fixed point in a process system per unit of time.

**FLOW SHEET:** A diagram, chart, or outline showing process material flow direction, successive operations, and equipment used, through which the materials flow during a manufacturing process.

**FLOW SPREADERS:** Paper machine approach flow design and arrangement that spreads stock from the header across the width of the headbox with an even velocity distribution.

**FLOWABILITY:** The ability of water to flow out of the stock furnish.

**FLOWBOX:** A flow control device consisting of a box-like chamber that receives the fluids and evens out the flow by allowing the fluid to flow between and over a set of baffles, and returns the excess. Once, it was a very common method used to control the flow of pulp stock slurries to paper and pulp mill process equipment such as washers, filters, save-alls, screens, and both fourdrinier and cylinder types of paper machines.

**FLOWMETER:** A device that measures the rate of flow or quantity of a moving fluid in an open or closed conduit.

**FLUE:** Any passage designed to carry combustion gases and entrained particulates away from the source.

**FLUE DUST:** Solid particles, usually smaller than 100 microns, carried in products of combustion.

**FLUE GAS:** A mixture of gases resulting from combustion and emerging from a chimney or stack.

**FLUE GAS SCRUBBER:** Equipment associated with power and recovery furnaces that removes fly ash and other materials from the flue gas by the use of sprays, wet baffles, or other means that require water or other liquids as a primary separation mechanism. Also referred to as *flue gas washer.*

**FLUE GAS WASHER:** *See* FLUE GAS SCRUBBER.

**FLUFF:** (1) Small, dust-like particles that break away from paper and collect on paper machine equipment such as rolls, frames, etc. (2) A soft, bulky, fibrous material produced by disintegrating fluff pulp into individual fibers.

**FLUFF PULP:** A thick sheet or batt of wood pulp fibers manufactured in roll or bale form and suitable for dry disintegration into individual fibers. *See* FLUFFING PULP.

**FLUFFER:** A mechanism used in oxygen delignification to break up or disintegrate high-density pulp before dropping it into the oxygen reactor.

**FLUFFING PULP:** A light, bleached, absorbent mixture of chemical and/or mechanical pulp used as a filler for sanitary products such as disposable diapers, bed padding, napkins, bandages, and other personal hygiene products. Also called *fluff pulp.*

**FLUFF-OUT:** A corrugated board defect caused by maladjusted equipment used in its manufacture.

**FLUID HEAD:** The pressure measured in relation to and expressed in terms of the height of a column of fluid of known density. It is computed by dividing the pressure by the density.

**FLUID POWER:** Energy transmitted and controlled through the use of a pressurized fluid.

**FLUIDITY:** That property of a liquid that manifests itself in its ability to conform to flow, which is opposite to viscosity.

**FLUIDIZED BED CALCINERS:** Lime recovery units that are alternates to rotary lime kilns, used for the calcination of lime mud to lime. They operate on the principle of fluidization.

**FLUIDIZED BED PROCESSOR:** A device in which solid particles are suspended in a gaseous medium and exhibit properties of a fluid.

**FLUIDIZED BED TECHNIQUE:** A combustion process in which heat is transferred from finely divided particles to combustible materials in a combustion chamber. The materials are supported and fluidized by a column of moving air.

**FLUIDIZER:** Material that when added to liquids increases its ability to conform to flow. (2) Process equipment used to maintain a mass of finely divided solids in a turbulent, dense state during treatment by dispersing it in an upwardly moving gas stream. This imparts to the mass a turbulence resembling that of a boiling liquid.

**FLUME:** A channel, either natural or man-made, that carries both liquids and materials suspended in the liquid.

**FLUORESCENCE:** That property of optical brighteners that increases the visual whiteness of the paper sheet under normal light when added to the papermaking process.

**FLUORESCENT BRIGHTENERS:** Types of dyes used to increase the apparent brightness of white paper by absorbing some of the invisible ultraviolet radiation and re-emitting it in the violet and blue portion of the spectrum. *See also* OPTICAL BRIGHTENERS.

**FLUORESCENT PAPER:** White paper made with synthetic dyes that produce a brighter appearance by increasing their ability to reflect light, or paper that is surface-coated with colored, light-emitting dyestuff materials (that reflect white light as a color).

**FLUORESCENT WHITE DYES:** Colorless dyes and pigments used to enhance the brightness of white paper and paperboard by absorbing ultraviolet light energy and re-emitting it at a higher wavelength of visible light. Same as *optical whiteners*.

**FLUORESCENT WHITENING AGENT:** A colorless paper dye that has the ability to convert ultraviolet light into additional visible blue light, thereby adding brightness or whiteness to the sheet.

**FLUTE:** One of the triangular-shaped configurations formed by an undulation of the inner corrugated liner in corrugated-type paperboard similar to and sometimes known as *corrugation. See* A-FLUTE, B-FLUTE, C-FLUTE, and E-FLUTE for undulations of differing dimensions and frequency.

**FLUTED BOARD:** A paperboard that has been passed through a corrugating machine. In the machine, the sheet is re-formed into a series of triangular-shaped undulations (*see* FLUTE) and is glued to one or two sheets of flat paperboard to produce either single- or double-faced corrugated paperboard.

**FLUTING:** An expression used to refer to corrugating medium. *See* CORRUGATING MEDIUM.

**FLUTTER:** The bouncing or flopping motion of a sheet of paper, usually coming off winding rolls, due to defects in the roll, such as flat spots, bumped edges, and baggy edges. Also called *slap.*

**FLUXING:** Cleansing by dissolving or melting a substance through chemical action.

**FLY ASH:** Entrained, partially burned dust, soot, and other materials and chemicals carried over with the flue gases emitted from the smoke stacks of power and recovery furnaces.

**FLY BARS:** Knives located on the surface of beater rolls.

**FLY ROLL:** A non-driven roll used on cylinder-type paper machines strictly to support the felt.

**FLYING BRIDGE:** A catwalk located between the wet ends of adjacent paper machines in a multi-machine paper mill. On the catwalk, the operator can observe the entire length of the machine. Normally it is movable so it can be swung out of the way, if necessary, when the wire (forming fabric) is changed.

**FLYING DUTCHMAN:** A large, high-speed, single-drum dryer section normally used on a tissue paper machine. Also called a *yankee dryer.*

**FLYING PASTER:** A fast-operating mechanism that is used to automatically glue or tape the ends of paper sheets together during production to make up large rolls. The large rolls can then be continuously run when used for converting into user products or making up printed material.

**FLYLEAF:** A medium-weight paper, made from chemical pulp and used to bind a book's content to its cover. *See* END LEAF.

**FOAM:** A mass of air bubbles formed in or on the surface of liquids and pulp slurries due to entrainment during agitation or other mechanical disturbances. Also called *froth.*

**FOAM BREAKER:** Mechanical devices normally installed at the top of liquor, stock, or any other storage area where stable foam may form. Its purpose is to break up the foam.

**FOAM KILLER:** Solid, paste, and liquid chemicals added to stock systems to reduce the development of troublesome foam and entrained air. *See also* DEFOAMER.

**FOAM MARKS:** Blemishes occurring on the surface of the wet web on the formation wire (forming fabric) of a paper machine. They are caused by foam in the stock slurry being carried over from the headbox. It can also be carried over into the coating operation and cause some type of defect in the coated surface of the paper sheet. Also referred to as *froth marks* and *froth spots.*

**FOAM SPOTS:** Small crater shaped defects on coated paper surfaces caused by entrained air bubbles in coating colors applied to the sheet surface.

**FOG:** A mixture of a gas and suspended liquid droplets generated by condensation from the vapor to the liquid state, or by breaking up a liquid into a dispersed state as a result of splashing, foaming, frothing, or atomizing.

**FOIL BOARD:** An even-surfaced, uniform, special type of paperboard made for lining with various types of metallic foil material.

**FOIL PAPER:** A general reference to types of even-surfaced, uniform paper made specially to be lined with metallic foil material.

**FOIL POLISH:** A term indicating that the brightness of a sheet is due to light reflection.

**FOLD OVER:** A defect in a paper sheet caused during the slitting or winding operation by the loose edge of a tear, cut, or hole folding back on itself.

**FOLD STRENGTH:** The ability of paper and paperboard to withstand repeated bending and folding with a minimum of damage.

**FOLDED:** A descriptive reference to packaged or quantities of paper sheets that are stacked in a folded fashion and may also be interleaved.

**FOLDED NEWSPRINT:** Leftover, unread newspapers that are usually collected from newspaper stores and newsstands and returned to deinking plants for reprocessing.

**FOLDER STOCK:** A durable, surface-sized paperboard or stiff, heavyweight paper with good tear and folding characteristics. It is made from chemical pulp, recycled wastepaper stock, and hemp fibers, and converted primarily to various types of business file folders.

**FOLDING BOARD:** A multi-ply paperboard that can be folded along a score line without damaging the outer or top liner of the bend.

**FOLDING BOXBOARD:** A variety of paperboards that can be scored and folded with a minimum of damage. These boards are used in the manufacture of folded boxes and containers known as folding cartons.

**FOLDING BRISTOL:** A highly-finished, stiff, heavyweight printing paper made of long, flexible-fibered pulp stock so that it can be folded with minimum crease damage.

**FOLDING ENAMEL PAPER:** High-strength, coated paper with excellent folding properties, making it particularly applicable in bookmaking.

**FOLDING ENDURANCE:** Refers to the ability of paper or paperboard to withstand repeated bending and folding with a minimum of damage.

**FOLDING ENDURANCE TEST:** A paper mill laboratory test made on a specific strip of paper. The test determines the number of double folds that can be made under specified tension conditions before it will break, and is a measure of the folding properties of the sheet.

**FOLDING STOCK:** A type of paper with chemical pulp containing long fibers. It is especially coated with pliable-type coating mixtures that will not be damaged by bending and folding.

**FOLIO:** (1) A sheet of paper folded once in bookmaking. (2) Sheets of paper 17 by 22 inches in size.

**FOLLOWER PLATE FOOT:** The pulpwood grinder hydraulic cylinder plunger base that forces the wood against the pulpstone.

**FOOD WRAPPERS:** A general reference to clean types of waxed, coated, or lacquered finish papers that have characteristics making them suitable for packaging food products.

**FOODBOARD:** A water-resistant, coated, sanitary type of paperboard made from bleached chemical pulps and used for packaging food products.

**FOOTAGE:** The amount of paper usually wound up in a roll, expressed as total length in feet, and used as the basis for determining the selling price.

**FOOTCANDLE:** A unit of luminance, based on a 1-square foot-surface area on which a time rate of flow or radiant energy of 1 lumen is uniformly distributed.

**FORCED AIR DRYER:** A type of paper dryer that uses forced hot air impinging on the surface of the paper through specially designed nozzles in order to enhance moisture removal.

**FORCED CIRCULATION:** The recycling of digester cooking liquor by withdrawing it from one or more points in the digester and pumping it, sometimes through a heat exchanger for heating, back into the digester at one or more locations.

**FORCED DRAFT (FD):** Positive low pressure created by air furnished to a power or recovery boiler by a fan or blower for improved combustion.

**FORCED DRAFT (FD) FAN:** A fan that is used to force air through an air heater and into a furnace.

**FOREST PRODUCTS LABORATORY (FPL):** Headquarters: U.S. Dept. of Agriculture, Forest Service, Madison, WI., (608) 231-9200.

**FOREST RESIDUALS:** Cuttings, cullings, branches, and residual unmerchantable trees left in the woodlands after it has been harvested for saw timber and pulpwood. When these residuals are harvested for use as power boiler fuels, they become part of the total biomass supply.

**FOREST RESIDUAL CHIPS:** Wood chips made from forest residuals that are used for the manufacture of pulp. *See* FOREST RESIDUALS.

**FORK LIFT TRUCK:** A material moving vehicle used primarily in the warehouse area of a paper mill. It is generally gas or electric powered, and has a two-prong fork type of lifting apparatus used for moving shipping platforms called *skids* as well as other materials.

**FORMATION:** The physical distribution and orientation of fibers and other solid constituents in the structure of a sheet of paper that affects its appearance and other physical properties. Also referred to as *look through* because the formation can sometimes be observed by looking through the sheet.

**FORMATION AIDS:** Materials added to stock furnishes going to the paper machine that improve the attraction of fibers and assure more uniform orientation and distribution in the sheet.

**FORMAZIN TURBIDITY UNITS (FTU):** A measure of the turbidity of water.

**FORMER:** Types of machines used primarily in the manufacture of multi-ply paperboards. They are adaptions of, and incorporate design features of, fourdrinier- and cylinder-type paper machines in that a cylindrical forming roll is used to support the forming wire mesh and stock is admitted to the wire surface through nozzles and special-shaped inlets.

It is also a general term referring to a variety of specialized wet end units that apply stock and form sheets for many grades of paper and paperboard.

**FORMING:** The process of paper fibers aligning themselves to form a sheet as the water drains from a thin slurry through the wire (forming fabric) of the paper machine wet end. It is commonly associated with the type of wet end referred to as a *former.*

**FORMING BOARD:** A plastic-covered, sometimes perforated board located under the paper machine wire (forming fabric) between the breast roll and first table roll. It is used to eliminate the sag caused by impact of the stock jet from the slice and to improve flow onto the wire.

**FORMING BOXES:** Light, stainless steel boxes covered by narrow strips of laminated plastic and located under the wire (forming fabric) of a paper machine. Internal suction is applied to the boxes to draw water from the wet web being formed on the wire.

**FORMING FABRICS:** Continuous screen belts made of plastic fibers and used on paper machine wet ends where the paper sheet is formed from a pulp stock slurry. Also called *forming wires, plastic wires,* or simply *wire.* At one time metal "wires" were commonly used (and some are still in use).

**FORMING FOIL:** A baffle with the leading edge contacting the wire (forming fabric) on a paper machine and the top sloping away from the wire at an angle of several degrees to create a suction that enhances drainage of water from the wet web formed on the wire.

**FORMING ROLL:** A cellular shell roll with a vacuum compartment around which the wires (forming meshes) of a former type of paperboard machine run. The sheet is formed on this roll.

**FORMING SECTION:** That part at the front (wet) end of a paper machine that consists of all the dewatering components and is where sheet *formation* occurs.

**FORMING WIRES:** The continuous screen wire belt on any wet-type paper machine on which the water is drained from the stock slurry as the sheet is formed. *See* FORMING FABRIC.

**FORTIFIED SIZE:** A rosin acid that has been specially processed to impart water-resistant properties to paper and paperboard. It also is used in conjunction with conventional rosin size to further enhance the water-resistant characteristics of the sheet.

**FORTIFYING TOWER:** An absorber used in the sulfite pulping process in which the free sulfur dioxide ($SO_2$) in the acid is increased by the absorption of sulfur dioxide from the relief gas or from a sulfur or pyrites burner.

**FORTRAN:** A high-level computer language commonly used in many process control computer systems in the pulp and paper industry. It is derived from the original name *FORmula TRANslation.*

**FORWARDING:** An expression used in pulpwood logging referring to the movement of cut tree segments from the stump to an area from which they are transferred to the mill.

**FOSSIL FUELS:** Coal, oil, and natural gas, so called because they evolve from the remains of ancient plant and animal life.

**FOULING:** (1) The impedance of the flow of gas or heat that develops in gas passages or on heat-absorbing surfaces in a furnace. (2) The hindrance to any process operation due to the interference by undesirable foreign material accumulation.

**FOURDRINIER BOARD:** Paperboard made on a fourdrinier-type machine, in contrast to that made on a cylinder-type machine.

**FOURDRINIER MACHINE:** The papermaking machine invented by Louis Robert, but named after Henry and Sealy Fourdrinier who financed its development. It is used interchangeably to refer to only the wet end section onto which the pulp stock is fed and formed as well as to the entire paper machine, including the press section, the dryer section, and the calender section. *See also* FOURDRINIER WIRE.

**FOURDRINIER WIRE:** A continuously traveling, endless, woven, metallic, or plastic screen belt located in the wet end section of a fourdrinier-type paper machine. The pulp stock is fed onto the wire so that the water can be drained from it as the fibers become oriented to form a continuous web. Also known simply as a *wire*. *See also* FORMING FABRIC.

**FOURDRINIER YANKEE MACHINE:** A fourdrinier-type paper machine with a dry end section consisting of a large, high-speed drying cylinder. *See* YANKEE DRYER.

**FOURTH HAND:** A member of the paper machine operating crew whose job is to assist the third hand in operating the winder.

**FOXING:** The process of paper being damaged by mold or mildew due to unfavorable environmental ambient storage conditions. It appears as specks, spots, or stain marks.

**FPL:** *See* FOREST PRODUCTS LABORATORY.

**FPM:** Abbreviation for feet per minute.

**FRACTIONAL WASHING:** A method of washing pulp fibers after the cooking process by a succession of washes with different concentrations of liquors usually in a decreasing direction.

**FRACTIONATOR:** A device that separates various length fibers in a pulp slurry. It is commonly used to determine the size of the long fraction pulp in groundwood.

**FRACTURE TOUGHNESS:** The physical property of materials, including paper and paperboard, that resists crack stresses in the plane of a sheet and withstands its propagation.

**FRAME:** The supporting structure of a paper machine.

**FRAME BUNDLING:** Packaging a stack of paper for shipping purposes by placing wooden frame-like retainers on the top and bottom and banding the entire bundle in both directions.

**FREE CHLORINE:** Elemental chlorine in the pulp bleaching process that is in solution and not compounded with lignin elements in chlorinated pulp slurries.

**FREE DELIVERY:** The theoretical operating condition when static pressure and resistance in an air system equals the forced draft.

**FREE ROSIN SIZE:** Rosin that is intentionally not completely saponified when making up rosin size by boiling rosin with caustic soda or sodium carbonate. It therefore contains rosin acids.

**FREE SHEET:** A sheet of paper that contains no mechanical pulp fibers or is made of pulp that has been subjected to the very minimum of refining or hydration, allowing the water to drain quickly when the sheet is formed on the fourdrinier wire (forming fabric). In actuality, some "free" sheets may contain small amounts of mechanical pulp. *See* GROUNDWOOD FREE. Also known as *woodfree*.

**FREE STOCK:** A common papermaker's reference to pulp stock slurries from which water readily drains from the fibers.

**FREE SULFUR DIOXIDE ($SO_2$):** That portion of the sulfur content of sulfite cooking liquor that is the sum of the actual free sulfur dioxide, plus one-half of the sulfur dioxide combined as bisulfites. Also referred to as *available sulfur dioxide*.

**FREE WATER:** Moisture in a sheet of paper that is not bound chemically or mechanically with the fibers.

**FREEBOARD:** That portion of storage chests, tanks, and chip bins above the level of stored material and not normally filled.

**FREENESS:** The ability of a pulp and water mixture to release or retain water on drainage. The ease or lack of ease with which the mixture or slurry will drain water is sometimes referred to as *slowness* or *wetness*. *See also* SLOWNESS, WETNESS, and CANADIAN STANDARD FREENESS (CSF).

**FREENESS TESTER:** An apparatus that measures the degree of hydration or refining of a pulp stock by determining how fast water will drain from it. Such an instrument is used to determine *Canadian standard freeness (Csf)*.

**FREQUENCY:** The number of cycles through which an alternating electrical current passes per second, that is, the reciprocal of the period. Also refers to other event cycles during various time frames.

**FREQUENCY CONVERTER:** A piece of electrical equipment that converts the power of an *A.C.* system from one frequency to another, with or without changes in the number of phases or in the voltage.

**FRESH WATER:** Water entering the pulp and papermaking process from the source for the first time, i.e., it has not been reclaimed or recycled from the process.

**FRICTION BOARD:** A type of dense heavyweight paperboard made from chemical wood pulp and wastepaper furnish for use in making paper pulleys.

**FRICTION CALENDER:** A type of calender used specifically to mechanically impart a high polish to paper finish by passing the sheet between three rolls revolving at slightly different speeds. It is made up of a cotton roll between two metallic rolls and produces a burnishing action.

**FRICTION GLAZED:** Paper whose surface has been given a high polish by passing the sheet between metallic rollers running at different speeds so that it mechanically burnishes the sheet.

**FRONT SIDE:** The side of a paper machine that is free of drives and other mechanical equipment, and makes the machine more accessible for operation by the machine tender and crew. Also referred to as the *tending side*.

**FROSTED PAPER:** Paper with a cloudy, frost-like appearing surface. It is obtained by adding colored or white fibers on the wet paper web while it is being formed on the wire (forming fabric) of the wet end of a paper machine.

**FROTH:** *See* FOAM.

**FROTH BONDING:** A technique used for the latex bonding of disposable non-wovens in which the binder is diluted with air and the resultant unstable foam is crushed into the sheet.

**FROTH PITS:** Small, depressed spots on the surface of a sheet caused by the bursting of entrained foam bubbles in the furnish coming over into the paper machine.

**FROTH SPOTS:** *See* FOAM MARKS.

**FROWN PROFILE:** A cross machine moisture scan measurement made on the felt in the press section of a paper machine. It produces a gradual negative to positive to negative shaped trace record. It is the opposite of *smile profile*.

**FROZEN-IN STRAIN:** The residual stress in a sheet of paper due to its resistance to shrinkage during the drying process and stabilization at atmospheric conditions.

**FROZEN SMELT:** Layer of smelt that builds up on the chrome ore refractory bottom of a sulfate black liquor recovery furnace. It is on this surface that the fluid smelt flows to the smelt spouts.

**FRP:** *See* FIBERGLASS REINFORCED PLASTIC.

**FRUIT FIBERS:** Papermaking fibers that originate from the fruits of plants.

**FRUIT WRAPPING PAPER:** An odorless, lightweight grade of special oil-treated paper, made with a high-gloss surface. It is used to wrap and package fruits. *See also* APPLE WRAP.

**FTU:** *See* FORMAZIN TURBIDITY UNITS.

**FUEL AIR RATIO:** The relation of fuel to air in the combustion mixture at the burner of a furnace.

**FUEL BED:** The layer of solid fuel or solid waste material on the grate or hearth of a power furnace.

**FUEL SACK PAPER:** *See* CHARCOAL BAG PAPER.

**FUELS:** Materials used to produce heat or power by burning, such as coal, coke, oil, gas, and wood. Black liquor is also burned in the pulp mill, but the primary objective in doing so is to recover chemicals from it, with the heat produced as a by-product.

**FUGITIVE COLORS:** Paper and paperboard colors that are susceptible to fading and damage by light and substances such as acids, caustics, and bleaching-type chemicals.

**FULL MACHINE DIRECTION TENSION BURST:** A roll defect occurring as a complete separation or rupture of the paper sheet with the fiber separation in the machine direction. It is usually caused by using excessive tension during winding.

**FULLER BOARD:** *See* INSULATING BOARD.

**FULLING:** An operation where the felt is wetted by a soap solution and then passed through a series of vertical and horizontal rollers. The mechanical action thickens and compacts the felt, and the pressures induce a mechanical interlock of the individual fibers, thereby increasing the strength and wearability of the felt.

**FULLY COOKED PULP:** A pulp that has been cooked to the extent that, when discharged from the digester, the mechanical action in emptying the digester reduces the chips to their fiber components, allowing them to be pumped without additional mechanical treatment.

**FULLY CREPED PAPER:** Paper that has been crinkled after the sheet is made on the paper machine and brought to another location. There it is moistened and passed over a roll fitted with a doctor blade. This produces a high percentage of crepe on the sheet.

**FUME:** Tiny, solid particles commonly formed by the condensation of solid matter vapors.

**FUNGI:** Small, often microscopic, plants without chlorophyll. They are useful for breaking down waste and stabilizing sewage. They also grow on wood and produce undesirable effects on paper and paper products under dark, damp storage conditions.

**FUNGICIDES:** Chemical substances used to prevent the potential growth of fungus on wood and paper.

**FURANS:** Highly toxic contaminants associated with the formation of *dioxin* in the bleaching of pulps with chlorine compounds.

**FURNACE:** The chamber of an incinerating system where drying, ignition, and combustion occur.

**FURNACE PRESSURE INTERLOCK:** An interlock system on a recovery furnace that senses pressure and, at a preset value, automatically initiates a sequence of events for shutdown.

**FURNISH:** The materials in a pulp stock mixture such as the various pulps, dyes, additives, and other chemicals blended together in the stock preparation area of the paper mill and fed to the wet end of the paper machine to make the paper or paperboard. Commonly called *stock*.

**FURNISH RETENTION:** The percentage of the original material fed to the paper machine appearing in the product and not lost from the process. It is calculated as the oven dry pounds or tons of furnish appearing in the deckled sheet production divided by the total furnish weight in pounds or tons, and converted to a percentage.

**FURNISH ROLL:** The roll in a gravure offset-type paper coater that picks up the coating material from the pan and transfers it to a metering and distributing roll. This roll, in turn, transfers it to the application or offset roll which applies it to the sheet surface.

**FURNISH SHRINKAGE:** The percentage of the original furnish to the paper machine that is lost in the papermaking operation.

**FURNITURE BOGUS PAPER:** Lightweight grades of paper used for making up filler material for protective wraps and pads for the packaging and transporting of furniture.

**FURNITURE WRAPPING PAPER:** Heavy, high-strength paper made from a mixture of sulfate chemical pulp, recycled wastepaper, and screenings used as a protective wrapping for transporting and storing furniture.

**FUSE:** An electrical element designed to melt or dissipate at a predetermined current value to protect against abnormal conditions of current.

**FUSION POINT:** The temperature at which a particularly complex mixture of minerals can flow under the weight of their own mass.

**FUZZ:** Fine fibrous extensions that occur on paper and paperboard surfaces. They are the result of improper refining in the stock preparation area, the use of excessive vacuum during the sheet formation phase, or the use of insufficient surface sizing material.

**FUZZING:** A paper sheet characteristic that makes it susceptible to the development of fibrous projections during its manufacture or in its use. Also called *fiber piling.*

**FUZZY LOGIC:** A branch of *artificial intelligence* used to enhance pulp and paper mill process control systems. Its rules are generally similar those operators are already using. However, its logic is not based on "either-or" (zero or one), but rather a more gradual scale such as "few," "many," and "most."

**FUZZY PAPER:** Paper with a hairy type of surface due to fine fibrous extensions out of the sheet.

# G

**GAFF:** A pole with a hook used in handling logs in a groundwood mill.

**GAGE PRESSURE:** The pressure in a system that is measured in relation to the pressure in the surrounding atmosphere.

**GALVANIC ACTION:** Type of metallic corrosion due to electrical current, usually between two dissimilar metals.

**GALVANIZED SHEET SURFACE:** A paper surface characterized by an uneven surface such that the variations in the smoothness and gloss creates a metallic galvanized appearance.

**GAMMA-CELLULOSE:** The pulp or cellulosic material that will dissolve in a 17.5 percent caustic solution at 20°C and remain in solution when the alkaline solution is neutralized.

**GAMMA RAYS:** Waves of radiant nuclear energy that have very high penetrating power.

**GANGUE:** The worthless portion of a slurry from which useful material was separated by flotation or some other type of fiber recovery process.

**GANTRY:** A crane mounted on two movable supports running on rails and high enough to allow vehicles to pass under.

**GAP FORMER:** A type of twin wire former in which the headbox slice jet impinges directly into the gap of the unit's two converging wires (forming fabrics). There is little or no traditional forming zone in these types of arrangements.

**GARBAGE BAG PAPER:** A good, wet-strength type of sulfate paper used to convert into garbage and trash bags. It is also treated with paraffin, wax, oil, or asphalt for lining garbage pails.

**GARMENT BAG PAPER:** A medium-weight, chemical pulp paper, sometimes mixed with hemp fiber, and made to be used in the garment industry for protecting clothing.

**GARNET PAPER:** Heavy, strong, hard-sized rope or kraft paper coated on one side with a varnish-resin adhesive and garnet abrasive material. It is used for smoothing easily abraded materials.

**GARTER STRAP:** Sealing collars with a molded rubber inside contact strip and attached to the ends of the vat on a cylinder-type paper machine.

**GAS:** (1) A common pulp mill acid plant reference to sulfur dioxide gas formed by the burning of sulfur in the sulfite pulping process. (2) Materials used in the pulp and paper industry that are in a gaseous state, such as chlorine and oxygen used in the bleaching of pulp.

**GAS BURNER:** A furnace component used to feed fuel gas and air with the necessary velocity and turbulence to produce proper ignition and combustion of the gas within the furnace.

**GAS CONTACT EVAPORATION:** A method of evaporation of liquor before delivery to the recovery furnace in a pulp mill. It is carried out by contact with the hot flue gases from the furnace.

**GAS COOLER:** Large, water-cooled banks of cylindrical conduits in which sulfur dioxide gas from the sulfur burner is cooled before being introduced to the absorption towers during the cooking acid-making process of a sulfite pulp mill.

**GAS FAN:** An air-moving device used to move or circulate gases in the process or expel obnoxious gases from a room or process.

**GAS IGNITER:** A supervised gas burner in a furnace, usually ignited by a spark, and used to continuously provide sufficient energy to produce proper operation of an associated hearth or other combustion burners under a variety of lighting off and firing conditions.

**GAS MASK:** A protective breathing device used where hazardous gases may exist to protect personnel in the area, such as bleach plants in a pulp mill.

**GAS-OFF:** A common pulp mill term referring to the removal of noncondensible gases formed during the cooking of wood chips in batch-type sulfate digesters.

**GAS PASS:** The path that furnace gases follow in moving through and leaving the furnace.

**GAS PHASE BLEACHING:** The bleaching of pulp by the addition of gaseous chemicals such as chlorine, ammonia, and chlorine dioxide.

**GAS PHASE CHLORINATION:** The chlorination of unbleached pulp by the direct addition of gaseous chlorine.

**GAS SCRUBBER:** Equipment used to remove recoverable components from a gas stream by the use of an atomized stream of process liquid or water.

**GASDOWN:** Relief of gas and steam to lower the pressure of a sulfite digester from maximum cook pressure to blow pressure and to recover sulfur dioxide gas at the end of a cook. Also known as *blowdown, side relief,* and *pressure blowdown.*

**GASIFICATION:** The process of converting a solid or liquid fuel into a gaseous fuel.

**GASKET BOARD:** A highly absorbent, unfinished chip or wood pulp board that has been chemically treated for use in making up gasketing materials.

**GASKETS:** Various materials shaped and fitted to pipe joints and other similar connections to make them leakproof. Many gaskets are made of paper treated with special additives such as phenolic resins.

**GATE:** An adjustable partition on a stock flow box that is used to regulate the rate of flow of stock slurry. Also referred to as a *slot.*

**GATE VALVE:** A type of valve used extensively throughout the pulp and paper industry, particularly as manually operated shutoff valves, although some are automatically operated. They are characterized by a flat, wedge shaped, sliding element that is positioned with a linear motion in the body and against the seat by a stem.

**GAUGES:** Instruments and devices used to measure and indicate product dimensions and process variables such as thickness, length, height, diameter, pressure, level, temperature, etc.

**G.E. BRIGHTNESS:** A U.S. measure of paper or pulp *whiteness* or *brightness* using a General Electric test instrument. It is the reflectance value relative to a standard (magnesium oxide) in the blue region of visible light, at a specific wavelength of 457 nm, expressed as a percentage of the standard's brightness. The test is described in TAPPI standard T452. It is conducted by illuminating a specimen at 45% and observing it normal to the plane of the specimen over a circular area about 12 mm in diameter (thus being sometimes referred to as "directional brightness"). *See also* ELREPHO BRIGHTNESS, which is the brightness measure used in most other parts of the world. There is no direct relationship between the two brightness measures, although *Elrepho brightness* tends to be 0.5% to 1.0% higher on the average.

**GEAR:** A machined element connected to the moving part of a machine in order to transmit motion. It usually consists of a wheel with teeth on its circumference that engage similar teeth on another wheel.

**GEAR DRIVE:** A shaft terminating in a gear used to transmit rotary motion from a driving device, such as a motor, to other devices through companion gear connections.

**GEARBOX:** A container housing a number of gears used to transmit, or reduce, or increase the speed of rotary motion of a driven shaft connected to a piece of process equipment.

**GEIGER COUNTER:** An electrical device that detects the presence and measures the intensity of radioactivity.

**GEL:** A jelly-like material formed by the solidification of a colloidal solution and characteristic of some pulp and papermaking materials.

**GELATIN:** A tasteless, odorless, glutinous material obtained by boiling animal bones, hooves, and other tissues or similar vegetable substances and used as paper adhesive coating and sizing. Also referred to as *animal size* or *glue.*

**GELATIN DUPLICATING PAPER:** A highly finished, medium weight, chemical pulp paper used to make master and copy sheets in a duplicating process. The aniline ink-typed or written material is transferred to a gelatin film and then picked up as copy on paper pressed into contact with the gelatin surface.

**GELATIN-PLATE PRINTING PAPER:** A good-strength, well-sized, smooth-surfaced paper used for printing from gelatin plates prepared with aniline ink.

**GELATINIZATION:** A process used in the preparation of starch for use in papermaking. It involves cooking starch in an aqueous suspension so that the granules swell to form a viscous solution.

**GENERAL MANAGER:** A common title for one who directs and is responsible for all operations of a pulp and/or paper mill.

**GENERAL SUPERINTENDENT:** One who is in charge of and supervises the operation of all departments of a pulp and/or paper mill.

**GENERATING TUBES:** The tubular elements in a boiler in which water is converted into steam.

**GENERATOR:** A machine that transforms mechanical power into electrical power.

**GENERATOR ROOM:** A room or area in the powerhouse of a mill where the electric generators are located.

**GEOGRAPHY PAPER:** A single coated book paper to be used for printing geography books.

**GEOTEXTILES:** Synthetic fabrics especially made to be used in contact with the soil.

**GIFT WRAPPING PAPER:** A light tissue-type of chemical pulp paper, sometimes colored and decorated. It is used extensively to improve the appearance of gift packages.

**GIFT WRAPPING THE WINDER:** A paper mill expression referring to a paper winder during a power disruption in which the roll on the unwind is allowed to coast to a stop rather than stopping it with a braking system.

**GILLING:** The process of applying a high-gloss finish on paper with a "gill" glazing machine.

**GIRDLING:** Cutting a ring around a tree trunk by removing the bark so that it can be treated with chemicals or allowed to die. This procedure facilitates bark removal when harvested for pulpwood.

**GLAND:** That part of a pump that stops the pressurized liquid from discharging out around the rotating shaft.

**GLARE:** The shiny appearance produced from the surface of paper. It reflects a considerable amount of light as the result of the way it is finished.

**GLARIMETER:** *See* GLOSSMETER.

**GLASS FIBER:** Glass in thin, thread-like, fibrous form. It is used in making specialty-type paper.

**GLASS FIBER FELT:** A thick sheet or pad made of bonded glass fibers in a synthetic resin.

**GLASS FIBER PAPER:** A special type of paper made from glass fibers. It is used to separate out very fine materials from liquids and gases.

**GLASS FLAKE:** An electrical insulating material used with various types of pulp as a bonding medium to make special papers.

**GLASS TRANSITION TEMPERATURE:** The temperature at which plasticized lignin in mechanical pulp changes to a brittle, glass-like state.

**GLASSINE:** Light, dense, translucent paper made from highly refined chemical pulp and possessing a high degree of hydration. It is used as envelope windows and in protective packaging of various types of foodstuffs, candy, tobacco products, chemicals, and metallic items.

**GLAUBER'S SALT:** *See* SALT CAKE.

**GLAZED BOOK PAPER:** Paper made to be used as covering for set-up boxes having a highly polished surface finish.

**GLAZED CASINGS:** A reference made to types of paper that have been friction-polished to a high gloss on both sides of the sheet.

**GLAZED PAPER:** Paper that has been surface-finished on one or both sides by methods utilizing friction on calenders, platens, and smooth-surfaced dryers, such as a yankee, to produce a high-gloss polish.

**GLAZER:** A machine used to friction-glaze paper surfaces. It consists of three rolls housed in an upright frame with the top and bottom rolls usually made of chilled iron and an intermediate filled roll.

**GLAZING:** The process of friction polishing paper or paperboard sheets on process equipment such as calenders, platens, and yankee dryers to produce a glossy surface finish. *See also* ROLLED.

**GLAZING CALENDER:** A paper machine calender designed with alternating metallic and filled rolls. It imparts a high glossy finish to paper primarily by friction.

**GLAZING CYLINDER:** A large, highly polished drying drum used for simultaneously drying and glazing machine glazed paper.

**GLOBE BOILER:** A stationary, spherical, batch-type digester used to cook rags in a rag pulp mill. It is rotated in order to load rags and dump pulp when the cook is completed.

**GLOBE DIGESTER:** A spherical-shaped, batch-type digester that sometimes rotates during the cooking process.

**GLOSS:** That property of a paper sheet surface that produces a shiny, highly-reflective appearance when light is reflected from it.

**GLOSS CALENDER:** A type of calender, fitted with highly polished and heated metallic rolls and rubber rolls, used to produce a shiny, lustrous appearance on the surfaces of coated paper and paperboard.

**GLOSSING AGENTS:** Chemicals added to a paper furnish to enhance the ability of the finished sheet to reflect light so as to produce a shiny, lustrous appearance.

**GLOSSMETER:** A paper mill laboratory test instrument used to measure the gloss or glare created by the surface of a sheet of paper. It operates by measuring the amount of light reflected from the sheet surface when exposed to a standard light source. Also called a *glarimeter.*

**GLUE:** A general adhesive material consisting primarily of gelatinous protein prepared from animal remnants of the meat packing industry. It is used extensively in the papermaking process for sizing and making up coating preparations. Sometimes called *animal size* or *gelatin.*

**GLUEABILITY:** The measure of paper or paperboard's ability to accept or "hold" glue as an indication of its receptiveness to being attached together by adhesives. Sometimes spelled as "gluability."

**GLUE-COATED:** A descriptive reference to paper that has been coated with a material made up with glue.

**GLYCERIN:** A sweet, viscous, colorless, noncrystalline liquid sometimes used as a plasticizer or softening agent in paper.

**GLYCERIN PAPER:** A moistureproof paper that has been treated with glycerin. It is used for protective wrapping and as a base material for the manufacture of oilcloth, draperies, and gaskets.

**GO-DEVIL:** A stiff piece of cardboard over which paper is usually looped and fed into the ingoing seal on slow-speed vacuum dryers.

**GOLD PAPER:** A decorative type of paper coated with bronze powder.

**GOLDENROD PAPER:** A special type of paper frequently used to produce opaqueness when arranging and positioning negatives for exposure. It is characteristically orange-yellow in color.

**GOOD SHEETS WASTE:** Acceptable quality papers thrown out with bad paper in sorting paper to waste because of defects.

**GOVERNOR:** A device for regulating the power usage or speed of paper and pulp mill process equipment such as a pulpwood grinder.

**GRAB:** A power shovel commonly used for materials handling in and around a pulp and paper mill. It consists of a bucket fitted to a boom or swing hoist and is attached to a mast or frame mounted on a truck, railroad car, or crawler tractor.

**GRAB SAMPLE:** A paper or pulp test sample obtained by random selection from the process stream.

**GRAB TRUCK:** Equipment employed to move rolls of paper from one location to another in a paper mill using two opposed clamps to grab and carry the rolls. Sometimes called a *clamp truck*.

**GRADE:** A type or class of paper identified as having the same fiber, color, additive, and chemical composition, and manufactured to have the same physical and mechanical characteristics.

**GRAIN:** The directional alignment of fibers in a paper sheet structure.

**GRAIN DIRECTION:** The direction of fiber alignment in a sheet of paper. *See* MACHINE DIRECTION.

**GRAIN LOADING:** The rate of emission of particulate matter from a polluting source. Measurements are made in grains (65 milligrams) of particulate matter per cubic foot of gas emitted.

**GRAIN LONG:** A sheet of paper cut with the long dimension parallel to the direction in which most of the fibers lie.

**GRAIN SHORT:** A sheet of paper cut with the short dimension parallel to the direction in which most of the fibers lie.

**GRAIN SHOVEL:** A special drag scraper-type shovel used to unload dry, granular material from cars by scraping, rather than by lifting.

**GRAINED PAPER:** A cover-type paper that has been embossed or otherwise decorated so its surface resembles wood grain, marble, or leather. The process of such texturing is known as *graining* or *pebbling*.

**GRAINED PAPERBOARD:** Paperboard whose surface has been embossed or printed to give the appearance of a wood-grained surface. It is used for making up various types of boxes.

**GRAINING:** The process of embossing or printing paper surfaces to resemble wood grain, marble, or leather. Also called *pebbling*.

**GRAINY:** Paper surface defect caused by uneven distribution of color throughout the sheet, uneven wire (forming fabric) and felt impression made on the sheet, and uneven shrinkage during the drying phase of papermaking.

**GRAINY EDGES:** A paper defect that appears as small, rough variations in the surface appearance at the edge of the sheet. It is caused by excessive cross direction shrinkage during drying.

**GRAMMAGE:** A very common paper and paperboard weight designation used in European countries that is generally the metric equivalent of basis weight. It is expressed in grams per square meter as compared with pounds per ream.

**GRANITE PAPER:** Paper made with a mottled surface by adding small amounts of heavily dyed fibers into a lighter colored furnish. It is used for special and fancy notepaper and other similar items.

**GRANITE ROLL:** A roll in the wet press of a paper machine made of granite and usually located in the top of the press.

**GRAPH PAPER:** Paper made to be ruled in both directions with parallel, equally spaced lines. It is used for making charts and graphs. *See also* ALIGNING PAPER.

**GRAPHIC PANEL:** A central control room operator's panelboard with a process flow diagram depicting process equipment and associated recording and control instruments located in relation to their actual point of control in the process.

**GRAPHITE PAPER:** Conductive, lightproof paper made by treating a sheet with an aqueous paste of colloidal graphite in various ways, including spraying, painting, and dipping.

**GRASS ROOTS MILL:** *See* GREEN FIELD MILL.

**GRATE:** The floor on which the burning fuel bed is formed in a solid fuel power furnace. *See also* BED-FIRED FURNACE.

**GRAVEYARD SHIFT:** The working hours of the operating personnel of a pulp and paper mill that usually start at 11:00 p.m. or 12:00 p.m., and last 6 or 8 hours in a 24-hour operating schedule.

**GRAVITY DECKERS:** A drum vacuum filter that is used to dewater stock slurries. The vacuum is created by the force of gravity on a falling column of water in a drop-leg connected to the inner portion of the wire covered screen on which the pulp pad is formed.

**GRAVITY DEWATERING:** The drainage from stock as it is projected onto the wet end wire (forming fabric) over an open table without any vacuum or positive suction.

**GRAVITY FLOW:** The movement of material from a higher to a lower height due to the force of gravity alone.

**GRAVITY SLUSHER:** A pulp slurry mixing apparatus into which the pulp and water fall and become a slurry due to the force of gravity.

**GRAVITY THICKENER:** A type of filter used to remove water from a pulp slurry by allowing the water to fall away from the slurry by the force of gravity as it passes over a stationary or rotating screen.

**GREASE RESISTANCE:** *See* GREASE PROOF and GREASE PROOFNESS.

**GREASE SPOTS:** Defects in paper that appear as small, transparent, scattered areas in the surface coating.

**GREASEPROOF:** A descriptive reference applied to paper and paperboard to designate that the stock it was made from was given a special mechanical refining or that they have been coated with casein, lacquer, hot melt, resin, or silicate-type coatings in order to make them resistant to oil and grease.

**GREASEPROOF PAPER:** Protective wrapping paper that is specially coated with materials to make it resistant to grease and oil, or the stock that it was made from was highly hydrated by mechanical refining of chemical wood pulp to make it greaseproof. Generally referred to as *glassine.*

**GREASEPROOF PAPERBOARD:** A multi-ply paperboard made with an outer lining of glassine or paper that was made resistant to oil and grease.

**GREASEPROOFNESS:** The property of paper or paperboard to resist the penetration of oil and grease. Also referred to as *grease resistance.*

**GREASY:** Type of pulp that has been highly hydrated by refining to an extremely low freeness. This causes it to feel greasy. Sometimes called *dead beaten pulp.*

**GREEN:** Reference made to wood logs and/or wood chips that contain an excessive amount of moisture or are incompletely seasoned. Also used sometimes in reference to environmental activism.

**GREEN CHIPS:** Wood chips made from freshly cut wood that has not had enough time to age and lose free moisture. Also called *greenwood.*

**GREEN FIELD MILL:** A new pulp and/or paper mill built on a site on which no other mill exists. Also called a *grass roots mill.*

**GREEN LIQUOR:** A liquid that is formed during the sulfate chemical recovery process by dissolving smelt from the recovery furnace in a dissolving tank. The clear liquid takes on a greenish tinge.

**GREEN LIQUOR CLARIFICATION:** The removal of suspended solids (dregs) from green liquor, prior to causticizing in a pulp mill, by settling it in any one of several types of sedimentation units after flocculation.

**GREEN PAPER:** Newly manufactured paper which has not had enough time to condition itself and equalize the moisture content.

**GREEN WOOD:** Freshly cut wood that has not had enough time to age and lose free moisture. Basically the same as *green chips.*

**GREETING CARD STOCK:** White or pastel color heavyweight paper or card stock with a fancy or vellum finish to be used to make greeting cards.

**GRID METHOD:** A laboratory procedure used for measuring fiber length and coarseness in which a predetermined sample is cast on a glass plate or slide and projected on a screen to determine the number of times the fibers pass a fixed grid pattern.

**GRINDER:** A mechanical pulping machine that reduces wood logs into their fiber components through abrasion by pressing them against a revolving grindstone. Also referred to as a *pulpwood grinder.*

**GRINDER BLEACHING:** The addition of bleaching chemicals such as dithionite or peroxide to the grinder pit in order to bleach the hot groundwood pulp.

**GRINDER PIT:** The area below the pulpstone of a grinder into which the pulp suspension formed from grinding the wood logs drops and collects. From the pit it flows into open flumes that direct the suspension to subsequent screening operations.

**GRINDER POCKET:** That part of a pulpwood grinder into which the wood for grinding is placed to make contact with the grinding stone.

**GRINDER ROOM:** The building or area of a mechanical pulp mill where the pulpwood grinders are located.

**GRINDERMEN:** Groundwood mill personnel who take care of the individual pulpwood grinders.

**GRINDING EXHAUST SYSTEM:** A ventilation system designed to remove hot, moisture-laden air from pulpwood grinders in a groundwood pulp mill.

**GRINDING FACTOR:** (1) The ratio of wood advance towards the stone in a grinder to average fiber diameter. Also called *initial grinding factor.* (2) The rate of pulp production by grinding in relation to the rate of the passing groove volume of the stone. Also referred to as *secondary grinding factor.*

**GRINDING PRESSURE:** The pressure existing in the cylinder operating the pressure foot which pushes the logs against the grindstone of a pulpwood grinder.

**GRINDING OF ROLLS:** The machining of press and calender roll surfaces to obtain the proper crown profile consistent with good operating performance.

**GRINDSTONE:** A grinding wheel made from natural or artificial stone and having a standard set of surface designs cut into it. When used in a grinder, it shreds wood logs pressed against it as it is rotated. Also referred to as *pulpstone.*

**GRINDSTONE DRESSER:** The device used to prepare a grindstone surface by means of the proper type of burr, depending on the quality of the pulp desired from the pulpwood grinding operation.

**GRIT:** (1) The particles in a grindstone that are abrasive and which are responsible for tearing away fibers from wood logs pressed against the grindstone surface while it is rotating. (2) Impurities entering the slaker with the lime in the causticizing system, but usually removed by the sloping classifier associated with the slaker.

**GROCERY BAG PAPER:** A general term referring to strong kraft paper made to be converted into grocery bags of a variety of sizes.

**GROOVED ROLL PRESS:** A paper machine wet press having a specially designed roll with circumferential grooves on the surface into which water is squeezed out by hydraulic pressure created at the nip, instead of the conventional suction roll.

**GROOVED TABLE ROLLS:** Rolls located immediately after the breast roll or the forming board on the wet end of a paper machine. Such a roll consists of a steel tube covered by a thick rubber sleeve in which grooves have been cut to provide gentle drainage while a paper web is being formed on the wire (forming fabric) running over it.

**GROOVING:** The process of partially scoring or indenting heavyweight types of paperboards so that they can be easily bent along a prescribed line.

**GROSS HEATING VALUE:** A measure of the calorific content of any fuel stream to a furnace as determined in the laboratory with an oxygen bomb calorimeter.

**GROSS WEIGHT:** The combined weight of the material and its container. It includes both net and tare weights.

**GROUND:** Wire or connection for electrical equipment and circuits that is or is supposed to be at earth potential, or the earth that it is eventually connected to.

**GROUND CHALK:** Deposits of natural calcium carbonate used as a source of fillers for making certain grades of paper, particularly in Europe.

**GROUND LIMESTONE ($CaCO_3$):** Limestone that is finely ground and used as a paper loading filler, providing faster drainage and a strong sheet.

**GROUND WATER:** A pulp and paper mill water supply obtained from deep wells and springs.

**GROUNDWOOD (GW):** Pulp and paper made up of mechanical fibers produced by the grinding of pulpwood.

**GROUNDWOOD FREE:** Any type of paper that contains less groundwood than the paper industry standard of 5 percent, determined by microscopic analysis of a properly dye-stained sample. *See also* FREE SHEET.

**GROUNDWOOD PAPER:** Any paper that is made from a furnish containing a large percentage of groundwood pulp. Newsprint is usually not included because groundwood content must fall within certain specified limits to be considered in this category.

**GROUNDWOOD PRINTING PAPER:** Soft, bulky, opaque printing paper with a smooth finish. It is made with groundwood or mechanical fibers in the furnish.

**GROUNDWOOD PULP:** A fibrous slurry produced by mechanically abrading the fibers from barked logs through forced contact with the surface of a revolving grindstone. It is used extensively in the manufacture of newsprint and publication papers.

**GROWTH RING:** A layer of wood growth that is added annually to the trunks and roots of trees and appears as a ring in a cross section view. It is used as an indication of the tree's age and growth.

**GUARD BOARD:** A scraping device placed on the upper couch roll of a paper machine to squeeze water and scrape off lumps of stock from the couch roll jacket.

**GUIDE EDGE:** That side of a paper sheet used to control the position of the paper as it is fed into and between presses or other papermaking equipment.

**GUIDE ROLL:** Any roll in papermaking machinery that can be adjusted to compensate for any misalignment of a paper or felt sheet running over it.

**GUIDE SHEET:** Paper used to check and adjust automatic sheet guides on a paper machine.

**GUIDING:** The procedures of maintaining the proper alignment of running wires, forming fabrics, felts, belts etc. in the direction relative to contacting paper machine component surfaces.

**GUILLOTINE:** A paper cutter or trimmer used in paper finishing operations. It has a long, heavy, mechanically driven knife to cut and trim stacks of paper sheet. Also known as a *trimmer.*

**GUM:** (1) A species of hardwood trees used as a source of pulpwood. (2) A high molecular weight, colloidal polysaccharide substance of plant origin. It becomes a gelatinous, viscous suspension when dispersed in water. It is used in stock preparation as a wet end strength enhancement additive, paper sheet surface sizing material, and to prepare adhesives that are applied to paper surfaces.

**GUM ARABIC:** Gum obtained from some types of accacia tree and used to make adhesives for paper labels, stamps, cigarette paper, etc.

**GUM ROSIN:** A substance produced by the distillation of natural gum obtained from trees. It is used as a sizing material in the stock preparation and wet end operations to add to the final sheet strength characteristics.

**GUMMED FLATS:** Gummed printing papers made into sheets or rolls used primarily by the graphic arts and business forms industry.

**GUMMED PAPER:** Paper whose surface is coated with adhesive gum material such as glue obtained from animal tissues and/or dextrin.

**GUMMED TAPE:** A general term applied to any variety of tapes made from a clay-filled, synthetic fiber-filled, hard-sized, supercalendered, asphalt-impregnated paper base. Sometimes combined with cloth and coated with a water-activated adhesive. It is used extensively for sealing purposes, especially in making up shipping cartons.

**GUMMING:** The coating of a paper sheet surface with gum or adhesive material.

**GUMMING PAPER:** A hard-sized, chemical pulp paper usually put up into rolls for use as a base for applying gum on one surface so that it can be made into labels and sealing tape.

**GUN:** A burner in a furnace used to introduce oil or gas fuel and air with the proper velocity, turbulence, and concentration to produce proper ignition and combustion of the fuel within the furnace.

**GUNK:** A colloquial expression referring to any undesirable deposition in pulp and papermaking process equipment that may consist of a combination of unidentifiable contaminants.

**GW:** *See* GROUNDWOOD.

**GYMNOSPERMS:** Cone-bearing types of trees whose seeds are not enclosed in an ovary, as distinguished from *angiosperms*. Used as a source of pulpwood. Often referred to as *conifers*.

**GYPSUM:** A naturally occurring calcium sulfate ($CaSO_4 \cdot 2H_2O$) pigment used as a filler in the manufacture of paper and paperboard. Also known as *crown filler* or *pearl filler*.

**GYPSUM BOARD:** A board made up with a high percentage of calcium sulfate material and lined with reinforcing sheets of paper. It is used in building construction for walls and ceilings.

**GYPSUM LATH:** A board made with a high percentage of calcium sulfate material and lined with unsized paperboard, enabling it to readily receive plaster when used to replace laths in building construction.

**GYPSUM SHEATHING:** A special water-resistant board made with a high percentage of calcium sulfate material and lined with waterproof oil or asphalt-impregnated paperboard. It is used for exterior siding in building construction.

**GYRATORY CHIP SCREEN:** A type of chip screen that uses a circular motion and inclined plates as a screening medium. *Also called* ROTARY SCREEN.

# H

**H FACTOR:** Integration of the relative rate of reaction over a period of time during the cooking cycle of wood to obtain a single number that can be used to determine the degree of cooking that has taken place. In the case of batch-type digesters, it is sometimes used to determine when the cook is finished and the time to "blow" the digester.

**HVAC:** Abbreviated reference for Heating Ventilation and Air Conditioning.

**HAIR CUT:** A paper web defect caused by fiber from felts, human hair, etc., cutting the sheet when pressure is applied by processing units such as presses and calenders.

**HALF-FINE METALLIC PAPER:** A special adhesive, metallic patchlined type of paper.

**HALF PLATE PAPER:** A high-quality paper with special softness qualities, suitable to use for woodcut purposes.

**HALF WRAPPING:** A method of wrapping paper bundles so that only the top or bottom and part of the ends and sides of the lid or base are actually covered.

**HALFSTUFF:** A term, normally used in conjunction with rag paper manufacture, referring to pulp prepared to be charged into a beater for mechanical refining.

**HALFTONE BLOTTING PAPER:** A type of blotting paper specially treated to have a smooth surface sufficient for halftone printing while also retaining the ability to absorb liquids.

**HALFTONE PAPER:** A type of paper manufactured to a particularly fine finish so that it can be used in the halftone printing process.

**HAMMERMILLS:** Types of high-speed equipment that use pivoted or fixed hammers or cutters to crush, grind, chip, or shred solid materials prior to their being used in a process.

**HAMPERS:** Storage containers used in paper recycling mills to receive incoming shipments of wastepaper for temporary storage before processing.

**HAND CONTROL:** *See* MANUAL CONTROL.

**HAND FEEL:** A term referring to the impression obtained by handling paper to judge its finish, flexibility, and general quality.

**HAND VALVE:** Type of flow control valve in process lines that is manually opened, closed, or adjusted by a direct-connected wheel.

**HAND WORK:** Any manual operation performed with the use of hand tools, as in the manufacture of paper boxes.

**HANDBILL PAPER:** Paper made to be used in printing announcements, advertisements, and other large quantity public pass-out documents.

**HANDLING DAMAGE:** Any type of damage done to paper and paperboard that can be attributed to the preparation, movement, shipment, and storage procedures.

**HANDLING PAPER:** The manual manipulation of paper in order to sense various properties such as surface feel, rigidity, etc.

**HANDMADE FELT:** Types of wet press felts made to impart handmade paper surface characteristics to a sheet of paper.

**HANDMADE FINISH:** Handmade paper surface characteristics applied to a sheet of paper by embossing rolls or specially designed plates.

**HANDMADE PAPER:** Paper made, usually one sheet at a time, by manually dipping special screen wire molds in vats of pulp stock and allowing the water to drain away by gravity. Sometimes called *mould made paper.*

**HANDS:** Members of the paper machine operating crew.

**HANDSHEET:** A single sheet of paper usually made in a laboratory by draining water from a pulp suspension on a screen-covered sheet mold. It is used in testing and examining the properties of pulp and paper. Sometimes referred to as a *test sheet.*

**HANG-UPS:** Pulp and uncooked chips that remain in a batch digester after it has been "blown."

**HANGING PAPER:** *See* WALLPAPER.

**HANGING RAW STOCK:** Groundwood paper before it is coated, printed, embossed, or otherwise processed to convert it to wallpaper.

**HARD BEATING:** Extended mechanical treatment of pulp in beaters and refiners in order to produce specific characteristics in the paper made from it.

**HARD BLEACHING PULP:** Unbleached pulp having a high K or KAPPA number, indicating that it contains a higher-than-normal amount of lignin and other noncellulosic material, thus requiring a greater amount of chemicals to bleach it to a comparable brightness value. *See* HARD PULP.

**HARD COCKLES:** A paper defect caused by a wrinkle resulting from nonuniform shrinking as a result of uneven drying or sheet formations passing through a calender.

**HARD COOK:** A mild pulpwood cook carried out under conditions resulting in minimum removal of lignin and other noncellulosic materials for such purposes as producing increased pulp strength and yields. *See also* RAW COOK.

**HARD COPY:** A computer-produced printed document such as a typed printout, paper tape, punched cards, or a video copy.

**HARD END:** A localized paper roll defect consisting of extra hardness caused by the concentration of higher sheet caliper.

**HARD FIBER:** A descriptive reference to types of highly compacted rigid paperboards.

**HARD FOLD:** The process of and/or the crease remaining in the sheet as the result of folding a quantity of paper. This is usually done for shipping purposes.

**HARD LUMPS:** Paper defects appearing as white, dense spots in the sheet that become crushed if the sheet is calendered. They are usually caused by bits of dried pulp or broke that have not been completely dispersed before sheet formation.

**HARD PAPER:** Paper having very firm, even surfaces or rendered rigid by impregnation with synthetic thermosetting resins so it can be molded into various consumer products. Also called *bakelite paper.*

**HARD PULP:** Pulp usually made by cooking chips under conditions that result in the removal of a minimum amount of lignin and other noncellulosic material. Referred to as having a high *k* or *kappa* number; also called *hard bleaching pulp.*

**HARD-SIZED:** Types of paper and paperboard that have been made resistant to water and ink penetration by subjecting them to a high degree of sizing treatments. Also called *well sized.*

**HARD SPOTS:** Hard areas on a reel of paper caused by uneven "across the machine" bulk winding on the reel. This action produces tight windings where the paper is thicker.

**HARD STOCK:** (1) Pulp slurries that have been subjected to extended periods of mechanical treatment in beaters or refiners in order to impart desired physical properties to the paper produced from it. *See* HARD BEATING. (2) Pulp slurries containing a high content of jute, rope, or rag fibers.

**HARD WATER:** Types of mill source water that contains dissolved calcium, iron, aluminum, magnesium, and other similar minerals generally related to its capacity to precipitate soap.

**HARD WRINKLE:** Creases in a paper sheet that occur when it is wound into a shipping roll at the mill.

**HARDBOARD:** A building panelboard made by subjecting lignocellulosic fibers and other ingredients to high pressure and temperature.

**HARDENING AGENT:** Chemicals used to treat animal glues used in papermaking for moisture resistance.

**HARDNESS:** (1) The physical characteristic of paper or paperboard that is related to its resistance to damage by impact. (2) The density of a rubber covering on press rolls and to the calcium, aluminum, iron, and magnesium concentration in water. (3) A water quality related to the calcium, aluminum, iron, and magnesium concentration.

**HARDWARE:** The actual physical, mechanical, and electrical equipment that makes up the process measurement, control, and management systems in a mill, or to a computer system in general.

**HARDWOOD:** Pulpwood from broad-leaved dicotyledonous deciduous trees.

**HARDWOOD PULP:** Pulp produced from the wood of broad-leaved dicotyledonous deciduous trees.

**HARDWOOD ROTS:** Types of fungi that cause extensive decay of hardwoods used for pulping.

**HARDWOODS BOTTOM LANDS:** Swampy pulpwood timberlands that are more suitable for growing hardwood than softwood trees.

**HARSH PAPER:** Defective paper that feels rough to the sense of touch.

**HATCH:** A cover used to provide access from the top of storage bins or towers.

**HAUL:** The distance that pulpwood must be transported by log trucks from the loading point in the woodlands to the unloading point at the mill site, or the act of accomplishing this.

**HAUL-UP:** A common means of reclaiming pulpwood logs from a river or log pond by the wood preparation plant of a mill.

**HAYING:** A paper mill expression referring to the process of making and gathering paper (broke) produced during a break in the sheet of paper on the paper machine.

**HEAD:** (1) The pressure produced by a column of fluid related to the fluid height and density. (2) The upper chip inlet end of a batch digester.

**HEAD GATE:** A flow control device used on sources of water-to-water turbine generators in hydroelectric plants and to water-powered wheel driving mechanisms.

**HEAD LOSS:** The decrease in pressure as a fluid flows through a pipeline and across fittings in a pipeline.

**HEADBOX:** A flow chamber located at the head end of a fourdrinier or cylinder-type paper machine. It receives the diluted pulp stock slurry and regulates the head or level to provide uniform and even flow across the width of the wire (forming section) and the correct volume of stock to the screens and mixing boxes before the cylinder machine vats. It is also known as a *mixbox* on cylinder-type machines.

**HEADER:** A common process pipeline or vessel from which a number of smaller branch lines are fed. It is used to carry fluid from a common source to various points of use.

**HEADS:** The vapor portion of long tube-type black liquor evaporators.

**HEARTH:** The bottom of a black liquor recovery furnace where the dried black liquor burns and forms a molten mass called *smelt.* The burner is known as a *hearth burner.*

**HEARTH BURNER:** *See* HEARTH.

**HEARTWOOD:** The dark-colored center of a tree, consisting of dormant wood.

**HEAT AGING:** Accelerating the aging of pulp by heating it in an oven at 105°C from 1 to 18 hours. It is performed in a laboratory to test the color reversion properties by determining the degree of yellowing.

**HEAT BALANCE:** The comparison of the heat input to the heat output around a process, process equipment, or furnace on a fixed time interval basis.

**HEAT CAPACITY:** The energy needed to increase a unit mass of a material by 1°C at a constant pressure or volume.

**HEAT CAPS:** Covers placed on the ends of dryer cylinders on a paper machine to reduce steam consumption.

**HEAT EXCHANGER:** An enclosed tube-filled chamber in which a flowing medium in or around the tubes is heated or cooled by another medium in or around the tubes, without allowing the mediums to mix.

**HEAT EXTRACTION:** The reduction of pressure and temperature of superheated steam from recovery and power boilers as it is passed through turbogenerators down to valves so that it can be used directly in the pulp and papermaking unit processes (such as digesters, evaporators, and paper machine dryers).

**HEAT INPUT:** The total of the gross heat inputs of all streams having any fuel value, the sensible heats of material streams and air, or other heat credits going to a power or recovery furnace.

**HEAT OUTPUT:** The total heat content of all output streams from a power or recovery boiler, including the heat transferred from the fire side to the steam side, sensible heat and latent heats of flue gas and smelt, heat reduction of makeup salt cake, additional gases, etc.

**HEAT RECOVERY:** The process of extracting heat from flash steam or any other steam that would escape to the atmosphere, usually by having it heat fresh water used back in the process for pulp washing or dilution purposes.

**HEAT RELEASE RATE:** The quantity of heat (Btu) per hour per cubic foot of the internal volume of the furnace in which a material is completely burned.

**HEAT SEALING:** Reference made to types of paper and paperboard whose surface has been coated or treated in such a manner that it becomes adhesive when subjected to heat.

**HEAT SEAL ADHESIVE:** An adhesive that can be converted to a dry film or coating; the dried film can be rendered tacky or fluid for bonding by the application of heat.

**HEAT TRANSFER:** The flow of heat from a warm area to a cooler one by conduction, convection, or radiation, such as the transfer of heat from one fluid or material to another fluid or material, usually through a metal wall. In the paper manufacturing process it is common in evaporators, heaters, and condensers, in which the heating medium is usually steam but can be water or other materials.

**HEAT TRANSFER PRINTING PAPER:** Paper that has been previously printed and from which the print dye can be transferred to another medium, such as textiles, by the application of heat.

**HEAT OF COMBUSTION:** The amount of heat (Btu) released per unit quantity of material (waste, liquor, or fuel) burned.

**HEAT OF CONDENSATION:** Heat given up when vapors such as steam condense and become available as heat energy. It is commonly used in multiple-effect black liquor evaporators. Also referred to as *latent heat of vaporization.*

**HEAT OF REACTION:** The heat produced or absorbed during chemical reaction. It occurs in unit processes such as the cooking of chips and the bleaching of pulp.

**HEAT OF VAPORIZATION:** *See* HEAT OF CONDENSATION.

**HEAT OF THE LIQUID:** The heat required to elevate a pound of a particular liquid from 32°F to the temperature of its-boiling point.

**HEAT-SEAL TEMPERATURE:** Dry adhesive activation temperature at which it will bond two surfaces, such as paper and paperboard, with proper pressure over a prescribed length of time.

**HEAT-UP TIME:** The time required to heat up cooking liquor and chip mixtures to cooking or "hold" temperature in chemical pulping performed in a batch-type digester. It also applies to other batch-type heating operations performed in a pulp and paper mill.

**HEATERS:** Units used to heat process fluids with a hotter fluid by transferring the heat via conductance, convection, or radiation.

**HEATING SEASON:** The seasons of the year during which mills in temperate weather areas must heat their building structures around process equipment. The heating season varies depending on the number of cold weather months characteristic of that particular area.

**HEAVY BLACK LIQUOR:** Spent cooking liquor that has been evaporated down to a high solids content of about 50 percent or more. The term is generally used when referring to black liquor after the evaporators, which is sent to the black liquor recovery furnace in a pulp mill.

**HEAVY SPAR:** A naturally occurring type of barium sulfate used as a source of filler and coating pigment for paper. *See* BARITE, BARYTA, and BARYTE.

**HEAVYWEIGHT:** Paper falling into the upper range of weights manufactured for a particular grade.

**HECTARE:** A land area measuring 10,000 square meters or 2.47 acres.

**HECTOGRAPH PAPER:** A smooth, heavily sized, highly calendered paper made especially to be used for duplicating work by the hectograph process, which employs spirits and uses a gelatin process.

**HELPER:** An unskilled mill employee working as an assistant to an operator or a skilled tradesman as a journeyman in the process of learning the particular job involved.

**HELPER DRIVE:** A mechanical drive located anywhere on a paper machine or as an extension to sectional or lineshaft drives. It is used wherever slippage or undue strain is placed on a felt or wire (forming fabric), or where it is inconvenient to drive a roll with direct mechanical power.

**HEMICELLULOSE:** The alkali-soluble, noncellulosic polysaccharide portion of the wood cell wall.

**HEMLOCK:** A softwood species of tree used as a source of pulpwood. It is especially suited for the manufacture of sulfite pulp.

**HEMP:** Fiber obtained from a plant found in areas with moderate temperatures and used in the manufacture of ropes and textiles. Hemp sometimes becomes a waste material fiber source for making certain types of paper and paperboard.

**HERITABILITY:** The ability or tendency in trees for traits such as disease or branch angles to be inherited.

**HERTY FOUNDATION:** Headquarters: Savannah, Ga., (912) 964-5541. Does contractual R&D and extensive pilot scale projects for the paper industry on a nonprofit basis.

**HEURISTIC:** The trial and error method of solving mill problems.

**HEXOSAN:** A carbohydrate component of pulpwood containing six carbon atoms in its chemical structure.

**HI-LO PULPER:** A type of pulper distinguished by having two individually motor-driven, side-mounted rotors with one driven at high speed for rapid defibering of pulp and the other driven at slow speed for circulating the slurry.

**HICKEY:** (1) A defect appearing as a white- or dark-centered spot on a printed sheet as a result of paper imperfections, ink imperfections, improper press operation, or the interference from small particles of extraneous matter. (2) A length of pipe or an extension handle used on a wrench to get greater leverage. Sometimes known as a *persuader.*

**HIDE GLUE:** Papermaking adhesive derived from animal hide extracts.

**HIDEOUT:** Silica in boiler water that has precipitated in inaccessible areas of a boiler and become mechanically loosened to cause a sudden increase in concentration.

**HIERARCHICAL CONTROL NETWORK:** Millwide process management and control with systems distributed at different levels, each being under the direction of the system on the level above it.

**HIERARCHICAL CONTROL SYSTEM:** A control system involving two or more computers in which a master computer receives the information it needs from the process or local computers, manipulates it, and sends instructions to local computers on how to control the process. This is similar to the system of delegation of authority found in company organizational structures.

**HIGH CONSISTENCY:** Once used as a reference to any pulp slurry with a consistency of 6 percent or more. But with the trend toward processing in the higher consistency ranges, the reference value median consistency is now generally in the 10 to 16 percent range or higher, depending on the process.

**HIGH CONSISTENCY REFINING:** The mechanical treatment of pulp at consistencies of up to 20 and 30 percent; sometimes performed on mechanical pulp screen rejects.

**HIGH DENSITY PUMP:** Pumps designed especially for use on high consistency pulp slurries.

**HIGH DENSITY STORAGE:** The storage of pulp slurries in a high consistency condition, usually after the bleaching process and just prior to the stock preparation.

**HIGH FINISH PAPERS:** Papers that have been surface-treated to impart a high glossy appearance.

**HIGH FREE-ROSIN:** Rosin size emulsions used in papermaking having a 75 to 90 percent free-rosin content.

**HIGH HAT:** A protective, cylindrical cover bolted on pulpwood grinder shaft flanges to prevent damage to the threads.

**HIGH IMPULSE PRESS:** Any of various types of presses designed to be highly loaded, sometimes above 2,000 pli. Some use special covers that deflect under high loadings, thus effectively extending the nip. Also referred to as *extended impulse press. See also* IMPULSE PRESSING.

**HIGH INTENSITY PRESS:** A type of wet press consisting of two rubber-covered rolls with a small diameter, hard-surfaced, grooved roll located off-center between them, creating a high intensity nip with the top roll to extrude water from the felt and sheet as it passes through.

**HIGH LEAD LOGGING:** A logging method that uses a cable system to lift logs partially off the ground and drag them to a loading area.

**HIGH LEVEL LANGUAGE:** A computer language that is characteristically easy to program due to its powerful statement makeup and more closely resembles common conversational languages.

**HIGH LOAD PITLESS GRINDING:** The removal of the pit dam in a pulpwood grinder to eliminate stone submergence and the provision of high pressure showers in order to increase groundwood production.

**HIGH PERFORMANCE CONTAINERBOARD (HPC):** Grades of *linerboard* and *corrugating medium* that meet new *Rule 41 edge crush test* specifications. Generally, these grades have very high compression strength values with reduced basis weights, compared with similar grades made to meet *burst strength* (*Mullen*) and basis weight specifications of older *Rule 41* tests. These grades have *ring crush* values that are at least 2:1/lb of basis weight, dramatically improving the stacking strength of corrugated boxes. Other reported advantages include reduced fiber and lower overall costs, as well as a lower tare weight.

**HIGH PLI PRESS:** A wet press on a linerboard machine that operates with what is considered high loading pressure above 800 pounds per lineal inch (PLI).

**HIGH PRESSURE FEEDER:** A special valving device designed to feed materials, such as chips or pulp, in a pressurized vessel in such a way that a seal is maintained without a loss in pressure.

**HIGH PRESSURE INTERLOCK:** Pressure-sensing device used to close gas header, fuel oil, and igniter valves when the furnace pressure exceeds a safe value.

**HIGH PRESSURE SHOWERS:** Pulpwood grinder showers using high pressure water. They are usually used when the grinder is run at high speed.

**HIGH SPEED PITLESS GRINDING:** Operating a pulpwood grinder at higher speeds than normal by removing the pit dam, so that the stone is not submerged, and with increased shower water in order to increase production of groundwood.

**HIGH TEMPERATURE BLEACHING:** Operating the bleaching stages (hypochlorite or chlorine dioxide) of a multi-stage pulp bleaching system at temperatures higher than considered conventional.

**HIGH TEMPERATURE CHLORINATION:** Operating the first bleaching stage (chlorination) of a multi-stage pulp bleaching process at higher temperatures (usually 110°F to 120°F) than considered conventional (less than 80°F).

**HIGH TORQUE CENTERWIND:** The procedure of winding rolls using a center axial drive with torque controlled so that it results in uniform tension and regulated roll hardness.

**HIGH VELOCITY DRUM DRYING:** The drying of a coated paper by the use of high velocity hot air impinging on the surface of a sheet of paper as it is supported by a dryer drum heated by steam.

**HIGH YIELD:** Descriptive reference to processes in which the yield is higher than the conventional yield, such as in pulping and bleaching under controlled conditions to minimize lignin and hemicellulose removal.

**HIGH HEAT WASHING:** The countercurrent washing of pulp with weak black liquor while still in the lower part of a continuous digester.

**HIGH SOLIDS COATING COLORS:** Kneaded, high-density, pigmented coating dispersions in the 50 to 65 percent solids range.

**HIGHWAY:** A data communications system that allows a number of computer systems to transmit and receive information from each other.

**HIGHWAY FOR PROCESS MANAGEMENT AND CONTROL:** A data communications system that allows a number of computer systems to transmit and receive information from each other and other devices included in an overall process management and control system.

**HOG:** A machine used to grind wood waste into small pieces so it can be burned in power furnaces.

**HOG FUEL:** Raw bark, wood waste, and other extraneous materials that are pulverized and used as a fuel for power boilers in a mill.

**HOG PIT:** A term used in some paper mills to refer to the couch save-all pit located under the wet end of a paper machine.

**HOIST:** Types of lifting devices that mechanically elevate material by the use of chain, cable, rope, etc. It usually consists of a block and tackle with multiple sheaves.

**HOLDOUT:** The ability of paper or paperboard to resist surface liquid penetration.

**HOLE:** A paper defect appearing as a small aperture in the sheet as a result of fiber aggregate in the stock, slime, splashing, etc.

**HOLEY ROLLS:** Hollow shell rolls, perforated with a regular pattern of holes, mounted on stub axles, and located in the approach flow of stock to the slice of a paper machine headbox. It is used to stabilize and condition the stock.

**HOLLANDER:** A common mill term referring to beaters because of their original Dutch association. *See* BEATING ENGINE.

**HOLOCELLULOSE:** The total polysaccharide fraction of caustic extractive-free wood.

**HOMESPUN FINISH:** A type of paper finish achieved by pressing paper between glazed boards and homespun cloth.

**HOMOGENEOUS:** Characteristic of materials which are of uniform structure or composition.

**HOMOGENIZE:** To reduce suspended materials in a slurry to small particles of uniform size so that they may be evenly distributed and blended into a smooth mixture.

**HOOD:** (1) A canopy located over process machinery, such as paper machine dryers and pulp washers, to collect and carry away vapors by evaporation from the process. (2) The front or firing end area of a lime kiln.

**HOOK:** A tool used in handling wood in a grinder room.

**HOPPER:** A funnel-shaped chute or receptacle with an open bottom. It is used to add material such as chips, coal, etc., into digesters, furnaces, etc. It is also used to receive ashes being discharged from furnaces.

**HORIZONTAL BELT WASHER:** A type of brown stock washer designed to use, and whose operation is based on, a moving horizontal belt.

**HORIZONTAL CENTRIFUGAL SCREENS:** Types of screens used for groundwood screening in which the stock is forced to spin around the inside of a horizontal screen drum and propelled from one end to the other by a winged rotor. The good fibers pass through the holes while the rejects move toward and exit out the opposite end.

**HORIZONTAL CHESTS:** Pulp stock storage tanks, usually of comparatively low height, in which uniform mixing and suspension are maintained by moving the slurry longitudinally by the use of midfeathers fitted with strategically located propellers, depending on the size and shape of the tank.

**HORIZONTAL POROSITY:** That property of paper related to the ability of a liquid to flow through a sheet in a general direction parallel to its plane.

**HORIZONTAL SIZE PRESS:** Type of size press in which the rolls are arranged next to each other in the horizontal plane, and the paper sheet to be sized runs vertically between them.

**HORIZONTAL SPLITTER:** A log cutting machine, used in the wood preparation area of pulp mills, in which large diameter logs are placed in a horizontal position to be split into smaller sizes so that they will slide through chipper chute openings and better fit pulpwood grinder pockets.

**HORN ANGLE:** The wedge-shaped section of pulpwood logs in the lower portion and up against the stone; the wood-to-stone pressure in a ring-type pulpwood grinder is created in this section.

**HORSEPOWER (HP):** The unit rate of doing work. For instance, 1 horsepower is equal to 33,000 ft/lb per minute, or 42.44 Btu per minute. Electrically, 1 horsepower is equal to 764 watts.

**HORSEPOWER-DAY PER TON:** The specific energy consumption used as a basis of control for a grinder, refiner, or similar equipment. It is calculated from the average driving motor load in horsepower divided by the average oven dry production or throughput of oven dry pulp in tons per day.

**HORSEPOWER-HOUR:** A measure of work done. For instance, 1 horsepower-hour is equal to 33,000 ft/lb x 60 minutes, or 1,980,000 ft/lb.

**HOSE END VALVES:** Water service valves that have fittings for hose connections.

**HOISERY PAPER:** Stocking wrapping paper made from lightweight paper and commercial tissue.

**HOT ACID SYSTEM:** A process used in sulfite pulp mills to reclaim all digester reliefs and to absorb the sulfur dioxide ($SO_2$) gas in hot cooking acid.

**HOT ALKALINE EXTRACTION:** The treatment of pulp (as a stage in the bleaching process) with caustic solution at elevated temperatures.

**HOT BLOW:** Discharging cooked pulp slurry from a continuous digester into the blow tank without cooling with black liquor.

**HOT CALENDERING:** (1) Using a steam shower at the calender stack of a paper machine to enhance smoothness and gloss properties. (2) A thermal bonding process for nonwoven fabrics in which a hot metal roll is used with felt, cotton, or special composition rolls in a calender to bond materials, including disposables such as diapers, sanitary products, and medical products. Also referred to as *hot roll calendering.*

**HOT GRINDING:** The making of mechanical pulp by grinding pulpwood under conditions in which the shower water is adjusted to maintain the pit water temperature between 70° and 95°C, thereby facilitating defibration of the logs.

**HOT LIME:** High temperature lime as it emerges from the lime kiln in a lime recovery plant of a pulp mill.

**HOT LIME-SODA PROCESS:** A method of softening a hard water supply by the use of lime and soda ash in a pressurized settling tank at boiling temperatures and without coagulants.

**HOT LOGGING:** A pulpwood operation in which logs are transported directly to the mill site from the tree-cutting operation without any intermediate layover or storage.

**HOT MELT COATING:** The coating of paper or paperboard with hot molten wax or plastics applied directly to the surface using methods not requiring any carrier solvent.

**HOT MELTS:** Paper and paperboard plastic or wax coating materials in a molten state applied to a sheet to produce a pleasing, fluid-resistant surface with a high gloss.

**HOT PRESSING:** (1) The use of increased temperatures in the press section of a paper machine to speed up drainage and improve water removal from the newly formed paper web. (2) The process of producing a smooth, glossy finish on paper surfaces by applying slight friction under pressure with the use of polished metallic plates or rollers, which increases the temperature. It is also accomplished on supercalenders.

**HOT REFINING:** Refining of stock slurries at temperatures higher than normal for standard mill conditions. *Also see* REFINING.

**HOT ROLL CALENDERING:** *See* HOT CALENDERING.

**HOT ROLLING:** The process of producing a high glossy finish or polish on the surface of paper by passing it through nips of steam-heated calender and supercalender cylinders.

**HOT STOCK SCREENING:** High temperature screening of pulp between the blow tank and the brown stock washers in a pulp mill with the use of pressure-type rotary screens.

**HOT WAXED BOARD:** Paperboard that has been waxed by applying molten paraffin directly to the surface and allowing it to solidify by natural cooling. This permits some of the paraffin to penetrate into the sheet.

**HOTWELL:** A chamber at the bottom of a condenser where the warm condensate can collect prior to being pumped back into the process, such as in black liquor evaporators and power and recovery boilers.

**HOUSE SHEATHING:** Asphalt-saturated, waterproof, felt-type sheet, coated on both sides and used extensively in the building industry.

**HOUSING:** The outer shell of a pump that encloses the impeller. Also referred to as a *casing.*

**HOUSEKEEPING:** A common expression used around a mill that refers to the maintenance of clean, neat, and orderly conditions in and around the process areas.

**HP:** *See* HORSEPOWER.

**HPC:** *See* HIGH PERFORMANCE CONTAINERBOARD.

**HUE:** Shade or tint of various papermaking dyes and colors.

**HULL FIBER:** Fiber obtained from the coverings of crushed cotton seeds and used to make specialty paper.

**HUMIDIFIER:** A piece of equipment used to raise the moisture content of a paper sheet during the manufacturing process, or to inject moisture into the air of a room where it is finished and stored. It is usually accomplished by the use of water sprays.

**HUMIDITY:** The amount of moisture in a gas or gases, whether isolated or as part of the atmosphere, and expressed in absolute or relative terms. In and around the mill, it is generally considered as the water vapor content of the air. *See also* ABSOLUTE HUMIDITY and RELATIVE HUMIDITY.

**HUNTING:** An undesirable oscillation produced in the process or control system operation, usually around the point of control or desired operation point, that persists after the cause has been removed.

**HYDRA-NIP:** A system located in the dryer section of a paper machine that uses a combination of moisture content, heat, and soft nips to pre-smooth the sheet of paper before it is calendered.

**HYDRATE:** A reference to substances that have had water added to their molecular structure.

**HYDRATED ALUMINA:** A paper-coating pigment prepared by the hydrolysis of sodium aluminate liquors. It is also used as titanium extenders in coating mixtures.

**HYDRATED LIME ($CaOH_2$):** Partially slaked lime produced by adding water to lime ($CaO$).

**HYDRATED STOCK:** Pulp slurry submitted to prolonged mechanical treatment until it becomes partly gelatinized. It imparts a "greasy" feel to greaseproof and glassine papers made from it.

**HYDRATION:** A phenomenon occurring during the mechanical treatment of pulp in refiners and beaters where there is an increase in the capacity of fibers to absorb and hold water. Once generally referred to as *slowness* or *wetness.*

**HYDRAULIC:** A general reference to mill equipment operated, moved, etc., by means of a liquid force.

**HYDRAULIC ACTUATOR:** A device used for converting hydraulic energy into mechanical energy.

**HYDRAULIC BARKER:** A type of pulpwood bark removal machine that utilizes high pressure water jets (at a suitable angle) to separate bark from logs.

**HYDRAULIC COOKING:** Cooking fibrous materials (wood chips) in a pressurized digester that has no vapor space, such as in continuous digesters, when the chip mass moves through in plug form.

**HYDRAULIC CYLINDER OPERATOR:** A cylinder used to provide mechanical motion by use of water or other liquid power.

**HYDRAULIC DEBARKING:** *See* HYDRAULIC BARKER.

**HYDRAULIC HEADBOX:** Type of paper machine pulp stock feeding chamber that operates completely full without any air pad.

**HYDRAULIC PISTON OPERATOR:** A piston that provides mechanical motion by use of water or other liquid power.

**HYDRAULIC PRESSING:** The removal of water from pulp laps (from wet machines) by the use of a hydraulically operated press, which is common in groundwood stock preparation.

**HYDRAULIC PRESSURE:** The force created by compression of a liquid in a closed structure, which may be transmitted to a moving component in the compartment.

**HYDROCELLULOSES:** Those portions of the product formed when cellulose is treated with acids and are insoluble in water.

**HYDRODYNAMIC LUBRICATION:** A lubricating phenomenon occurring between a pulpwood grindstone surface and the wood logs under proper pressure during the wood grinding operation.

**HYDROENTANGLEMENT:** A mechanical process for bonding a nonwoven fabric using high pressure water jets during its production.

**HYDROFOIL BLADES:** Various shaped boards used in place of table rolls to support the forming fabric carrying the draining layer of stock on larger and faster paper machines. It assists in dewatering, retention, and stock activity in the forming zone. Depending on the action desired, some of the shapes used are referred to as linear, shaped, stepped, or turbo.

**HYDROFOILS:** *See* HYDROFOIL BLADES.

**HYDROGEN BOND:** Interfiber cohesion in a sheet of paper that is attributed to the hydroxyl groups of the carbohydrates sharing a hydrogen atom in adjacent fibers.

**HYDROGEN PEROXIDE ($H_2O_2$):** An oxidizing type of bleaching chemical commonly used in bleaching and deinking paper stock as well as in bleaching other chemical and mechanical pulps.

**HYDROGEN SULFIDE ($H_2S$):** A gas having an odor of rotten eggs and commonly found in power and recovery furnace flue emissions.

**HYDROGEN-ION CONCENTRATION:** The hydrogen ion ($H+$) content of aqueous solutions that determines its degree of acidity or alkalinity. Generally expressed as pH.

**HYDROLYSIS:** The absorption of water by a chemical compound, causing it to form a new compound or compounds.

**HYDROMETER:** A measuring device used to determine the density or specific gravity of liquid solutions or mixtures as an indication of their concentration.

**HYDROPHILIC:** The physical property of substances that gives them a strong affinity for water. Also referred to as *easily wetted.*

**HYDROPHOBIC:** The physical property of substances that make them strongly repellent of water and said to be "not" *easily wetted.*

**HYDROPOWER:** Power or electricity produced by generators operated by water turbines, water wheels, or other water-actuated devices.

**HYDROSTATIC:** Relating to liquids in the static state or at rest and the pressure they tend to exert or transmit.

**HYDROSTATIC HEAD:** When used in reference to pressurized paper machine head boxes, it is the pressure produced by weight of stock at the slices.

**HYDROSULFITE:** Reducing-type bleaching chemicals, usually in the form of zinc hydrosulfite or sodium hydrosulfite, used in the bleaching of groundwood pulp and, to a lesser degree, recycled furnishes.

**HYGROEXPANSIVITY:** That property of a sheet of paper related to the dimensional change occurring during a change of relative humidity of the surrounding air. Also known as *expansion.*

**HYGROINSTABILITY:** A paper and paperboard characteristic that makes it susceptible to dimensional and shape changes as the moisture content varies.

**HYGROMETER:** A measuring device used to determine atmospheric relative humidity.

**HYGROSCOPE:** An instrument used to measure the relative humidity of air.

**HYGROSCOPIC:** The ability of a substance to readily take up and retain moisture from the air.

**HYGROSTABILITY:** The ability of paper and paperboard to resist dimensional changes as the moisture content varies.

**HYPERFILTRATION:** A filtration system in which pulp mill spent liquor effluents are passed, under pressure produced by a pump, between semipermeable membranes which retain polluting, toxic, and colored substances, and molecularly separate useful products such as lignins, wood sugars, etc.

**HYPOCHLORITE:** Reducing-type of bleaching chemical, usually in the form of calcium hypochlorite or sodium hypochlorite, used extensively in the bleaching of chemical pulps.

**HYPOCHLORITE STAGE:** The step or steps in a multi-stage bleaching process in which hypochlorite bleaching chemicals (usually calcium or sodium hypochlorite) are mixed, allowed to react, and washed.

**HYSTERESIS:** The return of a function to its origin through a different path.

# I

**IGT TESTER:** A paper mill laboratory test instrument used to determine the resistance of paper to picking by printing inks.

**ICE:** Abbreviation for industrial corrugated containers, also known as double-lined kraft containers.

**ICE CREAM BOARD:** A clean, moisture-resistant paperboard made from chemical pulp especially for use in manufacturing ice cream cartons.

**ID:** *See* INDUCED DRAFT.

**I.D.:** Abbreviation for inside diameter measurement as used with piping, tankage, etc. *See also* OUSTIDE DIAMETER (OD).

**IDEAL PIGMENT:** A plastic papermaking coating pigment that would meet certain criteria, including low specific gravity, high brightness, high refractive index, controlled particle size, easily dispersible, chemically inert, compatible with other pigments, nonabrasive, low adhesive demand, and high price and performance efficiency.

**IDLE TIME:** The time that any piece of process machinery or equipment is out of service.

**IDLER ROLLS:** Nondriven rolls on a paper machine that ride on paper and felt sheets to take up any slack that might occur.

**IEEE:** *See* INSTITUTE OF ELECTRICAL AND ELECTRONICS ENGINEERING.

**IGNITOR:** A supervised burner in a power or recovery furnace that is spark ignited. It continuously supplies enough energy to sustain stable operation of other burners or a hearth no matter what lighting off or firing conditions exist.

**ILLUSTRATED LETTER PAPER:** A specially treated bond paper with one side coated for artwork applications. The other side is used for typing, painting, or writing.

**ILLUSTRATION BOARD:** A rigid board made with a liner of drawing paper. It is used for making illustrations and other artwork. Also called *drawing board.*

**IMBIBED WATER:** Water that cellulose fibers will absorb on immersion after they have taken up all of the moisture they can from being exposed to saturated water vapor.

**IMBIBITION:** The absorption of water by fibers without an increase in volume.

**IMITATION ART PAPER:** An imitation coated paper sheet made with heavy loading and high supercalender finish.

**IMITATION HANDMADE PAPER:** Paper whose surface has been processed between metallic plates or embossed to give the appearance of having been made by hand.

**IMITATION KRAFT PAPER:** A brown-colored paper made from unbleached mechanical pulp and wastepapers, and used as a substitute for kraft paper where strength and durability are not of prime concern.

**IMITATION LEATHER:** A type of high-strength resin or rubber-impregnated heavyweight or laminated paper embossed to look and feel like leather.

**IMITATION PARCHMENT:** Paper made from sulfite pulp that has undergone prolonged mechanical treatment so that it is well-hydrated and the sheet made from it resembles a vegetable parchment.

**IMMERSION:** A size resistance test used on heavy paperboards. It consists of weighing a conditioned sample before and after total immersion in water to determine the amount absorbed.

**IMMERSION TEST:** *See* IMMERSION.

**IMMISCIBILITY:** That property of materials that resists mixing or attaining homogeneity.

**IMPACT DRYING:** *See* IMPULSE DRYING.

**IMPACT RESISTANCE:** The ability of paperboard shipping containers to withstand damage due to impact stresses.

**IMPEDANCE:** The opposition to the flow of electricity in an *alternating current (A.C.)* electrical circuit.

**IMPELLER:** (1) The rotating member of a pump or blower that moves the flowing material. (2) A fan-like propeller installed in baffled walls of stock storage tanks to maintain movement and suspension of slurries.

**IMPERFECTIONS:** A paper mill reference to defective sheets that are sold as a low grade or wastepaper.

**IMPERVIOUS:** The ability of paper and paperboard to resist liquid and grease assimilation.

**IMPINGEMENT AIR DRYING:** The drying of paper by use of high velocity and high temperature air jets perpendicularly directed to the surface of the wet sheet.

**IMPINGEMENT SCRUBBER:** A type of flue gas scrubber normally used in lime kiln operations. It consists of a vertical tower equipped with perforated horizontal plates having holes with baffles located over them through which gases pass and impinge as they are subjected to a liquid cleaning process.

**IMPREGNATING COMPOUNDS:** Chemical materials and agents used to penetrate paper, paperboard, pulp, wood, etc., in order to impart certain desired physical characteristics.

**IMPREGNATING PAPER:** Soft, porous grades of paper with good wet tensile strength, made to be saturated with various types of solutions and compounds.

**IMPREGNATION:** (1) The process of treating fibrous materials with penetrating liquids prior to chemical or mechanical pulping. (2) The application of liquids or molten materials to a paperboard sheet so that it will be absorbed to impart certain desired characteristics.

**IMPRESSED WATERMARK:** A watermark applied to the wet sheet as it passes through the presses of a paper machine. This is done by the use of engraved rubber rings impressing the design on the web.

**IMPULSE DRYING:** A process of removing water from a moist sheet in the paper machine section using a heated press roll in a high pressure nip. This creates steam that expands and drives water into the press felt, thereby reducing the load and requiring less energy in the dryer section.

**IMPULSE PRESSING:** Use of a press section with special covered rolls that flatten and extend the nip under high pli-loadings, thus improving press efficiency. Sometimes referred to as *high impulse pressing.*

**IMPULSE TO RUPTURE:** A paper mill laboratory strength test made on paper by integrating the product of rupture force and the time which the force has been acting on a prescribed sample.

**IN-CROWN:** The crown profile of a paper machine roll before being run on the machine.

**IN-DIGESTER WASHING:** The process of washing wood pulp in the digester vessel in which it was produced.

**INCH-HOUR:** A means of expressing a paper mill's operating rate. It is determined by multiplying the total inches of machine widths by the number of hours over a selected period of time.

**INCHING:** Moving sections or the wire (forming fabric) of a paper machine at very low speed, usually with a small, *alternating current (A.C.),* high torque gear motor, large enough to overcome static friction breakaway from the rest position. It facilitates inspection, changing of felts, startup, and clearing the cylinder of broke after a machine break. Sometimes known as *jogging* and *knock-up.*

**INCINERATION:** The controlled consumption by burning of solid, liquid, or gaseous materials or fuels in a power or chemical recovery furnace.

**INCLINED PRESS:** A horizontal, three-roll, wet press with the rolls arranged at about a 45° angle to the horizontal.

**INCLINED SCREW THICKENER:** A type of press normally used to thicken pulp from about 4 percent to 12 percent consistency, employing a screw-type compressor operating in an inclined shell housing.

**INCLINED SCREW WASHER:** A type of washer used on deinked stock after cleaning and screening to remove the dispersed inks, clays, and chemicals. It consists essentially of an inlet section supplying a number of inclined screws. The stock is pushed upward through a common perforated cylinder as the screws rotate. Water drains through the perforations as the stock is pushed out the port in the upper end.

**INCLINED TUBE DIGESTER:** A type of continuous digester whose main vessel consists of a tubular chamber mounted at an angle and usually fitted with an internal chip-moving mechanism, such as a drag conveyor, to move the chips being cooked down and up the digester.

**INCLINED WIRE SAVE-ALLS:** Devices used to reclaim fibers and fillers from paper machine white or filtrate water by the use of an inclined screen of fine mesh wire over which the water is allowed to flow.

**INDENTED BOARD:** A soft, cushion type of paperboard used for protective packaging, subliner for carpeting, and other softening purposes. The sheet, containing a high percentage of secondary fiber (sometimes 100 percent), is run through impression rolls during manufacturing without puncturing the sheet.

**INDENTED PAPER:** A soft, bulky type of paper used for packaging protection purposes. It is made with raised bumps and/or indentations by passing the sheet through impression rolls during manufacturing without puncturing the sheet.

**INDEX BRISTOL AND BOARD:** A surface-sized paper and paperboard made from hard beaten chemical wood and cotton pulps. It is used for converting into index record cards.

**INDEX CARD:** Paperboard cards of various sizes made from index bristol and board. They are used primarily for filing and other record keeping purposes.

**INDEX PRESSBOARD:** A paperboard made on a cylinder-type paper machine using chemical pulp and some rag fibers. It is used for converting into user items such as folders, pamphlet covers, and other filing or record keeping articles that require a stiff and fairly durable paper product.

**INDIA PAPER:** An opaque, lightweight printing paper made especially for use in bibles, prayer books, dictionaries, or wherever a low bulk sheet is required.

**INDIA PROOF PAPER:** A soft, absorbent, unsized printing paper made especially for use with steel engravings or woodcuts.

**INDIA TINT:** A light buff paper color usually obtained by mixing yellow ochre into the paper stock slurry before making the sheet.

**INDICATING GAUGE:** An instrument on which a measured process variable is visually displayed.

**INDICATOR PAPER:** Special unsized, treated paper with specific chemicals so that it will change colors when in contact with designated substances in fluids (in order to determine their presence).

**INDIRECT AIR HEATER:** A type of heat exchanger used on chemical recovery furnaces in which the combustion air is heated by passing hot gas in a countercurrent direction through the opposite end of a tubular chamber with no direct contact between the air and the hot gas.

**INDIRECT COOKING PROCESS:** The process of chemically pulping chips with cooking liquor heated by steam in a heat exchanger. There is no direct contact between the cooking liquor and the steam.

**INDIRECT HEATED DIGESTER:** A batch digester that uses heat exchangers to heat the cooking liquor, usually with steam. The steam never comes into direct contact with the liquor, which is circulated through the heater by a circulating pump external to the digester.

**INDUCED DRAFT (ID) FAN:** A fan located between a lime kiln, power furnace, or chemical recovery furnace, and flue or stack to create a negative pressure (induced draft) that will pull excess air and combustion products out of the kiln and furnace.

**INDUCTION MOTOR:** A motor in which the electrical energy delivered to a primary circuit is converted to mechanical energy by electromagnetic interaction between it and a closed secondary circuit.

**INDUCTION VOLTAGE REGULATOR:** A type of voltage regulator in which the positions of the primary and secondary windings can be adjusted by rotation.

**INDUCTIVE LOAD:** The type of load on an electrical system in which the current lags behind the voltage over the entire load.

**INDUSTRIAL BOILER:** A boiler used primarily to generate steam for process requirements but may also be used to provide steam for turbine generator operation.

**INDUSTRIAL PAPERS:** Grades of paper made specifically to be used for such purposes as industrial packaging, tissues, wrappings, impregnating, insulating, etc.

**INDUSTRIAL POWER BOILER:** A boiler used to produce energy primarily for use in an industrial plant.

**INDUSTRIAL UNION:** A type of labor organization found in a mill that admits all workers to membership regardless of their job, occupation, or craft. Sometimes referred to as a *vertical union.*

**INDUSTRIAL WASTES:** Process effluents originating in a mill and discharged into a treatment plant or into the environment such as ponds, lakes, or streams.

**INDUSTRIAL WIPES:** Supple, lint-free, highly absorbent papers with good wet strength characteristics. They are used to wipe up industrial oils, water, and other liquid spills.

**INERTIA:** The tendency of an object to remain at rest or in motion until acted upon by some other force.

**INFILTRATION AIR:** Air that leaks into chambers and ducts of lime kilns and power and chemical recovery furnaces.

**INFLUENT:** Mill wastes, water, and other liquids that are raw or partially treated flowing into a treatment plant, reservoir, basin, or holding pond.

**INFRARED DRYING:** Use of dryers that emit *infrared rays* for paper drying purposes, particularly for drying after a coater station. The infrared heating can be either electrical or gas fired, generally utilizing special ceramic elements.

**INFRARED (IR):** Invisible light rays that are longer than red rays in the visible spectrum and have a thermal effect that has been utilized in the drying of paper and other materials.

**INGRAINED PAPER:** Special finished paper made with surfaces treated to impart a blotchy, granite-like appearance.

**INITIAL ADHESION:** *See* AUTO ADHESION.

**INITIAL GRINDING FACTOR:** *See* GRINDING FACTOR.

**INITIAL TEARING RESISTANCE:** A physical property of paper related to the force required to start a rupture or tear at the edge of the sheet specimen under standard conditions.

**INJECTION COOKING:** Type of kraft wood cooking process that includes a continuous injection of white liquor during the cook to keep the alkali concentration at a constant level.

**INJECTOR:** *See* EDUCTOR.

**INJECTOR CELL:** A recently developed low energy flotation machine used in the deinking process.

**INK ABSORBENCY:** That ability of paper and paperboard to take up and retain writing, printing, and other inks.

**INK FLOTATION TEST:** A common paper mill laboratory test used to determine the size resistance of paper. It consists of floating a 2-inch square sample on standard writing ink and timing the period when the color shows through the upper surface.

**INK HOLDOUT:** The resistance of paper or paperboard surfaces to the penetration of various writing and printing inks.

**INK RECEPTIVITY:** The ability of paper and paperboard to take up and retain oil-based inks.

**INK RESISTANCE:** A paper property that prevents the penetration of inks used in writing, printing, etc.

**INK RUB-OFF:** The degree with which printing inks, especially oil-based inks used on newspapers, will rub off when subjected to the friction of handling or coming in contact with other objects.

**INK SHOW-THROUGH:** A characteristic of lightweight, absorbent-type papers in which ink is visible from the opposite side of the sheet due to penetration of the oil vehicle, thus making the paper more transparent.

**INK TRANSFER:** The quantity of ink that will adhere to paper or paperboard surfaces during printing by impression methods.

**INLET:** The piping system located just prior to the paper machine headbox through which paper stock furnish is fed and distributed to the headbox proper.

**INNERSOLE BOARD:** A high-strength, flexible, wet machine board made from rag, hemp, and sulfate chemical wood pulp. It is used especially in the shoe industry as a substitute for leather innersoles.

**INORGANICS IN WOOD:** The mineral constituents in pulpwood.

**INPUT:** (1) Any data or information furnished to a computer or instrumentation system for processing. (2) Devices that perform the operation in (1) above.

**INSIDE ROLL:** A roll located right after the suction press roll on a paper machine. Its purpose is to break the vacuum by opening up the felt, thus helping to keep the sheet from following the felt when the tail is fed over the machine.

**INSOLUBILIZERS:** Chemicals added to paper-coating materials to make them water-resistant.

**INSPECTION:** A unit operation in the finishing process of papermaking to detect and sort out products that are defective due to variations in thickness, dirt, lumps, holes, etc. The operation can be done manually or with sensors installed on automatic sheeting machines.

**INSTANT STARCH:** *See* PREGELATINIZED STARCH.

**INSTITUTE OF ELECTRICAL AND ELECTRONICS ENGINEERING (IEEE):** Headquarters: New York, NY, (212) 705-7900.

**INSTITUTE OF PAPER SCIENCE AND TECHNOLOGY (IPST):** Formerly the Institute of Paper Chemistry. Headquarters: Atlanta, GA., (404) 853-9500.

**INSTRUMENT:** A device that automatically measures, indicates, or records and controls process variables.

**INSTRUMENT SHOP:** An area or room in a pulp and paper mill where repairs on various types of instruments and automatic valves are carried out.

**INSTRUMENT SOCIETY OF AMERICA (ISA):** Headquarters: Research Triangle Park, N.C., (919) 549-8411.

**INSTRUMENTATION:** A group of applied devices used to detect, measure, indicate, record, compute, and control manufacturing and process variables in a mill.

**INSULATING BOARD:** Bulky paperboards made with a substantial amount of entrapped air so that it can be used to repel heat and sound. Electric insulating boards are dense and made of pressed boards or are laminated.

**INSULATING LATH:** A type of board made from fibers obtained by mechanically shredding chips. It is used as a base for plaster walls in building construction.

**INSULATING TISSUE:** Thin, lightweight, porous, long-fibered paper made from hemp and chemical pulps. It is used for making up electrical insulation coverings for various types of wires.

**INSULATION PAPER:** A high-strength, long-fibered paper made primarily from sulfate chemical pulp and generally used to make up linings for batt-type insulations.

**INSULATION RESISTANCE:** The ability of a paper sheet or tape to prevent the flow of an electric current across its surface.

**INTAGLIO PAPER:** Grades of paper suitable for use in intaglio printing. In this process, printing ink is transferred to the paper surface by the use of plates in which the printing areas are engraved well below the surface.

**INTAKE:** The inlet side of process equipment such as pumps, fans, etc.

**INTAKE SHAFT:** The positive mechanical means of transmitting power from the drive mechanism to the sections of electric motor-driven paper machines.

**INTEGRAL CONTROL ACTION:** A process controller mode in which the output is proportional to the time integral input, or the rate of change of the output is proportional to the input. Also referred to as *set control action.*

**INTEGRATED PULP AND PAPER MILL:** (1) A pulp and paper mill manufacturing complex in which all pulp and papermaking operations are conducted at one site. Sometimes converting operations, such as bag and tissue manufacturing, are also included. (2) Paper and paperboard mills that produce their own pulp.

**INTEGRATED SYSTEM:** A system that links several process control systems together so they will operate as a unit.

**INTEGRATOR:** A device used to continuously add up the quantity being measured over a period of time.

**INTELLIGENCE:** The capability of a device to manipulate incoming data to produce a related result.

**INTENSITY:** The degree of brightness or dullness of colors.

**INTERCALENDER DRYERS:** Cylinder dryers located between calenders when water or sizing is added to the paper sheet at the first calender stack in order to improve the finish.

**INTERFACE:** (1) The place at which an operator and control system, an operator and process, a control system and process, or one control system and another control system meet and interact with each other. (2) The adhered surfaces which come into contact with the adhesive or the area of contact between them when a multi-layer sheet is made by gluing several of the sheets together.

**INTERFIBER BONDING:** The cohesion produced between fibers, usually by mechanical treatment, which lends toughness and other strength properties to the paper and paperboard sheets produced from them.

**INTERIOR BOARDS:** Any type of board material, including special paperboards, fiberboards, and gypsum boards, used for interior work in building construction.

**INTERIOR PACKING:** Paper and paperboard protective items, such as padding, dividers, and lining material, used inside shipping packages to prevent damage to the contents in transportation.

**INTERIOR WRAPS:** Types of protective paper used to wrap packaged material to prevent damage from invasion of liquids, gases, and other deleterious fluids.

**INTERLAPPING:** In the packing of paper, where a specified quantity of paper is placed over another specified quantity so that only one half of each quantity is in contact with the other, and the overlapping ends are folded over to form a package.

**INTERLEAVING BLOTTING PAPER:** A thin, highly absorbent, unsized paper made to be used by the printing industry as inserts between freshly printed material to absorb excess ink and prevent offsetting.

**INTERLEAVING TISSUE:** A lightweight grade of paper used as a separating interleave protective medium in the handling of printed material.

**INTERLOCKING:** The arrangement of equipment or devices so that they will operate only in a proper sequence of events within limit or off-limit conditions. They cause a starting or stopping of related operations and are commonly used in multi-stage pulp bleaching, paper machines, and chemical recovery and power boiler operations.

**INTERMITTENT:** Refers to discontinuous or batch operations of unit processes in the mill during the manufacture of pulp and paper and includes beater operation, cooking of fibrous material, color makeup, and coating makeup.

**INTERNAL BONDING STRENGTH:** The sticking together of fibers within the paper or paperboard sheet structure that provides toughness and strength.

**INTERNAL ENERGY:** *See* INTRINSIC ENERGY.

**INTERNAL GEAR:** Any machine gear that has teeth located on the inner side of the outer circular rim.

**INTERNAL SIZING:** The process of adding suitable chemicals to a papermaking stock slurry so that they precipitate on the fibers in order to control the penetration of liquids into paper or paperboard made from it.

**INTERNAL TEARING RESISTANCE:** The force needed to continue tearing a paper specimen under standard conditions once the tear has been initiated. It is thus a measure of that specimen's internal resistance to tearing (after the initial tearing force has been exerted).

**INTERNAL THREAD:** The screw thread located on the inner surface of a nut or any other screw type fitting.

**INTERNAL TREATMENT:** The treatment of raw boiler feedwater by direct feeding of chemicals in order to precipitate impurities as a sludge rather than scale, to condition the sludge so it will not stick (allowing it to be removed by blowdown), and to make it noncorrosive.

**INTRINSIC ENERGY:** The amount of energy contained in a unit mass of a substance due to its temperature. Sometimes referred to as *internal energy*.

**INVENTORY:** A surplus of raw materials, pulp, paper, or paperboard accumulated and stored for future needs and/or sales.

**INVERSE PRESS:** A vertically aligned, two-roll, single-nip, wet press with the smooth roll contacting the wire (forming fabric) side of the paper sheet moving in the machine direction. Also called an *upside-down press*.

**INVERSION:** A weather situation that produces stagnant air conditions. Layers of cold and hot air prevent normal rising currents, thus preventing mill emissions from being properly dispersed upward into the atmosphere.

**INVERTED BLADE-TYPE COATER:** A type of paper coater used on paper machines that employs an inverted flexible blade to remove excess coating after it has been applied to the paper by an applicator roll.

**INWARD FLOW SCREENS:** Types of rotating, cylindrical screens used directly ahead of the paper machine to remove impurities from the stock by having them pass from the outside through screen-type plates toward the center and out the end while the tailings settle to the bottom and are drawn off.

**IPST:** *See* INSTITUTE OF PAPER SCIENCE AND TECHNOLOGY.

**IR:** *See* INFRARED.

**IRON-FREE ALUM:** A grade of aluminum sulfate that is used in the sizing of specialty papers for photography and other applications where it is essential that they be practically free from iron.

**IRON SPECKS:** Paper defects caused by inadvertent entrainment of iron and rust particles in the sheet.

**ISA:** *See* INSTRUMENT SOCIETY OF AMERICA.

**ISENTROPIC:** A property describing (mathematically) the constant portion of heat that is added to a process but cannot be converted to useful work.

**ISO 9000:** A reference to the 9000 series of quality standards as maintained by the International Standards Organization. Basically, these standards–adopted by national standards agencies and organizations in more than 50 countries–focus on manufacturing, installation, servicing, etc., rather than the finished products directly. The objective is to assure consistency at every stage of manufacturing and implementation. Generally, ISO 9001 certification applies to design, development, installation, and servicing stages. ISO 9002 covers production and installation, and ISO 9003 certifies final inspection and testing procedures only.

**ISO BRIGHTNESS**: A measure of paper and pulp brightness that is basically the same as *Elrepho brightness.* It is established by the International Organization for Standardization, ISO, in Geneva, Switzerland. The test is made using a Zeiss Elrepho photometer slightly modified by the addition of a gloss tap.

**ISOBARIC:** A term used when referring to any process under constant pressure.

**ITEM:** When a particular paper grade is made in only one set of specifications such as size, weight, finish, color, etc.

**ITEM 222:** Regulation affecting containerboard boxes used for shipping by trucks in the U.S. *See* RULE 41.

**IVORY BOARD:** A good quality, rigid paperboard with a clay-coated high finish used for halftone art printing purposes.

**IVORY PAPER:** Any paper with a light cream color.

# J

**JACK:** A device used to lift weights or exert heavy forces in a pulp and/or paper mill. It is usually portable and manually operated, using the principle of a screw, lever, toggle joint, or hydraulic pressure.

**JACK LADDER:** A mechanism used to haul pulpwood logs from a holding pond or flume onto a mill conveyor or into chutes feeding barking drums. It consists of two or more parallel chains fitted with wings mounted side by side so that they will pick up logs and carry them lengthwise along the chains. Sometimes called a *log haul-up.*

**JACK PINE:** A resinous tree species used extensively as a source of pulpwood, especially for the kraft or sulfate cooking process.

**JACKETS:** An old paper mill reference to cylindrical wool covers used over paper machine rolls, such as couch rolls.

**JACKSON TURBIDITY UNIT (JTU):** A measure of the turbidity of water that is proportional to the silicon content. It is expressed in parts per million (ppm).

**JACQUARD BOARD:** A high-strength, rigid paperboard or pressboard made primarily from jute fibers. Sometimes called *verdol paper.*

**JAMMING:** An undesirable condition that occasionally occurs during the grinding of pulpwood in which there is a sudden short period of high production followed by a stable production of low freeness pulp at a low rate. This is caused by partial binding of the stone surface by high wood pressure interfering with hydrodynamic lubrication.

**JAPAN PAPER:** A strong, mottled-surface, vellum-type paper used in printing up diplomas, greeting cards, and other similar specialty purposes. Sometimes called *Japanese paper.*

**JAPANESE COPYING PAPER:** A very light, high-strength paper that has a characteristic silky appearance and feel and is made from the long fibers of the bark from the paper mulberry tree.

**JAPANESE PAPER:** A type of paper used for printing engravings and originally made in Japan from long-fibered *pulp. See also* JAPAN PAPER.

**JARTOP PAPER:** Waxed, oiled, or varnished manila paper used as an inner liner for jar or bottle tops.

**JELLY GUMS:** Types of high-solid content adhesive coatings used in making foil papers, laminates, and automatic labeling paper.

**JELLY STRENGTH:** A rigidity measurement of chilled paper-coating adhesive solutions under standard conditions.

**JET:** (1) The form taken by the pulp stock as it is ejected from the headbox through an adjustable slot (slice) onto the moving wire (forming fabric) on which the sheet is formed on a fourdrinier paper machine. (2) The shape that pressurized pulp and paper liquids take when flowing into a lower pressure through an aperture.

**JET COMPRESSOR:** *See* THERMOCOMPRESSOR.

**JET CONDENSER:** Type of heat recovery system in which the steam flashed off of process equipment, such as a batch digester, is recovered and utilized for heating wash water. It consists of a means of transferring heat from blow steam condensate to fresh water by use of a liquid-to-liquid or vapor-to-liquid heat exchanger.

**JET COOKER:** A type of cooking device, such as that used on starch suspensions, in which a controlled flow of steam is admitted into a chamber, referred to as a jet, where it is mixed with a stream of starch slurry continuously fed through it under pressure by a pump.

**JET DECKLES:** The feathered edges of a paper sheet produced on paper and paperboard with jets of water or air which are located at both side edges of a wet web. *See also* AIR DECKLE.

**JET DRYER:** Paper dryer that uses high velocity and high temperature air jets impinging directly on the wet sheet.

**JET VELOCITY:** The speed at which the pulp stock is ejected from a paper machine headbox through an adjustable slot (slice) onto a moving wire (forming fabric) on which the sheet is formed. It is controlled at a specific relationship to the speed of the wire (drag/rush ratio) depending on the grade of paper being made.

**JEWELER'S BRISTOL:** A stiff, heavy paper usually made with a high rag fiber content and used primarily by jewelers to display merchandise.

**JEWELER'S TISSUE:** A lightweight, pliable paper made from chemical pulp and cotton fiber, with antitarnish properties. It is used to make wrappings for silver and other jewelry items.

**JIB CRANES:** Types of cranes used throughout a mill to move materials. They generally consist of a beam or beams that project from a supporting wall or post in such a way that they can be swung around in the vertical axis.

**JIG:** A machine commonly used in a mill machine shop to hold two pieces of material in the correct positional relationship while work is being done on them by another part of the machine or another tool.

**JIGGERMAN:** A specially trained person whose job is to carry out the grindstone dressing operation in a mechanical pulp mill.

**JIGGING:** *See* DRESSING.

**JOB LOT:** Out of specification, defective, or discontinued paper. It is made up in small quantities for special orders, and sometimes sold at a lower than standard price.

**JOBBER:** A paper merchant who buys paper in bulk or large quantities and resells it in smaller quantities.

**JOGGING:** (1) Making up a compact, even-sided pile of paper sheets by shaking and pushing them together against a flat surface. (2) Running a piece of machinery for short, intermittent periods of time in order to inspect it before turning it up to full speed. *See* INCHING.

**JOINTS:** The specially designed steam line connecting fitting on the hollow journals feeding steam to the individual paper machine dryer drums, or any line position connector that connects two or more lines.

**JORDAN CHEST:** A paper stock storage chest located just before a *jordan refiner*, and from which it is fed.

**JORDAN FILLINGS:** Wood spacers placed in slots between knives on the plug surface of conical stock refiners.

**JORDAN REFINER:** A type of fiber processing machine consisting of an adjustable conical plug, surface-lined with metallic bar knives, and fitted into a shell with the inner surface having similar knives. It is used to mechanically work pulp stock pumped between them as the plug is rotated in order to impart the desired physical properties to the stock and the sheet produced from it. It is a member of the *conical refiner* group. *See also* CONICAL REFINER.

**JOULE:** An electrical unit of work done by 1 watt in 1 second.

**JOURNAL:** The part of a shaft or axle of a paper machine or any other rotating piece of machinery that rests on a bearing.

**JOURNEYMAN:** A mill worker who has successfully completed an apprenticeship on his specific job.

**JTU:** *See* JACKSON TURBIDITY UNIT.

**JUDGE DYEINGS:** A procedure used during the testing of papermaking dyestuffs in which one edge of dyed sample sheets is cut on a paper cutter and compared for shade and strength.

**JUDGING SHADES:** The process of color-matching paper samples in a paper mill or paper mill laboratory under a constant light source or standard lamp.

**JUMBO MULLEN TEST:** A laboratory test made on multi-ply paperboard to measure the force required to burst it at its weakest layer. It is an indication of its plybond strength. The device used to perform the test is called a *jumbo Mullen tester.*

**JUMBO MULLEN TESTER:** A testing device used to measure the force required to burst multi-ply paperboard at its weakest layer in order to determine its plybond strength.

**JUMBO ROLL:** (1) A larger than normal roll of paper as it is slit and rewound. (2) A roll of paper usually greater than 12 inches in diameter and used for converting into user products.

**JUMP STAGE COUNTERCURRENT WASHING:** A countercurrent method of washing pulp during the bleaching process in which the filtrate from the latter bleaching stages is used serially for showers of the earlier bleaching stages, skipping over the extraction stages. This is useful in bleach plants that do not use stainless steel washers on their extraction stage to prevent corrosion.

**JUNIPER:** A species of tree used as a source of pulpwood.

**JUNK REMOVER:** A device located on rag pulpers or their dump chests to remove large extraneous materials.

**JURY RIG:** A temporary or emergency repair job on a piece of process equipment.

**JUST PERCEPTIBLE DIFFERENCE:** Paper color differences that are large enough to be perceived by the human eye or with the aid of optical detection devices.

**JUTE BAG PAPER:** A type of paper containing jute fiber and used for conversion into bags.

**JUTE BOARDS:** A general term used to refer to high-strength paperboards made mostly from kraft pulp mixed with reclaimed kraft and mixed paper pulp on cylinder machines. It is used to make shipping cartons.

**JUTE BRISTOL:** A type of high-strength, rigid paper containing jute fibers and having good printing surfaces. It is used to make tags and other items that must resist wear and tear.

**JUTE CONTAINER LINER:** A multi-ply paperboard made on a cylinder-type paper machine, the top liner made of kraft waste and the other layers made from reclaimed waste pulp.

**JUTE FIBERS:** Fibers originally derived from the hemp plant in India and used to make coarse sacks and bags. These, in turn, are used as a source of fibers for making reclaimed stock used in papermaking.

**JUTE LINERBOARD:** A multi-ply paperboard lined with waste sulfate pulp, which makes it better suited for printing purposes.

**JUTE PAPER:** Grades of paper made with jute fiber mixed with hemp, rope, stock waste, and small percentages of chemical pulp. It is converted primarily into wrapping paper, envelopes, bags, folders, and other specialty items.

**JUTE PULP:** Type of paper pulp made by chemically cooking waste burlap sacks and bags in a caustic liquor.

**JUVENILE FIBERS:** Fibers from *juvenile wood.*

**JUVENILE WOOD:** That portion of the wood structure of a pulpwood tree that is formed during the early years of its life and differs from the older structure by having short, weaker, and less dense fibers. Also referred to as *corewood.*

# K

**K:** The capacity of digital process computer memory. It means 1,024 bits, bytes, or words of digital data.

**K NUMBER:** A value, also called *permanganate number*, which is the result of a laboratory test for indirectly indicating the lignin content, relative hardness, and bleachability of pulps usually having lignin contents below 6 percent. It is determined by the number of milliliters of tenth normal permanganate solution (0.1 $KMnO_4$) absorbed by 1 gram of oven dry pulp under specified conditions. *See also* KAPPA NUMBER.

**K-B BOARD:** A rigid, water-resistant paperboard made from reclaimed wastepaper stock and asphalt emulsion on a cylinder or wet-type machine. It is used extensively for the manufacture of automobile and other decorative paneling.

**K-B SHEATHING:** A rigid, water-resistant building or insulating paperboard made from reclaimed wastepaper stock and asphalt emulsion on a cylinder or wet type of paper machine.

**KAMYR DIGESTER:** A type of continuous cooking system, characterized by tall, vertical, cylindrical cooking vessels. The chips or other materials to be cooked are fed into the top with the proper liquor make-up under pressure with a special rotary feeding device. The unit takes its name from its manufacturer, Kamyr, Inc.

**KAOLIN:** A naturally occurring white form of anhydrous aluminum silicate clay mineral that is refined and added to paper stock furnish, increasing the opacity of the sheet. It is also used to make a paper coating mixture. *See* CHINA CLAY.

**KAOLINITE:** *See* CHINA CLAY.

**KAPPA NUMBER:** A value obtained by a laboratory test procedure for indirectly indicating the lignin content, relative hardness, or bleachability of higher lignin content pulps, usually with yields of 70 percent or more. It is determined by the number of milliliters of tenth normal permanganate solution (0.1 $KMnO_4$) absorbed by 1 gram of oven dry pulp under specified conditions, and is then corrected to 50 percent consumption of permanganate. *See also* K NUMBER.

**KENAF:** A fibrous annual plant used as a source of papermaking pulp.

**KERF:** The width of a saw cut when cutting pulpwood trees into logs.

**KETTLE:** A process vessel heated by direct steam in a steam coil or a steam jacket. It is used frequently to make up materials such as rosin size and other papermaking materials that require heat for cooking or processing.

**KILN:** *See* LIME KILN.

**KINETIC ENERGY:** The energy forces of material bodies that relate to motion and which tend to keep a moving object in motion until counteracted by equal and opposing external energy.

**KING ROLL:** The bottom roll in a calender stack, and the only driven roll.

**KINK:** (1) A paper machine wire (forming fabric) defect that occurs as a bend or other deformation. It causes a blemish in the finished paper formed on it. (2) A sudden change in the radius of curvature of fiber caused during refining.

**KINKING:** An action that takes place during the refining of pulp in which the fibers are imparted with a sudden change in the radius of their curvature.

**KISS COATER:** A paper coater that applies the coating material on the sheet by the use of a rotating roll that picks up coating material from a pond and transfers it by lightly touching the sheet.

**KISS APPLICATOR:** *See* KISS ROLL.

**KISS COATING:** Paper coating procedure using a *kiss coater.*

**KISS ROLL:** A roll on a paper coater that leaves a coating film of closely controlled thickness on the paper by a single light touch. Sometimes called a *kiss applicator.*

**KITE PAPER:** A very light grade of paper made from chemical pulp and dried on a yankee dryer. It has a glaze on one side and is used to make kites.

**KLEINERT PROCESS:** *See* SOLVENT PULPING.

**KNEADER PULPER:** A type of pulper designed to fiberize wetted, shredded wastepapers prior to deinking. It consists primarily of a receiving hopper that feeds a conical chamber enclosure fitted with a driven rotating shaft equipped with kneader arms.

**KNIFE BARKER:** A machine used to remove bark from pulpwood logs, consisting of a set of rotating knives.

**KNIFE BLADE:** The sharp metal blade on paper cutters and trimmers that cuts the sheet to proper sizes.

**KNIFE COATING:** A method of applying a coating on a sheet of paper that uses a metallic blade to spread a controlled thickness of coating material on the surface.

**KNIFE EDGE:** The edge of a sheet of paper that has been made by slitting or cutting it with a knife.

**KNIVES:** Sharp blades, usually made of steel, for paper cutters, trimmers, pulpwood chippers, and other cutting machines used in the mill.

**KNOCKING BACK:** A common term used in pulpwood grinding mills referring to the process of dulling overly sharp grindstones, usually with a brick.

**KNOCKOFF SHOWER:** A high pressure shower located so as to remove any material adhering to a paper machine roll or other roll.

**KNOCK-UP:** *See* INCHING.

**KNOT DRAINER:** Type of screening device that separates uncooked wood materials from the pulp pumped from the blow tank.

**KNOTS:** (1) A hard lump in pulpwood logs where a branch grew out of the tree. They are difficult to cook in the manufacture of chemical pulp. (2) Lumps in pulp stock caused by fibers that have not been completely separated, or specks in rag content paper caused by knotted threads.

**KNOT SCREEN:** A flat screen used to remove uncooked chips in pulp coming from digesters in a pulp mill.

**KNOTTERS:** Either an inclined plate or a pressurized screen used to separate the knots and uncooked wood materials from pulp. They are located between the blow tank and the unbleached pulp washers in a pulp mill.

**KNUCKLE FREE FELTS:** Felts made with only a machine direction yarn structure. Also called *fillingless felts.*

**KNUCKLES:** The mesh locations in the weave pattern of a paper machine forming fabric where machine direction strands cross the cross-machine direction strands.

**KOLLERGANG:** A machine used in older paper mills consisting of two large stones travelling around in a vat to crush and pulp any rejected paper or to refine pulp.

**KRAFT-AQ PULPING:** Sulfate pulping process used primarily on sawdust in which anthraquinone is added to the cooking liquor to accelerate and improve yield of the cooking process.

**KRAFT PROCESS:** The sulfate chemical pulping process. Also any equipment used as well as any intermediate or final products derived from the process. It means "strength" in German, and is a common pulp mill name for the *sulfate process.*

**KRAFT BAG PAPER:** A high tear strength, usually brown paper used to manufacture shopping bags. It normally is made from unbleached sulfate chemical wood pulp.

**KRAFT BOARDS:** High-strength, bendable paperboards made from unbleached sulfate pulp.

**KRAFT COOKING LIQUOR:** A chemical mixture consisting primarily of sodium hydroxide (NaOH) and sodium sulfide ($Na_2S$). It is used to cook wood chips and convert them into wood pulp. Sometimes called *sulfate cooking liquor. See also* ALKALINE COOKING LIQUOR.

**KRAFT CORRUGATING MEDIUM:** A paperboard made from unbleached sulfate pulp mixed with straw or recycled paper stock. It is normally converted to corrugated board.

**KRAFT DIGESTER:** A pulpwood cooking vessel in which sulfate cooking liquor, consisting of sodium hydroxide (NaOH) and sodium sulfide ($Na_2S$) active chemicals, is used as the cooking medium.

**KRAFT LINER:** The outer layer of a multi-ply vat- or sheet-lined board made up of mostly unbleached sulfate chemical wood pulp mixed with small quantities of reclaimed stock.

**KRAFT LINERBOARD:** A paperboard made up of a high percentage of unbleached sulfate pulp mixed with recycled paper stock. It is used primarily as the outside layer of multiply corrugated container board.

**KRAFT MANILA:** Type of paper made with sulfate chemical wood pulp that does not have the typical dark brown color, but is yellow-colored to obtain a manila appearance.

**KRAFT PAPER:** High-strength paper made from sulfate pulp. It is usually made with a naturally brown color using unbleached pulp, but it can also be made of bleached pulp and dyed to other colors. Also known as *sulfate paper.*

**KRAFT PULP:** Wood pulp produced by the sulfate chemical process using cooking liquor made up primarily of sodium hydroxide (NaOH) and sodium sulfide ($Na_2S$). Softwood species of pulpwood are basically used to make this pulp. Also known as *sulfate pulp.*

**KRAFT PULPING LIQUOR:** A cooking chemical solution made up of sodium-based chemicals such as NaOH, $Na_2S$, $Na_2CO_3$, and $Na_2S_2O_3$.

**KRAFT RECOVERY BOILER POLLUTANTS:** Air contaminants emitted from kraft recovery boilers into the atmosphere, consisting of particulate material comprised of sodium sulfate ($Na_2SO_4$) and sodium carbonate ($Na_2CO_3$) and other solid materials. Also consists of TRS, comprised of gaseous material such as hydrogen sulfide ($H_2S$) and organic sulfur compounds.

**KRAFT RECOVERY CYCLE:** The series of unit processes in a sulfate pulp mill in which the spent cooking liquor is separated from the pulp by washing, concentrated by evaporation, supplemented to make up for lost chemicals, and burned to recover other chemicals. These recovered chemicals are converted to new cooking liquor by reacting them with fresh and recovered lime in a causticizing operation.

**KRAFT TWISTING PAPER:** A high-strength, sulfate chemical paper made especially to be slit into narrow widths and spun or twisted into yarn or twine.

**KRAFT WATERPROOF PAPER:** A highly moisture-resistant paper made from sulfate pulp and treated with moisture repellent materials such as paraffin or asphalt, and used extensively for wrapping purposes.

**KRAFT WRAPPING PAPER:** Grade of paper made from sulfate pulp with various types of finishes. It is used especially for general commercial wrapping purposes.

# L

**LABELABLE:** A reference to the property of a sheet of paper that allows a gummed label to adhere to its surface.

**LABEL CLOTH PAPER:** A high-strength, cloth-lined paper used to make durable labels and other items that require extra strength.

**LABEL PAPER:** A smooth-surfaced paper made from chemical pulp that will readily accept gum adhesives. It is used to make various types of labels.

**LABEL STOCK:** Normally printed, die-cut, pressure-sensitive, adhesive-coated paper or paperboard in a roll or sheet form. It is used in making labels.

**LABORATORY:** The area or areas in a pulp and paper mill where raw materials and products are tested and analyzed to determine if they meet the specifications established for them.

**LACE PAPER:** A medium-weight, well-sized paper made from bleached chemical pulp that has good tensile and tear strength. It is converted into such specialty items as paper lace doilies, lace packaging decorations, greeting cards, etc.

**LACQUER:** An organic solution with volatile solvents used for coating paper. It imparts such properties as high surface gloss, grease resistance, heat sealing, and improved surface appearance.

**LACQUERED PAPER:** Specialty paper whose surface characteristics have been enhanced by coating with a lacquer or similar plastic material.

**LADDER DRYER:** Conveyor-type dryer used to dry coated paper. Also called *flat dryer* and *tunnel dryer.*

**LAG:** A delay in any system output response with respect to a change in input.

**LAGOON:** A holding pond formed by mill effluents. The effluents are allowed to stand for further aeration and stabilization.

**LAID ANTIQUE PAPER:** Book, printing, or writing paper with a rough, ribbed, or lined surface appearance embossed by wires laid parallel to each other on the outer surface of the dandy roll on the wet end of the paper machine. *See also* LINEAR LAID PAPER.

**LAID DANDY ROLL:** A type of dandy roll on the wet end of a paper machine which has axially laid, parallel wires around its surface. It is used to emboss an identical ladder-like appearance on the surface of the paper being formed, which is known as *laid antique paper.*

**LAID FINISH:** The ribbed-surface appearance on *laid antique paper* imparted by a *laid dandy roll.*

**LAID PAPER:** Paper with a surface impression of laid mould wire produced by the use of a laid dandy roll on the paper machine.

**LAKE:** Bright, translucent, organic pigments consisting of a soluble dye adsorbed on or combined with an inorganic base. *See* COLOR LAKE.

**LAKE PIGMENT:** *See* COLOR LAKE.

**LAMELLAE:** Layers that make up the cell wall of a wood fiber and contain the fibrils.

**LAMINATED:** A descriptive reference made to multi-ply paper and paperboard consisting of firmly united, superposed layers that may be bonded with resin or other adhesives.

**LAMINATES:** (1) Adhesive layers between multi-ply sheets of paper or paperboard. (2) The multi-ply paper and paperboard made up of components joined by an adhesive layer.

**LAMINATING:** The process of firmly uniting superposed layers of paper or paperboard, or other materials, with or without adhesive, to form a multi-ply sheet of increased stiffness and thickness.

**LAND FILLING:** A method of disposing of solid mill waste by transporting it to a designated land area, dumping it into natural or man-made excavations, and covering it with earth.

**LANDING:** Cleared area in the woods where logs are stored for further handling during logging operations.

**LANGUAGE:** A system for representing and communicating information by digital process computers. It consists of a defined set of characters and rules for combining them into larger units, such as words or expressions. It also includes rules for word arrangement or usage to achieve specific meanings.

**LAP MACHINE:** A wet machine on which folded wet pulp sheets (*laps*) are produced from screened pulp slurries for storage or shipment.

**LAPPING:** The process of forming a wet pulp sheet by removing water from a screened pulp slurry, picking up the fibers on a felt, and transferring the fibers to a roll where a sheet is built up. The sheet can be removed and folded into a bundle for storage or shipment.

**LAPS:** Folded wet pulp sheets (approximately 30 to 45 percent consistency) produced on a wet machine (*lap machine*) and stored or shipped to another mill site for repulping. These sheets are used in making paper or paperboard. Also called *wet laps* or *wet pulp.*

**LAPWOOD:** Tops of pulpwood trees left in the woodlands after the logging operation is completed.

**LARCH:** A softwood tree species used as a source of pulpwood, especially suited for the sulfite pulping process.

**LASER WASTEPAPER:** Paper that contains inks used with laser printers. The term may also refer generally to papers containing others types of thermal-set inks used with various printers and copiers that are not "laser" units.

**LATE WOOD:** That portion of a tree that grows during the late summer and autumn seasons of the year. It is a slower growing wood and produces cells with thicker walls and narrower cavities. Sometimes referred to as *summer wood.*

**LATENCY:** The curl and spiral of individual pulp fibers produced during the mechanical action in refiners, especially in the thermomechanical pulping process.

**LATENCY CHEST:** A storage chest after the second stage refiner in a thermomechanical pulping process in which pulp is placed and agitated at a specified temperature for a predetermined time to remove the curl and spiral (*latency*) produced on the individual fibers by mechanical action in the refiners.

**LATENCY RELEASE:** The process of fibers straightening on reheating after they have been elastically distorted by mechanical action on grinding stones and refiners.

**LATENT HEAT:** The heat necessary to convert 1 pound of water at the boiling point temperature to 1 pound of steam at the same temperature. Generally, the energy needed to bring about a change of state or phase, but not a change in temperature (freezing and vaporizing).

**LATENT HEAT OF VAPORIZATION:** *See* HEAT OF CONDENSATION.

**LATERAL POROSITY:** The permeability of a sheet of paper or paperboard that exists in a parallel direction to the surface.

**LATEX:** A synthetic rubber-water emulsion made by polymerization of chemicals. It is used as an additive, coating material, adhesive, or impregnating medium in the manufacture of paper.

**LATEX AGGREGATES:** Types of contaminating specks occurring in the deinking process of wastepaper caused by rubbery particles that loosely stick to one another.

**LATEX BOARD:** Paperboard in which specific physical properties have been enhanced by the use of *latex* emulsion. It is either mixed with the fibers, coated on the surface, or impregnated into the sheet material.

**LATEX PAPER:** A type of paper that has been made from stock to which a *latex* emulsion has been added. It is used as a coating after the sheet is made or impregnated into the sheet.

**LATHE:** A device mounted on a pulpwood grinder consisting of a crosshead, lathe bars, and a mechanism to move a mechanically or electrically operated crosshead across the face of the pulpstone while it burrs the surface. Also called *fixture*.

**LATHE MICROMETER:** An adjusting mechanism on a pulpwood grindstone lathe used to control the depth of the burr impression made on the face of the stone.

**LATICES:** A soft, rubbery polymer used as an additive to papermaking pulp where toughness and high-tear strength are required in the sheet made from it. It is used for such products as sandpaper backing, various types of tapes, and imitation leather.

**LAWN FINISH:** A proper surface finish, resembling the texture of linen, applied to moisture-conditioned paper by embossing between sheets of linen and zinc plates under pressure, or passing between metallic engraved rolls with a linen design. It is used to make writing paper for correspondence and fancy-boxed stationery. Also known as *linen finish*.

**LAY-BOY:** The table on the front edge of a paper cutting machine located in the finishing room of a paper mill to which the cut sheets are automatically delivered, jogged into place, piled, and the load lowered as the pile builds up. Also known as a *delivery table.*

**LAYER BOARD:** A smooth-surfaced, usually brown-colored paperboard used between layers of packaged candy and baked goods.

**LAYFLATNESS:** The curl-resistant property of paper. Usually measured by visual comparison with a standard under controlled conditions.

**LAZY STEAM:** Pure, nonturbulent steam applied over vacuum areas on pulp and paper machines by the use of a specially designed injector (called a lazy injector) in order to increase production, improve paper quality, and save energy.

**LEACH CASTER:** A vertically arranged, cylindrical wooden tank used in older rotary digester pulp mills. The cooked chips were dumped into the tank and the excess liquor drained off through a perforated bottom.

**LEACH STAIN:** A chemical mixture that turns a blue-black color when mixed with mechanical pulp and a red-brown color when mixed with chemical pulp. It is used as a quick method of differentiating between the two types of pulp.

**LEACHATE:** Liquid containing dissolved chemicals picked up by flowing the liquid through a dry material, such as water through the combustion product (black ash) discharged from a rotary-type recovery furnace, to form green liquor in the soda pulping process.

**LEACHER:** A tank in which the combustion product discharged from a rotary-type recovery furnace (black ash) is mixed with water and drained to form green liquor in the soda pulping process.

**LEACHING:** In a soda pulp mill, the process of percolating water down through the combustion product (black ash) discharged from a rotary-type recovery furnace to dissolve out chemicals and produce green liquor.

**LEAD:** The slope angle of the teeth on the surface of a spiral-type pulpwood grindstone sharpening tool.

**LEAD DRYER:** The first, and usually a smaller, steam dryer drum in the dryer section of a paper machine following the press section.

**LEADING EDGE:** (1) The front edge of a sheet being processed or used. (2) The front of a suction box on the wet end of a paper machine.

**LEADING ROLL:** The first roll in a bank, series, or section of rolls on any mill process equipment.

**LEAF:** A sheet of paper with only one fold.

**LEAF FIBERS:** Fibers used in papermaking that are from leaves of plants, such as sisal, palm, manila, hemp, and flax.

**LEAFING:** Orientation during drying of a flake or plate-like pigment particle in a surface film so that the major dimensions of the particle is parallel with the surface of the dry film.

**LEATHER BOARD:** A paperboard used in shoe manufacturing. It is made from sulfate chemical, jute, manila, or wastepaper pulp mixed with scrap leather and usually formed on a wet machine.

**LEATHERETTE COVER PAPER:** Book and box cover paper made to resemble leather by embossing a leather grain design on the surface of the sheet.

**LEATHEROID:** A type of imitation leather made from paper and/or cloth with a coated or embossed leather-like surface design.

**LEDGER PAPER:** A strong, highly-sized paper made from bleached chemical wood pulp. It is used extensively to make accounting and record books and in accounting machines.

**LENGTH:** The dimension of paper or paperboard reels or rolls that run in the machine direction.

**LENS PAPER:** A soft, nonabrasive, lightweight, tissue-grade paper made from bleached sulfate pulp. It is used to make wrapping and polishing paper for eyeglasses, cameras, and other optical lenses.

**LETTER PAPER:** Any type of writing paper that is made to be cut into correspondence sizes.

**LEVAN:** The substance in liquid systems, commonly known as slime, that causes paper breaks on paper machines and creates spots, specks, and holes on the paper sheet.

**LEVEL:** The vertical distance between the surface of material (usually in storage) and some given reference point, which is commonly zero, in order to relate it to the quantity involved.

**LEVELING AGENT:** Paper-coating additives, including wax emulsions, stearate soaps, and modified oil, used to provide free-flowing properties to paper-coating mixtures.

**LICK-UP COUCH:** A grooved couch roll, usually associated with a rubber-covered top roll, commonly used for sheet transfer from wire (forming) section to press section on tissue paper machines.

**LICK-UP FELT:** A felt that is pressed against the sheet in the nip of a *lick-up couch* roll and rubber-covered top roll. It is used to pick up and carry the sheet from the wire (forming section) to the first press on a tissue paper machine.

**LIFE:** The length of time from initial installation of paper machine wires (forming fabrics) and felts to the time they have to be removed and replaced due to wear.

**LIFT:** (1) A stack or pile of paper or paperboard that is moved from one operation to another as a unit quantity. (2) The vertical distance between the level of a fluid in a reservoir and the centerline of a pump inlet.

**LIFTING:** The pulling away of the surface coating or fibers from a paper sheet by the ink during the printing process. It is usually due to the use of insufficient size when the paper was made.

**LIGHT AGING:** Accelerating the aging of paper in a laboratory by exposure to light. Performed as a test to predetermine the effects of natural aging on its brightness and color with respect to time.

**LIGHT SPOTS:** Paper defects caused at the wet end of a paper machine by a dandy roll throwing water drops out onto the wire (forming section), stock-jump created at the table roll, and improper settings of wet end equipment such as the dandy roll, forming board, slice, etc.

**LIGHTER BAR:** A lever that carries the roll spindle of a pulp beater. Its manipulation is the most common means of making a roll adjustment.

**LIGHTFASTNESS:** The resistance of paper, dye, ink, and other papermaking materials to color change when exposed to light.

**LIGHTPROOF PAPER:** Special paper made to prevent the transmission of light.

**LIGHTWEIGHT PAPER:** Paper sheets that have basis weights in the lower ranges specified for a particular grade.

**LIGHTWEIGHT COATED PAPER (LWC):** In North America, these papers are generally the No. 4 and No. 5 coated paper grades that range in basis weight from 32 lb to 40 lb. Some LWC basis weights have dropped below 30 lb. The No. 4 grade base sheet is primarily woodfree, while the No. 5 grade base sheet typically contains a high percentage of groundwood.

**LIGHTWEIGHT SANITARY TISSUE:** A soft, absorbent class of paper made from very lightly refined pulp, including such grades as table, napkins, toilet tissue, sanitary napkins, and toweling.

**LIGNIN:** A brown-colored organic substance that acts as an interfiber bond in woody materials. It is chemically separated during the cooking process to release the cellulose fibers to form pulp, and is removed along with other organic materials in the spent cooking liquor during subsequent washing and bleaching stages.

**LIGNIN-CARBOHYDRATE COMPLEXES (LCCs):** That portion of lignin that becomes chemically linked with carbohydrates (e.g., hemicelluloses, xylose, arabinose, galactose) in the latter stages of the kraft cooking (digester) stage. They tend to be alkaline stable, making further lignin removal difficult. Hemicellulose degrading enzyme treatment is now being used to partially and specifically hydrolyze the carbohydrate chains, making the pulp more bleachable in subsequent stages. *See also* ENZONE.

**LIGNIN CONTENT:** The amount of lignin present in the composition of the raw fibrous materials used in pulping and in the pulp after cooking and washing.

**LIGNOSULFONATE:** The material resulting from reaction of lignin with active alkali sulfur compounds in cooking liquor of the chemical pulping process.

**LIMBING:** The removal of limbs from felled pulpwood trees during the harvesting operation. Also called *lopping.*

**LIME (CaO):** A pulp mill chemical obtained by burning limestone ($CaCO_3$) and used to prepare cooking and bleaching liquors. It is also used in causticizing sulfate and soda cooking liquors, and to make up milk of lime [$Ca(OH)_2$] for the sulfite cooking process. *See also* LIMESTONE.

**LIME COOLERS:** Units designed to reduce fuel consumption of lime kiln systems by recovering heat from lime in preheated secondary combustion air.

**LIME KILN:** A refractory lined, open-ended, inclined steel cylinder located in the chemical recovery area of a kraft pulp mill and mounted on rollers. It is rotated about its longitudinal axis as lime mud ($CaCO_3$) is fed in the higher feed end and burned to form lime ($CaO$) as it travels to the lower discharge end. Also referred to as *rotary lime kiln* or simply *kiln*.

**LIME MILK:** The calcium hydroxide [$Ca(OH)_2$] formed by the reaction of lime ($CaO$) with water ($H_2O$). *See* MILK OF LIME.

**LIME MUD:** The primarily calcium carbonate ($CaCO_3$) sludge that settles out and is separated from the white liquor during the clarification operation in the causticizing process in a pulp mill recovery cycle prior to pumping over to the lime recovery area. Also called *white mud*.

**LIME MUD FILTER:** A continuous vacuum-type rotary filter located in the lime recovery area of a sulfate pulp mill. It is used for removing water from lime mud prior to feeding it to the lime kiln.

**LIME SLAKING:** The mixing of lime ($CaO$) with water ($H_2O$) to form lime water or calcium hydroxide [$Ca(OH)_2$]. In the causticizing process of a sulfate pulp mill recovery cycle, it is the mixing of lime with green liquor.

**LIME WATER ($CaOH_2$):** A pulp mill chemical solution produced by dissolving lime ($CaO$) in water ($H_2O$).

**LIMESTONE ($CaCO_3$):** A naturally occurring mineral that is heated to form lime. It is used by pulp mills in preparing cooking and bleaching liquors, causticizing of sulfate and soda cooking liquors, and other uses. *See also* LIME.

**LINE SHAFT DRIVE:** A type of paper machine drive consisting of a single prime mover with the line shaft connected to enclosed right angle gear units and input shafts at each drive point on the machine by a system of belts and cone pulleys.

**LINE SHAFT:** The main power links (shafts) that connect the electric motor, steam turbine, or steam engine prime mover on a paper machine drive to the various sections to be driven. Accomplished through couplings, pulleys, belts, and gears.

**LINEAR LAID PAPER:** Lightweight grades of writing paper made on a machine with a dandy roll fitted with surface wires that watermark the sheet with lines. These lines can be used as writing guides. *See also* LAID ANTIQUE PAPER.

**LINERBOARD:** A type of kraft paperboard, generally unbleached, used to line or face corrugated core board (on both sides) to form shipping boxes and various types of containers. Some linerboard may contain recycled fiber from small amounts to 100 percent.

**LINENS:** A reference to rags used as a source of flax fibers in making strong resistant papers such as bank notes, check paper, etc.

**LINEN BACK PAPER:** Paper made to be pasted on linen or other fabrics and made into bags, envelopes, tags, charts, etc.

**LINEN FIBERS:** Papermaking fibers originally derived from the flax plant and made into clothing, and from which rags and cuttings are obtained and used by paper mills to make paper.

**LINEN FINISH:** A paper surface finish resembling the texture of linen cloth. It is produced by embossing between sheets of linen and zinc plates under pressure or pressing between metallic engraved rolls with a linen design. Also known as *lawn finish*.

**LINEN PAPER:** Paper that has been made from pulp stock containing linen fibers obtained from rags and cuttings, or has a surface finish resembling linen cloth (*linen finish*).

**LINERS:** Light- to medium-weight papers that are treated and used as inner protection wrappers in food packaging, or the outer layers of multi-ply combination boards.

**LINING:** The tile, brick, or plastic refractory material used inside of a furnace, lime kiln, digester, and bleaching tower.

**LINING PAPER:** Paper used to paste on containers and the inside of book covers and pamphlets.

**LINT:** Minute quantities of surface-coating materials, additives, fillers, and fibers that separate from the paper sheet in the finishing operation, or during its use in printing and conversion into user items.

**LINTERS:** The shorter length cotton fibers used to make a higher quality pulp for cotton paper and cellulose derivatives.

**LIP:** The bottom edge of a paper machine headbox slice whose position determines the jet angle and thickness of the pulp stock being admitted onto the wet end wire (forming zone).

**LIPS:** The bottom and upper section of a paper machine headbox slice whose positions determine the jet angle and thickness of the pulp stock being admitted onto the wet end wire (forming zone).

**LIQUEFIER:** A chemical added to papermaking animal glue adhesive to lower the gelling temperature.

**LIQUID ADHESIVE:** Adhesive papermaking coating material usually made up of organic solvent solutions of film formers modified with plasticizers and resins.

**LIQUID ANIMAL GLUE:** Animal glue used to make paper adhesive coating material. It is in the form of a solution with added liquefiers and gel depressants to facilitate handling at room temperature.

**LIQUID CHLORINE:** The form in which elemental chlorine is received in the tank cars at a pulp mill bleach plant site for unloading with compressed air. It is evaporated in steam heaters before being introduced into the pulp slurry during bleaching.

**LIQUOR:** Chemical solutions used in the manufacture of pulp that change composition as they pass through process stages, e.g., those from digesting through washing, evaporation, incineration, recausticizing, and back to the digester, and those used in the bleach plant.

**LIQUOR DROPLETS:** Dispersed spray particles in the stream leaving liquor feed guns or burners in a pulp mill chemical recovery furnace.

**LIQUOR FLASH TANK:** Enclosed vessels in which hot pressurized liquor relieved from digesters is allowed to flash so that vapor heat can be recovered for reuse. The liquor is then sent to the evaporation and recovery operations.

**LIQUOR HEATERS:** Heat exchangers used on indirect heated batch digesters and continuous digesters to heat the cooking liquor with steam.

**LIQUOR ROOM:** The area or building where the liquor causticizing system is located.

**LIQUOR SPRAY GUNS:** Liquor-spraying nozzles through which concentrated pulp mill liquor is sprayed into a recovery furnace to be burned.

**LIQUOR-FILM COEFFICIENT:** A relation based on the condition at the liquor side of the tubes of a sulfate black liquor evaporator. It determines the possible rate of heat transfer.

**LIQUOR-TO-WOOD RATIO:** The relationship between the total liquid added to a pulp digester, plus the moisture in the wood, and the bone-dry wood fed or loaded into the digester. Sometimes referred to as *bath ratio.*

**LITHO PAPER:** Hard sized book papers suitable for lithography of offset printing.

**LITHOPONE:** A composite zinc-sulfide pigment made by mixing solutions of barium sulfide and zinc sulfate, and separating the precipitate formed. It is used in paper-coating mixtures.

**LITMUS PAPER:** Chemically treated, unsized paper used to detect the existence of specific substances in liquid solutions, gases, or vapors by changing to a distinct color.

**LIVE BOTTOM:** The bottom of a storage chest or bin that is designed so that mechanical motion, usually supplied by a screw or drag chain conveyor, can be imparted to aid in removal of stored materials.

**LIVE STEAM:** Full pressure steam as it comes from a boiler.

**LIVERING:** A pigment type coating or dye solution condition in which there is an increase in consistency that cannot be reversed.

**LOAD FACTOR:** An electrical term expressing the ratio of the average power to the peak power used by a mill, or a machine or system within a mill.

**LOADING:** Mineral substances such as clay, calcium carbonate, and other nonfibrous materials added to the fiber furnish of paper. Also, the process of adding these substances to impart certain desired properties such as increased opacity, brightness, bulk, weight, flexibility, softness, smoothness, and printability to the sheet made from it.

**LOCALIZED WATERMARKING:** A method of watermarking a sheet of paper with a dandy roll surface designed so that the watermarks will appear at specific locations in the sheet area.

**LOCUST:** A hardwood tree species used as a source of pulpwood, especially for the sulfite pulping process.

**LOFT DRIED PAPER:** A high quality, tub-sized paper dried in a *loft dryer.*

**LOFT DRYER:** A heated room in which wet sheets of paper are hung over poles to dry in a controlled atmosphere. The old method of doing this is known as *loft drying* or *pole drying.*

**LOFT DRYING:** *See* LOFT DRYER.

**LOG:** (1) Pulpwood lengths cut from felled trees in woodlands to convenient size for transporting to the pulp mill site. (2) A periodic operator generated listing or hard-copy summary of process operation information in the mill, automatically produced by a digital computer. (3) *See also* STICK.

**LOG FLUME:** A channel or chute filled with mechanically moved water used for conveying logs from a storage area to a debarking facility in the wood preparation operation of a pulp mill.

**LOG HAUL-UP:** A chain-type conveyor that picks up pulpwood logs from a holding pond and transports them to a storage pile at the pulp mill site. Also known as *jack ladder* and *wood conveyor.*

**LOG POND:** A natural or artificial pond of water where pulpwood logs in some mills are gathered and stored prior to being transported to the wood preparation area.

**LOG SPLITTER:** A log cutting machine used in the wood preparation area of pulp mills to split large diameter logs into smaller sizes so that they will slide through chipper chute openings and better fit the pulpwood grinder pockets.

**LOG TRANSPORTATION:** The movement of pulpwood logs by conveyors or flumes from storage area to debarking facilities in the wood preparation operation of a pulp mill.

**LOGGER:** An instrument that automatically scans certain measurement signals generated by a controlled process and records, usually in chronological order, the readings in a numerical format for future reference and other uses.

**LOGGING:** (1) The operation of harvesting trees from woodlands and processing them into barked and unbarked logs or chips so that they can be transported to the mill site. (2) The act of gathering process data and entering it on tabular reports for future referral, analysis, and similar uses.

**LOGIC:** The systematic method by which system information is processed with each successive step influencing the next.

**LONDON PAPER:** *See* BALSTON'S PAPER.

**LONG CORD:** A stack of pulpwood logs measuring 8 feet by 5 feet by 4 feet, equaling 160 cubic feet in volume.

**LONG DIRECTION:** The directional dimension of a paper sheet that runs parallel to the direction of movement of the paper machine on which it was made. *See* MACHINE DIRECTION.

**LONG FOLD:** A fold on paper or paperboard cut parallel to the direction of movement of the machine on which it was made.

**LONG GRAIN PAPER:** A type of paper that has been made with the longest sheet dimension running along parallel to the machine direction.

**LONG LOG:** A whole-tree length pulpwood log before shorter length logs are cut from it. Also known as *longwood logs*. *See also* TREE LENGTH.

**LONG LOG DEBARKING:** Removal of bark from whole-tree length pulpwood logs rather than cutting them up into shorter length logs.

**LONG STOCK:** Refined pulp stock where there is a minimum of cutting of the fibers during refining operations.

**LONG TUBE EVAPORATOR:** A type of spent liquor evaporator used in pulp mills characterized by long vertical tubes through which boiling liquor drives a mixture of vapor and liquor at a high velocity, depositing a thin film on the tube walls where water is further driven off to concentrate the liquor. Also referred to as *falling film evaporator* or *vertical film-type evaporator*.

**LONGEVITY:** That characteristic of paper that resists deterioration due to aging.

**LONGLEAF PINE:** A softwood tree species used as a source of pulpwood.

**LONGWOOD LOGS:** Sections of trees prior to being cut up into pulpwood lengths. Same as *long log*.

**LOOK-THROUGH:** The viewing of paper to observe formation by holding it up to a light source, or its appearance when it is observed in this manner. Also called *formation*.

**LOOP DRYING:** The drying of a wet paper sheet by looping it over poles traveling through a chamber in which the air is temperature and humidity controlled to achieve the desired paper-drying conditions.

**LOOSE COATING:** Paper-coating material that does not form a firm bond to the sheet during the coating operation, making it susceptible to lifting on subsequent use.

**LOOSE CORE:** A paper roll defect due to insufficient sheet tension being used at the beginning of a winding, or shrinkage of a core causing loose wound paper.

**LOOSE KNOT:** *See* ENCASED KNOT.

**LOOSE IN CASE:** An expression that refers to a paper packaging method where the reams are placed in wooden crates without wrapping.

**LOOSE PAPER:** A term referring to any extraneous pieces of paper caught up in a roll of paper that might cause the sheet to break. *See also* TRASH IN ROLL.

**LOOSE TABS:** Splice-marking flags that have not been placed in the proper location in a roll of paper or have been moved away from their proper position.

**LOOSE WINDING:** A paper roll defect showing up as slack, loose paper wound on the roll. It is caused by the use of insufficient sheet tension during the winding operation.

**LOPPING:** *See* LIMBING.

**LOST TIME:** The length of time that mill machinery and/or personnel are not productive due to accidents or any other types of malfunctions that interrupt normal production operations.

**LOT NUMBER:** An identifying designation assigned in a paper mill to a quantity of rolled or packaged paper of a specific grade used to make up a purchase order for a particular customer.

**LOUVER:** An air-damping device made up of metal slats located in the stream in such a way as to allow adjustment to regulate the air flow.

**LOW ANGLE GLOSS:** A paper mill laboratory measurement of the gloss of a paper sheet taken at an angle of 75 degrees.

**LOW CONSISTENCY:** Pulp slurry consistencies usually less than 6 percent, depending on what part of the process the reference is made.

**LOW COUNT:** The quantity of sheets in a standard ream of paper or stack of paperboard that does not have the minimum specified number for that particular classification.

**LOW DENSITY COOKING:** The cooking of wastepaper at 5 to 7 percent consistencies in a paper deinking process.

**LOW FINISH PAPER:** A paper sheet with a minimum of surface gloss or glaze.

**LOW FINISH:** A dull, matte paper and paperboard surface finish.

**LOW SLIP PUMP:** A type of pump that delivers nearly all of its internal displacement in each cycle of its impeller. Used where accurate metering of fluid is required.

**LOWERING TABLE:** An electrically or hydraulically operated platform in a roll handling system that can be raised and lowered for delivering paper rolls to floor level so that they can be moved.

**LUBRICANTS:** Chemicals added to paper-coating preparations primarily to improve flow properties. Their use results in a smoother coating with improved finish and printing qualities, along with other associated property enhancements. Also refers to chemical agents providing lubrication for dynamic machinery parts.

**LUMEN:** The center void portion of a wood fiber.

**LUMINOSITY:** The brightness effect produced on a paper sheet by a unit intensity of light.

**LUMINOUS FLUX:** Light measurements as related to the optical properties of paper; the time rate at which radiant energy is transmitted.

**LUMINOUS PAPER:** Decorative paper that has been prepared and treated to give off light or glow.

**LUMINOUS REFLECTANCE:** An optical characteristic of paper and paperboard sheet surfaces generally associated with *whiteness* or *brightness*.

**LUMP:** A paper defect due to incomplete separation of fiber bundles or the gathering together of fiber and other papermaking materials, causing raised, hard, localized spots in the finished sheet.

**LUMP BREAKER ROLL:** A rubber-covered roll that is sometimes mounted over the suction area of the couch roll of a paper machine to seal the wet sheet to the couch, thereby maintaining a higher, more consistent vacuum and improving water removal. It tends to consolidate the sheet and flatten any lumps that occur. Also called a *breaker roll* or a *presser roll*.

**LUSTER:** That part of the gloss property of paper related to the selective reflection of light as compared with its counter part, *gloss*, which is related to the reflection of excessively bright light. (2) The surface shine or sheen appearance of paper and paperboard as related to the amount of light reflected from it.

**LWC:** *See* LIGHTWEIGHT COATED PAPER.

# M

**MACERATION:** The softening and separation of wood chips or fiber bundles into their component parts by the application of physical mechanical treatment.

**MACHINE:** Very common paper mill expression referring to the paper machine.

**MACHINE BROKE:** Paper trimmings or paper damaged as a result of malfunctions or breaks in the sheet on the paper machine. The broke is either discarded or returned for reprocessing.

**MACHINE CALENDER:** A calender stack that is a part of the paper machine system normally situated between the last section of dryers and the windup reel.

**MACHINE CAPABILITY:** The maximum production of paper obtained from a machine regardless of its design capacity.

**MACHINE CHEST:** A large agitated storage tank located just ahead of the paper machine in which the stock furnish being pumped to the paper machine is stored. Sometimes called a "stuff chest."

**MACHINE CLOTHING:** Press and dryer felts on the dry end of a paper machine; now commonly includes the forming section *wire*, also known as a *forming fabric* or *forming wire*.

**MACHINE COATED:** Paper and paperboard that have a surface coating of adhesives and minerals applied while being made on the paper machine as an integral part of the papermaking operation.

**MACHINE CREPE:** A light, crinkly effect on paper produced by crowding the sheet on a roll on the last press or first dryer of the paper machine with a doctor blade. Also referred to as *primary crepe* or *semicrepe*.

**MACHINE DIRECTION (MD):** A term referring to the direction from the wet end to the dry end of a paper machine or to a paper sheet parallel to its forward movement on a paper machine. It is opposed to *cross direction* which runs from side-to-side of the paper machine or the sheet made on it. Also commonly referred to as *along machine direction*, *grain direction*, or *long direction*.

**MACHINE DRIED PAPER:** Paper that was dried on the paper machine dryer section, usually over steam-heated dryer drums. Also referred to as *cylinder dried*.

**MACHINE DRIED PULP:** Market pulp that was dried over steam heated dryer drums of a pulp dryer section.

**MACHINE DRIVE:** The mechanism used to impart motion to various sections of a paper machine. The motion can be obtained from one source, such as a motor, with each section connected to it, or from an individual source for each individual section.

**MACHINE EFFICIENCY:** The amount of paper produced on a paper machine in relation to the potential production capacity.

**MACHINE FILL:** The largest width of a paper sheet that can be manufactured on a paper machine.

**MACHINE FINISH (MF):** Surface finish produced on an uncoated sheet of paper as it is being made on the paper machine. It is usually accomplished with limited calendering on the machine calender stacks. Also referred to as *machine finish.*

**MACHINE GLAZED (MG):** A high-gloss surface finish produced on the wire (forming fabric) side of a sheet by passing it over a large diameter, highly polished, steam-heated roll as used on yankee-type paper machine dryers.

**MACHINE IMPRINTED:** Imparting a distinctive mark, symbol, or design (watermark) on a wet paper web while it is being formed on the paper machine with the use of metal or rubber impression plates or rollers.

**MACHINE LANGUAGE:** A language in binary form existing internally within a computer and executed by the computer for its operation.

**MACHINE LOADING:** The addition of mineral substances such as clay, calcium carbonate, and other nonfibrous materials to a sheet of paper at the size press while it is being made on the paper machine. This generally involves the use of a *pigmented size press.*

**MACHINE PIT:** The receiving chest located under the wet end section of some paper machines for draining white water from the sheet forming section of the wire where the tables, rolls, and foils are located.

**MACHINE ROOM:** The building or area of a paper mill where the paper machine is located.

**MACHINE SHOP:** A location in a mill equipped with tools for machining and repairing parts of machinery used in the manufacturing process.

**MACHINE SPEED:** The speed at which a paper machine is operating or the speed for which it was designed. In the U.S., it is usually expressed in terms of the speed of the sheet being made in feet per minute (fpm).

**MACHINE TENDER:** The person in a paper mill responsible for and in charge of the operation of the paper machine and the quality of the paper made on it.

**MACHINE WIRE:** A continuous, driven, endless wire belt screen on which the stock slurry is introduced. It allows the water to drain so that a web will form. The wires may be made of plastic. *See* FORMING FABRICS, FORMING WIRE, and WIRE.

**MACHINIST:** A machine shop employee who operates the tools to make or repair machinery parts in a pulp and paper mill.

**MAGAZINE GRINDER:** A type of pulpstone grinder used to grind pulpwood. It is fitted with a large log-holding container, called a magazine, that is located over the stone. This enables it to achieve greater capacity than other types of grinders.

**MAGAZINE PAPER:** Any paper made especially to be used for printing magazines, books, or similar periodicals.

**MAGAZINE STOCK:** Pulp stock made from reprocessed wastepaper, such as magazines, books, or similar periodicals. It is usually mixed with other pulps to make various grades of paper and paperboard.

**MAGIC EYE:** An electronic photocell commonly used on paper machines to detect a break in the paper sheet and sound an alarm to alert the operator.

**MAGNEFITE PROCESS:** An acid-type pulping process which uses magnesium bisulfite [$Mg(HSO_3)_2$] base chemicals as the cooking liquor.

**MAGNESITE ($MgCO_3 \cdot 3H_20$):** A precipitated *magnesium carbonate* compound used as a filler in the manufacture of cigarette paper. The native form is used to produce magnesium oxide (MgO) by heating to a high temperature. Sometimes referred to as *precipitated magnesite. See also* MAGNESIUM OXIDE.

**MAGNESIUM BASE:** A reference to wood cooking liquor containing magnesium (Mg) as its primary element.

**MAGNESIUM BISULFITE [$Mg(HSO_3)_2$]:** The main chemical constituent of the acid-cooking liquor used in the magnesium-based sulfite pulping process.

**MAGNESIUM CARBONATE:** *See* MAGNESITE ($MgCO_3 \cdot 3H_2O$).

**MAGNESIUM HYDROXIDE [$Mg(OH)_2$]:** A chemical solution used in making magnesium-base sulfite cooking liquor and in oxygen pulp bleaching. It is prepared by dissolving magnesium oxide (MgO) in water. Also called *milk of magnesia.*

**MAGNESIUM OXIDE (MgO):** The entrained chemical material recovered from the recovery boiler stack gases in a magnefite pulping process. It is converted back into magnesium bisulfite [$Mg(HSO_3)_2$] cooking liquor by slaking and mixing with sulfur dioxide ($SO_2$). It can also be prepared by heating magnesite ($MgCO_3$) to very high temperatures, and can be precipitated from seawater.

**MAGNESIUM SILICATE:** A material used in various forms, such as talc, asbestine, and agalite, as a filler in making paper to produce a soft, silky feel and enhance brightness and opacity.

**MAGNESIUM SULFATE (MgSO):** A chemical used as a stabilizer of peroxide solution used as a pulp bleaching solution by inhibiting the undesirable effects of iron, copper, and manganese.

**MAIN FRAME:** The central physical structure of a computer with no external devices connected to it by cabling.

**MAIN PRESS:** The press that drives each section of a cylinder-board machine, which is usually a suction press.

**MAIN SQUIRT:** The high-pressure water jet used on the wet end of a paper machine to cut a tail on the wet web as it is fed from the wire (forming fabric) to the wet presses and expands the web once it is fed through the presses.

**MAIN STEAM:** The high-pressure steam that comes from the primary superheater of boilers and is sent to the first high-pressure stage of the turbine generator.

**MAKE AN ORDER (MO):** A paper mill expression referring to the manufacture of paper specially for specific customer's order specifications that cannot be filled from inventory stock.

**MAKE AND HOLD ORDER:** An order for paper made to customer's specifications without any shipping instructions so that it must be held in storage.

**MAKEUP:** Water that is added to a power or recovery boiler to replace losses due to blowdowns and other causes not related to the actual conversion into steam.

**MAKEUP ALKALI CONSUMPTION:** Weight of chemical used to manufacture a ton of air-dried pulp by the sulfate process, and includes salt cake ($Na_2SO_4$), soda ash ($Na_2CO_3$), caustic soda (NaOH), and other sodium compounds expressed as $Na_2O$.

**MAKEUP CHEMICAL CONSUMPTION:** The weight of new salt cake ($Na_2SO_4$) or other sodium compounds expressed as $Na_2SO_4$ per ton of air dried pulp production that must be added to the recovery process to maintain an even sodium level in the cooking liquor in alkaline-type pulp mills.

**MAKEUP SALT CAKE:** The sodium sulfate ($Na_2SO_4$) added to the sulfate black liquor just before incineration in the recovery furnace to make up for the loss of sodium (Na) and sulfur (S) in the process.

**MAKEUP WATER:** The raw water that is pumped into a steam boiler system to make up for losses at various points, such as leakage through pump bearings, steam leakage through valve packings, etc.

**MAKING BOARD:** A baffle or baffles located in the stock approach system of a conventional vat-type cylinder paper machine immediately ahead of the vat entry.

**MAKING ORDER:** A special order that cannot be filled from stock and has to be custom-manufactured by the mill.

**MALE THREAD:** The external thread on pipes, fittings, valves, etc., for making screwed connections.

**MAN LIFT:** A type of elevator used in pulp mills. It consists of a continuous rubber belt with one-person passenger platforms that move slowly through the floor openings in the mill.

**MANAGED FOREST:** The managing of pulpwood forests to optimize wood and its production without adversely affecting other desirable aspects of the stand.

**MANAGEMENT INFORMATION SYSTEM (MIS):** An advanced computer network that provides data base information on millwide processes, production, shipments, etc. on a real-time and historical basis. It is used to make marketing and production strategy decisions.

**MANDREL:** The shaft or spindle that holds the bearings and spindle of the burr-holding assembly used to dress the stone on a pulpwood grinder.

**MANIFOLD:** (1) The stock line piping arrangement located just before the furnish enters the headbox of a paper machine. It spreads and distributes the large pipe flow to a more even flow along the width of the headbox. (2) Also a light weight, dense paper with machine glaze or unglazed finish used to make carbon copies.

**MANIFOLD PAPER:** Very thin regular bond paper with glazed or unglazed finish used to make carbon copies of letters.

**MANILA:** A light straw or yellow paper color and finish resembling that of paper made from manila hemp *(abaca)* stocks with no particular reference to fiber content. Also called *manila paper.*

**MANILA BOARD:** A manila colored paperboard made to be converted to folding boxes. Usually made from chemical-type pulps.

**MANILA ENVELOPE PAPER:** A manila colored paper usually made from chemical-type pulp and used to make writing paper and envelopes.

**MANILA FIBER:** *See* ABACA.

**MANILA HEMP:** *See* ABACA.

**MANILA PAPER:** *See* MANILA.

**MANILA TAG:** A strong, rigid, heavy buff colored paper made from a variety of fibers, heavily sized for use in making all types of tags.

**MANILA TISSUE:** A light weight, manila colored paper used for wrapping and toilet tissue.

**MANILA WRAPPING:** A manila colored paper made from sulfite and groundwood pulp for use in wrapping purposes.

**MANILA WRITING:** A yellow colored writing paper having a groundwood pulp content and used for notebooks, tablets, telegram forms, scratch pads, etc. Also referred to as *railroad writing.*

**MANIPULATED VARIABLE:** The quantity or condition in a controlled process that must be changed to maintain another quantity or condition (controlled variable) to a desired value.

**MANNOGALACTAN:** An additive obtained from plant seeds and mixed with the pulp furnish to enhance sheet formation and bonding of fiber on the paper machine.

**MANNOGALACTAN GUMS:** Water-soluble, amorphous carbohydrate material that is added to the paper stock furnish to improve the bonding strength of the fibers.

**MANPOWER:** Personnel that maintain the operations of a pulp and paper mill.

**MANUAL BACKUP:** An alternative method to automatic control of the pulp and paper process by means of manual adjustment of the final operator mechanism.

**MANUAL CONTROL:** Controlling process variables by manually adjusting the final compensating control devices. Sometimes referred to as *hand control.*

**MANUFACTURED FILLERS:** *See* ARTIFICIAL FILLERS.

**MAPLE:** A species of hardwood trees used as a source of pulpwood.

**MAP PAPER:** A high-strength paper made from bleached chemical pulp sometimes mixed with rag pulp, heavily sized with good folding properties, and used for lithographic and map marking purposes.

**MARBLE PAPER:** Paper colored to look like marble and used as book and box liners.

**MARKED ROLL:** Filled calender roll whose surface has been marred by wrinkled, creased, or color-lumped paper passing through nips of calender rolls. Other causes include the roping and double-folding of paper tails into the calender and the accidental passage of foreign objects, such as cleaning tools, through the calender roll nips.

**MARKET PULP:** Any pulp manufactured by a pulp mill and sold on the open market as a commercial product.

**MARKING FELT:** A paper machine wet end press felt especially manufactured to impress a specific pattern, symbol, mark, or design on the sheet of paper with which it comes into contact.

**MARKING ROLL:** A soft rubber roll, located at the wet end press section, whose surface is provided with raised patterns, designs, or distinctive symbols that are impressed on the wet sheet surface as the sheet passes over it.

**MARKS:** A distinctive symbol or design imparted on the sheet by the use of impression rolls and felts when the web is still wet. Also referred to as *watermarks.*

**MASON:** A member of the mill maintenance crew responsible for working with cement and concrete.

**MASTER CONTROLLER:** The controller that adjusts the set point of other controllers in a control system.

**MASTER MECHANIC:** A pulp and paper mill title for the employee who is the foreman of the repair and maintenance crew.

**MAT:** A thick layer or layers of fiber webs that commonly exist in the pulp and paper manufacturing process.

**MAT BOARD:** A thick board made by pasting and used for mounting displays, samples, and exhibits.

**MATCH STEM STOCK:** *See* BOOK MATCH BOARD.

**MATCHING:** The procedure of achieving the equivalent color and shade on a particular run of paper being made on a paper machine to an order previously made with the same color specifications.

**MATERIAL BALANCE:** Method of calculating the performance of pulp and paper equipment or unit process, such as washers, evaporators, etc., by comparing the total input of all materials to the total output of all materials.

**MATERIALS HANDLING:** The movement and transportation of raw materials and products from one part of a pulp and paper mill to another, including the preparation of the product for shipping out of the mill.

**MATRIX BOARD:** Plastic-coated paper or paperboard made for use as casting forms in stereotype printing.

**MATTE:** Paper and surface finishes that have very low *gloss* or *luster.*

**MATTE PAPER:** Paper made with a low and dull finish. *See* MATTE.

**MATURE FIBERS:** Fibers that make up *mature wood*.

**MATURE PAPER:** Paper that has been stored long enough for it to come to equilibrium with the surrounding atmospheric conditions, which may be controlled.

**MATURE WOOD:** That portion of the wood structure of a tree that is older and closer to the outer surface. The fibers are shorter, weaker, and less dense than those in the newer portion toward the center. Also referred to as *slabwood*.

**MAXIMUM DEMAND:** The greatest of all demands occurring over a given period on an electrical system.

**MC:** *See* MEDIUM CONSISTENCY.

**MCC:** *See* MODIFIED CONTINUOUS COOKING.

**MD:** *See* MACHINE DIRECTION.

**MD MICROSTRIATIONS:** *See* MICROSTRIATIONS.

**MEALY FINISH:** *See* MOTTLED FINISH.

**MEASURED VARIABLE:** The physical quantity, property, or condition of the pulp and papermaking process that is measured by an instrument and control system.

**MEASURING ELEMENTS:** Those devices in an instrument and control system that ascertain the value of the measured process variables.

**MEASURING TANKS:** Storage tanks located above the batch digester in a pulp mill into which cooking liquors are drawn and pumped in measured amounts into the digester for each cook. Also called *draw down tanks*.

**MECHANIC:** A member of the repair and maintenance crew who works on process machinery in a pulp and paper mill.

**MECHANICAL ADHESION:** The joining or bonding of surfaces by interlocking action and actual physical penetration of adhesive material as in the mechanical embedding of adhesives in the pores of paper in order to obtain a strong bond between sheets in the manufacture of laminates.

**MECHANICAL ATTACHMENT:** A method of filler retention imparted to papermaking fibers by physical force, such as wedging into surface imperfections.

**MECHANICAL BARKERS:** Types of pulpwood log bark removal equipment that operate on the principle of physical mechanical separation using abrasion, scraping, and abrading techniques.

**MECHANICAL BARKING:** The removal of bark from pulpwood logs by physical mechanical means including abrasion, impact, scraping, and abrading through the use of chains, wheels, knives, drums, etc.

**MECHANICAL DECKLE EDGE:** A feathery edge produced on paper by physically abrading the dried sheet.

**MECHANICAL DRAFT:** Movement of air by the use of a mechanical device such as a fan.

**MECHANICAL DRIVE:** A method of moving separate sections of a paper machine by connecting them together through a driving mechanism powered by mechanical means.

**MECHANICAL FOAM BREAKER:** A mechanism that reduces the foam in pulp suspensions, coatings, and other papermaking mixtures.

**MECHANICAL INTERLOCKING:** Changing properties, such as increasing the strength of a synthetic metal-fiber paper web, by calendering to a higher density.

**MECHANICAL PAPER:** General reference to a variety of medium weight papers made primarily from pulp produced by a mechanical pulping process.

**MECHANICAL PAPER DEFECTS:** A class of paper defects caused by physical structural formation faults as compared with those caused by chemical constituent faults.

**MECHANICAL PULP:** Pulp produced by reducing pulpwood logs and chips into their fiber components by the use of mechanical energy, via grinding stones, refiners, etc.

**MECHANICAL PULPING:** The converting of pulpwood logs and chips into pulp by the use of mechanical energy. *See* MECHANICAL PULP and MECHANICAL TREATMENT.

**MECHANICAL REFINING:** Imparting specific properties to pulp fibers by mechanical treatment between moving surfaces in papermaking machinery such as beaters and refiners.

**MECHANICAL TREATMENT:** (1) The development of certain desirable characteristics in pulp slurries by subjecting them to mechanical energy in equipment such as beaters and refiners. (2) The reduction of wood chips into their fiber components by subjecting them to mechanical energy in equipment such as beaters and refiners.

**MECHANICALLY BOUND WATER:** Water in a sheet of paper that is retained primarily by surface tension and exists wherever there is space in the fiber or fiber network for the water to accumulate.

**MECHANOCHEMICAL PROCESS:** A semichemical pulping process used on nonwoody plants, such as straws and reeds, in which chemical and mechanical actions are used at the same time under atmospheric pressure, producing a high-yield pulp.

**MEDIUM CONSISTENCY (MC):** A generalized reference used to describe pulp slurries having consistencies within the approximate range of 6 to 15 percent, although it may vary somewhat depending on where in the pulp and papermaking process the reference is made.

**MELAMINE FORMALDEHYDE RESIN:** A syrupy-type liquid used in paper mills to improve the wet strength of paper and the water resistance of paper coatings.

**MELLOW PAPER:** Types of paper that are characteristically soft and have been allowed to mature.

**MEMORY:** The storage device or section of a computer in which programs and data are stored and easily retrieved. It is the central processor unit for manipulation and execution.

**MEMORY COLOR:** Color of paper that, according to the judgment of the operator, would match the color of a previous order made by him.

**MENSURATION:** Measurement methods as applied to pulpwood.

**MERCHANT GRADE:** A general reference to premium and ultra-premium coated woodfree printing and writing papers. These grades are typically very high brightness and are used in annual reports and as special art paper.

**MERCHANT'S BRAND:** Types of paper made specifically for a particular paper merchant whose names have been devised and are owned by the merchant and not by the papermaker.

**MESH:** The number count of wires per inch of wet end screen (forming fabric) on a paper machine running in the machine- and cross-machine direction. *See also* COUNT.

**METAL FIBERS:** Thin, short strands of metal used in a sheet alone or in combination with other types of synthetic or cellulosic fibers.

**METAL TIPPED CORE:** Type of paperboard core fitted with metal rings on each end.

**METALLIC PAPER:** Paper whose surface has been given a metallic appearance by coating with a preparation made with aluminum and bronze powder and combining with metallic foil or depositing condensed, vaporized metal on the paper surface while under high vacuum.

**METERING NIPS:** The contact point between the metering roll and the distributing roll on some types of paper coaters where the coating material is admitted.

**METERING ROLL:** The roll on certain types of paper coaters that determines the thickness of the coating applied to the sheet by forming a nip with a second distributing roll and controlling the amount it picks up from a holding pan.

**METHANOL ($CH_3OH$):** (1) A common alcohol by-product of the sulfite process obtained by the fermentation of sugars in spent cooking liquor and subsequent distillation. (2) A chemical used as a raw material in the generation of chlorine dioxide ($ClO_2$) pulp bleaching solution in a pulp mill.

**METHYL MERCAPTANS ($CH_3SH$):** The reduced sulfur compound formed during the cooking of wood in a sulfate pulp mill and released in mill gaseous effluents. It is primarily responsible for the characteristic offensive odors associated with the sulfate mill environment.

**METRIC SYSTEM:** An international system of units of weights and measures. It is used extensively throughout the pulp and paper industry except in the U.S., where English units are still commonly used.

**MF:** *See* MACHINE FINISH.

**MG:** *See* MACHINE GLAZE.

**MGD:** *See* MILLIONS OF GALLONS PER DAY.

**MICRO KAPPA NUMBER:** The result of a bleachability laboratory test on wood pulp to determine the bleachability or the amount of lignin by a modification of the standard *kappa number* test. It is performed so that it can be used on very small samples and semibleached pulp with low permanganate consumption.

**MICROCOMPUTER:** A small computer that can be made of one or more microprocessors. It is generally designed to handle 4- to 8-bit words.

**MICROMETER:** A measuring device used to determine very small distances and used to accurately measure the thickness or caliper of paper and paperboards. Also known as a *caliper.*

**MICROPROCESSOR:** Generally a monolithic, Large-Scale Integrated (LSI) Central Processing Unit (CPU) on a single chip or semiconductor material.

**MICROSTRIATIONS:** Tiny ridges and valleys in the paper surface oriented in the machine directions. Sometimes referred to as *MD microstriations.*

**MIDDLES:** The inner layers of multi-ply paperboard or lower-quality paper boards used as a structural base for lining papers.

**MIDDLE LAMELLA:** The intercellular region in the structure of wood fibers that provides strength and rigidity to the wood by holding the fibers together. The majority of the lignin in the wood is found in this region.

**MIDFEATHER:** A partition located in the center of a beater chest or stock storage chest around which the pulp slurry is circulated to achieve fiber mixing and suspension.

**MIL:** A unit of linear measurement commonly used to express the thickness of paperboards. One mil equals 0.001 inch.

**MILEAGE:** The total amount of adhesive required for 1,000 square feet of bonded paper or paperboard surface.

**MILK CARTON BOARD:** A hygenically clean, high-strength, plastic coated food paperboard made from bleached chemical pulp. It is converted into cartons for milk, cream, and other dairy products and beverages.

**MILK OF LIME [$Ca(OH)_2$]:** The solution formed when lime ($CaO$) is treated with water (slaking). *See also* LIME MILK.

**MILK OF MAGNESIA:** *See* MAGNESIUM HYDROXIDE [$Mg(OH)_2$].

**MILL:** The building or buildings and area where the pulp and papermaking operations are carried out. Sometimes called a *plant* when referring to one area of the whole operation. It can also refer to rotating steel rolls used in mixing materials.

**MILL BLANKS:** Multi-ply blanks made on multi-cylinder machines in which the middle layers are made from lower quality pulp than the outer layers.

**MILL BOARDS:** Types of heavy paperboards extensively used for box manufacturing and in the binding of books.

**MILL BRAND:** Types of paper whose names have been devised and are owned by the paper mill making them. Also referred to as *mill line.*

**MILL BRISTOL:** Bristol-type paperboard made on a cylinder-type paper machine.

**MILL BROKE:** Paper trimmings or paper damaged as a result of breaks on the paper machine, in finishing operations, or anywhere in the process of paper manufacturing. It is usually returned for reprocessing.

**MILL COUNT:** The number of sheets in a shipped order as counted at the paper mill.

**MILL CUT:** The edge of a paper sheet made by the paper machine slitter and cutter.

**MILL EDGE:** The edge of a paper sheet before being trimmed in the finishing operation of a mill.

**MILL FINISH:** *See* MACHINE FINISH.

**MILL HEADS:** *See* MILL WRAPPER.

**MILL LINE:** *See* MILL BRAND.

**MILL LINING:** A boxboard manufacturing operation in which a lining is pasted to the base as it comes from the roll on the paper machine.

**MILL MANAGER:** The person in a pulp and paper mill management organization who has the responsibility of directing or controlling the manufacturing process at one specific mill.

**MILL PASTE:** *See* MILL SPLICE.

**MILL ROLL:** A roll of paper coming off the paper machine that is made to be cut into smaller rolls and sheets.

**MILL SIZE:** Rosin size made and used by the paper mill to provide moisture-resistant characteristics in the paper it manufactures.

**MILL SPLICE:** The joining of ends of a sheet of paper with tape or adhesives to make up a roll.

**MILL WRAPPER:** Types of paper used for wrapping purposes in a paper mill. Also referred to as *mill heads.*

**MILL WRAPPERS:** Types and grades of paper used to wrap paper rolls, reams, and other packaging.

**MILLING:** Mixing materials by passing them between rotating steel rolls called a *mill.*

**MILLIONS OF GALLONS PER DAY (MGD):** A very common expression used to denote the water consumption of a pulp and paper mill.

**MILLWIDE CONTROL:** Process automation and control implemented throughout all mill areas, involving upper as well as lower levels of hierarchical control applied to all phases of manufacturing in order to maximize efficiency and profitability. *See* PROCESS AUTOMATION.

**MILLWRIGHT:** The person in a pulp and paper mill responsible for improvements to or modifications of existing process machinery and the erection of new machinery. The millwright also looks after some types of repairs on machinery.

**MIMEOGRAPH PAPER:** Type of paper made with an absorbent surface so that it is suitable for use in mimeograph work.

**MINERAL EARTH PIGMENTS:** A class of insoluble pigments used for coloring in papermaking, represented by such products as oxides, ochre, and umbers.

**MINERAL FIBERS:** Nonwoody fibers such as asbestos, glass, etc., used in papermaking.

**MINERAL FILLER:** A nonfibrous material added to paper stock furnish before forming a sheet on a paper machine.

**MINICOMPUTER:** A small computer that can be made up of one or more microprocessors, generally designed to handle words up to 16 bits.

**MINING NOZZLES:** High-pressure water jets located in the lower portion of a high-density pulp storage chest. They are used to provide dilution water so that the stock may be more readily pumped out. The water used by these nozzles is known as "mining water."

**MIS:** *See* MANAGEMENT INFORMATION SYSTEM.

**MISCIBILITY:** The ability of process liquids to mix in any ratio without separating.

**MISPRINT BLEACHED SULFATE:** A grade of wastepaper that includes misprint sheets and bleached sulfate printed carton paperboards that are free from nonsoluble coatings, waxes, inks, adhesives, etc.

**MIST BOARD:** A multi-liner paperboard with the top or outer layers made with a mixture of black and white fibers.

**MITSCHERLICH PULP:** A high-strength easy bleaching chemical pulp made by the sulfite cooking process.

**MIXBOX:** A tank on the inlet of a cylinder machine vat where makeup stock, drainage, and overflow white water are added and mixed. Also called a *headbox*.

**MIXED BASE PULPING:** Adding sodium-containing solution to calcium bisulfite cooking liquor to improve cooking and pulp quality in a calcium-base sulfite pulping process.

**MIXED OFFICE WASTEPAPER:** Wastepapers collected from business offices, consisting of various coated and uncoated printing and writing papers, both *free sheet* and groundwood-containing, that have been printed with various types of inks in several printing processes and contain a wide assortment of mineral fillers and sizing materials.

**MIXED PAPER:** Wastepaper that includes office discards, boxboard cuttings, mill wrappers, and other similar wastes.

**MIXED RAGS:** A grade of papermaking rags that have not been sorted.

**MIXER:** A container fitted with an agitator in which pulp, chemicals, additives, and dyes are agitated. Steam is sometimes employed to achieve an intimately stirred mixture.

**MIXING CHAMBER:** A compartment on some types of paper machine headboxes that extends across the width of the paper machine and into which the manifold distributes stock furnish to achieve initial mixing and coalescence of the fibers.

**MIXING ROOM:** *See* COATING KITCHEN.

**MODIFIED CONTINUOUS COOKING (MCC):** A modified continuous digester system (Kamyr) in which chips cooked in the vessel's traditional upper portion (cocurrent cooking) pass through a second (countercurrent) cooking zone. In some *extended modified continuous cooking (EMCC)* systems, the chips are cooked in yet a third zone–the traditional "high-heat" wash zone where white liquor is added and upflows through the chip mass, resulting in the *EMCC* feature. Both modifications improve delignification efficiency. The "gentler" cooking resulting from extended times in the digester and multiple applications of cooking liquors (rather than in one harsh "shock" stage) reportedly produces pulps with higher strength and considerable lower kappa numbers than pulps cooked in traditional, single-cooking-stage continuous digesters.

**MODIFIED STARCH:** Papermaking starch that has been processed, generally by oxidation, to achieve lower viscosity or chemical characteristics beneficial to its use as an additive in papermaking.

**MOISTURE CONTENT:** The percent of water by its weight in paper, pulp, paperboard, chips, etc., that will vary according to atmospheric conditions because of the ability of these types of materials to absorb or emit moisture. *See also* DRYNESS, AIR DRY, BONE DRY, and OVEN DRY.

**MOISTURE PICKUP:** Moisture that is taken up by paper or pulp sheet from the atmosphere or other things that it may come into contact with.

**MOISTURE RESISTANCE:** The ability of paper and paperboard to resist the penetration of water and water vapor.

**MOISTURE VAPOR TRANSMISSION:** The ease or rate of water vapor permeation in a sheet of paper.

**MOISTURE WELTS:** A paper roll defect due to uneven moisture content expanding the surface, causing ridge-like bands and wrinkles aligned in the machine direction.

**MOISTUREPROOF:** The ability of paper and paperboard to resist the penetration of water vapor.

**MOLD MACHINE:** A wet machine used primarily to make paperboards. It uses wire-covered cylinders or molds on which the web is formed. These cylinders are partially immersed and are rotated in vats that contain a dilute stock suspension from which fibers are picked up to form a web on the surface, with water draining through and passing out at the ends. The wet sheet is then "couched off" of the cylinder onto a felt for possible combining with other sheets, pressing, or drying.

**MOLD:** Any type of frame that is an integral part of the sheet-forming machine over which a porous fabric or metallic screen is placed to provide a surface for water drainage and sheet formation from a pulp slurry poured over it.

**MOLDABILITY:** That property of any wrapping paper that enhances its fold crease retention.

**MOLDED PULP TRAYS:** End user domestic and industrial utility trays and other similar products that are made by molding directly from a thick liquid suspension of pulp fibers.

**MOLECULAR CHLORINE:** *See* CHLORINE.

**MONITOR:** (1) A visual display, commonly on a video screen, of conditions and measurements occurring in the papermaking process. (2) To sense, record, or alert personnel to a process condition requiring attention, usually without the capability of initiating corrective action.

**MONORAIL CONVEYOR:** Type of load-moving equipment guided by a single metal rail used to transport paper rolls in paper mills. There are two types: the overhead and the in-floor.

**MONOSULFITE PROCESS:** A pulping process used extensively to cook hardwood and straw in conjunction with a neutral cooking liquor. Also referred to as the *neutral sulfite* process.

**MONOSULFITE PULP:** Pulp usually made from hardwood by the *neutral sulfite* process, in which the cooking liquor is a single radical sulfite chemical made to produce a neutral pH solution condition.

**MORDANT:** Materials added to pulp slurries to improve the fixation of dyes to the fiber.

**MOTOR:** Rotating equipment that converts fluid or electrical energy to mechanical power to drive valves and machinery.

**MOTOR BOOSTER:** An electric motor-driven generator placed in series in an electric circuit to change the circuit voltage. *See also* BOOSTER.

**MOTOR CONTROL:** An electrical motor that governs the electric power delivered and starts, protects, and maintains operational speeds of large motors in the mill.

**MOTOR CONTROL CENTER:** The room in a mill where the electrical motor control equipment is located.

**MOTOR DRIVE:** An electric motor used in a mill machine drive system as the source of motive power.

**MOTOR-GENERATOR SET:** One or more motors coupled to one or more generators that serve as an electrical power system.

**MOTTLE:** (1) The random nonuniformity in printed gloss, or the visual density or color of a printed area caused by uneven absorption of ink by paper. (2) A surface effect produced by the addition of heavily dyed fibers of a different color in the stock furnish.

**MOTTLED FINISH:** Any paper surface that exhibits random, localized spots of high and low gloss and variations in thickness. *See* MOTTLE.

**MOTTLED PAPER COLOR:** A colored paper defect that results in varying shades due to nonuniform formation, dye addition, and pressure used on the calender stack.

**MOTTLING FIBER:** Heavily dyed fibers added to a different color stock furnish to produce characteristic surface effects.

**MOULD MADE PAPER:** *See* HANDMADE PAPER.

**MOUNTING BOARD:** A paper lined stiff type of paperboard used for displaying pictures, samples, exhibits, etc.

**MOUNTING BOARDS:** Multi-layered paperboards that are laminated with high-quality paper on both sides. They are commonly used to mount photographs and prints.

**MOUSTACHE:** A type of paper defect appearing as areas of excessive amounts of coating buildup caused by foreign matter on the blade coating roll pushing up a small portion of the blade. Sometimes called *butterflies. See also* BACKING ROLL SPOTS.

**MULLEN:** The bursting strength of paper and paperboards. It is usually determined on a tester that bears the name and is expressed in pounds per square inch (kilopascals per square meter). Sometimes referred to as *burst strength* or *pop strength.*

**MULTI-CYLINDER DRYER:** A paper machine dryer made up of a number of steam-heated cylindrical dryer drums.

**MULTI-CYLINDER MACHINE:** A cylinder-type paperboard machine consisting of a series of vats from which a single sheet is formed and then put together with others to form a multi-ply board.

**MULTI-FOURDRINIER MACHINE:** A paper machine that has more than one fourdrinier forming section arrangement in which a *multi-ply* sheet is formed by the multiple forming sections feeding together to form a single web.

**MULTI-GENERATOR DRIVE:** A paper machine drive system where the various sections of the machine are driven by individual motors drawing their power from separate generators.

**MULTI-LAYER FABRIC:** A forming fabric woven with two or three independent layers so that different surface characteristics can be achieved.

**MULTI-PASS HEADBOX:** Type of headbox designed so that the stock flow reverses direction around a baffle as it travels to the outlet. It is typically used on high-speed newsprint machines.

**MULTI-PASS INLET:** A type of flow distributor used to spread flow to the vat of a cylinder machine. Its length is designed to double back on itself two or more times as it is progressively reduced in size.

**MULTI-PLY:** A paper or paperboard sheet made up of two or more layers.

**MULTI-STAGE BLEACHING:** Any pulp bleaching process consisting of two or more stages of operation in continuous series, rather than in one single step.

**MULTI-STAGE SULFITE PULPING PROCESS:** A chemical pulping procedure in which two or more sulfite liquors of different pH or composition and concentration are added in tandem during the impregnation or cooking phase.

**MULTI-WIRE MACHINE:** A paper machine having more than one forming wire (forming fabric) in various arrangements, such as a *twinwire former, topwire former, multi-fourdrinier,* etc.

**MULTIPLE MANIFOLD:** A type of paper machine headbox inlet furnish-spreading piping arrangement in which the flow is repeatedly split in half.

**MULTIPLE BRANCHING MANIFOLD:** A type of stock furnish distributor used on the wet end of a fourdrinier paper machine. It is designed to be short-coupled to the headbox with a number of entry pipes.

**MULTIPLE EFFECT EVAPORATOR:** An evaporator system consisting of a series of individual evaporator bodies arranged so that the vapor generated from one evaporator body becomes the steam supply to the next evaporator in the series. It is commonly used to concentrate diluted spent cooking liquor.

**MULTIPLE USE FORESTING:** The managing of pulpwood forestlands in such a way as to protect its scenic beauty, wildlife, watersheds, and recreational aspects.

**MULTIWALLED:** A bag made with three or more layers of paper. *See* BAG.

**MUSIC PAPER:** A sized, machine-finished paper suitable for offset work in printing music sheets and books.

# N

**NAP:** The soft, downy surface produced on papermaking felts by raising the surface fibers of the sheet. *See also* NAPPING.

**NAPKIN TISSUE:** A grade of tissue paper made especially for converting into paper dinner napkins.

**NAPPING:** The process of raising the surface fibers on a papermaking felt by passing it against the rotating surface of a drum, fitted with barrel-shaped burrs (teasels), in order to produce a soft, downy surface (the *nap*).

**NATIONAL COUNCIL OF THE PAPER INDUSTRY FOR AIR AND STREAM IMPROVEMENT INC. (NCASI):** Headquarters: New York, NY, (212) 532-9000.

**NATIONAL FOREST PRODUCTS ASSOCIATION (NFPA):** Headquarters: Washington, D.C., (202) 463-2700. Merged in 1993 with the American Paper Institute Inc. (API) and the American Forest Council (AFC).

**NATIONAL PAPER TRADE ASSOCIATION (NPTA):** Headquarters: Great Neck, NY, (516) 829-3070.

**NATURAL BOILER WATER CIRCULATION:** The movement of water in a power or recovery boiler caused by a force due to the specific weight differential of water and saturated steam at equivalent pressure.

**NATURAL COLORED PAPER:** A type of paper whose color is that obtained only from the fiber used to make it. No dyes, pigments, or other coloring materials are used. Also known as *self-colored paper.*

**NATURAL DRAFT:** The negative pressure created by the height of a stack or chimney and the difference in temperature between flue gases and the atmosphere.

**NATURAL EARTH PIGMENTS:** A class of pigment-coloring material used in papermaking found in its natural state in the earth. The color usually ranges from a light yellow to dark brown.

**NATURAL FILLERS:** Those types of naturally existing inorganic substances used as loading materials in the manufacture of paper.

**NATURAL GAS:** A mixture of methane and other naturally occurring hydrocarbons. It is used as a fuel by many power plants in pulp and paper mills to produce steam and electrical energy.

**NAVAL STORES:** Pine wood by-products such as rosin, turpentine, and various pine oils, some of which are derived from the wood-pulping process.

**NCASI:** *See* NATIONAL COUNCIL OF THE PAPER INDUSTRY FOR AIR AND STREAM IMPROVEMENT INC.

**NCE:** *See* NONCONTACT EVAPORATOR.

**N.C.R.:** *See* NO CARBON REQUIRED.

**NDT:** *See* NONDESTRUCTIVE TESTING.

**NECK:** A common pulp mill term referring to the narrow, cylindrical section of a batch-type digester, located just below the chip inlet (head) and between it and the larger, cylindrical body of the digester.

**NECKING:** The coming together of the coating and the paper between the rolls (nip) of the coater.

**NEEDLE PAPER:** A grade of black-colored paper that does not contain any substance or chemical that would corrode or discolor metallic objects which it may contact or for which it is used as a wrapping material.

**NEEDLED FELT:** A paper machine press and dryer felt made by transversely punching and interlocking fibers in a loose mat with barbed needles, thereby imparting the various combined properties of openness, finish, and strength.

**NEEDLES:** The leaves or foliage of softwood or conifer species of pulpwood trees.

**NEEDLING:** Refers to a bonding method used in the manufacture of nonwoven papermaking felts, consisting of the mechanical interlocking of fibers by specially designed barbed needles.

**NEGATIVE CROWN:** The surface curvature of a paper machine roll when the center diameter is less than the end diameters, resulting in a concave surface profile.

**NEGATIVE PAPER:** A high-quality, glue-sized paper made with cotton fibers. It has a hard, smooth surface, and is used primarily by the photographic industry in making a special type of negative for reproduction work.

**NET POSITIVE SUCTION HEAD (NPSH):** The amount above saturation pressure at the inlet of a boiler feedwater pump that should not be above saturation pressure to prevent flashing and pump damage.

**NEURAL NETWORKS:** A branch of *artificial intelligence* used to enhance a pulp and paper mill's process control system and to find certain relationships in the mill data base (e.g., to correlate product and process quality variability). Unlike with *fuzzy logic* and *expert systems*, these relationships can be "learned" without specific models or rules by the network's ability to "map" a function simply by being trained on that function.

**NEUTRAL KRAFT:** An acid- and sulfur-free paper made from sulfate pulp with a pH of 7.

**NEUTRAL SIZE:** Rosin size used to impart water-resistance properties to paper and paperboard. The size is treated with caustic to produce a dispersion or solution with a pH of 7. Neutral size can also be synthetic, not requiring the use of acidic alum, thus allowing paper to be made at a pH of 7. Neutral sized paper is used in special documents, books, etc., where the effects of aging must be minimized. Neutral size is sometimes referred to as *alkaline size*.

**NEUTRAL SULFITE:** A chemical wood pulping procedure in which the cooking liquor is made up of sodium sulfite, with enough sodium carbonate to maintain a pH of 7 or slightly above during the cook. *See* MONOSULFITE PROCESS or NEUTRAL SULFITE SEMICHEMICAL (NSSC).

**NEUTRAL SULFITE SEMICHEMICAL (NSSC):** A chemical wood pulping process using a neutral sodium sulfite-sodium carbonate cooking liquor. The cook is carried out under slightly alkaline conditions, but with the final separation of the fibers made by unpressurized means after the initial cook has been completed. Also referred to simply as *semichemical pulping process.*

**NEW BROWN KRAFT ENVELOPE CUTTINGS:** A grade of wastepaper that includes new brown kraft cuttings or sheets, usually in bale form, free of contaminants such as tars, gums, waxes, asphalt, latex, colors, etc., and having a maximum of 1 percent out-throws.

**NEW CORRUGATED CUTTINGS:** A wastepaper grade consisting of trim and cuttings from corrugated boards with liners, made from sulfate or jute pulp, and having a maximum out-throw of 5 percent.

**NEW MANILA ENVELOPE CUTTINGS:** A wastepaper grade consisting of cuttings or shavings of manila envelopes or sheets of manila-colored paper in bale form, without any printing, and with a maximum of 2 percent out-throws.

**NEWS BOGUS PAPER:** Type of paper made from deinked newsprint wastepaper.

**NEWS GRADES:** Types of paper that contain a high percentage of groundwood pulp and are similar to newsprint.

**NEWS VAT-LINED CHIPBOARD:** News-lined chipboard used for making setup boxes.

**NEWSBACK:** *See* NEWSBOARD.

**NEWSBOARD:** Type of paperboard made primarily from newsprint waste for use in the manufacture of setup boxes and containers. The board may be lined with a higher grade of pulp during manufacture. Also referred to as *newsback.*

**NEWSPRINT:** A grade of paper combining high percentages of groundwood and/or mechanical pulps made especially for use in the printing of newspapers.

**NEWSPRINT MACHINE:** A paper machine devoted to making a *newsprint* grade of paper.

**NEWTONIAN FLUID:** Fluid in which the shear stress is directly proportional to shear rate, and viscosity remains constant when the rate of shear is changed.

**NFPA:** *See* NATIONAL FOREST PRODUCTS ASSOCIATION.

**NINE-POINT:** Paperboards having a thickness of 9 mils (0.009 inch).

**NINE-POINT CORRUGATING MEDIUM:** A paperboard, 9 mils (0.009 inch) thick, used as the fluted component in the manufacture of multi-ply combined board or wrapping.

**NIP:** The contact area between two rolls on pulp and papermaking machines such as wet presses, coaters, calenders, supercalenders, etc.

**NIP PRESSURE:** The pressure between two vertically aligned rolls, as in a wet press of a paper machine, expressed in pounds per lineal inch (pli) of roll face, and determined by the sum of any mechanical loading and weight of the top roll in pounds divided by the length of the contacting face at the nip.

**NIPPLE:** A short length of pipe or tube used for joining piping elements together.

**NITROCELLULOSE:** An explosive chemical substance formed from nitric acid and cellulose. It is used to make synthetic lacquers.

**NO CARBON REQUIRED (N.C.R.):** Reference to a special type of copying paper in which the impressed image is transferred from the original sheet to other sheets of paper by means of chemical coatings on the front and rear of the sheets. *See also* ACTION PAPER.

**NODULIZING:** A method of preparing pulp for bulk storage or transportation. The pulp slurry is dewatered on a press and the cake is shredded to produce grain-sized nodular particles called *noodles.*

**NOISELESS PAPER:** Paper that has been treated to minimize the property of rattle so that it can be used in making up such user items as theater programs, manuscripts, etc. Also referred to as *silent paper.*

**NOMINAL WEIGHT:** The standard designated weight for a ream or other specified quantity of paper of the prescribed sheet size. *See also* ACTUAL WEIGHT.

**NONAVAILABLE SULFUR DIOXIDE:** The difference between the total sulfur dioxide and the free or available sulfur dioxide in sulfite-type cooking liquors in pulp mills. Also called *combined sulfur dioxide.*

**NONBIBULOUS:** Papers and paperboards that are not water absorbent, as related to the high amount of size used in their furnish makeup.

**NON-CHLORINE BLEACHING:** The process of bleaching pulps to "market" brightness without the use of elemental chlorine or chlorine-containing compounds. Such sequences typically employ oxygen, *ozone, hydrogen peroxide,* and other peroxygens.

**NONCOMBUSTIBLE PAPER:** Special type of paper, usually made with asbestos, ceramics, or glass fibers, that will not burn or support combustion. It is used for fire protection purposes.

**NONCONDENSIBLES (NCs):** A pulp mill term referring to gases which, if not removed or relieved properly, collect at the top of a batch digester and cause a *false pressure* that prevents the temperature from rising to the proper cooking levels.

**NONCONTACT EVAPORATOR (NCE):** Type of evaporator in which the evaporating medium (steam or hot gases) does not contact the liquid being evaporated, such as used to concentrate black liquor with hot exhaust gases in a recovery furnace.

**NONDESTRUCTIVE TESTING (NDT):** The testing of paper or other materials without destroying the samples or component structures of a system. For example, a digester, tankage, etc., can be tested for corrosion or stress cracking by using various nondestructive methods such as ultrasonic imaging and magnetic imaging without disassembly.

**NON-FADING POSTER PAPER:** Book paper made from sulfite pulp to be light and weather resistant and used to make outside posters.

**NONFIBROUS MATERIALS:** Any component of furnish or paper stock that is not of a fibrous nature, such as rosins, dyes, fillers, chemicals, starches, additives, brighteners, etc.

**NONINDUCTIVE LOAD:** The type of load on an electrical system or equipment in which the current is in phase with the voltage across the entire load.

**NON-NEWTONIAN FLUID:** Fluid in which the shear stress is not directly proportional to the shear rate, and the viscosity does not remain constant when the rate of shear is changed. It is the opposite of *Newtonian fluid*.

**NONPOROUS:** Particular types of pulpwood that do not have pores or vessels, such as coniferous or softwoods. It also applies to other more or less solid pulp and papermaking materials that do not have pores.

**NONRETURNABLE CORE:** Type of paper roll core made to be used only once and not returned to the originating mill.

**NONTARNISH PAPER:** A type of paper that does not contain any chemicals that would discolor metallic objects with which it may come in contact.

**NONWARP ADHESIVE:** An adhesive that will produce laminations able to remain flat when subjected to low relative humidities.

**NONWOOD FIBERS:** Pulp and papermaking fibers obtained from plants, such as straws, bamboo, sugarcane (bagasse), reeds, rice, cotton, etc., and not from pulpwood trees.

**NONWOOD PULP:** Papermaking pulps such as bagasse, bamboo, esparto, straw, cotton, etc., not manufactured from pulpwood trees, but from other fibrous plants.

**NONWOODY:** *See* NONWOOD FIBERS.

**NONWOVEN FABRIC:** A sheet of cloth-like material made from long natural and synthetic fibers. It is formed from a slurry on a wire screen, such as a wet end of a paper machine, or by laying on a fine mesh screen from an air suspension.

**NOODLES:** A form of pulp prepared by dewatering a slurry to approximately 35 percent air dry and shredding. It is used primarily for storage and transportation. The method of preparing this form of pulp is commonly called *nodulizing*.

**NORDEN EFFICIENCY:** The number of ideal mixing stages in a counter-current washing system that are required to give the same performance as a reference washing stage or system.

**NO-SULFUR PULPING:** The process of chemically reducing wood into its fiber components with a cooking liquor consisting of chemicals that do not contain sulfur.

**NOTE PAPER:** Various grades of folded writing paper used for personal note-taking and correspondence.

**NOTION BAG PAPER:** Lightweight machine finished and glazed kraft paper used to make notion bags.

**NOVEL NEWS:** Grades of bulky low finished newsprint paper used in printing magazines and books.

**NOXIOUS GASES:** Objectionable gases such as those contained in the flue gases of a sulfate pulp mill. They consist mainly of sulfur dioxide ($SO_2$), nitrogen oxides (NO), and total reduced sulfur (TRS), made up of hydrogen sulfide ($H_2S$), methyl mercaptan ($CH_3SH$), dimethyl sulfide ($CH_3SCH_3$), dimethyl disulfide ($CH_3SSCH_3$), etc.

**NOZZLE:** A tapering constriction in a pipeline, projection, or spout used to increase the velocity of the fluid flowing through it.

**NOZZLE HEADBOX:** Type of paper machine headbox designed with nozzle-like inlets, with the entire channel from the pump to the slice completely filled with stock.

**NOZZLE SLICE:** A type of paper machine slice, with a nozzle-like design, that operates entirely filled with stock.

**NPSH:** *See* NET POSITIVE SUCTION HEAD.

**NPTA:** *See* NATIONAL PAPER TRADE ASSOCIATION.

**NSSC:** *See* NEUTRAL SULFITE SEMICHEMICAL.

**NUBBINS:** Unground pieces of wood produced during the grinding of pulpwood when the log projects over the edge of the stone while being ground.

**NUCLEAR GAUGES:** Measurement instruments having radioactive sources as the primary element. They are used to measure level, density, and basis weight in pulp and paper mills.

**NUMBER:** The quantity of sheets contained in 50 pounds of standard basic size paper.

**NUMBER 1 BOND:** Bond paper made entirely from white rag stock.

**NUMBER 1 BRISTOL:** A heavy solid sheet of ledger paper made from rag fibers on a fourdrinier paper machine. The heavier grades are also made by pasting sheets together.

**NUMBER 1 BROWN KRAFT PAPER:** A grade of wastepaper consisting of used brown sulfate paper in bales, and with a maximum of 5 percent out-throws.

**NUMBER 1 FLY-LEAF SHAVINGS:** A grade of wastepaper consisting of trim and shavings from magazines, catalogs, and similar types of printed, mostly sulfite-type baled papers with a maximum of 1 percent out-throws.

**NUMBER 1 MANILA PAPER:** A wastepaper grade primarily made of sulfite or sulfate manila-colored, baled paper with light printing and having a maximum of 5 percent out-throws.

**NUMBER 1 NEWS:** A grade of wastepaper made up of newspapers put up in bales, having less than 5 percent of other paper content, and with a maximum of 2 percent out-throws.

**NUMBER ONE (1):** A number that indicates the highest grade or quality. Higher numbers indicate lesser quality grades.

# O

**OAK:** A hardwood species of trees used as a source of pulpwood.

**OATMEAL PAPER:** A coarse type of wallpaper made by adding very fine sawdust to its furnish.

**OCC:** *See* OLD CORRUGATED CONTAINERS.

**OCHER:** A naturally occurring clay made up of high percentages of aluminum silicate, iron oxide, and chemically combined water. It varies in color from pure yellow to dull orange and red-brown and is used as a pigment in paper coloring.

**OCS:** *See* OUTSIDE CHIP STORAGE.

**OD:** *See* OUTSIDE DIAMETER.

**O.D.:** *See* OVEN DRY.

**ODD LOTS:** Paper and papermaking materials that do not comply with prescribed standards established for them.

**ODOR ABATEMENT:** The technology of reducing the nuisance caused by disagreeable odors of waste gases from pulp mills.

**ODOR THRESHOLD:** The tolerable and perceptible limits of disagreeable odors of waste gases from pulp mills.

**OEM:** *See* ORIGINAL EQUIPMENT MANUFACTURER.

**OFF COLOR:** Describes a dyed paper, the actual color of which does not exactly match the standard sample used on a specification guide. Also called *off shade.*

**OFF CUTS:** Paper sheets cut to size from trimmings and waste from cutting paper to make up standard reams of larger size sheets.

**OFF LEVEL SHEET:** A paperboard made on a counterflow-type cylinder machine with one side thicker than the other due to misadjustment of the wings that control the levelness of the board.

**OFF MACHINE COATING:** The process of applying coating material to a sheet of paper or paperboard in a location that is away from the machine on which it is made. Also called *conversion coating.*

**OFF SHADE:** *See* OFF COLOR.

**OFF-SPEC:** Shipments of paper received by customers that are not of the specifications ordered.

**OFF SQUARE:** Paper sheets cut with corners not at 90° angles. This causes a folding defect known as out-of-square.

**OFFSET:** A paper roll defect showing up as an abrupt change in position of the edge of the sheet. It is caused by variable tension at the winder, poor splicing procedure, or uneven roll loading. *See also* PARTIAL MACHINE DIRECTION TENSION BURST.

**OFFSET CORE:** A paper roll defect in which the core is not evenly aligned with and protrudes beyond the roll end.

**OFFSET PAPER:** Well-sized paper made with the characteristics most suitable for use in offset printing, in which the ink is transferred from the plate with the image on it to a rubber blanket and then to the paper being printed.

**OHM:** A unit of resistance offered to the passage of 1 ampere when impelled by 1 volt.

**OIL BURNER:** A device used to inject and atomize fuel oil into fine sprays, providing for proper air mixture on combustion in a furnace. Also referred to as *oil gun*.

**OIL GUN:** *See* OIL BURNER.

**OIL PENETRATION TEST:** A common paper mill laboratory test consisting of floating a small sample of paper on oil and measuring the time for oil to completely penetrate the sheet.

**OIL RESISTANCE:** A physical property of paper that makes it withstand the penetration of oil and oily substances.

**OIL SPOTS:** A type of paper defect appearing as translucent spots caused by the inadvertent introduction of oil into the furnish stock, especially when wastepaper and rags are used.

**OILED MANILA:** Oil-treated manila paper used for meat wrapping.

**OILED TYMPAN PAPER:** A uniform caliper paper with a high finish treated with light mineral oil and widely used in the printing industry.

**OILPROOF PAPER:** A type of paper that is specially treated to withstand the penetration of oil and oily substances.

**OLD CORRUGATED CONTAINERS (OCC):** A valuable source of fiber for recycled combination paperboards. *OCC* includes container plant cuttings as well as used corrugated containers.

**OLD NEWSPAPERS (ONP):** Newsprint wastepaper used as a source of secondary fiber for recycling paper mills. *ONP* includes old newspapers collected either from households or as overissues at newsstands or inplant. Also includes groundwood paper trim.

**OLD PAPER:** Wastepapers used as a source of fiber for the manufacture of some paperboards and chipboards. *See* WASTEPAPER.

**ONLINE:** The type of operation of a computer and other instruments that is actively monitoring and controlling a process or operation. Also refers to having been put into operation, such as a piece of equipment that has come *online*.

**ON MACHINE COATING:** Applying surface coatings on a sheet of paper while it is being made on the paper machine as an integral part of the papermaking operation.

**ON-OFF CONTROL:** A type of control in a mill in which the controller is multi-position and has two discrete values of output, fully on and fully off.

**ONE HUNDRED SHEET SEALED:** A package containing 100 sheets, usually of bristol or paperboard.

**ONE-SIDE COATED:** *See* COATED ONE-SIDE (C1S).

**ONE-SIDE FINISH:** Paper that has been finished only on one side.

**ONIONSKIN:** A thin, lightweight, transparent paper made especially to be used for producing typewritten copies of correspondence.

**ONP:** *See* OLD NEWSPAPERS.

**OPACIFIED:** A descriptive reference to paper made more opaque by addition of an *opacifier* to the pulp it is made from.

**OPACIFIER:** A substance used to mix with stock to make an *opacified* sheet of paper, i.e., one that is more opaque than normal. Also called *opacifying agent.*

**OPACIFYING AGENT:** *See* OPACIFIER.

**OPACIMETER:** An optical paper mill laboratory instrument with which the degree of light transmission through a sheet of paper can be measured.

**OPACITY:** That ability of substances such as paper, flue gases (smoke), and liquids to resist the transmission of both diffuse and nondiffuse light through it. It prevents *show through* of dark printing in contact with the backside of a sheet of paper.

**OPAQUE:** That property of a sheet of paper that prevents the transmission of light through it.

**OPAQUE PAPER:** Paper that has been loaded with pigments such as titanium dioxide and calcium carbonate to render it non-transparent.

**OPCO PROCESS:** **(1)** A wood pulping process in which pulp produced by mechanically separating the fibers of preheated chips in a pressurized refiner, known as thermomechanical pulp (TMP), is treated with sodium sulfite before further refining. (2) A modification of the thermomechanical pulping process in which the fiber is produced and then chemically treated either between the refining stages or after the secondary refining stage.

**OPEN CONSISTENCY REGULATOR:** A type of consistency regulator whose sensor consists of an element in an open flow box with provisions for stock inlet, outlet, and overflow.

**OPEN DRAW:** The tension existing in a paper sheet as it moves from one section or part of a paper machine to another, such as between the couch roll and the first press, between first press and second press, etc. *See also* OPEN TRANSFER.

**OPEN-FACE CALENDER:** A type of calender designed to allow the removal of calender rolls from the front or operating side, or upward between the upright frames of the stack.

**OPEN HEADBOX:** A type of headbox used in slower speed paper machines. It is fed on one end by a distributor and discharges onto the wire (forming fabric) at the other end with the top open to the atmosphere.

**OPEN IMPELLER PUMP:** A type of centrifugal pump designed with widely spaced impeller blades and used extensively to handle heavy pulp and stock slurries in a mill.

**OPEN PAPER MACHINE HOOD:** A canopy located over the dryer section of a paper machine with side panels extending down to the journals of the top dryers. It is used to confine and control vapor-laden air rising from the dryers so that the heat can be recovered for heating the air entering the machine room.

**OPEN TRANSFER:** A means of transferring the wet sheet web from the wire (forming) section to the wet press of a paper machine, and includes *open draw* and *poor man's pickup.*

**OPEN WASHING:** A type of pulp-washing sequence used in multi-stage bleaching of pulp in which the circuit has an effluent discharge.

**OPENER:** A pulp treating device usually located after the blowpit in older sulfite mills. It consists of a cylindrical wooden structure equipped with rotating wooden bars and stationary baffles in order to liberate the well-cooked fibers from the knots to facilitate easier removal by subsequent knotters in the process.

**OPENNESS:** A property of papermaking felts imparted during their manufacture by loose weaving.

**OPERATING CONDITIONS:** Conditions, such as ambient temperature, pressure, vibration, moisture, etc., to which a device or process is subjected during its normal use in process and manufacturing operations.

**OPERATING FLOOR:** The floor on which the operating crews are located in various sections of a mill.

**OPERATING ROOM:** Control room from which the operating personnel run process equipment in a mill.

**OPERATOR:** A pulp or paper mill employee in charge of the operation of a piece of machinery or unit process.

**OPTICAL BRIGHTNESS:** Fluorescent dyes or pigments that absorb ultraviolet radiation and re-emit it at a higher frequency in the visible spectrum (blue), thereby effecting a white, bright appearance to the paper sheet when added to the stock furnish. *See also* FLUORESCENT BRIGHTNESS.

**OPTICAL PROPERTIES:** Those properties of pulp, paper, and paperboard that are associated with light absorption and light-scattering.

**OPTICAL TESTS:** Tests made on paper to measure characteristics that are related to their light absorption and light-scattering properties, such as color, brightness, opacity, and gloss.

**OPTICAL WHITENERS:** *See* FLUORESCENT WHITE DYES.

**OPTIMIZATION:** The computation of process control parameters, and applying them to maximize control effectiveness.

**ORANGE LIQUOR:** A cooking liquor used in the polysulfide cooking process that is produced by partially oxidizing sodium sulfide ($Na_2S$) in white liquor in accordance with the following equation: $4Na_2S+O_2+2H_2O = 2Na_2S_2+4NaOH$. *See also* POLYSULFIDE PULPING.

**ORANGE PEEL:** A type of sheet surface that has a pebble-like texture resembling that of an orange.

**ORGANIC:** Any chemical compound containing carbon, of which there are many used in the pulp and papermaking process.

**ORGANOCELL PROCESS:** *See* SOLVENT PULPING.

**ORGANOCHLORINE:** *See* CHLORINATED ORGANICS.

**ORIFICE PLATE:** One of the most frequently used primary devices in the paper industry for flow measurement of fresh water, steam, clean chemicals, and uncontaminated gases. It consists essentially of a thin gauge metal plate installed in a pipeline. It has a characteristically shaped and located hole in it to produce a differential pressure as the fluid flows through it. This differential pressure is measured and converted into flow values.

**ORIGINAL EQUIPMENT MANUFACTURER (OEM):** The maker of major pulping and papermaking equipment and systems such as paper machines. The *OEM* establishes specific specifications for the equipment it manufactures, often subcontracting various components.

**ORP:** *See* OXIDATIVE REDUCTION POTENTIAL.

**ORSAT APPARATUS:** A test instrument used in many mills to determine the percentage composition of selected components in a gas mixture, such as $SO_2$ and $O_2$ by using the proper reagent for the component to be measured.

**OSMOSIS:** Diffusion of a fluid through a semipermeable membrane from the side of higher concentration to the side of lower concentration.

**OUT-CROWN:** The crown profile of a paper machine roll after being run in the machine.

**OUT-OF-JOG:** Sheets in a pile of paper that are not even with the edge. It is caused by improper jogging of the paper in the layboy.

**OUT-OF-ROUND:** A paper roll defect where the core is not in the center of the roll and has a flat section on the outside caused by impact damage or improper roll winding.

**OUT-OF-SQUARE:** A paper sheet defect (*Off-square*) due to corners not being at 90° angles, causing an overlap when folded in half.

**OUTFALL:** The mouth of conduit drains and other conduits from which a mill effluent discharges into receiving waters.

**OUTSIDE CHIP STORAGE (OCS):** A method of storing wood chips in open piles and feeding them from there directly to the digester house.

**OUTSIDE DIAMETER (OD):** The measure of piping, tankage, etc., that is taken or made on the outside circumference, as compared with *inside diameter (I.D.)*, which is made on the inside.

**OUTSIDE WASTE:** The loss in paper weight of a roll after the outside wrapping and damaged, unusable paper are removed.

**OUTSIDES:** Defective paper sheets that are usually put at the top and bottom of reams when wrapped.

**OUT-THROWS:** Paper that is not suitable for use in the grade for which it was intended.

**OUT-TURNS:** Paper samples of orders run on a paper machine and used for mill or customer reference records.

**OUTWARD-FLOW SCREEN:** A type of rotating cylindrical stock screen designed so that the direction of flow is from the inside to the outside of the screen plates. The accepted fiber passes through and the oversize and dirt particles settle and are continuously drawn off.

**OVEN DRY (O.D.):** Moisture-free conditions of pulp and paper and other materials used in the pulp and paper industry. It is usually determined by drying a known sample to a constant weight in a completely dry atmosphere at a temperature of 100°C to 105°C (212°F to 221°F). Also called *bone dry (B.D.)*.

**OVEN DRY WEIGHT:** Moisture-free weight of materials involved in pulp and paper manufacturing.

**OVERCOOK:** A chemical pulping process that produces pulp not meeting the strength or bleaching requirement specifications prescribed for it because of improper cooking conditions such as excessive cooking time, temperature, pressure, and liquor concentrations.

**OVERDRIED PAPER:** Paper from which too much moisture has been dried from the particular grade made on the paper machine so that it becomes too brittle and too weak for its intended use.

**OVERFIRE AIR:** Combustion air fed into a furnace just above the stoker grate.

**OVERGAS:** The relief gas ($SO_2$) from a sulfite chemical cooking digester that is not absorbed in the reclaiming tower of the liquor fortifying system.

**OVERISSUE NEWS:** A type of wastepaper used primarily by deinking plants. It is excess printed material and comes directly from the publishers before distribution.

**OVERLAP JOINT:** A type of adhesive joint used for some paper splices in which the surface of one sheet extends over the edge of the other sheet.

**OVERLOAD PROTECTION:** The effect of an electrical device operating on excessive current and causing and maintaining an interruption of current flow to the device governed, such as an electric motor.

**OVERRUN:** Paper made on a paper machine beyond the amount specified on a particular order. It is put into the mill inventory or shipped to the customer on his instructions.

**OVERS:** Extra sheets of paper supplied in an order to a customer to cover any anticipated spoilage.

**OVERSEER:** A term used in some pulp and paper mills for a foreman or a supervisor.

**OVERSHOOT:** The amount that a mill process output exceeds the desired value after there has been a change in input.

**OVERSIZE PAPER:** Paper intentionally made larger than ordered so it can later be cut or trimmed to order size.

**OVERWEIGHT PAPER:** Paper inadvertently made with a higher weight than specified on an order. It is put into inventory or shipped to the customer on special agreement.

**OXIDANT:** Any oxygen-containing substance that is capable of giving up its oxygen in a chemical reaction.

**OXIDATION:** A very common chemical process reaction in which oxygen combines with other elements or compounds.

**OXIDATION LAGOON:** A holding pond in which mill liquid effluents are retained in order that oxygen may be added by the use of air so that biological oxidation of organic materials takes place.

**OXIDATION-REDUCTION POTENTIAL (ORP):** The voltage developed in chemical solutions when oxidation-reduction reactions take place. It is used to determine residual chlorine in the chlorination of pulp, water, and the manufacture of hypochlorite bleaching solutions in a pulp mill.

**OXIDIZED BLACK LIQUOR:** *See* BLACK LIQUOR OXIDATION.

**OXIDATIVE EXTRACTION ($E_O$):** A pulp bleaching stage in which oxygen is added in the usual alkaline extractive stage, making it possible to lower bleaching chemical consumption in later stages and achieve the same brightness.

**OXIDIZED STARCH:** Starch slurries that have been treated with oxygen to lower the viscosity so that they can be used more efficiently as an additive in papermaking.

**OXIDIZER:** The vessel of a black liquor oxidation system in which the air or oxygen is intimately mixed with the black liquor.

**OXYCELLULOSE:** A structureless substance that is formed when cellulose undergoes drastic action from oxidizing agents used in some chemical wood pulping processes and pulp bleaching processes.

**OXYGEN BLEACHING:** A stage in a bleaching process that uses oxygen under alkaline conditions to purify pulp by removing lignins and altering other coloring matter.

**OXYGEN DEMAND:** The amount of dissolved oxygen consumed in biological processes breaking down organic matter in liquid mill effluents, usually under specified conditions over a period of time (5 days).

**OXYGEN DELIGNIFICATION:** The use of oxygen to further break down the lignin bond between fibers in wood. It generally is applied in a separate stage following chemical pulping (cooking) in a digester system. The oxygen is applied in a reactor vessel, followed by extensive washing. Kappa numbers of cooked chemical pulps can be reduced 10 to 15 points by this treatment with oxygen.

**OXYGEN SCAVENGER:** Chemical materials used in boiler feedwater conditioning to aid in the removal of low level dissolved oxygen during deaeration to decrease its corrosive effect.

**OZALID:** *See* DIRECT PROCESS PAPER.

**OZONE ($O_3$):** An allotropic form of oxygen produced by passing air or oxygen through an electrical discharge (*ozone generator*). It is being used in the bleaching of pulp to very high brightness, representing a significant environmental gain over the use of chlorine and chlorine containing compounds in the bleach plant (resulting in the reduction of *chlorotoxins*). Very little or no washing is generally required after an *ozone stage*, cutting down on the volume of effluent that has to be treated or recovered. Since chlorine is potentially eliminated from the bleaching process (and thus a reduction or elimination of sodium chloride buildup), the use of ozone as the primary bleaching agent may open the door to a *closed cycle mill*. *See also* NON-CHLORINE BLEACHING.

**OZONE GENERATOR:** A device or system of devices used to generate *ozone* ($O_3$) from air, or more efficiently, from oxygen. It consists of many banks of high voltage electrical discharge cells through which the oxygen passes for conversion to ozone. These generators, which much be erected "on-site" at a mill because of the relatively short "life" of ozone (which means that it cannot be practically shipped to the mill), consume large amounts of energy and must be cooled extensively.

**OZONE STAGE:** The application of ozone as a separate stage in the bleaching of pulp. *See* OZONE.

# P

**PACKAGING PAPER:** All grades of paper made especially for use in wrapping and making up packages for bundling and shipping purposes.

**PACKET:** A bundle of writing paper, usually consisting of 125 sheets.

**PACKING:** The compactness of the mass of chips in a pulp digester.

**PACKING FACTOR:** Bone dry weight of wood per digester in a batch cooking process, expressed as pounds per cubic feet in North America. Also referred to as *bulk density.*

**PACKING PAPER:** Type of high-strength paper used for wrapping and packing.

**PAD:** (1) Paper sheets glued on one side to make up a small writing packet. (2) Protective sheets of cut-to-size corrugated or solid paperboard used inside shipping containers to separate the contents.

**PADDING:** An operation in a paper-converting plant in which sheets of paper are glued on one side to make up pads.

**PAGE PAPER:** Paper or paperboard sheets used in bookmaking to support tied-up pages for storage or before placing in the printing frame on the press.

**PALLET:** A low platform on which materials such as paper are placed so that they can be moved from one location to another in the mill by the use of a forklift truck.

**PAMPHLET:** A type of booklet made up of a small number of pages, usually less than 8 inches wide and made with paper covers.

**PAMPHLET PAPER:** Printing paper made to be cut and assembled into thin paper-covered booklets.

**PAN LINER:** Chemical pulp paper made for use in bakeries to line pastry trays and pans.

**PANELBOARD:** A dense, heavy, stiff paperboard usually made on a wet machine and used in making suitcases and in building construction and automobile manufacturing.

**PAPER:** A homogeneous sheet of felted cellulose fibers bound together by interweaving and by the use of bonding agents, and made in a variety of types. It is used for a multitude of purposes such as writing, printing, wrapping, clothing, industrial, domestic, sanitary, etc.

**PAPER ADDITIVE:** Chemical or other material added to paper, paperboard, or their stock slurries to impart specific physical and visual properties to the sheet such as wet strength, water repellency, and fire resistance. *See also* ADDITIVES.

**PAPER-BASE LAMINATE:** A strong, thick, multi-ply sheet material made of sheets of paper impregnated with a thermosetting resin and firmly united while the resin sets. It is used for household, building, decorative, and industrial purposes.

**PAPER-BASE PLASTIC:** *See* PAPER-BASE LAMINATE.

**PAPER BOX MAKER:** One who converts paper and paperboard into consumer cartons.

**PAPER CLOTH:** Cloth-like fabric sheet material made with paper that has been woven from twisted paper threads, cloth with paper lining on one or both sides, or any type of specially treated paper that is used as a substitute for cloth. It is used for such purposes as household cleaning and book binding. Sometimes known as *paper textiles*.

**PAPER CORE:** A convolute-wound cylindrical tube made from paperboard, around which paper is wound in preparation for transporting and shipping. Metal tubes are also used for this purpose.

**PAPER CUTTER:** A type of paper-cutting machine designed to cut and trim rolls of paper into sheets of specified lengths suitable for the intended end use.

**PAPER DRILL:** A type of rotating cutting machine designed to make holes in paper.

**PAPER GRADE:** A type or class of paper identified as having the same fiber, color, additive, and chemical composition, and manufactured to the same physical and mechanical characteristics.

**PAPER HYGROSCOPE:** An instrument used to measure relative humidity of air around paper and between sheets in a pile to determine if it has reached equilibrium with ambient conditions and whether it will give up or absorb moisture.

**PAPER INDUSTRY MANAGEMENT ASSOCIATION (PIMA):** Headquarters: Arlington Heights, IL, (708) 956-0250.

**PAPER INSPECTOR:** An employee in the paper machine and finishing room whose duties are to examine rolls or sheets of paper and sort and set aside defective paper for resorting or reprocessing. Optical detecting devices have been developed and are used in some mills to perform some of this operation automatically.

**PAPER MACHINE:** The primary machine in a paper mill on which slurries containing fibers and other constituents are formed into a sheet by the drainage of water, pressing, drying, winding into rolls, and sometimes coating.

**PAPER MACHINE PIT:** *See* MACHINE PIT.

**PAPER MACHINE ROOM:** The area in a paper mill where the paper machine is located.

**PAPER MACHINE VENTILATION:** The removal of moisture-laden air and its replacement with warm, fresh air in the paper machine room of a pulp mill by the use of a hood over the machine.

**PAPER MANUFACTURER:** A concern involved in making one or various grades of paper, such as newsprint, bond, book, writing, and boards.

**PAPER MERCHANT:** A commercial establishment that acts as an intermediary between paper manufacturers and users by either carrying and warehousing supplies of various types of papers for wholesale distribution or making arrangements to have it shipped from the paper mill direct to the customer.

**PAPER MICROMETER:** An instrument used for measuring the thickness of paper or paperboard sheets.

**PAPER MILL:** A factory or plant location where various pulps in slurry form are mechanically treated, mixed with the proper dyes, additives, and chemicals, and converted into a sheet of paper by the processes of drainage, formation, and drying on a *paper machine.* Some paper mills also finish the paper in various ways.

**PAPER NAPKIN:** A type of soft, absorbent tissue paper, cut and folded to various sizes. It is usually creped, embossed, or printed with appropriate designs and used for household serving and table napkins.

**PAPER ROLL:** A form of packaging paper sheeting by winding it around a core or shaft so that it can be moved from one location to another within a paper mill, or transported to converters and other end users. Also refers to a paper filled roll used alternately with other rolls in super calenders to extend the nip. Also called *fiber roll* or *soft roll.*

**PAPER ROLL STOCK:** Any paper that makes up a roll for shipping purposes.

**PAPER SCALE:** A balance especially designed to weigh a sample of paper and calibrated to read the weight of a standard ream for that grade of paper.

**PAPER SCREENS:** A broad group of screen types used on stock in paper mills to remove unwanted materials and separate fibers from each other.

**PAPER SHREDDER:** A machine used in the mechanical treatment of wastepapers to open the pages of magazines, books, pamphlets, etc., and reduce the sheets to smaller pieces, making them more suitable for further processing.

**PAPER SIZE:** (1) Dimensions of paper which have been established by the paper industry according to their type. Also called *sheet size.* (2) Substances such as rosin, gelatin, starch, glue, wax, etc., added to paper stock slurries or coated on the sheet to make it resistant to penetration by water and ink. Often known simply as *size.*

**PAPER STOCK:** A water slurry of various pulp fibers, dyes, additives, and chemicals that will be pumped to the paper machine for forming into a sheet.

**PAPER TESTS:** Various paper mill physical test procedures conducted on samples of paper taken from rolls to determine whether the paper meets the specifications established for that particular lot or order.

**PAPER TEXTILES:** Cloth-like fabric sheets made from woven, twisted paper base threads or nonwoven fibers and formed on a paper machine. *See* PAPER CLOTH.

**PAPER TOWEL:** Highly absorbent, thick paper that has been cut into plain or appropriately printed sheets or rolls for use in industrial and household drying and cleaning purposes. The base material for the converted product is called *towelling paper.*

**PAPER TRUCK:** Truck used to raise and carry paper on skids, in cases, or in roll form from one place to another in a paper mill. It is usually hand- or motor-driven.

**PAPER TWINE:** Rope of various sizes made from threads or twisted, narrow strips of treated paper.

**PAPER WASTE:** (1) Printed or other used or excess paper commonly utilized by paperboard mills as a source of fibers in the manufacture of various types of paperboards such as chipboards and fillers for containerboards. (2) The total unusable weight in a roll of paper plus that damaged during shipment, storage, and handling in the preparation for use.

**PAPER WEB:** The basic structure of a sheet of paper consisting of a random network of intertwining fibers.

**PAPERBOARD:** A thick, heavy-weight, rigid, single, or multi-ply type of paper traditionally made on multi-cylinder paper machines, but now also made on fourdrinier-type machines, with and without dual headboxes or multi-former arrangements. Thickness and material vary, depending on its end use. It is used for wrappings, packagings, boxes, cartons, containers, advertising and merchandising displays, building construction, etc. Also known simply as *board*.

**PAPERBOARD PACKAGING COUNCIL (PPC):** Headquarters: Washington, D.C., (202) 289-4100.

**PAPERMAKER:** A paper manufacturer or the paper mill employee responsible for the operation of the paper machine.

**PAPERMAKERS' ALUM:** *See* ALUM.

**PAPERMAKERS' FELT:** *See* FELT.

**PAPERMAN'S FINGERS:** The fingers of employees who have been trained to detect paper qualities such as finish, smoothness, and roughness by sense of touch.

**PAPETERIE:** A chemical pulp and cotton fiber content paper made especially for conversion to a class of writing types. It is usually appropriately treated, decorated, cut, and packaged for use as correspondence stationery.

**PAPETERIE PAPER:** A reference to writing paper that is used primarily for boxed stationery. *See* PAPETERIE.

**PAPIER-MACHE:** A material made by mixing repulped wastepaper, virgin pulps, various fillers, adhesives, and sizes and treating the semi-dried product with linseed oil. It is generally used for molding purposes in conjunction with merchandising articles, models, etc.

**PAPRICAN:** *See* PULP & PAPER RESEARCH INSTITUTE OF CANADA.

**PAPYRUS:** A plant used by ancient Egyptians and others for writing and record-keeping purposes. It was made up into layers and matted into sheets. The term *paper* was eventually derived from the name of this plant.

**PARAFFIN PAPER:** *See* WAXED PAPER.

**PARAFFIN WAX:** A hard, glossy, nontacky, water-resistant material with a low melting point and viscosity, produced from naturally occurring paraffin oils, and used extensively in paper mills for sizing and coating paper and paperboard. Some types may be obtained from other sources such as bees, plants, etc.

**PARCHMENT:** Writing sheet material made from the skins of animals, such as goats and sheep.

**PARCHMENT FINISH:** A parchment-like, flat, undulating surface finish produced on firm paper by special treatment on a plater.

**PARCHMENT PAPER:** A parchment-like paper made from wood pulp and chemically treated to resist the penetration of grease. It is used to wrap greasy food items.

**PARCHMENT WRITING:** A durable, high-strength paper made from well-hydrated cotton and chemical wood pulps. It has the same appearance and texture as goat and sheet parchment.

**PARCHMENTIZING:** The process of treating a paper sheet with a 75 percent solution of sulfuric acid, followed by washing, in order to swell the cellulose such that the fibers are welded together into a continuous formation.

**PARENCHYMA:** The soft, thin-walled, cellular tissue of a woody plant that contains living protoplasm in growing parts of the plant.

**PARENT ROLL:** The original roll of paper made on a paper machine and put into inventory prior to being rewound, trimmed, or slit into smaller rolls that are wrapped and shipped to customers, or before being sent to the finishing operations.

**PARIS WHITE:** A natural form of calcium carbonate ($CaCO_3$) used as a filler or coating material in making paper. Also referred to as *whiting* and *chalk*.

**PARTIAL MACHINE DIRECTION TENSION BURST:** A paper roll defect manifested as a partial, irregular separation or rupture of the web. While the defect is in the cross-machine direction, the fiber separation is in the machine direction, hence the term "partial machine direction tension burst." It may be visible to a trained eye if the defect occurs at the edge of the roll. It will appear as a *stick-out* or *offset*.

**PARTIAL PRESSURE:** That portion of the total pressure in a gaseous mixture, such as the vapor that collects in the upper portion of a batch digester, which is contributed by any one component.

**PARTICLE BOARD:** A wood board formed from knife-cut flat chips and adhesives. Also called *flakeboard* or *waferboard*.

**PARTICLES:** Very small pieces of wood or bits made by mechanically breaking pieces or chips, usually by a hammermill or by a saw. Also called *sawdust*.

**PARTICULATE LOADING:** The introduction of finely divided solid or liquid particles into the surrounding mill air by gaseous emissions from various areas.

**PARTICULATES:** Finely divided solid or liquid particles in gaseous emissions from various areas of a mill, including chemical droplets, dust, mists, fogs, fumes, and smoke.

**PARTS PER MILLION (PPM):** The unit commonly used to represent the degree of pollutant concentration where concentrations are small. Larger concentrations are expressed in percentages.

**PASS:** A run of pulp or stock through a refiner. Also refers to the flow of stock by the baffles in paper machine headboxes.

**PASSIVITY:** The corrosion-resistant property of metals or alloys used in the paper industry, such as titanium, due to the formation of a protective film, such as oxides, on its surface.

**PASTE:** An adhesive, usually made from flour or starch, used in pasting sheets of paper or board together to produce multi-ply products.

**PASTEBOARD:** A rigid paperboard lined with sheets of paper pasted to its surface.

**PASTED:** Reference to a paper sheet or board that consists of two or more layers held together.

**PASTED BRISTOL:** A multi-ply type of bristol paper made by pasting two or more sheets together.

**PASTER:** A machine on which multi-ply paper or board is made by continuously pasting several sheets together. Also called a *pasting machine.*

**PASTING MACHINE:** *See* PASTER.

**PASTING PAPER:** A class of papers made with properties suitable for adhesion to other materials such as boards and various paper products.

**PATCH MARKING:** A method of producing a watermark on a sheet by sewing a wire mark patch into the wire surface of the dandy roll on the wet end of a paper machine, or sewing it into the wire surface of a cylinder-type machine.

**PATENT COATED BOARD:** A type of paperboard made and lined with white or colored fibers on a multi-cylinder machine. It possesses a high surface finish suitable for use in making cartons.

**PATTERN:** Any undesirable surface nonuniformity visible on paper.

**PATTERN BOARD:** A highly calendered, compacted, lightweight paperboard made from waste paper and sulfate pulp to be used by the clothing and shoe manufacturing industry for making patterns.

**PATTERN PAPER:** Paper used by tailors and the clothing industry for making patterns for the designing and making of clothes.

**PC:** *See* PROGRAMMABLE CONTROLLER.

**PCBs:** *See* POLYCHLORINATED BIPHENYLS.

**PCC:** *See* PRECIPITATED CALCIUM CARBONATE.

**PEAK POWER:** A power plant and user mill expression referring to the average power during a time interval of specified duration occurring within a given period of time, with the interval being selected when the average power consumption is the greatest.

**PEARL FILLER:** A naturally occurring, anhydrous calcium sulfate compound ($CaSO_42H_2O$) employed as a filler in making writing-type paper. Also referred to as *anhydrate,* and sometimes known as *gypsum.*

**PEARL WHITE:** *See* BLANC FIXE.

**PEBBLING:** The process of embossing or printing paper surfaces to resemble wood grain, marble, or leather. Also referred to as *graining.*

**PECKERWOOD:** A reference to small pulpwood logs.

**PEEL ADHESION:** The force required to break the bond between an adhesive paper tape and the surface to which it is applied. It is usually expressed in ounces per inch width when pulled at a 180° angle.

**PEEL WASTE:** Damaged outer wraps of containerboard rolls.

**PEELED LOGS:** Pulpwood logs from which the bark has been removed by hand.

**PEELING:** (1) A paper defect manifested by surface scaling. (2) Process of removing bark from pulpwood logs by hand.

**PELLETIZER:** A rotating, perforated drum positioned on a vertical axis, through which dewatered pulp is passed to form grain-like pellets for bulk storage or transportation.

**PELLETIZING:** A method of preparing pulp for bulk storage or transportation in which pulp slurries are dewatered on a press and the cake is converted to grain-like pellets.

**PEN AND INK TEST:** A common paper mill laboratory test for checking the water resistance of paper by the observation of the spreading or "feathering" of an ink line on a sample.

**PENCIL POINT:** A paper roll defect in which the paper slips off its core, causing a cone-shaped end. Also referred to as a *slipped roll.*

**PENESCOPE TEST:** A laboratory test to measure water resistance of paperboard by placing a circular sample in a hollow screw cap on a brass chamber fitted with an upright piece of glass tubing (penescope). The chamber is filled with test liquid and excess is added to fill the glass tubing to a 1-foot height. The time it takes for the test liquid to penetrate the board is interpreted as a measure of the water resistance of the board.

**PERCENT ACTIVE TIME:** The ratio, in percent, of the actual operating time of a piece of machinery in a mill to the theoretical time operation if no unforeseen circumstances to prevent its operation are encountered.

**PERCENT SOLIDS IN BLACK LIQUOR:** The percentage of solids by weight in black liquor fluid in a recovery system in a sulfate pulp mill. For example, if black liquor to a recovery furnace is 70 percent solids, then 30 percent is water.

**PERCENT STRETCH:** An expression of the degree of elongation of paper to the point of rupture as determined in a tensile strength test, and expressed as a percentage of the original length.

**PERCENT WET TENSILE STRENGTH:** The percentage ratio of water-saturated paper tensile strength to its air dried tensile strength.

**PERCENTAGE HUMIDITY:** A humidity expression computed by dividing the absolute humidity of moist air by the absolute humidity of moist air saturated at the same temperature and multiplied by 100.

**PERCOLATION:** Downward flow or infiltration of liquids through pores or spaces in soil.

**PERFECT PAPER:** One hundred percent defect-free paper.

**PERFECT SPLICE:** *See* SAFETY SPLICE.

**PERFORATED ROLLS:** Cylinder rolls covered with holes and located in the headbox of a paper machine just ahead of the slice. They are used to defloc and even the flow of pulp slurry.

**PERFORATING:** A paper-converting process in which a row of small holes is punched in a sheet to allow easy tearing.

**PERFORATING PAPER:** Paper made with special characteristics so that it will cut cleanly when holes are punched in it.

**PERIPHERAL SPEED:** The surface speed of any rotating, cylindrically-shaped piece of papermaking equipment such as a pulpstone and a dryer drum. It is expressed in feet per minute (FPM).

**PERMANENCE:** *See* AGING RESISTANCE.

**PERMANENT WHITE:** *See* BARIUM SULFATE.

**PERMANGANATE NUMBER:** A value, also known as *k number*, that indicates the relative hardness or bleachability of chemical pulp usually having lignin contents below 6 percent. It is determined by the number of milliliters of one-tenth normal potassium permanganate solution ($KMnO_4$) that is absorbed by 1 gram of oven dry pulp under specified and carefully controlled conditions.

**PERMEABILITY:** That property of paper and paperboard directly related to the ability of a fluid to penetrate and spread throughout the sheet.

**PEROXIDE:** A short name for sodium peroxide ($Na_2O_2$) or hydrogen peroxide ($H_2O_2$), which are used to make up bleach liquor for bleaching mechanical-type pulps and for "brightening" pulps made from recycled fibers.

**PEROXIDE BLEACH:** A bleaching chemical made with sodium or hydrogen peroxide and used to bleach mechanical-type pulps and to "brighten" pulps made from recycled fibers.

**PEROXIDE BLEACHING STAGE:** A sodium or hydrogen peroxide bleaching step or steps sometimes used in a multi-stage chemical-bleaching sequence as one of the operations making up the complete pulp-bleaching system.

**PERSUADER:** A length of pipe or an extension handle used on a wrench to get greater leverage. Also called a *hickey*.

**PGW:** *See* PRESSURIZED GROUNDWOOD.

**pH:** A symbol that represents the acidity or alkalinity of a solution, and is a measure of the key ingredient of aqueous solutions of all acids and bases–the *hydrogen ion concentration*. It is actually the negative logarithm to the base 10 of the hydrogen ion concentration, with 7 being the value of a neutral solution. It drops toward 0 as the acidity increases and rises toward 14 as the alkalinity increases.

**PHLOEM:** That circular portion of a tree located just below the bark, which forms a soft porous path for the sap to flow through.

**PHOTO FOLDER PAPER:** A thick paper used in the photographic industry for mounting photographs and cut into sizes so that it can be folded to form a protective cover.

**PHOTO MOUNT BOARD:** A heavy weight paperboard made for back supporting of pictures and photographs.

**PHOTOCOPYING PAPER:** A chemically clean, high-grade paper made as a base to apply light-sensitive emulsions. It is used for making copies of correspondence, records, drawings, etc. *See also* PHOTOSTAT PAPER.

**PHOTOGRAPHIC PAPER:** A chemically clean, high-grade paper made to be used as a base to apply sensitized photo emulsions. It is used in the photographic industry.

**PHOTOSTAT PAPER:** Paper that has been coated with a light or chemically sensitive material so that it can be used to make copies of printed, written, and photographic records, drawings, and other information. *See also* PHOTOCOPYING PAPER.

**PHOTOSYNTHESIS:** The formation of carbohydrates by the natural process of combining water and carbon dioxide in trees, which produces the wood substance used to manufacture pulp in pulp mills.

**PHYSICAL/CHEMICAL WASTE TREATMENT:** The treatment of mill waste waters by unit processes involving physical and chemical means, such as clarification, filtration, ion exchange, and carbon adsorption, to remove objectionable organic and inorganic contamination.

**PICAROON:** A tool used in a groundwood mill with a long, axe-type handle equipped with a hook and used for moving wood.

**PICK:** (1) The phenomenon of pulp or fibers pulling away and sticking to the paper machine parts they come in contact with, such as rolls, in the wet end and dry end sections. (2) A paper mill control test to determine surface adhesion properties of paper. (3) The small particles of paper that loosen from the surface of paper, especially during printing.

**PICK RESISTANCE:** The ability of a sheet of paper to withstand the lifting of particles and/or fibers from its surface on contact with other objects. *See also* DENNISON WAX.

**PICK STRENGTH:** The resistance of a sheet of paper to the pulling away of fibers or coating from its base. It is directly related to the force with which the fibers adhere to each other and the coating.

**PICK TESTER:** A paper *pick resistance* measurement instrument for wet or dry paper.

**PICKING:** The lifting of fibers and very small pieces of coating from the paper sheet surface during the printing process due to the paper being too soft and/or the ink being too sticky. *See also* PLUCKING.

**PICKOUTS:** Defects consisting of lumps made up of coating and fibers that have been lifted out of a sheet of paper.

**PICKUP:** The process of lifting and removing the wet, formed sheet from the forming wire (fabric) of a paper machine so that it can be transferred to a moving press felt. The felt carries it into the press section for further water removal.

**PICKUP FELT:** The felt on a paper machine that contacts and lifts a wet web of paper from the surface of a sheet-forming wire (fabric) or another felt during the papermaking process.

**PICKUP PRESS SECTION:** That portion of the wet end press on a fourdrinier paper machine that transfers the wet web from the forming wire (fabric) and removes some of the water from the sheet.

**PICKUP ROLL:** A paper machine roll located at and in contact with the couch roll and wire (forming fabric) of the wet end. It aids in transferring the wet web to the press felt that passes underneath it.

**PIGMENT:** An insoluble mineral or organic powder used as a dye to color paper and as an additive to impart specific properties, such as bulk, porosity, and opacity, to the sheet.

**PIGMENT DISPERSION:** The technique used in making up pigment slurries for paper coating. The process involves the wetting of the dry solids into the liquid with intensive high shear mixing to reduce agglomerates into individual particles and maintaining good surface motion.

**PIGMENTED SIZE PRESS:** Any of various types of size presses designed or modified to apply pigments such as calcium carbonate or clay along with sizing materials such as starch to the paper sheet, usually in or near the paper machine dryer section. *See also* SIZE PRESS.

**PIGTAILS:** The accumulation of pulp fibers in the twisted-together fibrous cover on the surfaces of papermaking felts. Sometimes referred to as *shankers*.

**PIKE POLE:** A long pole used to pull or push logs from a distance in a stream. It is fitted with a steel point and a hook.

**PILE CURL:** The curling experienced in a similarly oriented stack of paper that is apt to occur when the top is unrestrained.

**PILOT DRYER:** An isolated dryer drum in a paper machine dryer furnished with an independent steam and condensate system. It is used as part of a paper moisture measuring and control system.

**PIMA:** *See* PAPER INDUSTRY MANAGEMENT ASSOCIATION.

**PIN CHIPS:** Undesirable, small sized chips usually removed from the chip supply to a pulp mill during the screening operation. However, some inadvertently get through to the digesters.

**PIN HOLES:** Paper defects consisting of tiny holes in the sheet or coating due to extraneous gritty material such as sand and froth caused by air bubbles entering the papermaking process.

**PIN PAPER:** A specialty rust- and tarnish-resistant manila paper used to package metallic pin fasteners.

**PIN SEAM:** A thin monofilament wire used to connect the ends of paper machine dryer fabrics.

**PIN TICKET BOARD:** A stiff, white, bristol-type paperboard made to be converted to merchandise price tags.

**PINE:** Evergreen softwood trees used as a major source of pulpwood.

**PINK LIQUOR:** Weak, spent magnesium-base cooking liquor, washed from pulp made by cooking wood with magnesium bisulfite in a sulfite pulp mill, which becomes *red liquor* when water is evaporated from it.

**PIPE FITTERS:** Mill employees who construct and maintain pipelines in the mill.

**PIPE SHOP:** A room or area where pumps, pipelines, and pipe assemblies are made up and plumbing-type repairs are made.

**PIPELINE:** A conduit connected to pumps, valves, and control devices used for transporting finely divided solid material, pulp slurries, and other process fluids from one location to another in pulp and paper mills.

**PIRSSONITE:** Deposits occurring primarily in kraft pulp mill recausticizing smelt dissolving, green liquor clarifier storage tanks and pipe lines. The deposits are made up of sodium and calcium carbonate double salts.

**PIT:** Mill reference to receiving vessels and tanks for drainage water and white water from screening and sheet formation operations involving pulp stock slurries.

**PIT-TYPE GRINDING:** Normal operation of a pulpwood grinder in which the stock outlet dam under the stone is in a raised position, forming a pond (pit) into which the lower part of the stone is immersed.

**PIT VENTILATION:** Heated fresh air pumped into the space between the sole plates of paper machines having pits.

**PITCH:** (1) A resinous substance that is a combination of unsaponifiable organic chemicals mixed with resin and fatty acids that exists in mechanical and unbleached chemical pulps. When conditions are right, it will become separated and contaminate paper machine wet end components. (2) The angle at which the thread of a screw or bolt runs with respect to the body. (3) The shrillness level of noises produced by various pulp and papermaking machinery and the process from which the workers must be protected. (4) The number of threads or number of teeth per inch on a grindstone burr surface.

**PITCH CONTENT:** The amount of *pitch* in pulp, paper, or paperboard.

**PITCH CONTROL AGENT:** Materials used to solubilize and emulsify pitch in pulp stock during papermaking.

**PITCH SPOTS:** A paper defect consisting of specks on the sheet caused by the separating of resinous materials in the pulp stock before the sheet is formed on the paper machine.

**PITH:** A continuous, cylindrical-shaped section of spongy-like tissue, oriented axially in the center of pulpwood trees and other plants, whose primary function is the storage of liquids.

**PITLESS GRINDING:** The operation of a pulpwood grinder in which the stock outlet dam under the stone is lowered so that there is no (or a minimum) pond (pit) formed. The stone runs free above the surface level at increased surface speeds and loads.

**PLAIN MILL BLANKS:** A class of uncoated, single-ply or multi-ply, rigid paperboards with good printing surfaces. They are made from chemical pulp, mechanical pulp, deinked pulp, and clean broke. The blanks are used for making up posters, merchandising displays, and other types of signs.

**PLAIN PRESS:** A type of paper machine wet press in which the wet sheet and a felt are passed through the nips of two solid rolls with the squeezed water flowing backward from the nip, opposite the direction of the sheet, and along the surface of the bottom roll and into a pan.

**PLAIN PRESS ROLLS:** Rolls on a linerboard press having smooth surfaces with no grooves.

**PLANER HEAD BARKER:** *See* BARKER-CUTTER HEAD.

**PLANT:** A building or group of buildings housing equipment, machinery, tools, and people to carry out a process or manufacturing operation, such as pulp and papermaking. Collectively, the plants are usually referred to as a *mill.*

**PLANT CAPACITY:** The maximum processing, production, or output potential of a pulp or paper mill, usually expressed in tons per day (TPD).

**PLANT FACTOR:** The ratio of the average load to the rated capacity of a power plant in a mill. The rated capacity is the aggregate rating of all the electrical generators.

**PLANTATION:** Woodlands that are planted and managed for maximum production by essentially one man.

**PLASTIC WIRES:** *See* FORMING FABRICS.

**PLASTICIZER:** (1) A chemical agent used in the manufacture of glassine-type papers and in making up protective coating materials for paper sheets. (2) Paper-coating additive used to make paper coatings more flexible.

**PLATE:** A paper mill reference to the rotating disc in refiners.

**PLATE BOARD:** A stiff, moldable type of paperboard used as a base from which paper food serving plates are made.

**PLATE CLEARANCE:** The distance between the surfaces of refiner discs through which pulp is passed for mechanical treatment.

**PLATE FINISH:** A smooth, polished paper surface achieved by treating the sheets between polished zinc or copper plates and subjecting the stack to high pressure and a small amount of friction between the rollers of a plater. Also referred to as *plate glaze.*

**PLATE GLAZE:** *See* PLATE FINISH.

**PLATE PAPER:** A smooth-surfaced, lightly sized, bulky, printing-type paper used to take impressions from engraved plates. It is used in gravure printing.

**PLATEN DRYER:** A paper-drying machine made up of a pair of closely spaced, parallel, heated surfaces, between which the wet paper is passed.

**PLATEN ROLL:** A roll on a rewinder made up of highly polished, hardened steel sleeves butted together on a core shaft over which the paper passes to be cut by circular knives pressing against it on the opposite side.

**PLATER:** A machine for applying a surface finish on paper by passing piles of sheets back and forth between rolls under pressure as they reverse rotational direction.

**PLATER BOOK:** A number of sheets of paper stacked for pressure finishing between rolls of a plater by inserting surface impression sheet material such as fabrics, metallics, etc., between them.

**PLATER FINISHED:** Paper whose surface finish was obtained by passing between metallic rolls under pressure on a plater.

**PLATING:** The operation of finishing paper by pressing a stack of sheets piled alternately with impression sheet materials between rolls on a plater-type finisher.

**PLAYING CARD STOCK BOARD:** A slightly stiff, coated paperboard made of two pasted chemical pulp sheets. It is used especially for making playing cards.

**PLC:** *See* PROGRAMMABLE LOGIC CONTROLLER.

**PLI:** *See* POUNDS PER LINEAL INCH.

**PLIABILITY:** The property of paper and paperboard that permits it to be bent and folded without breaking.

**PLOW:** (1) A groundwood mill device located on top of some pulpstones to prevent the accumulation of shims. (2) A diverting device used to remove wood from a conveyor.

**PLUCKING:** The lifting of bundles of fibers and small areas of coating from the paper surface during the printing process. It is caused by the use of unusually high tackiness inks. *See* PICKING.

**PLUG:** (1) The inner conical portion of a Jordan-type refiner. (2) Wooden inserts installed in the ends of paper roll cores to keep the rolls from flattening.

**PLUG VALVE:** A type of valve used in pulp and paper mills having a rotary motion-operated cylindrical, conical, or spherical segment-shaped closure element housed in an appropriately shaped body.

**PLUGGER:** *See* BROKE BOY.

**PLUGGING THE FELT:** The process of wet end press felts filling up with materials that are used to make up the sheet of paper, such as rosin, clay, starch, etc., by coming away from the web and being picked up by the felt.

**PLUGGING THE STACK:** A paper machine calender malfunction caused by the outer liner of multi-ply paperboards pulling away and wrapping around the rolls.

**PLUME:** The visible emission that rises from mill flues and chimneys and is mostly a form of evaporated water in steam.

**PLUNGER CHIP FEEDER:** A type of continuous digester chip feeder in which the chips are compressed while coming in contact with the digester steam so as to form a wet plug, compact enough to seal-in the digester pressure.

**PLY:** A layer that makes up a multi-layered/pasted or a multi-cylinder/formed paperboard.

**PLY BOND STRENGTH:** Refers to the ability of layers or plies of paperboard to adhere together and not separate or delaminate.

**PNEUMATIC CHIP CONVEYOR:** A chip transporting system sometimes used to move chips from outside storage to the pulp mill digester house. It uses air as a carrying medium, and consists of a large positive displacement blower discharging through a venturi tube into sheet metal piping to which the chips are fed by a star-feeder.

**POCKET CHIP FEEDER:** A type of continuous digester chip feeder having a star- or conical-pocketed, revolving plug into which wet or dry chips fall and are then fed into the digester feeding system.

**POCKET COUNTER:** A unit mounted on a pulpwood grinder to record and total the number of times a pocket is charged with wood. It is used as a guide in the production of pulp.

**POCKET GRINDER:** A type of pulpwood grinder designed with containers in which logs are loaded and mechanically pressed against a revolving grindstone by a plunger.

**POCKET VENTILATION:** The introduction of pressurized conditioned air in the area between the dryer rolls of multi-drum dryer sections on paper machines in order to minimize moisture profile non-uniformities.

**POINT:** A mill term used to express the thickness of a sheet of paper or paperboard. One point is equal to one one-thousandth of an inch.

**POINT SOURCE:** Stationary source of large mill contaminating emissions such as a digester house, chemical recovery area, and power house.

**POINTS PER POUND:** A term commonly used to express the bursting strength of paper. It is determined by dividing the strength test results by the basic weight of the sheet in pounds.

**POLE DRYING:** *See* LOFT ROOM.

**POLE MARK:** *See* BACK MARK.

**POLISH:** A paper surface finish that possesses a high gloss or shine.

**POLISHED DRUM COATING:** A paper-coating procedure in which the coated side of the sheet is dried by contact with a highly polished surface of a rotating, heated cylinder.

**POLLUTANT:** Any undesirable gaseous, liquid, or solid material released from the pulp and papermaking process to the surrounding mill environment and natural bodies of water.

**POLLUTION:** The contamination of the surrounding mill environment and natural bodies of water with undesirable gaseous, liquid, and solid materials released from the pulp and papermaking process.

**POLLUTION ABATEMENT:** The process of implementing means of reducing or eliminating the contamination of the surrounding mill environment and bodies of water by minimizing the release of undesirable gaseous, liquid, and solid materials from the pulp and papermaking process.

**POLLUTION INDEXES:** Values determined by various test methods and used to express the extent of release of undesirable process materials and waste to receiving streams and other natural bodies of water.

**POLLUTION LOAD:** The solids and oxygen demanding characteristics contributed to natural water bodies and streams by mill waste material.

**POLYCHLORINATED BIPHENYLS (PCBs):** A group of highly toxic *chlorinated organics* whose main source in the pulp and paper industry has been electrical transformer fluids, although they do exist and are produced in limited amounts in various mill other areas and process segments.

**POLYELECTROLYTES:** Negatively and/or positively charged materials used as a coagulant or a coagulant aid in mill water and waste treatment operations.

**POLYMER:** Long chain, high molecular weight chemical substances made up of molecules of the same substance, such as cellulose and lignin.

**POLYMERIC IMPREGNATION:** The process of impregnating paper and board to improve their mechanical properties.

**POLYURETHANE:** A material used as paper machine roll coverings instead of conventional rubber coverings.

**POLYSULFIDE (PS) PULPING:** A sulfide pulping process in which dissolved sulfur in the white cooking liquor is used for sulfur makeup.

**POLYVINYL ACETATE (PVAc):** A synthetic latex binder used in coating formulations.

**POLYVINYL ALCOHOL (PVA):** A papermaking chemical used as a surface sizing medium for printing and writing grades of paper.

**POOR CORE START:** *See* BAD CORE START.

**POOR EDGES:** A paper defect caused by malfunctions in the paper machine, producing edges on the sheet that are too thin or too heavy.

**POOR MAN'S PICKUP:** A means of open transfer of the wet web from the fourdrinier wet end section to the wet press. It consists of an arrangement where a wire-turning roll is employed. The sheet follows the wire (forming fabric) after leaving the couch roll and is pulled off midway between the couch roll and the wire-turning roll. *See also* OPEN DRAW and OPEN TRANSFER.

**POOR ROLL START:** A paper roll defect occurring when loose paper is the cause of wrinkles on starting the roll on a winder, sometimes resulting in a misaligned core due to slippage.

**POOR SLITTING:** A paper roll defect caused by dull slitters which cause a rough or fuzzy paper edge appearing as nicks and dents in the roll.

**POOR SPLICE:** A type of paper roll defect referring to paper sheet splices that do not give satisfactory performance in later processing or use.

**POP STRENGTH:** A slang reference to the maximum uniformly distributed pressure applied at right angles to its surface that a test piece of paper will stand before it breaks under conditions defined in the standard test method. Also called *burst strength* and *Mullen.*

**POP TEST:** Rupture test, whose name was derived from the popping sound made on a testing machine to determine the bursting strength of paper and board. Also referred to as *burst test.*

**POP VALVE:** A spring-loaded, quick-opening safety valve that opens automatically when pressure exceeds limits for which the valve was set. It is commonly used on power and recovery boilers and other equipment to protect them from damage due to excess pressure.

**POPE REEL:** A type of reel used for winding up rolls of paper as they come from the calenders. It consists of a large diameter drum whose peripheral speed is synchronized with the paper machine. There are steel arms at each end of the drum with saddles into which the bearings of the reel spool fit for continuous, uninterrupted change of rolls.

**POPLAR:** Deciduous, broadleaf tree used as a source of pulpwood.

**PORES:** The larger openings or cross sections of vessels found in the wood of broadleaf trees.

**POROSITY:** That property of paper and paperboard related to the ability of fluids to pass through them. It is an indication of the size, shape, and distribution orientation of the pores in a sheet and the compactness of the fibers. Sometimes incorrectly referred to or related to *air permeability* and *vapor permeability*.

**POST COLOR NUMBER:** A value determined in the measurement of pulp and paper color reversion before and after heat aging.

**POST SCREENS:** Plate-type and vortex-type screens located after the main centrifugal screens in a groundwood stock screening system to rescreen the accepted stock.

**POSTCARD BRISTOL:** Bristol-type paper that is converted to postcards.

**POSTCARD STOCK:** Coated paper and paperboard made with a variety of finishes and used for making postcards.

**POSTER PAPER:** A machine-finished, durable, non-fading type of paper, usually made from chemical pulps. It is used for general poster applications, especially billboard poster work.

**POUND-CALORIE (LB-CAL):** The heat required to raise the temperature of 1 pound of water 1°C.

**POUNDS PER LINEAL INCH (PLI):** The measure of the loading pressure on paper machine presses, particularly linerboard presses, in reference to the *cross direction* length (in inches) of the specific nip.

**POUNDS PER POINT:** A term commonly used to express the density of paper. It is determined by dividing the basis weight of the sheet, in pounds, by the thickness, in mils. *See also* APPARENT DENSITY.

**POWDER PAPER:** A specially treated, manila paper made for use by the mining industry to wrap blasting powder.

**POWDERING:** The breaking away of pigment from the paper surface in the finishing and converting operations.

**POWER:** The rate at which work is done.

**POWER BOILER:** That portion of a steam-producing unit in which water is converted into steam for use in the process or to produce electricity.

**POWER CONSUMPTION:** The energy used by a process or piece of pulp and papermaking machinery and other equipment.

**POWER DISTRIBUTION:** A system that spreads moderately high voltage power from the main incoming transformer bank of a mill to unit substations, where it is reduced to workings voltages. The unit substations are located close to the motors and loads they serve.

**POWER FURNACE:** That portion of a steam-producing unit where the combustion of fuel occurs to produce heat required to convert water into steam in the boiler portion.

**POWER GENERATION:** The production of steam and electricity in the power house for process use and running machinery in a mill.

**POWER HOUSE:** The area of a pulp and paper mill where steam and electricity are generated to supply heat and power for pulping and papermaking processes and machinery. It is usually located in a separate building. Also called *boiler house* or *power plant.*

**POWER PLANT:** *See* POWER HOUSE.

**POWER PUMPS:** The types of reciprocating pumps that are driven from the outside through a crankshaft or other device.

**POWER SUPPLY:** The auxiliary energy supplied to a device, system, or process.

**POWERED RELIEF VALVE:** An electrically or pneumatically operated, quick-opening safety valve on a power boiler which can be manually opened from a remote control panel.

**PPC:** *See* PAPERBOARD PACKAGING COUNCIL.

**PPM:** *See* PARTS PER MILLION.

**PRECIPITANT:** Substances such as alum, sodium aluminate, resins, etc., added to stock slurries to precipitate size onto the fibers during papermaking.

**PRECIPITATE:** A solid that separates from a solution because of some chemical or physical change, or the formation of such a solid.

**PRECIPITATED CALCIUM CARBONATE (PCC):** A wet end filler used in alkaline papermaking and also in some coating formulations. It generally has good opacity and bulking characteristics and can be produced in several particle morphologies and sizes. Most *FCC* being used in the paper industry is made in on-site satellite plants that utilize available mill carbon dioxide streams as well as mill power and water resources. *See also* CALCIUM CARBONATE.

**PRECIPITATED MAGNESITE:** A compound used as a filler in cigarette paper. *See* MAGNESITE ($MgCO_3 \cdot 3H_2O$).

**PRECIPITATION:** The separation of dissolved substances from a solution by transformation into a solid form and settling. It is sometimes induced by the addition of chemicals.

**PRECIPITATORS:** Mechanical and/or electrical separating devices used to collect fly ash and solid particulate matter from gaseous emissions emitted by power and recovery furnaces to prevent discharges into the atmosphere. *See also* ELECTROSTATIC PRECIPITATOR.

**PRECIRCULATION:** The pumping of cooking liquor continuously for a period of time from accumulator storage to the digester and back again in a sulfite-type batch cooking process before starting to bring it up to cooking temperature and pressure.

**PRECOAT FILTER:** A type of filter used in the recausticizing area of a pulp mill to wash and dewater lime mud before going to the lime kiln.

**PRECOATING:** (1) The application of a thin layer of coating material on a sheet of paper prior to the final coating operation. (2) Applying a thin coating of material on the screen of a filter before beginning the actual filtering operation.

**PREFORMS:** Resin-filled paper plastics made in either flat sheets or shaped to conform with the final molded product. They are used as automobile panels.

**PREGELATINIZED STARCH:** Papermaking starch used to make up a paste of adhesive that has already been cooked and dried. Sometimes called *instant starch*.

**PREHYDROLYSIS:** An initial, pressurized, presteaming stage prior to the sulfate chemical cooking of chips in a digester in order to subject the chips to a mild acid hydrolysis before the cooking liquor is added. It is usually a stage in the production of *chemical cellulose* or *dissolving pulp*.

**PREMETERED SIZE PRESS:** *See* FILM PRESS.

**PRESCREENS:** Drum and vibrating flat screens located just ahead of the main centrifugal screens in a groundwood stock screening system. They reduce the shive content of the feedstock to improve the screening operation.

**PRESERVATURE:** A chemical agent used in the paper industry to discourage the growth of harmful microorganisms.

**PRESS:** A paper mill term for a pair of rolls, usually located between the paper machine wire (forming) section and the dryers, through which the newly formed paper sheet is passed. It is usually installed in three sets and used for pressing out water and smoothing the newly formed paper sheet. It is also used for the application of surface coatings on the paper machine. *See also* PRESS SECTION.

**PRESS DRYING:** Sheet drying accomplished by the application of both pressure and heat.

**PRESS FABRIC:** A woven, continuous belt of wool, cotton, and/or synthetic fibers used to transport the sheet of paper between the rolls of the presses on a paper machine. Also referred to as a *press felt* or a *wet felt*. *See also* FELT.

**PRESS FELT:** *See* PRESS FABRIC.

**PRESS MARK:** Paper defect caused by faulty rolls or felts used on the paper machine.

**PRESS ROLLS:** A pair of rolls made up of an upper metal or granite roll over a lower rubber-covered roll, through which the newly formed paper is passed to have water pressed out and the surface smoothed or coating material applied.

**PRESS SECTION:** That part of a paper machine located between the wet end section and the dryer section where water is removed by passing the wet web between rolls and felts while applying a combination of pressure and vacuum. *See also* PRESS.

**PRESS TRANSFER:** The location on a paper machine where the wet web moves from the press section to the dryer section.

**PRESS TRIMMING:** The cutting of paper to accurate size and shape on a paper trimmer in the finishing operation.

**PRESSBOARD:** A high-strength laminated-type paperboard that has been glazed and used in the building industry for construction and insulating purposes.

**PRESSER ROLL:** *See* LUMP BREAKER ROLL.

**PRESSINGS:** Heavy, high-strength paperboard, generally made from chemical wood and cotton pulps on a cylinder-type paper machine, with a machine-glazed finish produced by drying over one single dryer. It is primarily used for making covers for books and box coverings.

**PRESSMARK:** A mark made on a paper sheet while passing through the press section (somewhere after the first press). The impression design is generally contained on a rubber collar.

**PRESSURE:** The action of one force against another force. It can be a thrust, either distributed over a surface, such as in a press or calender, or acting against a given surface within a closed container, such as a digester or a paper machine drum. It is usually measured as force per unit area and commonly expressed as pounds per square inch.

**PRESSURE BLOWDOWN:** Relief of gas and steam in a batch sulfite cooking process to lower digester pressure from the hold pressure to the blow pressure in order to recover excess $SO_2$ at the end of the cook. Also referred to as *blowdown* and *gasdown.*

**PRESSURE FOOT:** That part of a hydraulically operated ram that forces the pulpwood logs against the revolving pulpstone in a pulpwood grinder.

**PRESSURE FORMING:** The formation of a sheet from the head box onto the wire (forming fabric) of a paper machine under conditions in which the head in the headbox is greater than the wire speed and is referred to as *rush.*

**PRESSURE JOINT:** A specially designed rotary steam fitting used on paper machine dryers to admit steam to the individual drums and remove the condensate formed by the condensing steam.

**PRESSURE PULP WASHER:** An enclosed rotary drum pulp washer that operates under pressure at higher temperatures, permitting washing closer to the boiling point.

**PRESSURE REDUCING VALVE (PRV):** A valve used to regulate either upstream or downstream pressure in steam distribution systems in a power plant and in other pressurized flow vessels and lines.

**PRESSURE SENSITIVE ADHESIVE:** Type of dry adhesive paper coating that is tacky at room temperature and will stick firmly to the surface to which it is applied with only a small amount of pressure.

**PRESSURE-SENSITIVE PAPER:** A high-strength paper coated with a pressure-sensitive type of adhesive. It is used for conversion to tapes and labels.

**PRESSURIZED CENTRIFUGAL SCREENS:** Enclosed, centrifugal screens that are completely filled on both feed and accept sides, and are kept from plugging by vanes that move around the inside of the cylindrical screen and which create a +pressure wave ahead of them and backflow after them.

**PRESSURIZED FURNACE:** A combustion unit in which the furnace pressure is higher than atmospheric pressure. It is usually produced by the use of a *forced draft (FD)* fan only and sealing of the furnace casing.

**PRESSURIZED GROUNDWOOD (PGW):** Mechanical pulp made on a stone grinder where the whole grinder casing is pressurized and increased shower water temperature is used.

**PRESSURIZED HEADBOX:** An enclosed tank at the wet end of a paper machine in which the head or level of the stock is regulated by loading it with air or applying vacuum to provide uniform flow on the paper machine wire (forming fabric). Also called *air cushion headbox. See also* AIR PAD.

**PRESSURIZED REFINER MECHANICAL PULP (PRMP):** Pulp made by mechanical separation of fibers from chips in refiners under pressurized conditions and with no preheating.

**PRESSURIZED REFINING:** Mechanically treating pulp in a closed refining system in which the refiner operates under positive pressure. *See also* PRESSURIZED REFINER MECHANICAL PULP (PRMP).

**PRESTABILIZING AGENT:** In making up a latex paper coating mixture, it is the material added to the latex before it is added to the coating mix, such as casein or protein.

**PRESTEAMING:** Treating wood chips with low pressure steam to remove air. This action promotes cooking liquor penetration during a subsequent cooking process.

**PRETREATMENT:** The preliminary treatment of wastes in the mill prior to discharging them into the waste treatment plant where they are processed further.

**PRIMARY AIR:** That part of the air to a power or recovery boiler which enters with the material to be burned and amounts to about 15 percent of the total required combustion air.

**PRIMARY AIR FAN:** An air fan located in the incoming air duct and upstream of a coal pulverizer on a power furnace.

**PRIMARY BLACK LIQUOR HEATER:** A heater located between the discharge of the black liquor nozzle pump and the black liquor nozzles in a recovery furnace of a sulfate pulp mill. It can be of either the indirect or direct steam injection type.

**PRIMARY COLORS:** The red, yellow, and blue colors used in the dyeing of paper, two of which can be mixed together to obtain secondary colors (green, orange, and violet).

**PRIMARY CREPE:** *See* MACHINE CREPE.

**PRIMARY FINE SCREENS:** Pulp screens located immediately after the *bull screens* and used to remove *shives*, fiber bundles, and coarse fibers in a groundwood pulp mill. These screens include both horizontal and pressurized centrifugal types.

**PRIMARY HEADBOX:** The first and main headbox on a fourdrinier wet end of a paper machine making a multi-ply board, with the liner being applied by a subsequent (secondary) headbox.

**PRIMARY PRESSES:** *See* BABY PRESSES.

**PRIMARY SLUDGE:** The settlings removed from the first stage of a waste treatment plant which consists of a sludge settling tank. The sludge is normally dried over vacuum filters and disposed of in landfills or dried and burned in the power furnace.

**PRIMARY TREATMENT:** The removal of suspended matter from mill waste water by sedimentation. It is usually the first stage in a multi-stage waste treatment process, where substantially all floating or settleable solids are mechanically removed by screening and sedimentation.

**PRIMARY WALL:** The very thin outer wall of a papermaking fiber, comprised of layers of cellulose fibers or strands oriented at different angles from one layer to the next.

**PRIMER COAT:** With reference to adhesive-coated paper, it is a facing of a thin layer of adhesive-like material that serves as a bonding agent between the adhesive and the backing on the paper.

**PRIMING:** The passing of water from a power or recovery boiler with the steam.

**PRIMING THE PUMP:** Filling a centrifugal pump with the liquid to be pumped so that the pump will start to work.

**PRINT BONDING:** A technique used for the latex bonding of disposable nonwovens in which the binder is applied on discrete areas of the sheet.

**PRINT-ON COATING:** A method of applying high density coating material to a paper sheet by the use of applicator rolls, which distribute an even film on the surface of a sheet.

**PRINT SHOW-THROUGH:** The extent to which printed material can be seen from the other side of a sheet of paper, and is a function of the transparency of the sheet.

**PRINT SNAP:** The perceived difference in gloss between printing and the coated sheet surface that contains it. The greater this difference, the more *snap* the printing (usually photos or other graphics) is said to have. This quality has becoming highly desirable, and is being accomplished in some cases by using special coating formulations that provide maximum ink holdout while producing a matte or semi-matte coated surface. The ink formulation is obviously an important factor in achieving maximum *snap*. Also called *delta G* or simply *snap*.

**PRINTABILITY:** The ease with which paper can be printed to high quality standards with the least amount of spoilage.

**PRINTED GLOSS:** The reflectivity of a printed surface of paper to light, as if seen through glass or in a mirror.

**PRINTED GLOSS MOTTLE:** The random nonuniformity in the visual density and/or color of a printed area of paper.

**PRINTING PAPER:** All types of paper that have characteristics suitable for printing purposes.

**PRIVATE BRAND:** A brand name of paper that has been assigned to a specific grade of paper and is considered to be the property of a particular paper dealer or user.

**PRMP:** *See* PRESSURIZED REFINER MECHANICAL PULP.

**PROCESS:** The operation of pulp and papermaking in which physical or chemical changes in matter occur; or a conversion in energy, such as when a change in pressure, temperature, or speed takes place.

**PROCESS AUTOMATION:** Implementing *process control* in a pulp and paper mill by the use of computers and advanced electronic instrumentation. It minimizes direct operator involvement with the process itself, while maximizing data availability for more informal decision making. Process automation exists on at least three levels–loop level, supervisory level, and management level (or millwide level). Although the goal is to automate control of the process from every level, operator interface is theoretically broadened, rather than diminished, by having information available to exercise more options as needed.

**PROCESS COATED PAPER:** Paper whose surface coating was applied while it was being made on the paper machine as an integral part of the papermaking operation.

**PROCESS CONTROL:** The regulation or manipulation of variables influencing the conduct of any pulp or papermaking operation in such a way as to obtain the maximum amount of product within prescribed quality specifications. *See also* PROCESS AUTOMATION.

**PROCESS LAG:** *See* DEAD TIME.

**PROCESS WATER:** Any water in a pulp and paper mill that is used to dilute, wash, or carry raw materials, pulp, and any other materials used in the process of making pulp and paper.

**PRODUCTION:** The actual amount of paper made on a paper machine or in a paper mill, usually expressed in tons per day.

**PRODUCTION COST:** The fixed costs plus the raw material and labor costs to manufacture paper, usually expressed in dollars per ton.

**PROFILE:** Continuous cross-machine basis weight and caliper records across the paper web.

**PROFILE PAPER:** Paper that has been horizontally and vertically ruled and is used in the engineering and architectural professions.

**PROFILED SCREEN CYLINDERS:** Types of pulp screens designed to induce high intensity pulses at the apertures so that it can run at higher consistencies, lower pressure drops, and with less fractionation of the longer fibers.

**PROFILERS:** On- and off-paper-machine measuring and recording systems used to obtain continuous cross-machine basis weight and caliper variations across the paper web.

**PROGRAMMABLE CONTROLLER (PC):** A solid-state device used in pulp and paper mills to replace conventional electrical and mechanical logic equipment such as relays, stepping switches, timers, and cam and drum programmers. They perform sequence and logic control by accepting binary (on-off) inputs coming from such devices as pushbuttons, limit switches, and other contact devices in the process. They generate binary outputs to such devices as motor starters, stepping switches, and solenoid valves. They can also be used to accept and transmit analog signals that can be used for control purposes. Often referred to as a *programmable logic controller (PLC)*.

**PROGRAMMABLE LOGIC CONTROLLER (PLC):** *See* PROGRAMMABLE CONTROLLER (PC).

**PROOFING PAPERS:** Types of high-quality, smooth-coated and lower grade book and newsprint paper used in the printing industry for making engraver's and galley proofs.

**PROPORTIONERS:** Various types of units or systems used to mix the various components in making multi-pulp furnishes in a paper mill. They can consist of multi-compartment stock regulating boxes with adjustable dams to complete automatic in-line control systems using a combination of ratio controllers all responding to a signal representing paper stock demand. Sometimes referred to as *stock blending systems*.

**PROTECTED BUNDLE:** A shipping ream or bundle of paper packaged with an outside protective wrapping of heavy, stiff news or paperboard.

**PROTOCOL:** A set of rules programmed into digital process computers for converting signals that allow them to communicate with each other.

**PRV:** *See* PRESSURE REDUCING VALVE.

**PRY BAR:** A wooden tool designed and used in paper mills to move rolls of paper.

**PS:** *See* POLYSULFIDE PULPING.

**PSEUDOPLASTIC:** A material, such as some coating materials, rheologically defined as showing no yield value but whose viscosity falls with increasing stress, such that the consistency curve approaches a straight line that intercepts the stress axis at a finite point.

**PSI:** Abbreviation for pounds per square inch pressure.

**PSIA:** Abbreviation for pounds per square inch, absolute pressure.

**PSIG:** Abbreviation for pounds per square inch, gauge pressure.

**PSYCHROMETER:** A measuring instrument used to determine the concentration of water vapor in air to ascertain the humidity.

**PSYCHROMETRIC CHART:** A convenient graph generally used by paper mills to determine the composition and thermodynamic properties of moist air. It includes dry bulb, wet bulb, relative humidity, absolute humidity, and dew point readings.

**PUCKER:** A paper defect appearing as a wavy, "cockled" surface caused by the uneven contraction of the sheet during the drying process on the paper machine.

**PUCKER WRINKLE:** A paper roll defect appearing as a small, unopened fold in the paper, caused by a blister entering the nips of rolls on a paper machine.

**PUG MILL:** A unit used to treat lime mud in the lime recovery area of a pulp mill before being processed in fluid bed calciners.

**PULLEY:** A wheel or combination of various sized wheels with grooved rims in which a rope, cord, or chain is run to amplify the force applied at one end for lifting heavy weights or pulling heavy loads in a pulp and paper mill.

**PULLEY BOARD:** *See* FRICTION BOARD.

**PULLOVER:** A pulp mill term for the cooking liquor mixed with relief gases and emitted from a sulfate batch digester during the cooking process. Also called *carryover*.

**PULP:** A fibrous material produced by mechanically or chemically reducing woody plants into their component parts from which pulp, paper, and paperboard sheets are formed after proper slushing and treatment, or used for dissolving purposes (*dissolving pulp* or *chemical cellulose*) to make rayon, plastics, and other synthetic products. Sometimes called *wood pulp.*

**PULP & PAPER RESEARCH INSTITUTE OF CANADA (PAPRICAN):** Headquarters: Pointe Claire, Que., (514) 630-4100.

**PULP BALES:** Sheets of dried, fibrous material made up in bundles for shipment.

**PULP BLEACHING:** The process of purifying and whitening pulp in a pulp mill by chemically treating it to alter the coloring matter and to impart a higher brightness to the pulp.

**PULP BOARD:** A coarse paperboard made on a multi-cylinder type of paper machine from mixed wastepaper and/or mechanical pulps.

**PULP CONTENT:** The amount of fiber in paper and paperboard as compared with nonfibrous clay, fillers, and coating materials. It is determined by laboratory analysis.

**PULP COOKING:** The process of reacting fiber-containing materials with suitable chemicals, usually under high temperature and pressure, in order to reduce them into their component parts with the fiber portion separated in the form of pulp. More commonly known as *pulping.*

**PULP FURNISH:** The pulps used to make up a multi-material stock mixture or blend used for manufacturing a particular grade of paper on a paper machine.

**PULP GRADES:** Various types of pulps made for specific end uses such as dissolving pulp, bleached pulp, glassine or greaseproof pulp, newsprint pulp, corrugating medium pulp, etc.

**PULP LAP:** Folded, wet pulp sheets, approximately 30 to 45 percent consistency, produced on a wet machine and stored and shipped to another mill site for repulping. It is used in making paper and paperboard.

**PULP MILL:** A plant in which pulp is mechanically or chemically produced from fibrous materials such as woody plants, together with other associated processes such as pulp washing and bleaching. Chemical preparation and cooking chemical recovery operations are also conducted there.

**PULP PRESS:** A wet press designed specifically to perform high grade dewatering of pulp.

**PULP SCREENS:** Various types of flat and rotary screens used in pulp mills to remove shives, slivers, and knots from diluted pulp slurries.

**PULP SHEET:** Bleached or unbleached pulp made into sheet form on a cylinder, wet, or fourdrinier-type paper machine for ease of shipment to off-site paper mills to be mixed with other pulps, additives, dyes, and chemicals for conversion into paper.

**PULP SLURRY:** A mixture of various types of cellulose fibers and additives, chemicals, dyes, or fillers that make a particular grade of paper.

**PULP SUBSTITUTES:** Grade of unprinted wastepaper consisting of brown and colored kraft, white, and semibleached sheets, cuttings, tabulating cards, and trim shavings.

**PULP SUSPENSION:** A mixture of one or several varieties of cellulose fibers in water.

**PULP WASHER:** A piece of pulp mill equipment designed to separate soluble, undesirable components in a pulp slurry from the acceptable fibers, usually by some type of screening method combined with diffusion and displacement with wash liquids, utilizing vacuum or the natural force of gravity.

**PULP WASHING:** The process of separating soluble, undesirable components in a pulp slurry from acceptable fibers by the use of suitable wash liquids and screening procedures.

**PULPER:** A disintegrator made up of a vessel fitted with a suitable agitator and used in a paper mill to break up and defiber dry intramill broke, purchased pulp, wastepapers, etc., in water to form a slurry. This slurry is further treated mechanically and chemically (to improve its quality) prior to converting to paper and paperboard. Also called a *repulper.*

**PULPER STOCK:** Pulp stock slurry made up of dry broke, purchased pulp, wastepapers, or other fibrous materials that have been mechanically disintegrated in water in a pulper.

**PULPING:** (1) In a paper mill, the mechanical reduction of dry mill waste or old papers and dried and lapped pulp into a slurry form with enough suspension water to make it satisfactory for beating or refining. (2) In a pulp mill, the reduction of fibrous material to its component elements through mechanical means, such as pulpwood grinders, or chemically by the use of various types of cooking liquors (*pulp cooking*).

**PULPING PROCESSES:** Processes for converting fibrous raw material into pulp. They are usually classified by either the nature or degree of the chemical and/or mechanical treatments used in the pulping action.

**PULPSTONE:** A grinding wheel of abrasive segments mounted on a reinforced concrete core, sized, and used on pulpwood grinders to mechanically reduce logs to their fiber components by abrasion. Also known as GRINDSTONE.

**PULPSTONE BEVEL:** A flat area made on the edges of the sides of a pulpwood grindstone face. Also called *edge of stone* or *bevel.*

**PULPSTONE BINDER:** A material used to hold the abrasive grits or particles in position in making up the sections of a pulpwood grindstone.

**PULPSTONE DRESSER:** A sharpening tool made from surface toothed, seamless steel tubing, and used in a groundwood mill to condition pulpstone surfaces.

**PULPSTONE GRADING:** The assignment of letters to pulpstones, indicating the degree of hardness. It progresses from "A" to "Z" as the hardness increases.

**PULPSTONE SEGMENT:** One of the abrasive blocks that is mounted on a concrete core to make up a pulpstone.

**PULPSTONE STRUCTURE:** The relative spacing of the abrasive grits or particles in a pulpstone segment.

**PULPWOOD:** Wood cut or prepared primarily for making logs and chips for converting to wood pulp to be used for subsequent manufacture into paper, fiberboard, and other products.

**PULPWOOD GRINDER:** A mechanical pulping machine that reduces pulpwood logs into their fiber components through abrasion by pressing them against a revolving grindstone. Also called a *grinder.*

**PULPWOOD PRODUCER:** One who manages a pulpwood harvesting crew and sells wood either directly to a pulp mill or to a pulpwood dealer.

**PULPWOOD PRODUCTION:** All of the operations used to produce pulpwood, such as cutting the wood from the stumps of trees to delivering it to the pulp mill woodyard.

**PULVERIZED COAL:** Coal that is mechanically crushed or ground to a powdery consistency and blown into a power furnace where it is burned as a fuel in suspension.

**PULVERIZER:** A mechanical grinding and crushing machine used to reduce solid materials to a powdery consistency, such as for coal fed to a power furnace.

**PUMPAGE:** The fluid being moved by a pump.

**PUMPING:** The addition of energy to a fluid to move it from one location to another, usually through a pipeline.

**PUNCTURE RESISTANT:** The property of a paper sheet that makes it capable of withstanding piercing by sharp objects.

**PURGE:** (1) A flow of air, water, gas, or any other fluid used to maintain a process measurement connection free of buildup or plugging with the process medium. (2) In a power or recovery furnace it is the flow of air through the furnace boiler gas passages and associated flues, ducts, evaporators, and precipitators to remove any combustible gases.

**PURITAN WHITE:** Ground gypsum ($CaSO_4 \cdot 2H_2O$) used to make a filler for various writing, papeterie, and specialty grades of paper.

**PURITY:** The depth of color in colored paper. It is sometimes expressed as a colorimetric quantity mathematically related to what is referred to as "excitation purity."

**PVA:** *See* POLYVINYL ALCOHOL.

**PVAc:** *See* POLYVINYL ACETATE.

**PYRITES:** A sulfur bearing ore that has been used as a source of sulfur dioxide ($SO_2$) gas to make cooking liquor in sulfite pulp mills.

**PYRITES BURNER:** A rotary-type furnace used by sulfite pulp mills to burn pyrites to make $SO_2$ for use in making cooking liquor.

**PYROLYSIS:** The chemical decomposition of a material by heat in the absence of oxygen.

**PYROMETER:** Temperature-measuring system that operates on the radiation effect of a heated body without physical contact with the body. It is used to measure temperatures of objects that are inaccessible by any means other than visual, such as molten lime in lime-burning kilns, recovery and power boiler interiors, and moving paper sheet surfaces on paper machines.

**PYROXLIN PAPER:** A lacquered paper made from sulfite pulp and additives to produce an irridescent, bright, metallic appearance.

# Q

**QUADRAFOS:** Another name for sodium tetraphosphate, an inorganic dispersing agent used in paper mills to reduce the viscosity of pigment suspensions for the surface treatment of paper and paperboards.

**QUADRANT SCALE:** A type of balance used to determine the ream weight of paper. It has a graduated scale and is in the shape of a quarter circle arc.

**QUADRILLE FINISH:** A paper finish which is watermarked or embossed with small squares. It is usually found on fine writing paper and fancy table specialties.

**QUALITATIVE ANALYSIS:** Mill laboratory procedure used to determine and identify the makeup of various materials.

**QUALITY:** A reference to the "goodness" of paper and paperboard which is a sum of those properties and physical characteristics of importance to the user.

**QUALITY CONTROL:** Maintaining, during the production of paper and paperboard, the conditions necessary to produce a level of properties and physical characteristics in the final product within variable limits as established to be satisfactory for its particular end use.

**QUANTITATIVE ANALYSIS:** Laboratory testing method used in mills to determine amounts of various substances, such as glue, casein, starch, rosin, etc. in paper.

**QUARTO:** Eight pages of paper obtained by folding or cutting a sheet.

**QUEEN ROLL:** The second roll from the bottom and the first undriven roll on a typical paper calender stack in a paper mill. It is smaller in diameter than the bottom, driven roll (*king roll*), but larger than the rest of the undriven rolls above it.

**QUENCH CIRCULATION:** Withdrawing and returning cooking liquor from a Kamyr-type continuous digester to a point where reaction is almost completed. It is done to produce a countercurrent flow in the digester that cools the chip column and essentially stops further reaction.

**QUENCH EXTRACTION:** The removal of reaction products and spent cooking liquors from the pulp after reaction is almost completed in a Kamyr-type continuous digester.

**QUENCH TANK:** A water-filled tank used to quickly cool incinerator residues and other hot materials.

**QUICK LIME (CaO):** A pulp mill chemical obtained by burning limestone ($CaCO_3$), which forms milk of lime [$Ca(OH)_2$] when dissolved in water. It is used to prepare bleach and pulping liquors, and in recausticizing sulfate and soda cooking liquors in pulp mills.

**QUICK OPENING VALVE:** A valve (usually of the gate type) that has a sliding stem and lever so that it can be opened and closed quickly.

**QUICK SETTING:** The ability and speed with which papermaking materials, such as sizes, coatings, etc., establish themselves to the desired constant, stable conditions after being added to the stock, paper sheet, etc.

**QUICK-SETTING INKS:** Special resin-oil combination inks which are most effective on certain coated papers. The oily material penetrates into the stock, leaving a dry resin film on the surface after printing.

**QUICK STEAMING BOILER:** A small capacity, water tube boiler designed for rapid steaming. It is generally made up of small, coiled steel tubes, with forced circulation and intense heat release. Fired by a gas or oil burner, it can come to full capacity and pressure within minutes. It is sometimes used for standby purposes in mills.

**QUICK STICK:** *See* AUTO ADHESION.

**QUIRE:** A quantity of paper that is equal to 1/20th of the standard ream for that grade.

**QUIRED PAPER:** Paper sheets that have been packaged in twenty sheet piles, folded in half.

**QUIRED REAM:** A ream of paper in which the sheets are folded in half, rather than laying flat.

**QUIREFOLD:** A package of paper made by folding a quantity of paper equal to 1/20th of a standard ream for that grade of paper.

# R

**RACEWAY:** (1) A channel that contains electrical wiring and runs throughout the mill. (2) A channel for routing water to various parts of the mill.

**RADIAL:** Along or in the direction of the radius of the pump impellers.

**RADIAL FLOW MIXER:** A type of mixer developed especially for use in medium density, upflow bleach towers. They are designed to mix the bleach chemical with the pulp and direct the flow in such a manner as to provide uniform movement up the tower.

**RADIAL-FLOW TURBINE:** A turbine in which the steam flows radially from the center and out through concentric, contra-rotating rows of blades located on the surface of its rotor.

**RADIANT DRYERS:** Electric or gas infrared heaters consisting of a bank so located as to provide high-density heating for selective drying of paper on a paper machine.

**RADIANT DRYING:** A method of selectively drying paper on a paper machine with a bank of infrared heaters placed close to and aimed at the sheet in such a manner as to provide high-density heating.

**RADIANT HEAT TRANSFER:** The electromagnetic transfer of heat between body surfaces that are separated by a vacuum or by a transparent media.

**RADIANT SUPERHEATER:** A type of heat exchanger installed in a furnace so that it is usually a part of the wall or gas pass, allowing it to absorb the heat of radiation from the hot gas.

**RADIATION-CURABLE LAMINATING ADHESIVES:** Laminating adhesives that are cured by exposure to ultraviolet or electron-beam radiation.

**RADIOACTIVE:** A descriptive reference to elements, such as krypton, strontium, radium, etc., that emit alpha, beta, or gamma rays. They are used as a means of making basis weight, density, and level measurements in the paper industry.

**RAFFOLD:** A mixture of calcium carbonate and magnesium hydroxide once used extensively by pulp and paper mills as a filler material in the production of uncoated papers for magazines.

**RAG BOILER:** A rotating, spherical or cylindrical pressure vessel used to cook rags at elevated temperatures with caustic soda.

**RAG BOOK PAPER:** A high-quality grade of paper, with varying cotton fiber content, used to make high-grade book paper.

**RAG CONTENT PAPER:** Paper containing from 25 to 75 percent cotton or rag fibers; usually includes bond, ledger, and specialty papers.

**RAG CUTTER:** A machine used in a rag pulp mill to cut sorted rags into strips prior to cooking. It generally consists of a revolving lawnmower-type knife assembly.

**RAG CUTTINGS:** Linen and cotton cuttings and trimmings from clothing manufacturing sent to rag mills to be converted into rag pulp and paper.

**RAG DUSTER:** *See* DUSTER.

**RAG OVERLOOKERS:** Rag pulp mill employees who inspect rags to be used for papermaking and remove any objectionable materials.

**RAG PAPER:** Paper normally made up entirely of cotton or rag fibers for use in high-grade paper products, but sometimes used to refer to paper made partly from rags.

**RAG PLATE PAPER:** Paper having a rag or cotton fiber content with a high finish applied on a supercalender and used in bookmaking.

**RAG PRINTING PAPER:** Any paper having rag or cotton fiber content used for printing purposes.

**RAG PULP:** Pulp made by cooking rags with caustic soda under elevated temperature and pressure. It is shipped to paper mills that make rag content paper but do not have rag pulping facilities.

**RAG SORTING:** A rag pulp mill operation in which rags are sorted and classified according to color and quality, based on the type and quantity of undesirable materials to be removed.

**RAG THRASHER:** A piece of rag pulp mill machinery consisting of two rotating, cylindrical drums to which old rags are fed for a teasing treatment prior to passing into the dusting operation.

**RAG THRASHING:** The first cleaning steps in the preparation of rags for papermaking. The rags are submitted to a beating action by a revolving cylindrical drum fitted with surface-mounted hooks or pins. A coarse screen beneath the drum allows the dirt to drop through.

**RAG WASHER:** An open tube-type container in which rags and wash water are circulated by a large roll as it washes them in preparation for the cooking and papermaking processes.

**RAG WRAPPING:** An inferior grade of heavyweight paper made from a mixture of waste pulp and used to wrap bales of rags for shipment to rag pulp mills.

**RAGS:** Generally, cotton or linen trimmings and cuttings from the textile industry that are sorted, graded, and cooked to make pulp in a rag pulp mill.

**RAILROAD BOARD:** A colored paperboard made on a cylinder type of paper machine from old newspaper stock. It is used for ticket stock (originally for railroad tickets), showcards, tags, signs, etc.

**RAILROAD DUSTER:** A type of rag-dusting machine used in a rag pulp mill prior to the cooking process. It consists of a series of revolving drums with pins or teeth helically arranged around them which carry the cut rags over screens through which the dust falls and is removed.

**RAILROAD TRACKING:** Certain types of paper imperfections caused by indiscriminate shrinkage in a lateral direction during drying and improperly adjusted rolls during coating. It appears as a small series of consecutive ridges running parallel to the machine direction.

**RAILROAD WRITING:** A writing paper made primarily from groundwood pulp. It has a canary yellow or manila color and is used extensively for making telegram forms, tablets, and pads.

**RAMIE:** A native tropical plant of Asia which has been a fiber source in papermaking.

**RANDOM LENGTH:** *See* TREE LENGTH.

**RANGE:** The operational scope of variations of the pulp and papermaking process and equipment as defined by the upper and lower limits.

**RAPID-DISPLACEMENT HEATING (RDH):** A modified kraft batch-pulping method (Beloit Fiber Systems Div.) that involves the displacement of spent liquor and recovered heat used in a cook so that they can be used in a later cook. Initially, the objective of RDH was to reduce the amount of heat used in order to save energy and reuse chemicals. But more recently, bleached pulp mills have employed the method to extend delignification in the pulp cooking stage and reduce subsequent bleach chemical demand. The method is similar to SuperBatch (Sunds Defibrator), with which it shares common developmental origins.

**RATE-OF-SET:** A property used as a basis in selecting an adhesive emulsion for a particular application. It is a measure of the minimum time required for the emulsion system to create a bond between the two surfaces of paper sheets such that the paper would fail before the bond.

**RATIO CONTROL:** A control scheme based on relative proportions between two variables, usually flow, and extensively used in adding and mixing materials in the process of pulp and paper manufacturing, especially in stock blending.

**RATTLE:** The crackling noise made by shaking, crumpling, or snapping a sheet of paper. It is used as a casual indication of its stiffness.

**RAW ACID:** Sulfite cooking acid (liquor) leaving the acid towers in a pulp mill acid plant after $SO_2$ absorption, filtering, and storage in pressureless "raw acid" storage tanks prior to fortification with gases in accumulators.

**RAW COOK:** A pulpwood cook in which not enough lignin and other noncellulosic materials are removed to meet the established mill standards for that particular grade of pulp being produced. It is usually caused by insufficient time, steam, cooking liquor, and cooking liquor strength used in the wood cooking operation. *See also* HARD COOK.

**RAW EFFLUENT:** Untreated wastewater being discharged from pulp and paper mill operations.

**RAW MATERIALS:** Any materials, such as wood, water, chemicals, dyes, additives, fuels, etc., brought into a pulp and paper mill that are required and used in the production of pulp and paper or used in producing products for use in pulp and paper manufacturing.

**RAW STOCK:** A general reference made to paper and paperboard before they are coated or used for other converting purposes.

**RAW WATER:** Source water supplied to pulp and paper mills that has not been given any type of a purification treatment, except for possible screening.

**RAW WEIGHT:** The weight of a sheet of paper before it is coated. It is usually expressed in pounds per ream or 1,000 square feet. Same as *uncoated weight.*

**RAY CELLS:** The short cells generally located in the structure of wood which make up the tissue running radially across the growth rings of pulpwood logs.

**RAYON:** A fine, smooth filament made from modified cellulose solutions. The solution is extruded through spinnerets into a chemical bath and warm air. The basic raw material for rayon is known as *dissolving pulp,* or chemical cellulose.

**RDH:** *See* RAPID DISPLACEMENT HEATING.

**R.E.A.:** Abbreviation for residual effective alkali.

**REACTION WOOD:** Wood whose physical characteristics have been caused by tension in hardwoods and compression in softwoods. It is due to its location in crooked stems and branches of trees.

**REAGENT:** Materials that take part in chemical reactions during pulping and papermaking processes.

**REAL TIME:** The actual time for a physical process to take place. Also refers to a system in which the time taken by a computer to process information from the mill operations and to send out signals to the mill operations is related to the duration of the events controlled by the calculations made. It generally applies to present time situations.

**REAM:** A stack or package of paper containing a number of sheets designated as standard for that grade, usually 480, 500, or 1,000 sheets.

**REAM LABEL:** An identification slip pasted on the exterior of a wrapped ream package of paper. It contains information about the paper therein.

**REAM MARK:** Pieces of colored paper placed in a stack of paper to indicate where the ream divisions are located.

**REAM WEIGHT:** The weight of a ream of paper considered standard for that grade.

**REAM WRAPPED:** A package or bundle of paper made by wrapping up the required number of sheets designated as a standard ream for that grade.

**REAM WRAPPER:** A class of chemical pulp and jute or rope fiber content paper, usually laminated or coated to provide resistance to moisture transmission, used to wrap standard ream size packages.

**REBURNED LIME:** *See* QUICK LIME.

**RECAUSTICIZING:** *See* CAUSTICIZING.

**RECEIVING WATERS:** Natural bodies of water, such as rivers, lakes, oceans, etc., into which treated or untreated wastewaters are discharged.

**RECEPTIVITY:** The ability of a paper sheet surface to accept liquids such as oil, water, ink, etc.

**RECESSES:** Spaces formed and occupied by voids or spaces in a sheet of paper. Sometimes called *voids in paper. See also* VOID FRACTION.

**RECHIPPER:** A machine used in the wood preparation process in a pulp mill to reduce to acceptable size large wood chips and slivers removed during the screening operation so that they can be returned for processing.

**RECIPE:** The list of pulp chemicals, additives, and dyes blended together to make a particular grade of paper.

**RECIPROCATING PISTON PRESS:** A type of press used to dewater and compact rejects in some mills; it uses a piston in a press chamber to form a compressed plug.

**RECIPROCATING PUMP:** A type of liquid-moving device that uses a piston, plunger, diaphragm, and other devices moving in a back-and-forth manner to positively displace a given volume of liquid during each stroke of the unit within a suitable fixed casing.

**RECIRCULATION:** (1) The flow around and behind an impeller before exiting through the pump discharge. (2) The removal and return of a portion of the process material during the manufacturing operation, such as cooking liquor during a batch cooking operation and during a pulping cycle or from a storage tank.

**RECLAIMED FIBER:** Any fiber used for papermaking that was obtained from fibrous waste materials.

**RECLAIMED PULP:** Fiber obtained primarily from waste newspapers, magazines, boxboard, and discards from offices and print shops.

**RECLAIMING TOWER:** A vessel in a sulfite mill cooking acid makeup system where the digester gas, during its discharge after cooking, is collected in order to recover the entrained cooking acid for reuse.

**RECORD PAPER:** Type of paper made to be very durable and of high quality so that writings and other records made with it will last a very long time.

**RECORDER:** A record keeping instrument that receives information from process measurement elements or transmitters and produces a continuous record of their values.

**RECOVERABLE RESOURCES:** Any used process materials that still have additional useful purposes and are recycled back into the process, such as water and paper broke in a paper mill.

**RECOVERY:** The process in a pulp mill where the active chemicals are derived from spent cooking liquors and then used to make up fresh cooking liquors.

**RECOVERY BOILER:** A combination boiler unit in a pulp mill used to recover the spent chemicals from cooking liquor and to produce steam.

**RECOVERY FURNACE:** The unit in a sulfate pulp mill in which concentrated spent cooking liquor (*black liquor*) is burnt to a smelt to recover inorganic sodium salts and to generate steam.

**RECOVERY ISLAND:** *See* RECOVERY PLANT.

**RECOVERY LOOP:** *See* RECOVERY PLANT.

**RECOVERY PLANT:** The area, building(s), and all of the process units considered to be included in the chemical recovery cycle of a pulp mill. Sometimes referred to as *recovery island* or *recovery loop.*

**RECOVERY TOWER:** A vessel in a sulfite mill acid system in which blowoff gases are recovered for reuse.

**RECTIFIER ROLL:** A roll located just ahead of the slice in some paper machine headboxes that is used for stabilization and conditioning of the stock flow.

**RECTIFYING:** A paper-finishing room operation consisting of putting sheets in a stack of paper to bring it back into standard ream size for that particular grade if sheets had previously been removed during sorting or other finishing operations.

**RECYCLABLE PAPER:** Paper that can be reused in the papermaking process. *See* RECYCLED PAPER and WASTEPAPER.

**RECYCLED PAPER:** Any used paper returned to the paper mill as a source of fiber. It is usually mixed with other unused or virgin fibers to make up various grades of paper. The term is sometimes used interchangeably with *wastepaper* or *old paper.* When reprocessed into pulp, recycled papers are known as *recycled fiber.*

**RECYCLED FIBER:** That component of a paper or paperboard furnish that is derived from *recycled paper* or *wastepaper.*

**RECYCLED PAPERBOARD:** Any grade of paperboard made primarily from fibers obtained from used materials, for example, waste newsprint and paper.

**RECYCLING:** The return of processed or used materials, such as fiber, paper, water, and some chemicals, back to the original process to make a new product.

**RED ALDER:** A hardwood species of tree found in the northwestern area of the United States and Canada, and used as a source of pulpwood by pulp mills.

**RED GUM:** A hardwood species of tree found in the eastern area of the United States, and used as a source of pulpwood by pulp mills.

**RED LIQUOR:** Spent liquor (condensed *pink liquor*) that is separated from the pulp after cooking with magnesium-base liquor (magnesium bisulfite) and contains the dissolved constituents of the wood. It is subsequently sent to an incineration process to recover chemicals for reuse in making up fresh cooking liquor.

**RED MAPLE:** A hardwood species of tree found widely scattered throughout the United States and Canada, and used as a source of pulpwood by pulp mills.

**RED ROPE PAPER:** *See* RED WALLET PAPER.

**RED SPRUCE:** A softwood species of tree used as a source of pulpwood by pulp mills.

**RED WALLET PAPER:** A strong, flexible paperboard made from a mixture of rope, jute, chemical pulp, and mechanical pulp and used to make file folders, expanding envelopes, and wallets. Also referred to as *red rope paper.*

**REDUCE:** The dilution of a paper finishing material with volatile liquids of a lighter specific gravity in order to thin its consistency or viscosity.

**REDUCER:** A pipeline connector having a smaller size on one end than on the other.

**REDUCIBLE SULFUR:** The amount of sulfur and sulfur compounds existing in paper. It is used as an indication of tarnishing properties because the compounds can be reduced to hydrogen sulfide ($H_2S$) when coming in contact with certain metals.

**REDUCING AGENTS:** In pulp bleaching, it refers to chemicals, such as sulfurous acid and its salts, dithionites, borohydrides, etc., which brighten pulp by the reduction of lignins and other coloring impurities.

**REDUCTION:** A chemical reaction occurring in a sulfate recovery furnace in which the sodium sulfate ($Na_2SO_4$) in, or added to, the black liquor as salt cake is reduced to sodium sulfide ($Na_2S$). This becomes part of the green liquor to the causticizing process. It is measured by analyzing green liquor and is expressed as a percentage, obtained by dividing $Na_2S$ by $Na_2SO_4 + Na_2S$, all expressed as $Na_2O$.

**REDUCTION EFFICIENCY:** An expression used in a pulp mill chemical recovery plant to indicate how well the conversion of the sulfur to sodium sulfide ($Na_2S$) has taken place in the recovery furnace. This is usually determined by dividing the sodium sulfide by the total sulfur in the smelt, and expressed as a percentage.

**REDWOOD:** (1) A species of tree, sometimes used as a source of pulpwood. (2) A reddish-colored abnormality in softwoods occurring in areas where the wood has become compressed.

**REEL:** (1) A machine located after the dry end section of a paper machine. It consists of a revolving frame, drum, and core or shaft, around which the sheet is wound. (2) The roll of untrimmed paper that is wound upon a core or shaft after leaving the dry end of a paper machine.

**REEL BOY:** An apprentice on early paper machine crews who did general cleaning, housework, and miscellaneous jobs around the winder and rewinder. Sometimes called a *broke boy.*

**REEL DRUM:** A variable-speed, motor-driven, revolving cylinder on a reel in a paper mill used to wind paper into a roll around a core by friction between it and the roll.

**REEL SPOOL:** A shaft or core around which paper from the paper machine is wound on the reel in a paper mill.

**REELING:** A paper mill operation of winding a roll of paper coming off the dry end of a paper machine around a revolving core or shaft.

**REFINER:** A machine for mechanically treating fibers in pulp and paper mills when rubbing, brushing, crushing, fraying, or cutting is desired to process or impart certain properties to the finished pulp slurry and the sheet web formed on the paper machine.

**REFINER BLEACHING:** The process of adding bleach liquor to a mechanical refiner in which chips or pulp are being defiberized.

**REFINER GROUNDWOOD:** Mechanical pulp made with a grinder and put through a rubbing, brushing, crushing, fraying, or cutting treatment in a pulp mill processing machine called a *refiner.*

**REFINER MECHANICAL PULP (RMP):** Pulp made by processing untreated wood chips in mechanical atmospheric refiners.

**REFINING:** Pulp and paper mill operations conducted on fiber suspensions to rub, brush, crush, fray, or cut the fibers as desired in order to impart such characteristics as increased capacity to absorb water and improved sheet formation.

**REFINING AGENTS:** Materials added to pulp slurries prior to refining to facilitate the separation and conditioning of fibers undergoing mechanical treatment.

**REFLECTANCE:** The light-repelling ability of the paper surface, also referred to as *gloss*, and expressed as the ratio of the intensity of the light repelled to that repelled from a standard reference.

**REFLECTIVITY:** The amount of light or energy repelled from a surface as compared with the total amount directed or incident to it.

**REFLECTOMETER:** An instrument used to measure the amount of light that a sheet of paper is capable of reflecting by comparing it with that reflected by a standard.

**REFRACTIVE INDEX:** A measure of the ability of materials to bend a light beam from a straight course. Liquids' ability to bend light varies with their density and is used as a method to determine this variable.

**REGENERATED CELLULOSE:** Cellulose that is produced by changing its form in a chemical solution and reproducing it in another form or derivative.

**REGISTER PAPER:** A lightweight writing grade of paper made from chemical pulp and possessing good tensile and tearing strength. It is usually put up in rolls especially for use in automatic register machines, and in flat, folded, snap-apart packages for use as computer printouts and other copying purposes.

**REGRINDING:** A pulpwood grinding phenomenon in which fibers freed from the wood logs pass through the grinding zone and are exposed to a secondary mechanical action by being transported back into the grinding zone on the surface of the grindstone. It is sometimes used as reference to all free fibrous material passing through the grinding zone.

**REGULAR NUMBER:** The quantity of standard size paperboard sheets in a standard weight bundle as established for that particular type.

**REGULAR SIZE:** A standard size of any sheet of paper as established for that particular type, grade, and form, such as rolls or flat sheets.

**REGULAR WEIGHT:** The standard weight reference established for a particular type or grade of paper.

**REGULATING BOX:** A constant level, baffled flow chamber provided with a consistency sensing element. It is used to add water in order to maintain the consistency at a constant value and is usually found on the stock feed lines to the paper machines.

**REGULATOR:** A device that senses a process variable or condition, compares it with some preset value, and initiates action to correct any deviation between them. Consistency, pressure, and temperature regulators are examples of the wide variety used in the pulp and paper industry.

**REINFORCED PAPER:** A paper sheet whose strength has been improved by combining several sheets with asphalt or other adhesive, by combining with other materials such as cloth and thin, flexible metallic coatings, or by filling the sheet with string, yarn, glass, or synthetic fibers.

**REINFORCING AGENT:** A pigment of larger size than that existing in a film, which is added to increase its strength.

**REJECT REFINING:** An expression originally used in groundwood pulp mills to refer to the practice of refining rejects from *bull screens* and primary screens, but can also be applied to the refining operation of rejects of any screening and cleaning system used in pulp and paper mills.

**REJECTS:** The materials or that portion of a chip or pulp mass that are removed when processed through a screening or cleaning system.

**REJECTS REFINER:** A type of refiner used to mechanically treat and refine rejected fiber from stock screening and cleaning operations in a pulp and paper mill.

**RELATIVE HUMIDITY (RH):** The actual amount of water vapor present in the air as compared with the maximum amount of water vapor the air could hold at that temperature. It is a measure of its ability to moisten or dry materials. *See also* ABSOLUTE HUMIDITY.

**RELEASE AGENTS:** An antisticking substance, such as soap, ammonia, glycols, etc., added to moistening fluid. It is used in the gloss-coating of paper on supercalenders to facilitate the release of the coating from the polished drying cylinder surface.

**RELEASE COATING:** A paper surface coating applied to the backing of pressure-sensitive tape, the side opposite the side the adhesive was applied. It is used in making rolls of pressure-sensitive tapes to provide for ease of subsequent unwinding.

**RELEASE PAPER:** A type of paper made especially to be easily removed from and having very little attraction to sticky surfaces.

**RELIEF:** The reduction of pressure in a pressurized vessel or process equipment by releasing some of the enclosed material. It is performed on such equipment as digesters, dryer drums, steam boilers, etc. for proper process operation or for safety purposes.

**RELIEF GAS:** Vapors commonly removed from batch-type pulp cooking digesters, such as noncondensible gaseous components, in order to maintain proper temperature-pressure relationship during the cooking process.

**RELIEF VALVE:** A valve used to reduce pressure in a pressurized vessel or process by releasing some of the enclosed material. It can also be a self-operated, quick opening valve.

**REMOTE CONTROL:** A control system for automatically operating devices and valves located remotely on the process.

**REP FINISH PAPER:** Paper with a ribbed surface produced by embossing the sheet between grooved steel rolls.

**REPELLENCY:** The refusal of paper to take or accept ink.

**REPRODUCTION PAPER:** A type of high quality, single-side-coated paper suitable for fine screen printing, color printing, and reproduction proofs.

**REPTISSUE:** Lightweight tissue grades of paper with a ribbed surface.

**REPULPABLE TAPES:** Paper tapes that can be recycled to the process without contamination of the broke pulp.

**REPULPER:** *See* PULPER.

**REPULPING:** The process of mechanically disintegrating and rewetting dry pulp and paper to form a slurry, which can be treated further and reformed into a sheet of paper.

**RESERVOIR:** A natural or artificial pond, lake, tank, or basin used for storage, regulation, and control of the mill water supply.

**RESET CONTROL ACTION:** In process control, it is the number of corrections the control system makes per unit of time, usually expressed in minutes. *See also* INTEGRAL CONTROL ACTION.

**RESIDUAL HYPOCHLORITE:** The excess of hypochlorite bleaching solution maintained after the retention time of pulp in the hypochlorite bleaching stage of a multi-stage bleaching process. It prevents possible color reversion in the hot, moist environment.

**RESIDUALS:** (1) Valuable by-products and leftovers that can be reused in other processes or as finished products. (2) Refers to materials remaining in a product after it has gone through a process to remove them, such as lignin, extractives, and noncellulosic carbohydrates in bleached pulps.

**RESIDUE:** Material that remains after all gases, liquids, and solids are removed from mill effluents.

**RESILIENCY:** The ability of paper and paperboard to return to their original shape after being bent or warped.

**RESIN:** The water-insoluble vegetable excretion or exudation that results in an amorphous, lustrous-appearing, solid, organic substance that is soluble in organic solvents. Also, it can be an artificially produced synthetic material. Resin is used in making both paper and ink.

**RESINOUS WOOD:** Softwood obtained from coniferous-type trees that is used as a source of pulpwood.

**RESISTANCE:** Generally, that which opposes an acting force and is usually equal and opposite to the force. Electrically, it is the opposition offered by any material to the passage of a steady electric current through it.

**RESISTANT COLOR:** *See* FAST COLOR.

**RESOURCE RECOVERY:** The process of obtaining materials or energy, particularly from solid waste materials such as wood, sawdust, and bark.

**RETENTION:** The retaining of dyes, additives, fillers, fiber fines, and other materials added to the stock slurry furnish in the sheet formed from it. It is expressed as a percentage of the total amount originally added.

**RETENTION AIDS:** Materials such as vegetable gums, cationic starches, potato starch, sodium aluminate, colloidal animal glue, acrylamide resin, etc., added to the papermaking process at the paper machine headbox, fan pump, or other location close to the wire. They are added in small amounts for the express purpose of maximizing the retention of fillers by altering their electrical charge or bonding.

**RETENTION TIME:** The elapsed period of time between when a material enters a process stage until it leaves it, such as wood chips through a continuous digester and pulp passing through various stages in a bleaching process.

**RETREES:** Substandard and defective watermarked papers that are sorted at the mill and sold at a discount.

**RETROGRADATION:** The reversion of starch that has been cooked to form paste back into an insoluble condition. It is caused by hydration which, in essence, is a reverse of gelatinization.

**REVERSE ROLL COATER:** A type of paper-coating machine consisting of a metal metering roll that picks up and transfers a prescribed amount of coating material to a metal coating roll. This roll then applies the material to the surface of the paper moving in the opposite direction with a rubber-covered backing roll located on the opposite side of the sheet, and around which it runs for support purposes.

**REVERSION:** *See* BRIGHTNESS REVERSION, COLOR REVERSION, and YELLOWING.

**REWATER:** Water previously used to dilute and transport the slush pulp through the papermaking process, but is later removed and returned for reuse. Called *white water* when done at the paper machine.

**REWETTING AGENTS:** Materials, such as sulfonated polyesters of dicarboxyle acid, used in the treatment of tissues, toweling, and other wet-strength grades of paper to improve absorbency.

**REWINDER:** A paper-winding machine located away from the paper machine. Paper rolls from the paper machine winder are sent to it for re-rolling, slitting, and removal of defective paper. It is also used to make smaller, tighter rolls for shipping to users or for finishing operations in the paper mill.

**RH:** *See* RELATIVE HUMIDITY.

**RHEOLOGY:** The study of the deformation and flow of matter as related to its behavior to stress-strain forces.

**RHEOMETER**: An instrument used to measure viscosity and flow behavior of coatings at very high shear rates. Its value is an indication of flow beneath the blade tip in a blade-type sheet coating process.

**RHEOSTAT:** An electrical device provided with means to vary its resistance to electrical current. It is used extensively in the mill to adjust motor speeds.

**RIBBON PAPER:** Strips of paper used to support and protect textile ribbon by interwinding with it.

**RICE PAPER:** A nonfibrous sheet that technically is not paper. It is made from the pith of trees cut up in thin layers with a sharp knife to produce an ivory-like textured surface.

**RICE STRAW:** Type of straw sometimes used as a source of fiber in the manufacture of certain grades of chip and news boards.

**RICK:** A cordwood stave bolt pile and other materials split from short pulpwood logs.

**RIDER ROLL:** A small, sometimes counterweighted paper machine roll that is let down on a running sheet of paper to help maintain a pressure, especially on winders, after the paper has been turned around the core to give the sheet a better start.

**RIDGE:** (1) A paper roll defect appearing as rings around the circumference of the roll. It is caused by the nonuniform flow of stock from the headbox, plugged wet felt, plugged suction roll, faulty dryer canvas, buildup or filling of the calender rolls, etc. Sometimes known as *bars, buckles, chain marks, rope marks* and *crow's feet.* (2) A synthetic forming fabric problem appearing as a raised line on the surface and usually removed by the use of a household iron while still on the paper machine.

**RIFFLER:** *See* SAND TRAP.

**RIGGING:** A system of cables, blocks, and hooks used in the logging of pulpwood.

**RIGHT SIDE OF PAPER:** A paper mill expression referring to watermarked paper to indicate the surface of the paper from which the watermark can be properly viewed. It is normally the upper or felt side of the sheet on a paper machine. *See also* WRONG SIDE OF PAPER.

**RIGIDITY:** *See* STIFFNESS.

**RING:** (1) The crackling sound made by crumpling or crushing a piece of paper. It is used to obtain a quick, general evaluation of its quality. (2) The cross section presentation layer of annual growth or increment of a pulpwood tree.

**RING CRUSH:** The ability of paper and paperboard, made up into cylinders, to withstand deformation forces applied to the edges. Also called *ring stiffness.* In reference to *Rule 41* affecting U.S. shipping boxes made from containerboard, a specific *edge crush test* and/or *concora crush* test are used.

**RING GRINDER:** A type of pulpwood grinder used to make groundwood pulp. The logs are fed to the grindstone surface by a chain-driven, ring-type chamber and set eccentric to the grindstone, whose speed determines the rate of feed.

**RING MARK:** (1) Watermark made on less expensive grades of paper using rubber rings with raised images or designs. (2) Paper defects appearing as cloudy rings or blotches, particularly on pigment-colored papers, caused by the bursting of bubbles formed while the web is on the paper machine wire (forming fabric).

**RING POROUS:** A characteristic of pulpwood logs from hardwood trees in which the annual spring growth rings are made up of large pores and vessels.

**RING STIFFNESS:** *See* RING CRUSH.

**RING WATER:** Dilution water injected in a number of locations around the periphery of pulp stock storage tanks and bleaching towers. It is injected just above the agitators and scraper arms to facilitate the removal of the pulp stock from the bottom by stock pumps.

**RINGELMANN CHART:** A chart of gray-shaded color standards used to evaluate the opacity and density of smoke in order not to produce excessive emissions.

**RIPPLE FINISH:** An undulating surface finish produced on paper by passing the paper between embossing rolls or by plating.

**RISE TIME:** The elapsed time required to achieve top operating temperature and pressure by steam-heating a batch-type digester in a pulp mill.

**RISER TUBES:** Heated tubes in a steam boiler containing the steam-water mixture that is displaced by the relatively heavier water in the downcomers.

**RMP:** *See* REFINER MECHANICAL PULP.

**ROAD MAP PAPER:** Long fiber paper made from chemical pulp. It has good printing and folding properties and is used for making up road and other types of maps.

**ROD MILL:** A defibering machine made up of a revolving, rugged, lined, metallic, horizontal drum, partially filled with iron rods. It was once used extensively by pulp and paper mills to refine pulp screenings by passing the stock from one end to the other.

**ROE CHLORINATION NUMBER:** A value determined by a pulp mill laboratory test on pulp to indicate the quantity of gaseous chlorine absorbed by a specified sample. It is a measure of its bleachability.

**ROLL:** (1) A continuous sheet of paper or paperboard wound up around a core or shaft, usually to specified diameter and width to best suit subsequent finishing and converting operations. (2) A rotating, cylindrically shaped, solid or hollow structure made from a variety of materials. It is normally used in a horizontal position to support or carry paper or other sheet materials, such as felt, on a paper machine. They may be driven or rotated by the sheet they support, or they may be independently driven.

**ROLL-A-PACK:** An expression referring to a method of wrapping paper rolls with stretch film and end headers made of corrugated board.

**ROLL A WAD:** A paper mill expression referring to the operation of transferring a sheet of paper from one reel to another at the paper machine winder.

**ROLL BOX:** A small, vacuum suction box sometimes located just behind the breast roll on the wet end of a paper machine. It is used to remove air and water from the returning wire (forming fabric) in order to aid in stabilizing the stock flow jet from the slice, and to minimize the tendency to form bubble marks in the sheet.

**ROLL BREAKER:** A wooden paddle designed and used to smooth the sheet of paper in a paper mill while it is being rewound.

**ROLL-COATED:** A descriptive reference to paper that has been surface coated in a type of coating operation which applies the coating material with a rubber applicator roll. The material is picked up in metered amounts from another feed roll.

**ROLL COATER:** Any type of paper-coating machine that uses cylindrical rolls to pick up, meter, and apply coating material to the surfaces of a paper sheet.

**ROLL COCKLE:** A paper roll defect occuring as a buckle on the surface of a finished roll of paper.

**ROLL CONVEYORS:** A pulp and paper mill materials moving system made up of a series of rollers with end bearings or loose wheels on spindles, spaced at regular intervals, and mounted on parallel supports.

**ROLL CORE:** A metal, wood, or, more commonly, a convolute-wound fiber tube used to wind paper or paperboard for transporting and shipping.

**ROLL CROWN:** The surface curvature shape or diameter profile of the press and calender rolls to compensate for deflection in order to obtain a uniform nip pressure distribution across the full width through which the paper passes.

**ROLL CURL:** The curvature a sheet of paper assumes due to becoming set when wrapped into a roll around a core or shaft. Also referred to as *roll set* and *wrap curl*.

**ROLL DOCTOR:** A scraping device in the form of a thin blade in contact with and extending across the face of paper machine rolls, usually to keep the surface clean. Also called a *doctor* or *doctor blade*.

**ROLL EDGE DAMAGE:** A paper roll defect consisting of any damage occurring at the ends, such as nicks, cuts, and indentations.

**ROLL END:** A piece of circular paper or paperboard placed on the side end of a roll of paper (*roll head*) prepared for shipment. It is used to prevent damage to the edges. *See also* ROLL HEADING.

**ROLL FINISHING:** A paper mill operation in which paper rolls from the paper machine winder are sent through a series of handling steps. These steps range from removing their shafts, stripping the outer layers, mechanically conveying to automatic wrapping, crimping, heading, weighing, labeling, and printing, to final conveying to the shipping area.

**ROLL GRINDING:** The machining of paper machine press and calender roll surfaces to obtain the proper crown profile consistent with good performance operation.

**ROLL HANDLING:** A transporting system used to move paper rolls from one location to another in a paper mill. Sometimes includes automatic cutting into smaller rolls and wrapping for shipment.

**ROLL HEAD:** The side end of a roll of paper or the circular pieces of material (*roll end*) placed on it for protection purposes during shipment. Also referred to as *end cap*. *See also* ROLL HEADING.

**ROLL HEADING:** A manual or automatic operation during roll wrapping of paper in which the folded end wrapping is covered with a piece of material (*roll end*) and attached to it for protection during shipment. Also referred to as *end capping*.

**ROLL PAPER:** Any paper that is packaged into rolls of specified dimensions.

**ROLL POSITION:** The location on the paper machine where the paper roll is made and taken in relationship to the front or back side of the paper machine.

**ROLL-OUT TABLE:** Roller-type conveyors used in structural board manufacturing to transport the wet, pressed sheet to the dryer.

**ROLL SET:** The tendency of a sheet of paper to assume the curvature of the roll on which it is wound. Also known as *roll curl* and *wrap curl*.

**ROLL SETTING:** The manual or automatic positioning of a beater roll in relation to the bedplate as the beating process proceeds on the pulp fiber mass.

**ROLL STAND:** *See* BACKSTAND.

**ROLL TICKET:** A paper or paperboard card, usually color-coded, inserted in a roll during winding or rewinding for purposes of identification.

**ROLL TISSUE:** Lightweight tissue paper put up into rolls appropriate for the intended end use, such as toilet tissue.

**ROLL WEIGHT:** The gross weight of a roll of paper (which includes the core and wrapping), or the net weight (which does not include the core and wrapping).

**ROLL WRAPPING:** A paper roll finishing operation, done manually or automatically, in which finished rolls are packaged by wrapping and sealing for shipment from a paper mill to converters and other users.

**ROLLED:** Paper that has been surface-finished on one or both sides by friction-glazing between metallic and filled rolls on a machine called a *glazer*, or on calenders and supercalenders.

**ROLLED EDGES:** The curled edges of a sheet of paper.

**ROLLER BACK:** An operation performed in a groundwood mill in which a blank or dull burr is run across the grindstone face to knock off the high points.

**ROLLER BEARING:** A friction support assembly commonly found in pulp and papermaking machinery. It consists of a journal-type socket fitted with rollers on its interior perimeter.

**ROOF AIR SUPPLY:** Warm air introduced under the roof of a paper machine room to prevent condensation and dripping.

**ROOFING FELT:** A base sheet of porous, soft paper made from used rags and paper fibers. It is made especially for saturating and coating with asphalt to make some types of roofing papers.

**ROOFING PAPERS:** Any type of paper sheet material that has been treated and coated with asphalt, crushed slate, and other materials. Roofing papers are used for the roofs of buildings, and are sometimes known as "roll roofing."

**ROPE CARRIER:** A system developed to safely thread the sheet of paper through the dryer section of a paper machine. It consists of two ropes running in grooves on the tending side of the dryer drums. They run together and carry the sheet through the dryers when fed into the nips of the rope.

**ROPE MANILA PAPER:** A high strength paper used for wrapping purposes. It is usually made of fibers from manila and hemp ropes and cables.

**ROPE MARKS:** A type of paper roll defect appearing as welts. It is caused by soft areas produced during the reeling operation and results in nonuniform hardness. *See also* BARS, BUCKLE, CHAIN MARKS, CORRUGATION MARKS, RIDGES, and TIRE TRACKS.

**ROPE PAPER:** Paper containing manila hemp fiber content obtained from old ropes.

**ROPE TAG:** Types of paper ranging from light paperboards to heavy weight paper made from manila rope and chemical wood fibers.

**ROPE WRAPPING:** A high-strength, usually brown paper made with hemp fibers obtained from rope and chemical pulp.

**ROPES:** (1) All types of cords made of hemp, jute, sisal, flax, cotton, and other plant fibers. They are used as raw materials for making certain grades of bristols, tag, folder stock, etc. (2) Wet-lay nonwoven paper machine fabric defect caused by gatherings of fiber, with unaligned ends, that are more agglomerated than in normal dispersions.

**ROPING:** A type of paper defect appearing as a machine direction wrinkle in the sheet. It is caused by the sheet tending to move transversely on the dryer drum combined with the pulling forces exerted by the draw as it passes over the drums.

**ROSIN:** A material made up of a suspension and used for internal sizing of paper and paperboard. It is obtained as a residue from the distillation of gum from resinous southern pines. Sometimes called *colophony*.

**ROSIN SIZE:** Rosin made up as a suspension and used for internal sizing of paper and paperboard to enhance its ability to repel moisture and water.

**ROSIN SIZED:** Any type of paper or paperboard whose water and moisture resistance was produced by the addition of rosin to the pulp furnish from which it was made.

**ROSIN SOAP:** *See* SAPONIFICATION.

**ROSIN SPECKS:** A paper defect due to the addition of rosin size to the stock that has not completely emulsified and has agglomerated and fallen out of the furnish before it is completely dispersed. It causes yellow-colored spots in the sheet.

**ROTAMETER:** A fluid meter device consisting of a vertically oriented, tapered tube in which a floating element rides in a stream flowing through it, rising and falling in accordance with the velocity of the moving stream. The tube is calibrated with a scale so that the position of the floating element is indicative of the flow rate.

**ROTARY BURNERS:** *See* SULFUR BURNERS.

**ROTARY CUTTING:** A paper-cutting machine used in the paper mill finishing room to continuously convert rolls of paper into untrimmed sheets by the use of a rotating cutting element fitted with suitable knives.

**ROTARY DIGESTER:** A cylindrical or spherical drum often used to cook rags. It is mounted on trunnions resting on bearings at each end, about which they slowly revolve during the cooking process. Also called a *bleach boiler*.

**ROTARY DRAINER:** Revolving, cylindrical screening equipment through which freshly cooked pulp is passed so that the spent cooking liquor drains away and is followed by a dilution and washing operation.

**ROTARY FURNACES:** Large, horizontal, rotating furnaces once used extensively to burn black liquor from soda and sulfate pulping processes to recover chemicals.

**ROTARY LIME KILN:** *See* LIME KILN.

**ROTARY NEWSPRINT:** Newsprint paper put up into rolls especially for use on rotary printing presses.

**ROTARY PRESSURE JOINT:** A type of steam supply connection designed for getting steam into rotating vessels, such as rotary digesters, or getting condensate out of paper machine dryer drums and other similar equipment.

**ROTARY PUMPS:** Types of fluid-moving devices commonly used in mills for moving chemical solutions and other liquids through pipelines. They are characterized by having a rotating cam, screw, gear, or vane as the moving element, fitted into a suitable fixed casing.

**ROTARY SCREEN:** (1) A stock-cleaning device used just ahead of the paper machine to remove dirt by passing the slurry slowly through a rotating cylinder mounted with fine-slotted screen plates, which are constantly scoured by a shower. (2) Any type of pulp screening equipment in which the screening chamber or drum is rotated or revolved. (3) A type of screen used on wood chips that is known as a *gyratory chip screen.*

**ROTARY SYPHON:** A specially designed piping system designed to scoop condensate from the inside of rotating paper machine dryer drums and to remove it through a *rotary pressure joint.*

**ROTARY VACUUM FILTER:** A type of pulp washer consisting of a revolving, wire-covered, cylindrical drum partially immersed in a vat of low consistency stock slurry. The slurry forms a mat on the outer surface of the drum as liquor drains away from the mat and down a drop leg to create a vacuum while wash liquor is sprayed on the drum.

**ROTATIONAL GRINDING:** A method of pulpwood grinding that employs a series of grinding wheels so arranged that they impart a rotational movement to the pulpwood log and impel it forward while grinding off the fiber.

**ROTOGRAVURE PAPER:** Sized and unsized paper whose printing smoothness has been maximized, usually by supercalendering the sheet. It is used for rotogravure printing of catalogs, magazines, newspaper supplements, etc.

**ROTOR:** The heavy, motor-driven, pulp-defiberating element of a pulper, located in the bottom area. It consists of various designs of impellers, depending on the manufacturer.

**ROUGH FINISHED:** Reference to paper and paperboard that have not been given any, or have been given a minimum of, calender treatments.

**ROUGH SURFACE:** A type of paper defect caused by poor drying procedures, too fast drainage on the wire (forming fabric), coarse stiff fibers, improper headbox consistency, poor sheet formation, too coarse wires (fabrics), and rough spots.

**ROUNDING:** The preparation of a roll of paper that is to be shipped directly to the user. Also called *rewinding.*

**ROUNDWOOD:** Logs as delivered to a pulp mill, with bark attached and cut to specified lengths, usually up to 10 feet.

**RPM:** Abbreviation for rounds per minute that is used to measure the speed of rotating equipment.

**RUBBER LATEX:** A sizing material consisting of liquid rubber milk. It is also used to impregnate the paper sheet to strengthen its properties.

**RUBBER MARK:** A translucent mark made on paper of lesser grades by means of rubber rings with raised designs that are impressed on the damp sheet.

**RUBBER RINGS:** Rubber watermarking devices with raised images or designs. They are used on the wet web of less expensive types of paper.

**RUBBER SPOTS:** Undesirable spots in paper caused by the inadvertent pickup of tiny particles of rubber during the manufacturing process.

**RUBBER STAMP MARK:** A type of *watermark* produced in a wet web of paper by the use of rubber bands fitted on press rolls of paper machines.

**RULE 41:** Corrugated shipping box regulation for U.S. railroads, corresponding to *Item 222* that regulates similar shipping boxes for trucking in the U.S. Revisions to the regulations in 1991 included a new *edge crush test* that can be substituted for separate *burst strength* and basis weight tests on *linerboard* and *corrugating medium.* This shift toward compressive stacking strength (*ring crush* and *concora crush*) rather than bursting strength (*Mullen*) has opened a new market for *high performance containerboard.*

**RUN OF PAPER:** A scheduled order of any given grade of paper being made on a paper machine.

**RUNNABILITY:** (1) In the paper mill, how well pulp stock furnish to the paper machine forms a sheet on the wire (forming fabric) and passes through the drying and finishing operations. (2) Used by customers in reference to how well the paper performs in their converting operations, such as on printing presses. (3) A general reference to how efficient a paper machine or any of its segments, or other pulp and paper mill equipment and systems, are operating.

**RUNNING SCHEDULE:** Paper mill operating plans for running paper machines most efficiently, taking into consideration changes in grades, colors, and weights made on each paper machine, with a minimum of downtime and lost material during production.

**RUPTURE PICK:** A surface strength failure of paper that consists of a continuous removal of the surface during manufacture or use.

**RUSH/DRAG RATIO:** *See* EFFLUX RATIO. Also referred to as *push/drag ratio.*

**RUSH:** The phenomenon of "pushing" created on the fiber suspension being ejected on the paper machine wire (forming fabric) from the headbox. It is caused by the slice jet velocity of the stock being greater than the wire speed. It is functionally opposite to "pulling" of the suspension, know as *drag.*

**RUST SPOTS:** Undesirable reddish-brown specks in a paper sheet caused by inadvertent pickup of extraneous iron particles.

**RUST-PREVENTIVE PAPER:** Paper made especially to be free from acidity and reducible sulfur compounds so that it can be used for wrapping silverware, razor blades, needles, and other metallic items without rusting or discoloring them. Also called *antitarnish paper.*

**RUTILE:** A titanium dioxide ($TiO_2$) mineral containing iron.

# S

**S-FACTOR:** A sulfite cooking parameter that gives an integral measure of the "pulping work" done by the three main pulping parameters: time, temperature, and partial pressure of $SO_2$.

**S PAPER:** An abbreviation for paper that has been sized.

**SC PAPER:** Paper with supercalendered surfaces. Sometimes written as "S/C paper."

**SSC:** An abbreviated term used to distinguish paper that has been both sized and supercalendered.

**SACK KRAFT:** A kraft paper with good stretch, high tensile energy absorption, and high tear resistance.

**SACK PAPER:** A strong, heavyweight paper usually made from rope and/or sulfate pulp especially for conversion into large, heavier type sacks rather than lighter bags.

**SAD PAPER COLORS:** Types of paper colors which have been given a duller or flatter finish by the use of other chemicals and colors.

**SAFETY BASE STOCK:** Paper made especially to be surface-treated with chemicals so that any erasures or alterations become very obvious and are easily detected. *See* ANTIFALSIFICATION PAPER.

**SAFETY SPLICE:** *See* PERFECT SPLICE.

**SAFETY TYPE PAPER:** Any grade of paper so treated that it will produce an obvious discoloration, smudge, or reveal a hidden surface design when mechanically tampered with by attempting to alter the surface with erasures or other types of alterations. *See* ANTIFALSIFICATION PAPER.

**SAFETY VALVE:** A self-operated, quick-opening, spring-loaded valve used for fast relief of excessive pressures in process equipment and enclosed vessels, such as on power and recovery boilers, and to relieve pressure during emergency conditions to protect drums, boilers, etc.

**SAL SODA:** Sodium carbonate ($Na_2CO_3$).

**SALES BOOK PAPER:** Any grade of paper, such as bonds, manila, tissue, etc., made specifically for use in the manufacture of sales or order books.

**SALT:** Technically, it is a product of reaction between an acid and a base, but usually refers to sodium chloride (NaCl) when used in the mill.

**SALT CAKE:** A form of natural sodium sulfate ($Na_2SO_4$) added to the thick black liquor just prior to incineration in a sulfate recovery furnace where it is converted to sodium sulfide ($Na_2S$) to provide one of the active chemicals in the subsequent makeup of raw cooking liquor in the sulfate pulping process. Also referred to as *glauber's salt.*

**SALT CAKE MAKEUP:** The sodium sulfate ($Na_2SO_4$) that has to be added to the alkaline black liquor during the chemical recovery cycle to provide the desired level of active chemical. It is expressed in pounds per ton of air dry pulp produced.

**SALT CAKE MIX TANK:** A specially agitated tank through which black liquor is passed so that sodium sulfate can be added to it just before incineration in a chemical recovery furnace.

**SALTED PAPER:** Type of paper that has been treated with a weak salt (NaCI) solution and dried in preparation for use as a sensitizing base for silver nitrate ($Ag_2NO_3$) for printing purposes.

**SALTING OUT:** The formation of a layer of fat and resin acids present in sulfate black liquor as soap due to the increasing salt content caused by the evaporation process. It is removed so that it will not plug the evaporator tubes.

**SAMPLE:** Representative portion of raw materials, pulp, paper, and any other materials obtained during the manufacturing process and used for analysis, testing, or to make any other measurements or evaluations on it to determine whether it meets the prescribed specifications or as a guide to proper process operation. Sometimes called a *specimen.*

**SAMPLE CARD:** Paperboard used for mounting samples of textile cloth for exhibition purposes.

**SAMPLE DEPARTMENT:** A room or area in the paper-finishing room of a paper mill where paper samples of every order filled are kept for reference purposes in case the order is repeated or a complaint is made regarding lt.

**SAMPLER:** A device used to obtain periodic or continuous fractional portions of raw materials, process materials, and products of a pulp and paper mill for use in analysis, testing, and evaluation. It is usually associated with a flow measurement when fluids such as chemicals, water, and waste are involved.

**SAMPLING:** The process of obtaining representative portions of raw materials, pulp, paper, and any other materials during the manufacturing process in order to analyze, test, or make other measurements on it.

**SAMPLING PAPER:** Kraft paper of high strength used as wrapper for packaging of wool, cotton, and other textile samples.

**SAND TRAPS:** A pulp-cleaning system used on some older paper machines. It consists of a long trough with baffle plates over which pulp stock slurries are allowed to flow so that heavy foreign material, such as dirt and sand particles, will settle out to be removed. Also called *rifflers.*

**SANDPAPER:** Heavy, strong, hard-sized rope or kraft paper coated on one side with glue, sand, or sand-like gritty material such as silicon carbide in varying degrees of fineness to suit the type of surface smoothing and polishing work required of it.

**SANDSTONE:** Natural abrasive material from which grinding stones were made for early groundwood pulp mills.

**SANITARY LANDFILL:** Areas in which solid wastes from the pulp and papermaking process are disposed.

**SANITARY TISSUE:** Highly absorbent, bulky types of lightweight tissue papers used to make sanitary, disposable items such as napkins.

**SAP:** The fluid part of a tree that moves up from the roots through the outer portion of the trunk and branches and contributes to its growth.

**SAP-PEELING:** The process of bark removal from pulpwood. It is done on-site as soon as it is cut, usually in the late spring or early summer because it is easier to remove at that time.

**SAPONIFICATION:** The conversion of fatty acids to soaps by treatment with alkalies, as used in the manufacture of rosin size employed in papermaking. Also called *rosin soap.* This same phenomenon occurs when the alkali in sulfate cooking liquor reacts with the fatty acids in wood during the cooking process.

**SAPPY:** A descriptive pulp mill reference to pulpwood or chips having a lot of sap or sapwood.

**SAPWOOD:** The lighter colored, outer portion of a tree, containing the cells through which water is conducted from the roots, through the trunk, and into the stems, branches, and leaves.

**SATELLITE CHIP MILL:** Mills remotely located from pulp mills at the source of wood (the woodlands) which process woodchips from treelength stems. The chips are subsequently transported to the pulp mill site.

**SATIN FINISH:** A type of smooth paper surface finish applied to the sheet in such a manner that it resembles satin.

**SATIN WHITE:** A white aluminum hydroxide and calcium sulfate pigment mixture made by reacting calcium hydroxide with aluminum sulfate. It is used in the preparation of paper coatings for high white grades of enameled finished sheets.

**SATURATED AIR:** An air-water vapor mixture in which the partial pressure of the water vapor equals the vapor pressure of liquid water at the temperature of the mixture.

**SATURATING FELTS:** A soft, porous, absorbent, fibrous sheet made from inferior grades of rag fiber mixed with small amounts of other fibers. It is used to make waterproof building-type papers by impregnating it with tar, asphalt, or bituminous products.

**SATURATING PAPERS:** Types of paper made with an open, porous formation so that they will absorb impregnating liquids such as asphalt and tar.

**SATURATING PROPERTIES:** Those properties of paper that influence the amount and rate of impregnation of liquids into the sheet.

**SATURATION:** The impregnation of unsized paper, such as felt paper, with waterproofing or preserving liquid to the point that it will not absorb any more.

**SATURATION BONDING:** A technique used for the latex bonding of disposable nonwovens in which the web is immersed in a binder bath that completely covers each individual fiber with binder solution.

**SAVE-ALL:** Separation devices used in a paper mill to reclaim fibers and fillers from white water by the use of screening, flocculation, sedimentation, flotation, etc.

**SAWDUST:** Small particles of wood formed during slashing and chipping of pulpwood logs. It is screened from the chips and disposed of or burned in power furnaces. Some mills will also bring in this material from sawmills to use as a source of fuel for power furnaces. Also, it is sometimes "cooked" in specially designed digester vessels for use in certain paperboard furnishes.

**SBK:** *See* SEMIBLEACHED KRAFT.

**SBLO:** *See* STRONG BLACK LIQUOR OXIDATION.

**SBS:** *See* SOLID BLEACHED SULFATE.

**SCABBING:** The adherence and buildup of foreign material on the surfaces of paper machine calender rolls. This material is usually scraped off and the rolls kept clean by a *doctor blade*.

**SCALDER SCREEN:** A screen used to remove oversized chips in the chip manufacturing process of a pulp mill.

**SCALE:** (1) A device used to determine the basis weight of paper and the total weight of paper rolls, or any other material in the mill. (2) A range of values assigned to instruments used for reading measurements of variables made in and around the pulp and papermaking process. (3) An imperfection in a sheet of paper caused by particles of paper and/or surface coating sticking to calender rolls, breaking off and adhering to other portions of the sheet, or any small, localized aggregate of foreign material, such as rust, dust, etc., becoming embedded onto the sheet. (4) Sometimes refers to the scale-like material that builds up in pipelines, black liquor evaporators, etc.

**SCALPING SCREEN:** A type of screen used to remove frozen clusters of wood chips in a flow stream of chips from storage to digesters in a pulp mill.

**SCARIFICATION:** A paper defect showing up as the breaking up and loosening of the surface coating and fibers.

**SCFH:** *See* STANDARD CUBIC FEET PER HOUR.

**SCHOOLMARM:** A standing tree that has split part way up into two separate trunks.

**SCHOPPER FOLDING MACHINE:** A widely used apparatus to test the folding endurance of paper. It was one of the first such instruments to be used for this purpose. *See* FOLDING ENDURANCE.

**SCION:** A shoot removed from a superior tree and grafted onto root stock to produce superior seed.

**SCMP:** *See* SEMICHEMICAL MECHANICAL PULP. Also is used with SULFONATED CHEMIMECHANICAL PULP.

**SCOOP TRUCK:** A material mover used in paper mills that is similar to a *forklift truck* except that it has a semicircular scoop instead of forks. It is used especially to handle rolls and boxes by picking them up from the floor in any position and depositing them in any other position.

**SCORE:** To put an indentation on the surface of a board or heavy paper so that it is easier to fold. Also referred to as *crease.*

**SCORING:** The process of producing an indentation on the surface of a board or heavy paper formed by passing between a grooved metal surface and a depressor, or with metal rollers, to facilitate folding the sheet along the line produced.

**SCOTCH PINE:** A species of deciduous tree used as a source of pulpwood by pulp mills.

**SCRAP:** Extraneous bits of paper such as trim that inadvertently become wound into a roll.

**SCRATCHES:** Surface defects that occur during the coating process of paper in a paper mill.

**SCREEN:** Equipment that separates desirable material from undesirable material by the principle of filtering through a wire mesh. Examples are screening sawdust from chips and dirt from pulp suspensions. It is used for white water cleaning, dewatering, and fiber recovery processes.

**SCREEN ROOM:** The area in a pulp mill where unwanted particles called tailings or rejects are separated from the accepted fibers with the use of equipment such as knotters, rifflers, refiners, separators, thickeners, and flat or rotary screens.

**SCREENED PAPER:** *See* CONDITIONED PAPER.

**SCREENING:** The process of separating undesirable sawdust, fines, and oversize chips from acceptable chips, or separating unacceptable dirt, fibers, shives, and knots from acceptable pulp in a pulp and paper mill by the use of screens.

**SCREENINGS:** Rejected materials such as knots, shives, and large bark particles from the screening operations of pulp suspensions in a pulp mill. This material is either dried and burnt or used in the manufacture of coarse wrapping papers or insulating boards.

**SCREW CHIP FEEDER:** A chip-feeding mechanism used to feed chips into continuous-type digesters under pressure. This is accomplished by the formation of a plug of chips compacted with a screw-type impeller.

**SCREW CONVEYOR:** A short-distance material mover used in a pulp and paper mill. Its primary moving element is a rotating spiral that pushes solid material such as pulp, chips, sawdust, coal, etc. from one point to another. Also called *spiral conveyor.*

**SCREW PRESS:** Liquid extraction equipment used to squeeze chips prior to refining in the manufacture of mechanical pulp. It is also used to thicken pulp from approximately 10 percent consistency to about 45 percent consistency by the use of a tapered compression screw rotor.

**SCRIP:** A name sometimes used for *currency paper.*

**SCRUBBER:** A large steel vessel used in some sulfite pulp mill acid plants in which the combustion of sulfur (S) to sulfur dioxide ($SO_2$) is completed, intimately mixed, and dust particles removed.

**SCRUBBING:** The removal of impurities from gaseous mill effluents by the use of a wash liquid, usually in the form of a spray.

**SCUFF RESISTANCE:** The ability of paper and paperboard to withstand the lifting of fiber when rubbed on another surface. It is measured by a laboratory instrument consisting of an element of fixed weight and size, and rotates at a fixed speed on one spot of a sample sheet. *See also* SURFACE SCUFF.

**SCUFFING:** The process of lifting the surface fibers when paper and paperboard are rubbed together or against other rough surfaces, especially when wet.

**SEAL PIT:** *See* SEAL TANK.

**SEAL TANK:** A receiving tank located beneath vacuum-type washers and filters. The water drops into it through a pipeline and forms a seal to create a vacuum in the sheet-forming cylinder portion of the unit. Sometimes referred to as a *seal pit.*

**SEAL WATER:** A water purge added to the packing gland of a pump or other rotary joints for the purpose of lubrication, cooling, and to prevent process fluids from coming in contact with the packing gland.

**SEALING PAPER:** Sheets of strong kraft or manila paper used in wrapping packages of paper and paperboard.

**SEALING TAPE:** Long strips of strong kraft or manila paper with an adhesive applied to one side so that it can be used for sealing boxes, packages, bundles, etc.

**SEAM MARK:** A paper defect caused by poor formation in the area of a defective or broken seam on the paper machine wire or on the dryer canvas. It appears as adjacent light and dark streaks on the sheet, running in the cross machine direction.

**SEAM PICKING:** A synthetic forming fabric problem in which a seam end may be pulled from its locked position in the seam joint and projected above the plane of the fabric surface causing picking of fibers and pinholes in the paper web.

**SEASONED PAPER:** *See* CONDITIONED PAPER.

**SEASONING:** Allowing the paper and paperboard to stand long enough for the moisture to come to equilibrium with the humidity of the surrounding air.

**SECOND GROWTH:** Timber that has grown up after a virgin stand was cut or destroyed by fire or insects.

**SECOND PRESS:** The second set of press rolls through which the wet sheet of paper passes after leaving the first press section of the paper machine.

**SECONDARY AIR:** That portion of the combustion air in some pulp mill chemical recovery furnaces that is admitted tangentially through registers in the upper side walls above the primary air but below the burners. It aids in the completion of black liquor burning in the upper furnace area.

**SECONDARY BLACK LIQUOR HEATER:** A heater in the black liquor feed system of a sulfate recovery furnace located between the discharge of the black liquor nozzle pump and the black liquor spray nozzles. It is used to raise the temperature of the black liquor, usually by direct contact injection steam or by indirect heat exchanging.

**SECONDARY COLORS:** Colors, such as green, orange, and purple, which are obtained when two of the primary colors (red, yellow, and blue) are mixed together.

**SECONDARY CONDENSATES:** Condensates from the process in a pulp and paper mill such as evaporators, digesters, etc., that are treated and reused.

**SECONDARY CREPE:** A heavy, crinkly effect on paper produced by rewetting the dry sheet and crowding it as it is passed over a doctor-fitted roll after it is made on a paper machine. This is usually done off-site in a converting-type operation. Sometimes referred to as *water crepe.*

**SECONDARY FIBER:** Any type of paper- and paperboard-making fiber obtained from wastepapers and other used, reclaimable fiber sources.

**SECONDARY GRINDING FACTOR:** *See* GRINDING FACTOR.

**SECONDARY HEADBOX:** A type of headbox designed to sit across the wet end wire of a fourdrinier type of paper machine, downstream from the normal primary headbox, for the purpose of applying and spreading a top layer of stock when running certain grades of paperboard.

**SECONDARY PULP:** Fiber suspensions obtained from repulping wastepaper broke and other reclaimable fiber sources (called *secondary fiber*).

**SECONDARY STOCK:** Paper machine stock furnishes made from pulp obtained from wastepapers, broke, and other reclaimable fiber sources (known as *secondary fiber*).

**SECONDARY WALL:** The inner shell of a wood fiber consisting of layers of cellulose fibrils or strands.

**SECONDARY WASTE TREATMENT:** Biological treatment of some pulp and paper mill effluents after sedimentation in a primary waste treatment plant.

**SECONDARY WASTEWATER TREATMENT:** A biological mill wastewater treatment operation that follows a primary sedimentation treatment.

**SECONDS:** Inferior, mill-sorted paper and paperboard with lower qualities than the standards established for specific grades and types, but which can be and are sold at reduced prices.

**SECTION:** A bank of dryer drums on a paper machine usually getting their steam from the same header and having a common drainage system. A typical paper machine dryer arrangement will incorporate several sections.

**SECTIONAL DRIVE:** The use of individual electric motors to run the various sections of a paper machine. They are adjusted and regulated to achieve the necessary related speeds for proper operation.

**SECURITY TYPE PAPER:** Grades of paper that have been prepared or treated to produce properties that would make it difficult to counterfeit. It is used for stocks, bonds, and currency papers.

**SEDIMENT:** Any material that settles out of pulp slurries, liquid solutions, treated water, wastewater, and other fluids.

**SEDIMENT VALUE:** A means of evaluating the settling rate of various types of fibers in suspension.

**SEDIMENTATION BASIN:** A large container in which wastewater is retained so that any suspended solids will settle by gravity and then can be removed.

**SEDIMENTATION TANK:** A vessel in which any precipitate, such as liquor containing calcium carbonate from a causticizing tank in a sulfate pulp mill, is left to settle out so that the clear liquid can be decanted from the top portion.

**SEDIMENTATION:** The settling of suspended solids from pulp slurries, liquid solutions, treated water, wastewater, and other fluids. It is usually accomplished by reducing the velocity of the liquid below the point where it can transport the suspended material.

**SEED HAIR FIBERS:** Types of papermaking fibers attached to and obtained from seed hulls of plants such as cotton.

**SEEDING:** The process of forming pigment aggregates from individual particles in paper coating mixtures.

**SELF-COLORED PAPER:** *See* NATURAL COLORED PAPER.

**SELF-DOCTORING ROLLS:** *See* SELF-SKINNING ROLLS.

**SELF-PEELING ROLLS:** *See* SELF-SKINNING ROLLS.

**SELF-SEALING PAPER:** Paper with a special adhesive coating that allows it to adhere to other sheets or to other surfaces by the use of pressure.

**SELF-SKINNING ROLLS:** Paper machine rolls characteristically possessing the property of easy sheet release due to special microporous neoprene coverings that introduce air between the roll surface and the sheet. Sometimes referred to as *self-doctoring roll* or *self-peeling roll.*

**SEMIBENDING:** A descriptive reference to paperboards, especially chipboard, that can be creased to a 90° angle bend without damage.

**SEMIBLEACHED KRAFT (SBK):** Sulfate pulp that has not been bleached to the extent that normally fully bleached pulp has. It is used to make up end products considered of lower quality.

**SEMIBLEACHED PULP:** Pulp that has been only lightly bleached to what is ordinarily considered a very low brightness range.

**SEMICHEMICAL:** A descriptive reference to any type of paper or paperboard made from fibers produced by a combination of chemical and unpressurized mechanical processes.

**SEMICHEMICAL MECHANICAL PULP (SCMP):** Pulp made by pretreating the chips at a temperature over 100°C and then refining them at atmospheric pressures.

**SEMICHEMICAL PULP:** A lower quality pulp made by cooking fibrous materials in a neutral sodium sulfite/sodium carbonate cooking liquor followed by a final separation of the fiber using unpressurized mechanical means. Also referred to as *neutral sulfite semichemical (NSSC) pulp.*

**SEMICHEMICAL PULPING PROCESS:** A two-step pulping process that uses a mild liquor, such as neutral sodium sulfite/sodium carbonate solution, for partial softening of chips, followed by a final separation of fibers by mechanical action. In this case, it would be referred to as a *neutral sulfite semichemical (NSSC)* pulping process.

**SEMICREPE:** *See* MACHINE CREPE.

**SEMICREPE TISSUE:** Lightweight paper used for toweling, table napkins, tablecloths, bathroom tissues, etc., having a surface finish that looks like crepe but without the physical characteristics associated with it.

**SEMIDULL FINISH:** A paper and paperboard surface that is smooth and even, but has low gloss or luster.

**SENSIBLE HEAT:** The energy required to change only the temperature of a material.

**SENSITIVITY:** The ratio of a change in output to the change in input of a device or process after a steady state has been reached.

**SENSITIZED PAPER:** (1) Chemically treated paper sensitive to light. It is used for photographic and blueprint reproduction. (2) Chemically treated paper responsive to elements such as heat, moisture, erasures, and other chemicals.

**SENSITIZING PAPER:** A base paper made especially to be treated with a sensitizing solution so that the surface becomes reactive to light, chemicals, or gases, and then used to manufacture such items as blueprint and photographic papers.

**SENSOR:** An element or device directly responsive to the value of a process measurement quantity.

**SEPARATOR:** Any filtering or sedimentation type of equipment that divides material mixtures into separate components, such as a solid from a liquid, a solid from a solid, and a liquid from a liquid.

**SEPTA:** Cross walls in pulpwood fibers.

**SEPTATE FIBER:** Pulpwood fibers that contain cross walls.

**SEQUESTERING AGENT:** Any chemical used to stabilize other chemical solutions, such as using magnesium sulfate ($MgSO_4$) to stabilize hydrogen peroxide bleaching liquor or to control the brightness and reversion effects of iron salts formed during the pulp-bleaching process. In effect, it is a type of *chelating agent.*

**SERIM:** The weave in plain woven felts used on paper machines.

**SERVIETTE PAPER:** *See* NAPKIN TISSUE.

**SET:** The transformation of adhesive paper coating material by either chemical or physical means to a fixed or hardened condition in which it becomes sticky enough to adhere to other surfaces.

**SET OF ROLLS:** A complete complement of all types of rolls used on a paper machine.

**SET POINT:** A manually or automatically set input variable to a process control system that sets the desired value to which the measured process variable is to be maintained.

**SETUP BOXBOARD:** Type of paperboard made especially to be used in the manufacture of rigid cartons in the shape that they are to be used.

**SETUP BOXES:** Rigid cartons made from paperboard in the size and shape that they are to be used.

**SETTLE TIME:** Time needed for a cyclic twin motor sheeter to synchronize to material speed after the cutter motor is cycled to accommodate the length against the circumference distance.

**SG:** *See* SPECIFIC GRAVITY.

**SGW:** *See* STONE GROUNDWOOD.

**SHADE:** The color depth, strength, and hue in the comparison of papers of the same color.

**SHADOW MARKS:** Paper defects made by low spots in a roll, such as couch roll hole patterns, leaving thicker, more translucent areas that produce a darker appearance at these points.

**SHAFT:** (1) A rotating, cylindrically shaped, solid rod or bar equipped with rotating pieces or pulleys for the purpose of transmitting power or motion in mill machinery and operations. (2) A metal, wood, or convolute-wound fiber tube used to wind paper or paperboard into rolls for transporting or shipping purposes. Also called a *core.*

**SHAFT PULLER:** The unit of a roll-handling system used to remove the winder shaft from a paper roll in a paper mill.

**SHAKE:** An eccentric stroking device driven by a variable speed motor and connected to the wire section of slower speed paper machines to move the wire section from side to side, thus improving the orientation of fibers during the formation of the paper web.

**SHANKERS:** *See* PIGTAILS.

**SHARPENING LATHE:** A pulpwood grindstone sharpening device fitted on grinders so that the stone can be resurfaced by the application of a properly surface-designed burr or sharpening tool.

**SHATTER SPRAY NOZZLE:** A fishtail-like nozzle used on sulfate pulp mill chemical recovery furnaces to discharge a heavy stream of recirculated green liquor into the stream of molten smelt issuing from the smelt discharge spouts. It is used to break up and disperse the molten smelt before it hits the surface of the liquor in the receiving smelt dissolving tank. *See also* STEAM SHATTER NOZZLE, which it is sometimes called.

**SHAVING BOARD:** A type of board made by bonding wood shavings together with a suitable adhesive.

**SHAVINGS:** Narrow strips cut from the sheet on the paper machine and during the finishing operations. They are recycled back to the process to become part of the broke pulp. Also refers to thin slices of wood pared off with a knife or other cutting equipment.

**SHEAR:** (1) The hydraulic cutting blade on a mechanical pulpwood harvesting machine. (2)

A refining action that results in two contiguous parts of the fiber sliding past each other, in a direction parallel to their plane of contact. (3) The cutting action of the impeller of a pump as it passes through the material trapped within the casing, particularly at the cut-water point.

**SHEAR CUT WINDER:** A type of winder used in a paper mill to wind paper as it passes from the unwinding stand, over tension rolls, between upper and lower slitter disks, then over other rolls and spreader bars or rolls, to two matched winding drums where the cut sheet is wound around a core shaft.

**SHEAR THICKENING:** The apparent increase in volume of slurries, such as clay and other high solid content coating materials, when a shear is applied. It is often experienced in pumps used to move the slurry. *See also* DILATANT.

**SHEAR THINNING:** *See* THIXOTROPY.

**SHEAR-TYPE AGITATOR:** A type of agitator that not only mixes various emulsions, solutions, and pulp suspensions in pulp and paper mills, but also breaks up the particles.

**SHEATHING:** Various sizes and thicknesses of structural and insulating fiberboard used in building construction. Also refers to book or newsprint paper used as a liner for chip and straw board.

**SHEAVE:** A grooved wheel, usually found on a pulley, or the wheel in which the rope or belt rides in a paper machine drive assembly.

**SHED:** A descriptive number of machine direction strands used in a paper machine forming fabric weave pattern before the pattern repeats itself.

**SHEET:** A flat piece of any type of pulp, paper, and paperboard of a variety of characteristics, sizes, and finishes.

**SHEET BOW:** Edges of sheets that are not perfectly straight, but slightly curved, as the result of the type of cut used on the paper.

**SHEET CALENDERED PAPER:** Paper sheets whose surfaces are finished or glazed by passing them separately through a calender rather than in the conventional continuous web.

**SHEET CUTTER:** (1) A cutting machine consisting of a single sharp guillotine knife blade. It is used in the paper-finishing operation to cut and trim reams of paper to order sizes. (2) The roll-cutting machine used to slit paper rolls into smaller widths and then crosscut the web into sheets before the trimming-to-size operation.

**SHEET DRYER:** Type of drying equipment that dries individual sheets of paper laid flat or hung up on racks. A *loft dryer* is an example.

**SHEET FORMATION:** The degree of fiber dispersion and orientation of a web of paper developed on the wet end of a paper machine as the water is drained away. Also, the act of forming a mass of fibers into a sheet.

**SHEET LINED:** A type of paperboard to which paper has been pasted on either one or both sides.

**SHEET LINER:** A liner applied to paperboards after they are cut into sheets.

**SHEET MOLD:** An integral part of a sheet-forming device over which a porous fabric or metallic screen is placed to provide a surface for water drainage and individual sheet formation from pulp slurry poured over the mold.

**SHEET SIZE:** Dimensions of standard sheets that have been established by the industry for the various grades of paper. Sometimes referred to as *paper size*.

**SHEETER:** A paper finishing machine that automatically cuts and slits a roll of paper to produce and deliver sheets in cartons, reams, or on skids.

**SHEETING:** The paper mill finishing room operation in which the paper web is cut up into individual sheets.

**SHELF PAPER:** Paper treated to resist oil and water, and made specifically for converting into decorative, protective coverings for shelves.

**SHELL MARK:** Paper defect caused by rough edges in the holes of the couch roll or suction pickup roll and the use of excessive vacuum on these rolls on the paper machine.

**SHELTERWOOD LOGGING:** A method of harvesting pulpwood leaving scattered selected trees to provide seed for regeneration and shelter for seedlings.

**SHIFT:** A period during a day in which a crew of operating personnel work. In a 24-hour pulp and paper mill operation, there may be 3 or 4 shifts of 8 or 6 hours in length, with a new crew taking over each period. Also called a *tour*.

**SHINERS:** Transparent spots in a sheet of paper caused by squeezing partly cooked fiber bundles and splinters during the calendering operation of papermaking.

**SHIPPING ROOM:** The area of a paper mill from which paper to be shipped to customers is loaded onto trucks and railroad cars.

**SHIPPING SACK PAPER:** A very high-strength paper made from kraft pulp according to strict specifications. It is used for converting into *multiwalled* protective shipping bags.

**SHIRT BOARD:** A moderately rigid chip or news type of paperboard used as support for packaging shirts after laundering.

**SHIVES:** Uncooked or undisintegrated bundles of fibers or splinters in wood pulp that may show up as imperfections in the finished sheet of paper made from it, if not previously removed. It is also found in groundwood pulp. Sometimes called *slivers*.

**SHOE BOARD:** Paperboard made from waste and chemical pulps and sized with rosin, rubber latex, wax, etc., to impart characteristics suitable for use by shoe manufacturers to make counters, stiffeners, and in building up heels.

**SHOE PRESS:** (1) A type of paper machine press section arrangement that utilizes a curved "shoe" piece to force the *wet felts* and sheet against a top press roll. The shoe section usually has thick belts that conform directly to the shape of the roll. By expanding the contact area between the sheet and press roll, the nip is proportionally extended, dramatically improving overall press section performance. (2) The last press in a series of presses used on the wet ends of board, pulp, and paper machines. *See also* EXTENDED NIP PRESS.

**SHOOT:** Same as *blow* when used to mean the discharge of contents of a digester at the end of a cook. *See also* BLOW.

**SHOPS:** Rooms or areas in a pulp and paper mill where repairs are carried out, such as machine, pipe, electrical, and instrument shops.

**SHORT-DWELL COATER:** A coating head developed primarily for thin papers to provide better runnability than conventional blade coaters by utilizing an optimum combination of forces on the blade, water transport from color to paper, and flow characteristics.

**SHORT GRAIN PAPER:** Paper made with the machine direction in the shortest sheet dimension.

**SHORT-SEQUENCE BLEACHING:** The process of obtaining fully bleached pulp brightness with acceptable strength and cleanliness using a minimum number of bleaching stages, usually three or four.

**SHORT SHEETS:** Sheets of paper found in a packaged ream that have lesser dimensions than the standard for the grades of paper involved.

**SHORT STOCK:** Pulp consisting of natural short fibers or fibers that have been mechanically reduced to considerably shorter lengths by cutting in beaters and refiners.

**SHORTENING:** *See* CUTTING.

**SHORTWOOD:** Wood logs that have been cut into lengths suitable for use in pulp mills.

**SHOT-PEENING:** The bombardment of the surface of metal components of pulp and paper making equipment to induce residual compressive stress, thereby counteracting the destructive effects of tensile stress and resulting metal fatigue.

**SHOW-THROUGH:** A reference to the resistance of a paper sheet to transmission of both diffuse and non-diffuse light so that any printing on one side cannot be seen from the other side. *See also* OPACITY.

**SHOWCARD BOARD:** Type of coated paperboard usually made from newsprint and sometimes lined with paper. It is used for window and counter merchandising displays.

**SHOWER WHIPPER:** A type of conditioner used on cylinder machine pickups or top felts. It consists of a high pressure shower located within the felt loop and a whipper equipped with bars that beat the felt approximately four times per foot immediately after the showers.

**SHOWERS:** Water jets or sprays used throughout pulp and paper mills to wash wire mesh screens, wires, wet felts, and pulp pads on paper machines, cylindrical-type washers, pulp screens, pulp drainers, etc.

**SHREDDER:** A machine used in the mechanical treatment of wastepaper to open the pages of magazines, books, pamphlets, etc., and to reduce the sheets to smaller pieces, making them suitable for further treatment.

**SHRINK:** (1) The total furnish actually lost during the manufacture of paperboard. (2) The drop in a steam boiler liquid level occurring on load drop when a decrease in volume caused by an increase in drum pressure and a decrease in firing rate results in a "shrinkage" effect.

**SHRINKAGE:** Dimensional decrease of a paper sheet, loss of moisture, or decrease in weight between stock furnish to a paper machine and the paper made from it. It is also used to define a decrease in sheet width from couch (wet end) to reel (dry end), expressed in percent.

**SHRUNKEN EDGE:** An undesirable phenomenon occurring when just the edges of paper sheets in a pile have contracted due to moisture loss, causing a dishing effect in the sheet which raises the corners. Sometimes called *tight edges.*

**SHUNT:** A resistor connected in a circuit to be measured that is parallel with the current circuit of the measuring instrument.

**SHUT IN:** To shut down and bottle up a pressurized, continuous-type digester in a pulp mill for a period of time without any loss of pressure.

**SHUTDOWN:** Planned or unplanned stoppage of the paper machine, another process, or the entire pulp and paper mill for repairs, alterations, washing, cleaning, lack of orders, or any other reason.

**SHUTE WIRES:** The individual wires or fabric strands (also known as *weft wires*) that make up the wire screen or forming fabric on a fourdrinier wet end. They run from side to side, or in a cross-machine direction, and cross the wires or fabric strands that run parallel to the machine direction called *warp wire.*

**SIDE CUT:** A paper roll defect in which the edges of the sheet are cut by markings or other causes.

**SIDE HILL DECKERS:** A screening device used to thicken rejects in groundwood mills.

**SIDE HILL SCREEN:** An steeply inclined screening device commonly used in deinking systems. As pulp slides down the large screen deck by gravity, ink is drained off with the water.

**SIDE RELIEF:** The removal of gas and steam to lower the pressure in a sulfite batch digester from hold to blow pressure in order to recover excess sulfur dioxide ($SO_2$) at the end of the cook. Also referred to as *blowdown, gasdown,* and *pressure blowdown.*

**SIDE ROLL:** A roll of paper made during a side run to utilize the full deckle of the paper machine on a particular order.

**SIDE RUN:** The making of a roll of paper of shorter length than the others, slit on the rewinder for a specific order to utilize the full deckle of the paper machine. It is sometimes sold at a discount price.

**SIDERUN NEWS:** Odd size rolls of newsprint paper.

**SIDE TRIM:** Combined board trimmed as waste at the slitter stage in the processing of paperboard.

**SIDES:** The surfaces of felts, wires, and a paper sheet associated with distinguishing references to the upper and lower sides.

**SIENNA:** A natural form of ferric hydrate used to make a lightfast paper dye that is resistant to acids, alkalies, and chlorine.

**SIEVE TEST:** An analysis made on pulpwood chips to determine the size proportions by running a representative sample through different size screens.

**SIFTING:** A method of producing dry-formed paper by forcing fiber through a sieve onto a moving wire with suction applied under the bottom surface.

**SIGHT GLASS:** A direct visual indicator of liquid level, consisting of a transparent glass or plastic tube attached to vessels in such a way that the liquid height on the tube is equal to the level of liquid in the vessel.

**SIGN BOARD:** A thick paperboard, usually made from used paper and chemical pulp, with properties so that it can be used for making outdoor advertising poster signs.

**SIGN PAPER:** A machine-finished, durable, nonfading type of paper, usually made from bleached chemical pulps and surface-sized for general poster and billboard advertising purposes.

**SILENT PAPER:** *See* NOISELESS PAPER.

**SILHOUETTE PAPER:** Black paper base used to cut out silhouettes.

**SILICA:** Objectionable silica compounds that volatilize and boil off with the steam produced in a boiler. Under certain conditions they will condense out and glass-coat turbine blades and process equipment.

**SILICATED PAPER:** A type of paper whose surface has been hardened and finished by coating it with a sodium silicate mixture.

**SILK BOARD:** A thick pasted chip board that is pasted and lined with colored paper and used as a core to wind silk into bolts.

**SILK PAPER:** Special writing paper containing silk threads that serve to make it difficult to duplicate.

**SILK WRAPPING PAPER:** A duplex-colored paper used to wrap silk cloth. It is usually made on a multi-cylinder paper machine from a mixture of chemical pulps.

**SILO:** (1) A tall, narrow storage tank for solid materials, such as chips, clay, and other papermaking substances, that are fed into the top and withdrawn from the bottom. (2) A holding tank in the white water system of the wet end of a paper machine, located under the breast roll end of the wire, that receives the white water as it drains through the wire, away from the forming paper web.

**SILVER LABEL PAPER:** A printing-type paper used for labels. It is made by either mixing aluminum powder in the furnish or by coating the sheet with aluminum powder. *See also* ALUMINUM PAPER.

**SILVER PAPER:** *See* TINFOIL PAPER.

**SILVER TISSUE PAPER:** Lightweight paper treated with copper salts or other inhibitors so that it will not tarnish silver items. It is used as a wrapping or packaging material. *See* ANTITARNISH PAPER.

**SILVICHEMICALS:** Chemical by-products of the wood-pulping process and other chemicals derived from wood.

**SILVICULTURE:** The science of forest management.

**SINGLE COATED PAPER:** Paper that has been surfaced with only one layer of coating material on one or both sides.

**SINGLE CYLINDER MACHINE:** A type of papermaking machine commonly used to make paperboards. It is characterized by having only one wire-covered, partially immersed cylinder in a vat of stock suspension on which the matted sheet is formed.

**SINGLE FACE CORRUGATED BOARD:** A type of corrugated paperboard with only one single liner pasted to the corrugated medium. It is used primarily as protective wrapping in packaging fragile materials.

**SINGLE FACED ROLL:** A roll of corrugated paperboard lined on only one side of the corrugated medium, and used for protective packaging.

**SINGLE FELTED PRESS:** A type of press used on a linerboard machine which has only one felt.

**SINGLE LOG BARKER:** A type of pulpwood bark removal equipment that removes the outer bark from logs, one at a time.

**SINGLE MOTOR PAPER MACHINE DRIVE:** The use of one big adjustable or fixed speed electric motor to run the various sections of a paper machine through a shaft or rope-type drive.

**SINGLE PLY:** A descriptive reference to any solid paper or paperboard sheet made up of only one layer.

**SINGLE SHEET CUT:** A paper finishing operation in which paper sheets are cut individually.

**SINGLE STAGE CENTRIFUGAL PUMP:** A type of pump used in pulp and paper mills in which the total head is developed by one impeller.

**SINGLE STAGE HYPOCHLORITE BLEACHING:** The simplest chemical pulp bleaching process, consisting of only one operation, using either calcium or hypochlorite bleaching solution.

**SINGLE TIER DRYING:** A dryer system arrangement in which the sheet passes over only a single row or bank of traditional dryer drums, rather than serpentine fashion over alternating top and bottom dryer drums. This reduces draw in the dryers and reportedly boosts overall drying efficiency.

**SINGLE WHITE VAT-LINED BOARD:** A two-ply paperboard made on a cylinder machine with one white layer made from chemical pulp and white wastepapers, with a backing made from old used papers.

**SISAL HEMP:** Fiber obtained from a plant found in areas with moderate temperatures. It is used in the manufacture of ropes and textiles which become a waste material fiber source for the manufacture of certain types of paper and paperboard. It is sometimes used directly for papermaking.

**SIXTH HAND:** A member of the paper machine crew who is under the supervision of the machine tender but generally receives instructions from the back tender and the third hand. His duties include keeping the general floor areas around the machine clean, emptying the broke trucks, maintaining general neatness, and being knowledgeable of the work and duties of the third and fifth hands on the paper machine crew.

**SIZE:** A substance, such as rosin, gelatins, glues, starch, waxes, etc., added to paper stock furnish or to the surface of a sheet made from it. It is used to impart ink- and water-repelling properties. Also called *paper size.*

**SIZE HYDROLYZATE:** Waxy substances formed when a synthetic size used in the wet end of alkaline paper machines reacts with water, in addition to carbohydrates in the fiber. Wet end deposits, press picking, and problems in converting the paper can be caused by the buildup of these substances.

**SIZE PRECIPITANTS:** Chemicals, such as alum, copper, sulfate, and other copper salts, added to papermaking stock furnishes to precipitate rosin size out of solution and onto the fibers to impart water-resistant properties to the paper.

**SIZE PRESS:** A paper mill processing unit consisting of two, usually rubber-covered, rolls located between two dry end sections of the paper machine. It is used to apply size solution to the surface of the paper sheet.

**SIZE SPECKS:** Paper defects appearing as transparent spots in the sheet. They are caused by agglomerated size material in the stock furnish which did not disperse in the papermaking process.

**SIZE TUB:** A long, shallow vat used in applying size materials to the surface of a paper sheet by passing it through a glue or starch solution and squeezing off the excess through two press rolls. *See also* TUB SIZE PRESS.

**SIZING:** (1) A paper mill operation in which materials are either added to the stock furnish before going to the paper machine or to the surface of the paper sheet on the paper machine in order to enhance its resistance to liquid penetration and improve its abrasive resistance, creasability, finish smoothness, and surface characteristics. (2) The resistance of paper to water and vapor penetration.

**SIZING AGENTS:** *See* SIZE.

**SKATING:** A web-forming phenomenon occurring on the paper machine wet end wire caused by the localized stock flow interferences at the slice. It results in streak-like formations on the sheet surface which appear to move from side to side in the across machine direction.

**SKETCHING PAPER:** Types of paper especially made with a surface finish for use in making rough pencil and crayon drawings.

**SKID:** (1) A paper-moving and shipping platform fitted with runners of appropriate height and separation width to enable the use of forklift trucks to move them. (2) The operation of dragging tree lengths of logs from the cutting site to a collecting point.

**SKIDDER:** A special type of tractor used to drag pulpwood logs from the cutting area to the loading area.

**SKIM:** (1) The removal of the top layer or scum from the surface of a liquid. (2) The thin top or bottom layer of paper on a lined sheet of paper or board.

**SKIM TEST:** A paper test procedure used in a paper mill as a means to determine the water resistance of paper by the amount of curl when wetted on one side.

**SKINNING STICK:** A wooden tool designed and used for "skinning" layers of paper off rolls in a paper mill.

**SKIPPED COATING:** Imperfections in the surface coating of paper caused by small, random, localized disturbances of the application process.

**SL:** *See* SPENT LIQUOR.

**SLAB:** A multi-sheet slice of paper off the outer surface of a roll of paper that is removed to reduce the size of a roll, or to break down the roll for reprocessing or removal of certain defects.

**SLAB WOOD:** *See* MATURE WOOD.

**SLABBING:** A pulp-forming operation performed on a wet pulp machine in which a thick sheet of pulp is formed around the outer surface of a roll and cut off in sheets for shipment.

**SLABBING DOWN:** The act of removing a multi-sheet slice of paper off the outer surface of a roll of paper for the purpose of reducing the size, breaking down the roll for processing, or removal of certain defects.

**SLACK AREAS:** Paper defects consisting of localized areas of high moisture content, producing a web with nonuniform draw and designated by terms denoting their position in the sheet. *See also* BAGGY PAPER.

**SLACK CENTER:** Paper defect consisting of localized high moisture content areas situated in the center of the sheet causing a nonuniform draw on the web. *See also* BAGGY PAPER.

**SLACK EDGES:** Warped or wavy edges on sheets of paper, usually caused by the edges being thinner than the center, resulting in soft ends when wound up into rolls. It also occurs during the coating operation if the middle of the sheet is under higher tension than the sides.

**SLACK END:** A paper defect referring to a web with nonuniform draw and caused by high moisture content areas at the ends of the sheet. *See also* BAGGY PAPER.

**SLACK ROLL:** A reel of paper that has been wound with insufficient tension, resulting in a loose roll.

**SLACK SIZED PAPER:** Paper made from stock furnish with less than the usual amount of sizing material. This produces a soft, slightly water-absorbent sheet having a below-normal finish.

**SLACK START:** Inadequate application of sheet tension at the start of a roll-winding operation, causing an abnormally loose wound roll.

**SLAG TANK:** A water-filled sump beneath a coal burning power furnace. It is used to sluice ashes from the wet bottom or receive the melted ash with low fusion temperatures.

**SLAKED LIME:** A lime slurry produced by the hydration of lime (CaO) when mixing it with water, forming calcium hydroxide [$Ca(OH)_2$].

**SLAKING:** Generally, the mixing of water with lime (CaO) to form calcium hydroxide [$Ca(OH)_2$]. Specifically, in the paper industry it is the reaction in the chemical recovery cycle of an alkaline pulp mill.

**SLAKING/CAUSTICIZING:** A two-stage chemical process in the causticizing plant of an alkaline pulp mill in which the sodium carbonate ($Na_2CO_2$) in the green liquor is converted to sodium hydroxide (NaOH) to produce white liquor. The first stage is slaking, which consists of the addition of lime (CaO) to green liquor where it reacts with water to form calcium hydroxide [$Ca(OH)_2$]. The second stage is causticizing, in which the calcium hydroxide reacts with the sodium carbonate to form sodium hydroxide. Both stages overlap.

**SLAP:** *See* FLUTTER.

**SLASH:** A pulpwood logging term referring to the debris, such as limbs, tops, culls, etc., left over after the tree has been cut down and removed from the woodland.

**SLASHER DECK:** That section of a pulp mill's woodyard than cuts lengths of wood (sometimes tree length) into lengths suitable for processing in the subsequent debarking and chipping operations. It consists of a large metal deck with regularly spaced circular saws. Logs slide down the deck by gravity and are "slashed" to size by the saw blades in the process.

**SLAT DRYING:** The drying of paper by the use of air in some paper mills. It is done by suspending the sheets over moving wooden staves, supported by and carried on metallic frames with radiating arms. Also called *spar drying.*

**SLICE:** A long, flat outlet located on the bottom side of the headbox from which the stock furnish is fed to the wire (forming fabric). It extends across the entire width of the wire and can be adjusted to regulate the flow of stock onto the wire and thereby control the thickness of the sheet of paper produced.

**SLICE HEAD:** The total pressure behind the slice on a paper machine headbox made up of the height of the stock plus any added padding pressure. It affects the speed at which the stock is delivered onto the wire.

**SLICE MARK:** Paper defect caused by a malfunction or misadjustment of the slice on the paper machine. It appears as a random translucent area and surface unevenness of the sheet.

**SLICE VELOCITY:** The speed at which a paper machine headbox slice delivers the stock onto the wire of the wet end. Often referred to as *spouting velocity.*

**SLICK FINISH:** A smooth, glossy surface finish on paper.

**SLIME:** An undesirable, slippery, glutinous film formed by microorganisms and the agglomeration of nonbiological matter. It is found throughout the pulp and paper stock flow and storage system.

**SLIME HOLES:** Small undesirable holes with translucent edges formed on the paper. They are usually caused by the breaking through of dried, entrained slime in a sheet as it is subjected to high pressure when passing between the rolls of a calender stack.

**SLIME SPOTS:** Dark grayish defects in paper showing up as smear spots in the sheet as a result of fungi and microorganism growth in the pulp stock furnish carrying over through the paper machine.

**SLIME TOXICANTS:** Chemical poisons added to stock slurries to prevent the bacterial growth that produces slime.

**SLIMICIDE:** Toxic chemical substance added to the pulp and paper process to inhibit the growth of undesirable microorganisms that cause slime.

**SLIP:** (1) Small, narrow strip of colored paper inserted in the ends of a roll of paper as it is wound. It is used to identify locations of defects or breaks in the sheet. (2) A semi-solid liquid such as clay and pigment-water mixtures used in papermaking.

**SLIP HOLE:** A light sheet defect made on a cylinder-type paper machine, appearing as a hole, and caused by centrifugal force at speeds of 300 to 400 feet per minute.

**SLIP RATIO:** The ratio, expressed as a percentage, of the difference between operating speed and synchronous speed to the synchronous speed of an induction-type motor.

**SLIPPED ROLL:** A paper roll defect in which one end is cone-shaped because of slippage of the core. Also referred to as *pencil point* and *telescoped roll.*

**SLIPPING:** A cylinder-type paper machine operating problem caused by insufficient differential of head inside and outside the cylinder mold and the use of a stock that is too free.

**SLIPS:** A common paper mill reference to a clay slurry with proper dispersant used to make up paper sheet coating preparations.

**SLIT SCORE:** A cut made in a sheet of fiberboard that does not go all the way through. It is used to facilitate bending.

**SLITTER:** A paper web cutting assembly located on the winder or rewinder of a paper machine. It consists of suitably arranged mating pairs or a single set of rotating circular knives riding on a drum between which the paper is threaded and cut to make up large rolls into shorter rolls of specified widths.

**SLITTER DUST:** Tiny, coated fiber fragments that break off the edge of a sheet of paper during the slitting operations, and may cause problems in finishing and converting operations later.

**SLITTER EDGE:** The ends of paper produced on the sheet when rolls are cut on a slitter on a paper machine.

**SLITTER EDGE CURL:** Rippled- or ruffled-like paper edge resulting from improper adjustment of the paper slitter knives, caused by the knives being too far into the sheet.

**SLITTER RINGS:** Colored or shadow-like rings that appear on the roll edge in a concentric pattern due to defective or misaligned slitter knives. *See also* SLITTER RUNOUT.

**SLITTER RUNOUT:** A roll defect caused by a wobble in a slitting component, appearing as a wave on the edge of the sheet. *See also* SLITTER RINGS.

**SLITTING:** Converting rolls of paper as they are produced on a paper machine into shorter rolls of specified widths. It is done by cutting the web in the machine direction using circular, flat-faced, rotary knives during the winding operation on a paper machine, or as an off-machine finishing room operation.

**SLIVER DETECTOR:** A device used in groundwood mills to indicate the presence of bundles of fibers and splinters in the pulp by catching them and causing an overflow condition. It consists of a coarse screen or wire mesh.

**SLIVER SCREEN:** A pulp-cleaning, flat type screen or classifier used in groundwood pulp mills to remove undisintegrated bundles of fibers and splinters from the pulp stock.

**SLIVERS:** *See* SHIVES.

**SLOT:** *See* GRATE.

**SLOT PAPER:** A dense paper with a high, glossy finish made from cotton pulp. It is used in making formed electrical insulation.

**SLOW SHEET:** A sheet of paper made from pulp stock furnish which has been given an extended mechanical treatment in beaters or refiners so that water drains from it slowly.

**SLOW STOCK:** Pulp stock which tends to slowly release water on drainage as the result of mechanical treatment in beaters or refiners.

**SLOWNESS:** The ability of a pulp and water mixture to retain water on drainage. The opposite condition is called *freeness* or *wetness* and refers to the ease with which the mixture or slurry will drain water.

**SLUDGE:** Solid material filtered out of mill wastewater. It is either disposed of in landfill operations or burned in power boilers.

**SLUDGING:** A paper blade coating defect consisting of a viscose, pastelike buildup at the nips and along the inside of the blade, which breaks away and adheres to the surface of the paper web as it passes through the nip.

**SLUG HOLE:** A paper imperfection caused by an aggregate of fibers in the sheet that is lifted during the paper machine drying or pressing operation, leaving a hole in the web.

**SLURRY:** (1) A liquid mixture made up of suspended fibers, fillers, coating pigments, and other solid material in water or adhesive used in papermaking. (2) Any pumpable mixture of solids and fluids in the pulping and papermaking processes.

**SLUSH:** A pulp stock-water suspension thin enough to flow or to pump through a pipeline, usually running about 1 to 6 percent consistency.

**SLUSHED PULP:** A pulp/water suspension that is transported from the pulp mill to the paper mill by pumping through pipelines.

**SMASHED BULK:** The compactness property of a stack of sheets of compressed book paper (specific number of sheets) with essentially no air between them.

**SMELT:** The molten chemicals that collect at the bottom of the sulfate recovery furnace *(hearth)* in a pulp mill. They consist primarily of sodium sulfide ($Na_2S$) and sodium carbonate ($Na_2CO_3$).

**SMELT BED:** The accumulation of chemicals that form a pile at the bottom of the sulfate recovery furnace. The pile is drawn off as it becomes molten for further processing in the pulp mill chemical recovery cycle.

**SMELT DISSOLVING TANK:** A metal vessel located under a sulfate recovery furnace into which the molten chemicals flow. There they are quenched and dissolved in water or weak green liquor from the causticizing area to form fresh strong green liquor in the chemical recovery cycle.

**SMELT FURNACE:** A furnace used in the soda recovery process in which the combustion of black ash is completed and converted to molten chemicals that are reused in making up cooking liquor.

**SMELT SPOUT:** A water-cooled, jacketed trough through which the molten smelt is drained from a sulfate black liquor recovery furnace into a dissolving tank.

**SMELT SULFIDITY:** Sodium sulfide ($Na_2S$) content in a sulfate chemical recovery furnace smelt. It is expressed as percent $Na_2O$ and calculated by the equation:

$$\left(\frac{Na_2S}{Na_2S + NaOH + Na_2CO_2} \times 100\right)\%$$

**SMILE PROFILE:** A cross-machine moisture scan measurement made on the felt in the press section of a paper machine. It produces a gradual positive to negative to positive shaped trace record. It is the opposite of *frown profile.*

**SMOKE:** The mixture of gaseous and particulate emissions from power and recovery boiler stacks. It contains solid particles generated as a result of incomplete combustion of materials containing carbon.

**SMOKE STACK:** A vertical pipe, flue, or chimney designed to exhaust gases and suspended particulate matter into the atmosphere from a power and recovery furnace in a pulp mill.

**SMOOTHING PRESS:** A paper machine press consisting of a set of hard-surfaced rolls located close to the wet end section or between the dryer sections. It is used for ironing out and densifying the sheet while it is still damp.

**SMOOTHING ROLLS:** Press rolls on a paper machine wet end that are also located between dryer sections to iron out the sheet of paper while still damp.

**SMOOTHNESS:** The surface property of a paper sheet related to its degree of uniform evenness and flatness. It is sometimes measured by the rate of flow of air between two or more sheets or between one sheet and a smooth standard surface.

**SMOOTHNESS TESTER:** A common paper mill laboratory instrument used to determine the degree of flatness and evenness of a paper sheet. Instruments have been developed to continuously make this measurement on the paper machine.

**SMOTHERED WATERMARK:** A type of paper sheet watermark design that covers the entire surface area.

**SNAILING:** Paper machine wet end markings or streaks caused by the accumulation of air bubbles and water at the slice.

**SNAKING:** Streaks of higher moisture that meander back and forth in a sheet of paper. They are caused by a drainage disturbance or by a too slack or too stiff paper machine wire (forming fabric), resulting in excessive water beneath the paper sheet.

**SNAP:** (1) The ability of paper and paperboard to immediately return to their original shape when bent and released. (2) A reference to the difference in gloss between a coated sheet surface and the ink that is printed on it. Also called *delta-g* or *print snap.*

**SNAP-OFF:** Paper sheet breaks occurring inside a roll of paper during the winding operation due to cuts, wrinkles, and spots that are too dry or too hard, or the use of excessive sheet tension. They cannot be seen from the ends of the roll.

**SNOW:** A reference to fluffy, low density particles issuing from the stack of a sulfate recovery furnace.

**SNOWSTORM:** The appearance of paper with a wild, nonuniform formation. Also called *wild formation.*

**SOAKING ZONE (SZ):** The upper part of a continuous-type vertical digester where chips are soaked in cooking liquor.

**SOAP:** The product of reaction between the alkali in sulfate cooking liquor and fatty acid portions of wood, which precipitate out of the spent black liquor when water is evaporated from it. The soap is separated for making tall oil as a by-product.

**SOAP SKIMMING TANK:** A storage tank used with black liquor sulfate evaporators. It is designed so that the partially evaporated black liquor between the effects is allowed to stand while the soap floats to the top, where it is scraped off and sent to a tall oil plant for processing. The remaining liquor proceeds to subsequent evaporator effects for further evaporation.

**SOAP WRAPPER PAPER:** An alkali-proof paper made from chemical pulp that is treated to prevent the growth of fungi and other microorganisms, and is used to make a variety of wrappers for bars of soap.

**SOAPSTONE:** A naturally occurring magnesium silicate mineral that is pulverized and used to make a softening filler material called *talc* or *Spanish chalk.* It is used in some grades of cheaper writing papers, envelope papers, papeteries, etc.

**SODA:** (1) Sodium carbonate ($Na_2CO_3$). (2) A word used to designate pulp made by the soda process.

**SODA ASH:** The commercial form of anhydrous sodium carbonate used in pulp mills.

**SODA PROCESS:** A chemical pulping process that consists of the reduction of chips to their individual fiber components by use of cooking liquor made up of caustic soda solution, the recovery and preparation of this liquor, or the treatment of pulp and paper produced from it.

**SODA PULP:** Pulp made by the cooking of chips from deciduous or broadleaf trees in a sodium hydroxide or caustic soda solution.

**SODA STRAW PAPER:** A high strength paper made from bleached chemical wood pulp that is treated with paraffin and used to make drinking straws.

**SODIUM ALUMINATE ($Na_2Al_2O_4$):** (1) A chemical used in a paper mill in conjunction with alum to control pH of stock to the paper machine for more efficient sizing and for better filler and fiber retention. (2) A chemical used as a sizing material in papermaking to provide water and ink penetration resistance in the sheet.

**SODIUM BISULFITE ($NaHSO_3$):** An acid-type cooking liquor chemical component sometimes used to neutralize chlorine residual in the pulp during the bleaching process.

**SODIUM CHLORATE ($NaClO_3$):** A raw material used by bleach plant chlorine dioxide generating plants. It is treated with sulfuric acid ($H_2SO_4$) to produce chlorine dioxide gas ($ClO_2$) for the bleaching process.

**SODIUM DITHIONITE:** *See* SODIUM HYDROSULFITE.

**SODIUM HYDROSULFITE ($Na_2S_2O_4$):** A chemical used as a bleaching agent for mechanical and deinked pulps. It may be generated on-site from sodium borohydride ($NaBH_4$).

**SODIUM HYPOSULFITE:** *See* SODIUM THIOSULFATE ($Na_2S_2O_3$).

**SODIUM HYDROXIDE (NaOH):** A strong alkali-type chemical used in making up cooking liquor in alkaline pulp mills. It is commonly referred to in the mill as caustic, caustic soda, or lye.

**SODIUM HYPOCHLORITE (NaOCl):** A chemical used as one of the bleaching agents in some multi-stage pulp mill bleach plants.

**SODIUM LOSSES:** The total sodium expressed in the equivalent of $Na_2O$ lost during the alkaline pulping process. It is usually calculated as a percentage of the white liquor to the digester or the actual loss per ton of air dry pulp produced.

**SODIUM MAKEUP:** All sodium chemicals, such as sodium sulfate, caustic, sodium carbonate, etc., that have to be added to the alkaline pulping process to maintain the desired chemical balance, usually expressed in pounds of equivalent $Na_2O$ per ton of air dry pulp produced.

**SODIUM OXIDE ($Na_2O$):** A standard chemical formula established in the pulp and paper industry in North America to express equivalent weights of chemicals used and existing in the process.

**SODIUM PEROXIDE ($Na_2O_2$):** An oxidizing-type chemical used in groundwood and chemical pulp mills as a bleaching agent.

**SODIUM RECOVERY:** The difference between all of the sodium lost during an alkaline pulping and recovery process and the original total sodium, expressed as a percentage of the total.

**SODIUM RECOVERY INDEX:** A sulfate pulp mill chemical recovery performance value determined by dividing the sodium in the smelt in the recovery furnace by the total sodium entering the unit. It can be expressed as a percentage.

**SODIUM SILICATE ($Na_2SiO_3$):** A chemical used as a sizing material in papermaking to provide water and ink penetration resistance in the sheet. It is also used to make up an adhesive in corrugated paperboard and for the lamination of solid fiberboard. Commonly known as *water glass*.

**SODIUM SULFATE ($Na_2SO_4$):** *See* SALT CAKE.

**SODIUM SULFIDE ($Na_2S$):** A common inorganic chemical compound that makes up one of the active chemicals in sulfate cooking liquor used in the pulp mill.

**SODIUM THIOSULFATE ($Na_2S_2O_3$):** A chemical salt used in pulp bleaching to neutralize residual chlorine that remains in the pulp after treatment.

**SOFT CALENDER:** On-machine calender with a soft-nip created by using rolls with polymeric covers or special fillings opposite steel rolls.

**SOFT END:** A paper roll defect caused by paper thickness and weight variations at the edge of a sheet, resulting in a spongy-like winding at the ends of the roll.

**SOFT FOLD:** A paper-finishing operation in which a stack of large sheets of paper is folded in such a manner that no crease marks remain.

**SOFT LUMP BREAK:** Paper breaks, usually occurring in the dryers or calenders of paper machines, caused by the presence of *soft lumps*.

**SOFT LUMPS:** Agglomerates of improperly disintegrated fibers and stock collecting at various points in the stock delivery system, such as inlets, that break away from time to time to form small clotted masses that become embedded in the sheet. This can result in a *soft lump break*.

**SOFT PAPER:** Paper made from stock furnish to which very little or no sizing material has been added. The sheet is pliant and moderately absorbent with low finish.

**SOFT PULP:** Pulp made by cooking chips under severe conditions that remove more than the usual amount of lignins and other noncellulosic material. Soft pulp is referred to as having low *k* or *kappa* numbers or as *easy bleaching pulp*.

**SOFT ROLL:** A reel of paper that is wound with insufficient tension or contains paper with uneven thickness in the cross machine direction, resulting in a loose roll.

**SOFT SIZED PAPER:** Paper made from stock furnish with less than the usual amount of sizing material, thus producing a soft, somewhat water-absorbent sheet having a below-normal finish.

**SOFT WHITE SHAVINGS:** *See* BOOK SHAVINGS.

**SOFTENERS:** (1) Substances added to paper stock furnish that will promote softness in the final sheet. (2) Chemicals added to hard water to convert it to soft water, making it more potable for use by the mill.

**SOFTENING POINT:** The temperature required to soften papermaking rosin to a specified viscosity.

**SOFTNESS:** That quality in a sheet of paper achieved by breakdown of fiber bonds through calendering, called *surface softness*. When it is obtained by producing a tight crepe through a high Yankee dryer adhesion, it is called *body softness*.

**SOFTWARE:** The collection of programs and routines associated with computers used to control various pulping and papermaking processes.

**SOFTWOOD:** Wood obtained from evergreen, cone-bearing species of trees, such as the pines, spruces, hemlocks, etc., that are characterized by having needles.

**SOFTWOOD PULP:** Pulp produced from the wood of evergreen coniferous species of trees such as pines, spruces, hemlocks, etc.

**SOLENOID VALVE:** A valve that is opened or closed by means of an electric solenoid.

**SOLID BLEACHED SULFATE (SBS):** Paperboard made from bleached kraft pulp that is sometimes coated with clay for printing purposes and/or coated with polyethylene plastic to make it more moisture resistant.

**SOLID BOARD FRAME:** A packaging structure used for shipping paper. It consists of solid wooden boards placed on flat top and bottom sides of a stack of paper and suitably bound in both directions.

**SOLID BOARD:** Single-ply, homogeneous types of paperboards, made from the same stock throughout the sheet structure.

**SOLID BRISTOL:** A grade of single, homogeneous, stiff, heavy paper made from one type of stock throughout the sheet body.

**SOLID FELT:** A very heavy grade of paper used by the building trade as environmental and sound insulation.

**SOLID FRACTION OF PAPER:** The volume of solid material in a sheet of paper, such as fibrous materials, fillers, sizes, dyes, pigments, and other chemicals, divided by the total volume of the sheet.

**SOLID STATE:** Electronic pulp and paper process measurement, control, and regulating equipment that has no electronic tubes, but is made up primarily of integrated circuit chips and boards.

**SOLID WASTE DISPOSAL:** The ultimate disposition of solid refuse that cannot be salvaged or recycled at the mill.

**SOLID WASTES:** Any solid material produced during the manufacturing process in a pulp and paper mill, such as sawdust and sludges, that must be disposed of by transporting it off the site, recovered, recycled into the process, or utilized to produce energy by incineration.

**SOLUBLE PAPER:** A special type of paper made to quickly dissolve and become completely dispersed when put into water.

**SOLVENT:** Any liquid used to form a solution by dissolving, liquefying, or dispersing one or more other materials.

**SOLVENT PULPING:** The use of organic solvents such as methanol and ethanol to chemically pulp chips in a "cooking" vessel. Some solvent processes such as the *Organocell process* use caustics with methanol, such as sodium hydroxide and anthraquinone. The *Alcell process* and the much earlier (1931) *Kleinert process*, utilize only ethanol. The *ASAM process* (alkaline sulfite, anthraquinone, methanol) allows high-yield, kraft-type pulp strength and bright pulp that can be bleached in a non-chlorine process.

**SOOT:** Chemical dust and other suspended materials that will deposit on tubes in the superheater, boiler, and economizer banks of power and recovery furnaces. It must be periodically removed for proper operation of the machinery.

**SOOT BLOWER:** A manually or automatically operated, probe-type device used to blow dust, soot, and ash from the fireside of boiler tubes in a furnace by employing steam or compressed air.

**SORBENT:** A material that takes up moisture or liquid when a physical or chemical change takes place in the process. Examples are calcium chloride and sulfuric acid. Also referred to as *adsorbent*.

**SORTING:** A manual or mechanical inspection operation carried on in the finishing room of a paper mill to remove imperfect sheets from an order. In a rag mill it refers to separating the different types of rags received for processing.

**SORTS:** Acceptable sheets of paper of various grades, types, and sizes put aside to be used to replace imperfect sheets removed from orders during the sorting operation in a finishing room of a paper mill.

**SOUND ABSORPTION COEFFICIENT:** *See* ACOUSTICAL ABSORPTION COEFFICIENT.

**SOUNDING THE REEL:** The method used by paper machine backtenders in evaluating the operation of the paper machine by slapping the reel of paper with their hand or a billy stick to obtain an indication of its hardness by the sound produced.

**SOUR COATED PAPER:** Reference to paper with a casein base coating that has decomposed, causing a distinctive, disagreeable odor.

**SOURED COATING:** A paper defect caused by the use of deteriorated casein in preparing the coating material, detected by the emission of a disagreeable odor. *See* SOUR COATED PAPER.

**SOURING:** The cleaning of wet end components, such as the wire and rolls, with the use of acid solutions.

**SOYBEAN PROTEIN:** That part of the soybean used as a source for making adhesives, sizing, coating materials, etc., for use in papermaking.

**SPANISH CHALK [$H_2Mg_3 \cdot (SiO_3)_4$]:** A naturally occurring silicate of magnesium (*soap-*

*stone*) with some magnesia replaced by alumina and used as a filler to improve the texture and feel of paper. Commonly called *talc.*

**SPAR DRYING:** *See* SLAT DRYING.

**SPARE PARTS:** Extra, critical-type pulping and papermaking equipment components that are most likely to break down or wear. They are put into storage to be used as emergency replacements when required so that the undesirable downtime of the equipment is minimized.

**SPARGER:** A type of liquid agitator that uses compressed air or gas to create and disperse bubbles throughout the mass to keep it in motion and mixed.

**SPAR POLE:** A tower constructed in the woodlands to do high-lead logging.

**SPC:** *See* STATISTICAL PROCESS CONTROL.

**SPEAR POLE:** (1) A wooden pole designed and used for spearing wraps and removing broke from the dryer section of a paper machine. Also called *dryer spears* and *dryer spear pole.* (2) Poles sometimes made from aluminum and used in the hot pond area of the woodyard to steer logs into jack ladders.

**SPECIAL MARKED ORDER:** A quantity of paper produced by a paper mill to specifications made up and supplied by the user.

**SPECIALTIES:** Papers and paperboards not considered as one of the standards by the paper industry, but made with unique characteristics and properties suitable for specific purposes or in accordance with special order specifications.

**SPECIES:** A class of trees, wood types, and designations of kinds of pulps made from these sources by pulp mills.

**SPECIFIC ADHESION:** The joining or binding of surfaces by valence forces resembling the type giving rise to the adhesion.

**SPECIFIC GRAVITY (SG):** The ratio of a fluid's density to the density of water at a standard specified temperature, or the comparative weight of a given volume of material with an equal volume of water at a specified temperature. *See also* APPARENT SPECIFIC GRAVITY.

**SPECIFIC HEAT:** The relationship of the amount of energy required to raise the temperature of a unit mass of material 1°C as compared with raising an equivalent quantity of reference material, such as water or air.

**SPECIFIC SURFACE:** Surface area of fibers per unit weight of dry fiber, expressed in square centimeters per gram.

**SPECIFIC VOLUME (SV):** The volume per unit mass of paper when measured under standard conditions. It is usually expressed in cubic centimeters per gram, which essentially is the reciprocal of density.

**SPECIFICATIONS:** A detailed statement of particulars listing the specific physical and chemical properties or characteristics to which pulp and paper is to be made to be accepted for the required end use.

**SPECIMEN:** *See* SAMPLE.

**SPECK:** A very small area of a sheet of paper that has a different appearance than the surrounding area. It is caused either by extraneous materials or discolored particles of paper components.

**SPECKLE:** Used in the printing industry to refer to a small area of low ink density.

**SPECTROPHOTOMETER:** An optical instrument used in the paper mill laboratory to measure the reflectance or transmittance of light from a paper sample.

**SPENT LIQUOR (SL):** Used cooking liquor in a chemical pulp mill that is separated from the pulp after the cooking process. It contains the lignins, resins, carbohydrates, and other extracted substances from the material being cooked. Usually, this liquor is processed through a recovery cycle to produce fresh cooking liquor and steam for process use and/or power generation.

**SPHERICAL ABERRATION:** An image imperfection caused by light passing the edge of an optical lens with spherical surfaces coming to focus in a different plane from that passing through the center of the lens.

**SPHERICAL BOILER:** A round, globe-shaped, batch-type digester that is used in pulping rags. It is usually held stationary during the cooking process and rotated only to fill with rags and to dump pulp at the end of a cook.

**SPHERICAL DIGESTER:** A round, globe-shaped vessel used to cook rags and wood chips. It may be either held stationary or rotated during the pulping process.

**SPIRAL BURR:** A pulpwood grindstone sharpening tool whose surface teeth are aligned diagonally to its axis.

**SPIRAL CONVEYOR:** *See* SCREW CONVEYOR.

**SPIRAL GRAIN:** The grain of a standing pulpwood tree in which the fibers are aligned spirally.

**SPIRAL SEAM:** Thick wire used to join ends of a dryer felt or screen on paper machines.

**SPITS:** Tiny, elongated, tear-shaped areas of excessive coating on the surface of paper sheets caused by the splattering of coating material during the coating operation. *See also* BLADE BLEEDS.

**SPLICE:** A joint made in a continuous sheet of paper with a glue or adhesive-type tape when there is a break in the web caused by winding or rewinding into a roll in the paper mill.

**SPLICE OFFSET:** An offset occurring at or near the splice in a roll of paper.

**SPLICE TAG:** A paper or paperboard marker placed at, and extending beyond, the edge of a roll of paper. It is used to identify the location of a joint or splice made on the web. Also referred to as a *flag.*

**SPLICING TISSUE:** A thin, narrow strip of paper, cloth, or plastic adhesive-coated material, usually put up in rolls and used in paper mills to join ends of a continuous paper sheet whenever there is a break in the web during roll makeup operations.

**SPLITTER:** A log-cutting machine used in the wood preparation area of a pulp mill to split large diameter logs into smaller sizes so that they will slide through the chipper chute openings and fit better into the pulpwood grinder pockets.

**SPLITTING:** (1) A multi-ply paperboard defect in which the layers that make up the sheet tend to separate from each other. (2) A refining action causing a physical separation of a fiber in a direction that is parallel to its axis.

**SPOILER BARS:** A series of axial bars placed on the inner circumference of the paper machine dryer to promote turbulence at high speeds and increase drying efficiency.

**SPONGY:** A descriptive paper mill term referring to types of absorbent, bulky, loose body paper made with little or no sizing material added to the pulp stock.

**SPOT:** Any small area in a sheet having a different appearance from the overall web. It may or may not be considered objectionable, depending on the quantity and intended use of the paper.

**SPOT COATING:** A paper-coating operation in which only certain areas of the sheet are coated.

**SPOT WEAR:** The localized wearing of a paper machine wire (or forming fabric) caused by dents, jams, stock lumps, or any foreign material passing through a nip between rolls.

**SPOUTING VELOCITY:** *See* SLICE VELOCITY.

**SPRAY BONDING:** A technique used for the latex bonding of disposable nonwovens in which the dilute latex is sprayed onto the moving sheet.

**SPRAY COATING:** Applying coating material on a paper sheet surface by employing a series of spray guns to deposit successive layers.

**SPRAY DYEING:** The application of dyes onto the wet end web of paper by the use of spraying devices or brushes.

**SPRAYER:** Devices used throughout a pulp and paper mill to convert liquids into small droplets or mists.

**SPREAD SHAFT COATER:** A type of paper coater that uses a 1-½-inch to 3-½-inch diameter rod rotating at high speed and in the opposite direction to the travel of the web to spread the coating material on the surface of the sheet.

**SPREADER:** A device, usually a slightly bent metal bar or roll, used in paper mills to produce uniform tension across a sheet of paper entering a winder, rewinder, calender, supercalender, or other similar sheet processing and converting operations. Also called *spreader, bowed bar,* or *curved bar.*

**SPREADER OFFSET RING:** Offsets on a roll that occur in one direction.

**SPREADER STOKER:** A type of solid fuel burning system in which a traveling grate is fed by a distributor that throws the fuel through flames onto the top of the fire, with the ash removed continuously by a slow, forward movement opposite to the stream of fuel and into a receiving pit.

**SPREADERS:** Stock flow distributors configured to provide a uniform spreading of stock slurry flow from the stock entry system to a cylinder-type paper machine.

**SPRING ROLL:** Spring-mounted, sometimes counterweighted, cylindrical rolls located throughout a paper machine or other sheet handling machinery to smooth out the tension by riding on the web of paper. Sometimes called *dancing rolls.*

**SPRINGBACK:** (1) The tendency of flat sheets of paper and paperboard to return to their original flat shape after bending or folding. (2) The extent to which a pad of pulp fiber will return to its original thickness after pressure is applied and released from the surface.

**SPRINGWOOD:** The wood of a tree produced early in the growing season of each year. It is characterized by wider, thinner walled cells than produced later in the growing season, which is called *summerwood* or *latewood.* This early growth is also called *earlywood.*

**SPRUCE:** Species of evergreen resinous softwoods used as a source of pulpwood by pulp mills.

**SPRUCE TURPENTINE:** A by-product of the sulfite cooking process obtained by condensing the relief gases. Also called *cymene.*

**SPUR SHOOTS:** Short, stocky branches of pulpwood trees with little or no internodal development.

**SQC:** *See* STATISTICAL QUALITY CONTROL.

**SQUARE SHEET:** Paper or paperboard having strength and tear resistance properties distributed uniformly in both machine and cross machine directions throughout the sheet.

**SQUARED PAPER:** Paper sheets that have been cut on a guillotine type of trimmer with all sides at right angles to each other or between one end and a side.

**SQUEEZE ROLLS:** Types of small cylindrical rolls used on multi-cylinder paperboard machines to press the layers of webs together to form a multi-ply sheet.

**SQUIRREL CAGE INDUCTION MOTOR:** A type of motor used in mills in which the secondary circuit consists of conductors suitably disposed in slots in the secondary core and connected at both ends of the core.

**SQUIRT TRIM:** A fine, high-pressure water jet located on each edge of the paper machine wire (forming fabric). It is used to trim the web to the desired width.

**STABILITY:** *See* AGING RESISTANCE.

**STABILIZED PAPER:** Paper in which moisture is uniformly distributed throughout the sheet and has reached equilibrium with the air.

**STABILIZERS:** Materials added to emulsions such as paper coating preparations to keep them from coagulating.

**STACK:** (1) Calenders used on paper machines. (2) A pile of paper sheets or pulpwood logs, or the act of putting paper sheets or pulpwood logs into a pile. (3) A vertical pipe or flue designed to exhaust gases and suspended particulate matter into the atmosphere from power and recovery furnaces and lime kilns in a pulp mill.

**STACK BARRING:** A papermaking problem occurring during the calendering operation. It results in cyclic caliper variations in the machine direction of the finished sheet.

**STACK EFFECT:** The movement of hot gaseous mixtures up a pulp mill power and recovery boiler or lime kiln flue due to the temperature difference between the mixture and the atmosphere.

**STACK GAS ANALYZERS:** A group of gas analyzers that measure the presence and concentration of objectionable gases in the flue gases of a recovery boiler.

**STACK SAMPLING:** The process of obtaining representative samples of gaseous mixtures in ducts and flues of power and recovery furnaces or lime kilns in a pulp and paper mill for the purpose of making analytical measurements on the gaseous mixtures.

**STACKER:** Equipment used to pile up packages of paper in a paper mill or pulpwood logs in a pulp mill yard.

**STACKING THE SHEET:** A paper mill expression referring to the operation of "poking" the tail of a paper sheet through the nips of the calenders.

**STAGGER:** A paper-handling operation in a paper mill finishing room in which the stacks of paper are arranged in a zig-zag fashion.

**STAINED PAPER:** Those types of colored paper in which the dye solution is applied to the surfaces of the finished sheets during the calendering operation, or by passing through a color bath.

**STALACTITES:** A paper blade coating defect consisting of a feathery, whisker-like accumulation of coating material extending under the blade, which occasionally drops off and causes coating spots on the paper sheet.

**STALAGMITES:** *See* BLADE BLEEDS.

**STANDARD ATMOSPHERE:** A generally accepted standard of pressure equivalent to 14.696 psia, 760 mm Hg, or 29.92 inches Hg.

**STANDARD BUNDLE:** A paperboard standard referring to a unit of sheets weighing 50 pounds.

**STANDARD CONDITIONS:** Unless otherwise specified, it is assumed to be 29.92 inches of mercury and 70°F. For paper testing, it is 73°F (plus or minus 2°) and 50 percent relative humidity (plus or minus 3 percent).

**STANDARD CUBIC FEET PER HOUR (SCFH):** Used in denoting flow of gases referenced to a temperature of 68°F, a relative humidity of 36 percent, and under a pressure of 14.7 pounds per square inch absolute (*psia*).

**STANDARD PACKAGE:** A paperboard standard referring to a 500-sheet ream unit.

**STANDPIPE:** A water storage tank standing on the ground with the height greater than the diameter, used in a distribution system. It is also used to afford relief from pressure surges in the system.

**STAPLING:** The matting of fibers during pulp screening operations in pulp and paper mills.

**STARCH:** A type of papermaking adhesive and sizing material made primarily from corn and potatoes, but also extracted from tapioca, wheat, and other sources. It is used to produce a higher degree of rigidity in a sheet and to improve the finish by causing the fibers to lie flat.

**STARRED ROLL:** A star-like, loose wound paper roll defect that appears on the end of the roll. It occurs when the outer winding is tighter than the inner winding.

**STARTING BURNERS:** Auxiliary fuel burners located on a sulfate black liquor recovery furnace. They are used to ignite and sustain the ignition of the black liquor and are not normally used for continuous firing.

**STARTUP:** The initial operation of new process equipment, especially pertaining to a new paper machine in a paper mill.

**STARVED JOINT:** A paper joint made with insufficient adhesive coating on the surface due to excessive applied pressure or penetration in the body of the sheet.

**STATIC ELECTRICITY:** A charge of electricity that accumulates on the surfaces of insulating materials such as paper when in frictional contact with other materials, especially in warm, dry air. This condition exists particularly around the final dryer and calender areas of a paper machine.

**STATIONER:** One who purchases stationery classes of paper products from converters and retails them to users.

**STATIONERY:** A paper-converting reference to a whole class of papers, including writing, typewriting, book, and other office and business-type papers used for communication purposes.

**STATISTICAL PROCESS CONTROL (SPC):** A quality control program implemented by many pulp and paper mills that defines their "statistical process realms" so that production quality can be maintained within specific, accepted limitations. It usually involves extensive data recording, charting, and graphing of process parameters to establish and analyze a mill's realistic production capabilities. This information is normally shared with the mill's customers and each production lot is accompanied by *SPC* process data. Sometimes referred to as *statistical quality control (SQC)*.

**STATISTICAL QUALITY CONTROL (SQC):** *See* STATISTICAL PROCESS CONTROL (SPC)

**STEAM:** Water in the form of a colorless vapor that becomes visible when it comes in contact with air, forming a mist or spray.

**STEAM ATTEMPERATION:** A process whereby steam temperature is lowered through partial desuperheating by spraying water into the steam (direct contact), or by passing all or a portion of the desuperheated steam through various types of heat exchangers.

**STEAM BLISTER:** Paper and paperboard defect formed during the drying process when steam becomes locally entrapped under the sheet when drying too fast. On multi-ply paperboards, it may occur in small, local areas between layers.

**STEAM CHEST:** Part of a black liquor evaporator that contains steamheated tubes with which the liquor comes in contact to boil off the water.

**STEAM FINISHED PAPER:** Paper in which the surface is treated with a steam spray as it enters the calender or supercalender on a paper machine. It improves smoothness and gloss, and increases the density.

**STEAM JOINT:** A special rotary fitting on paper machine dryer drums through which steam is admitted and condensate is removed.

**STEAM PLANT:** A plant in a mill complex that generates steam for use in the pulp and papermaking processes.

**STEAM SHATTER NOZZLE:** A duck-billed pipe jet installed above and in front of the discharge end of the smelt discharge spout on a sulfate chemical recovery furnace. It projects a steam jet on the issuing smelt as an additional aid to the breaking up of the smelt stream before it comes into contact with recirculated green liquor shatter sprays. *See also* SHATTER SPRAY NOZZLE, which it is sometimes called.

**STEAM STRIPPING:** A method used by some sulfate pulp mills to purify contaminated condensates by removing the total reduced sulfides (TRS) and biochemical oxygen demand (BOD) by a controlled direct steam injection process.

**STEAM TRAP:** A device used to automatically drain condensate from steam systems such as found on dryers and heat exchangers.

**STEAM TURBINE:** A source of prime movement consisting of a series of blades on which steam jets are allowed to impinge in such a way that the action rotates a shaft on which they are mounted and which is connected to an electric generator, pump, or other machinery to be driven.

**STEAMED MECHANICAL PULP:** A brown-colored, unbleached, groundwood pulp made by steaming the pulpwood prior to grinding.

**STEM FIBERS:** Types of fibers obtained from the stems of plants, e.g., bamboo, bagasse, straw, or the main trunk of a tree.

**STENCIL BOARD:** A high strength glazed kraft paperboard made to be treated with oil or varnished so it can be used to make stencil cutouts.

**STENCIL PAPER:** High-strength type tissue paper made especially to be wax- or oil-treated. It is used as a base for conversion into master sheets for use on mimeographs and similar duplicating machines.

**STENCIL SHEET:** A type of coated waterproof tissue paper used to make duplicating stencils.

**STICK:** (1) A common name used in the wood preparation area of a pulp mill in reference to a pulpwood *log.* (2) A roll of paper wound up on a winder shaft after being made on a paper machine. (3) Wet-lay nonwoven paper machine fabric defect caused by bundles of fibers with aligned cut ends not dispersed in manufacture.

**STICK MARK:** *See* BACK MARK.

**STICK OUT:** A paper defect manifested as a partial irregular separation or rupture of the web. *See* PARTIAL MACHINE DIRECTION TENSION BURST.

**STICKIES:** Sticking conditions occurring on paper machines using secondary recycled fiber containing materials such as ink, tars, latex, adhesives, and other organic compounds. Also refers to sticky materials called *tackies* and *depositable pitch.*

STIFFNESS: That ability of paper or paperboard to withstand bending or crushing forces. Also called *rigidity.*

**STIPPLE:** Reduce the gloss of a sheet of paper by embossing a cross-grain design on its surface with use of imprinted rolls.

**STOCK:** (1) The fibrous mixture in a paper mill that is ready to make into paper. It may consist of one or more types of beaten or refined pulps, with or without suitable fillers, dyes, additives, and other chemicals. Also called *furnish* and *stuff.* (2) Paper suitable for a particular use, such as coating raw stock, milk bottle stock, tag stock, towel stock, etc.

**STOCK AGITATOR:** A type of equipment or method used to keep the fibers well-dispersed in a stock slurry storage vessel by the use of air or mechanical stirring action.

**STOCK BLENDING:** The process of mixing a variety of types of pulps, additives, dyes, and chemicals to make up a stock slurry satisfactory to meet the specifications of a particular grade of paper. *See also* PROPORTIONERS and STOCK PROPORTIONING.

**STOCK CHEST:** Most often refers to a large, agitated tank in a paper mill that is used for receiving and holding the stock during the stock preparation phase of papermaking.

**STOCK CUTTING:** A stock preparation process of cutting papermaking fibers to the desired length by mechanical treatment in beaters and refiners.

**STOCK ENTRY:** That portion of a conventional cylinder-type paper machine which takes the stock flow from the stock supply piping system and spreads it across the width of the machine, reducing the velocity, and delivers it to the entry pond.

**STOCK HOURS:** A paper machine production term used in some mills to refer to the actual time the stock ran on the wire (forming fabric) after deducting for other delay times.

**STOCK ITEMS:** Types of paper consisting of specific grades, sizes, weights, colors, and other characteristics that are kept in inventory by papermakers and/or merchants to fill orders for immediate shipment.

**STOCK ORDER:** An order for paper that can be filled from standard types of paper, called *stock items,* in a paper mill and/or merchant's inventory storage.

**STOCK PREPARATION:** The area of a paper mill where pulp is received from an on-site or off-site pulp mill, prepared for storage in slurry form, mechanically treated in beaters and refiners, mixed with other pulps, additives, dyes, and chemicals, and then cleaned and generally processed prior to sheet formation on the paper machine. Once known as the *beater room.*

**STOCK PROPORTIONING:** The process of mixing a variety of types of pulps, additives, dyes, and chemicals to make up a stock slurry satisfactory to meet the specifications of a particular grade of paper. Also called *stock blending.*

**STOCK PUMP:** A pump that moves pulp slurries through pipelines. It is sometimes specially designed for this purpose.

**STOCK REGULATOR:** A process control unit usually located in the stock preparation area just ahead of the paper machine to control the consistency of stock by adding water to it.

**STOCK ROOM:** (1) The section or room in a paper mill in which the finished reserve paper is stored. (2) The area of the mill where tools, spare parts, supplies, etc., are kept and distributed as needed by mill operation personnel.

**STOCK SIZES:** Paper and paperboard sizes established as standard for specific grades by the paper industry, and usually carried in inventory stock by paper mills and distributors.

**STOCK TROUGH:** A long, narrow, wooden conduit used as a means of transporting paper stock flow in older paper mills, especially to the screens and paper machine.

**STOCK WEB:** A roll defect in which the layers tend to stick together due to poor adhesive coating, water, or some other sticky substance being caught up between the sheets.

**STOCK WEIGHTS:** Paper and paperboard weights established as standard for specific grades by the paper industry, and usually carried in inventory stock by the paper mills and distributors.

**STOCKLIGHT:** The time it takes for a red color to appear when a sample of paper is floated on thiocyanate solution, and a drop of ferric chloride is placed on its upper surface in testing for its water resistance.

**STOCKMAN:** A traditional reference to the employee in a paper mill who is in charge of the stock preparation operation and maintaining the stock supply to the paper machines.

**STOKER:** A series of stationary or moving grates at the bottom of a solid fuel burning furnace on which a bed of solids and ash is maintained for proper combustion.

**STONE:** A grinding wheel of abrasive segments mounted on a reinforced concrete core, sized to fit on pulpwood grinders to mechanically reduce logs to their fiber components by abrasion.

**STONE DRESSER:** A special tool used to sharpen pulpwood grinder stones. Also referred to as a *burr* or *bush roll*.

**STONE GROUNDWOOD (SGW):** Pulp made by abrading wood logs against a revolving stone, usually at atmospheric pressure but sometimes done under pressurized conditions.

**STONE IMMERSION:** A groundwood pulp mill reference to the distance below the surface of the pond of pulp under the grinder where the grindstone runs. Also called *stone submergence*.

**STONE NUMBER:** An identification system used on pulpwood grinder room operating reports and log sheets with a number of the operating line and a number of each pulpstone on the line.

**STONE SHARPENING:** A groundwood pulp mill operation in which the surface of the grindstone is renewed by removing dull and ineffective abrasive grains with a burr or sharpening tool.

**STONE SUBMERGENCE:** *See* STONE IMMERSION.

**STORAGE ACID:** Raw cooking acid held in reserve for transfer to storage tanks to be fortified with sulfur dioxide.

**STORAGE CHESTS:** Storage vessels used throughout the mill to hold stock slurries and other liquids during the pulping and papermaking processes.

**STORAGE TANKS:** Large, metallic, or wooden vessels used throughout the mill to hold raw and finished materials in the pulping and papermaking processes.

**STORE ROOM:** An area usually located near the shop area of a mill where machine equipment parts and supplies are kept in storage for use in repairs, rebuilding, and replacements.

**STOREHOUSE:** A dry, well-lighted area of a paper mill where orders not immediately shipped to a customer are held for a period of time.

**STRADDLE TRUCK:** A material mover used in pulp and paper mills. It runs astride loads to be moved, fastening to them by slings or clamps and then hoisting them off the floor so that they can be moved in a horizontal direction.

**STRAIGHT BURR:** A pulpwood grindstone sharpening tool whose surface teeth are aligned parallel to its axis.

**STRAIGHT GRAIN:** The grain of a standing pulpwood tree in which the fibers are aligned in the vertical direction.

**STRAIN:** The ratio of the change in dimensions of a paper sheet under external force to the original dimensions.

**STRAINED FORMATION:** Small, light spots in a sheet of paper caused by displaced fibers forming a lump at the forward edge of the spot as a result of running too tight of a draw at the couch roll.

**STRAW:** Dry stalks of grasses and plants that contain enough fiber to be an economical source of pulp for use in making paper with particular characteristics.

**STRAW PULP:** Fibrous mass produced by cooking *straw.* It is used in the manufacture of corrugating medium and less expensive multi-ply filler-type papers.

**STRAWBOARD:** An inexpensive type of paperboard used primarily as a filler in multi-layered boards or as a base for corrugating medium. It is made from straw that has been chemically treated, generally with lime.

**STRAW WRAPPING:** A coarse yellow meat wrapping paper made from unbleached straw.

**STREAK DETECTOR:** A laser or other type of inspection system used to detect streaks and scratches on the surface coating of paper.

**STREAKS:** *See* BLADE MARKS.

**STREET ELL:** A pipeline elbow fitting having threads on the outside of one end (male) and threads on the inside of the other end (female).

**STRESS:** The internal force per unit area of a paper sheet that resists external load, expressed in dynes per square centimeter or pounds per square inch.

**STRETCH:** The elongation experienced by paper before rupture when tension is applied. It is measured as a percentage of the original length.

**STRETCHER ROLL:** Any adjustable roll in contact with the web of paper on a paper machine. It can be moved up and down or back and forth in order to increase or decrease the tension of the sheet.

**STRIATIONS:** Lines and streaks appearing on an adhesive-coated paper. It is caused by uneven application of the coating.

**STRIKE:** When used in association with paper coating slurries and inks, it refers to their penetration into the paper sheet.

**STRIKE THROUGH:** Penetration of ink through a sheet of paper so that it shows through on the unprinted side.

**STRING CATCHER:** A rake-like device lowered in one channel of a pulp breaker beater to remove any strings or rags, and thus prevent operating problems further down the processing line.

**STRINGINESS:** The tendency of paper adhesive coatings to form fine threads and filaments when the adhered sheet is pulled away from another joined surface. Also known as *webbing.*

**STRIPPLING:** Embossing paper to remove the high gloss of a sheet by running the sheet between rollers with countergrained surfaces.

**STRONG ACID:** A sulfite pulp mill term for the fresh sulfite cooking liquor used to cook chips in a digester.

**STRONG BLACK LIQUOR OXIDATION (SBLO):** The transformation of unstable sodium sulfide ($Na_2S$) to the more stable sodium thiosulfate ($Na_2S_2O_3$) in sulfate black liquor using air or oxygen after evaporation to 55 percent or more solids concentration. This minimizes hydrogen sulfide ($H_2S$) and organic sulfur compound formation during combustion in a chemical recovery furnace of a pulp mill.

**STRONTIUM 90:** A radioactive material used to make some basis weight, level, and density measurements in pulp and paper mills.

**STRUCTURAL CURL:** A mechanical deformation of a paper sheet resulting from being in a curved position over a prolonged period of time. It appears in the layers of a paper roll that are close to and wrapped around the small curvature of the core.

**STRUCTURED COATINGS:** Paper coatings that have good flow characteristics until applied to the sheet, at which time it no longer flows and does not strike further into the base sheet. *See also* COATING IMMOBILIZATION.

**STUB ROLL:** The roll of paper that remains after one of standard diameter is partially used.

**STUCK END:** A paper roll defect in which the web layers at the ends stick together at the edges.

**STUFF:** The fibrous mixture of mechanically treated pulps mixed with other papermaking ingredients such as fillers, dyes, additives, and other chemicals ready for making it into a sheet on a paper machine. Also called *furnish* and *stock*.

**STUFF BOX:** A compartment in a stock-metering box used on some paper machines to ration out enough furnish supply to the wet end and return the excess to the machine storage chest over a constant head baffle to meet the required sheet weight being made.

**STUFF CHEST:** *See* MACHINE CHEST.

**STUFF PUMP:** Specially designed pumps used to move pulp slurries and stock furnishes through pipelines.

**STUMP WOOD:** The bell-shaped base portion of a tree just above the roots.

**STUMPAGE:** A pulpwood timber stand before cutting takes place.

**STYRENE-BUTADIENE LATEX:** A chemical employed as a coating binder and saturating agent for paper and paperboards.

**SUBSTANCE:** The weight of a standard ream of paper. It is expressed in pounds for the number and size of sheets of paper in the established ream for that particular grade.

**SUBSTRATA THERMAL MOLDING:** A hot paper supercalendering process in which the thermal molding action of the calender nip is allowed to penetrate deep enough into the paper web so that the smoothness and gloss are obtained with minimum densification of the interior of the web.

**SUBSTRATE:** Any material used as an intermediate layer between an adhesive or bonding coating and the sheet surface to which it is applied.

**SUCTION APRON:** A suction plenum box with a top perforated with holes and located between the tapes of a cutter in a board mill. It is used to keep the sheets from leaving the cutter as they proceed to the collection boxes for proper alignment and bundling.

**SUCTION BOX:** *See* VACUUM BOX.

**SUCTION BOX MARKS:** Paper defects showing up as light and dark areas running in the machine direction, caused by uneven suction produced by the suction boxes.

**SUCTION COUCH ROLL:** A type of paper machine couch roll that is essentially a rotary suction box used to remove water from the wet web after it is formed on the wire (forming fabric), and to pick up the web so that it can be transferred to the wet press section.

**SUCTION DECKLE EDGE:** Sides of a paper sheet produced by removing the stock from the edge of the paper machine wire (forming fabric) by the use of a vacuum.

**SUCTION PRESS:** Type of paper machine wet end press that uses a roll having a vacuum applied to it. It aids in the removal of water from the wet paper web.

**SUCTION PRESS ROLL:** A roll located in the wet end press section of a paper machine fitted with an internal suction box. It is used to draw out water through a perforated rubber cover as the wet sheet of paper passes over it on a carrying felt.

**SUCTION PUMP:** A pump connected to the suction boxes and couch roll on the wet end of a paper machine to create a vacuum that helps remove water from the wet web forming on the wire (forming fabric).

**SUCTION ROLL MARK:** Paper defects caused by the use of excessive vacuum in the suction couch roll or suction roll on the wet end press.

**SUGARCANE BAGASSE:** The residual of sugarcane after the stalks have been crushed and the sugar has been extracted from them. It is used as a source of fiber in the manufacture of certain pulps.

**SULFATE COOKING LIQUOR:** *See* KRAFT COOKING LIQUOR.

**SUIT BOXBOARD:** A type of paperboard colored or lined with a good printing surface and the ability to be folded in making boxes for clothing.

**SUITCASE BOARD:** A pasted chipboard with graining on one side and lined with a leather embossed paper on the other side. It is used in making briefcases, suitcases, sample cases, and other leather appearing carrying cases.

**SULFATE PAPER:** A high-strength paper made up of or having a high percentage of *sulfate pulp* or *kraft pulp*. Same as *kraft paper*.

**SULFATE PROCESS:** An alkaline pulp manufacturing process in which the active components of the liquor used in cooking chips in a pressurized vessel are primarily sodium sulfide ($Na_2S$) and sodium hydroxide ($NaOH$), with sodium sulfate ($Na_2SO_4$) and lime ($CaO$) being used to replenish these chemicals in recovery operations. Often referred to as the *kraft process*.

**SULFATE PULP:** Fibrous material used in pulp, paper, and paperboard manufacture that has been produced by the *sulfate process* or *kraft process*. Also referred to as *kraft pulp*.

**SULFATE TURPENTINE:** Chemical by-product of the *sulfate process*. *See* TURPENTINE and TURPENTINE RECOVERY.

**SULFIDITY:** The $Na_2S$ content of sulfate cooking liquor divided by the active alkali content, expressed in percent as $Na_2O$:

$$\left(\frac{Na_2S}{NaOH + Na_2S} \times 100\right)\%$$

**SULFITE BOND:** Beater and tub-sized *sulfite paper* used for office stationery, forms, memos, circulars, etc.

**SULFITE MANILA:** Manila colored paper made from *sulfite pulp*.

**SULFITE PAPER:** A higher quality paper made up of or having a high percentage of *sulfite pulp* fibers. *See also* SULFITE PROCESS.

**SULFITE PROCESS:** An acid pulp manufacturing process in which chips are reduced to their component parts by cooking in a pressurized vessel using a liquor composed of calcium, sodium, magnesium, or ammonia salts of sulfurous acid.

**SULFITE PULP:** Pulp made by the *sulfite process*.

**SULFONATED CHEMIMECHANICAL PULP (SCMP):** Pulp made by sulfonating chips prior to subjecting them to two stages of atmospheric refining for final mechanical separation of the fibers.

**SULFONATION:** The principal reaction in the sulfite pulping process in which lignin combines with sulfur dioxide to form lignin sulfonates.

**SULFUR:** A yellow-colored, naturally occurring element that is burned in acid plants of sulfite pulp mills. It produces sulfur dioxide ($SO_2$) which is used in making up calcium, sodium, magnesium, and ammonia base sulfite cooking liquors (*sulfite process*).

**SULFUR BURNERS:** Rotary or spray type furnaces used in acid plants of sulfite pulp mills to burn molten sulfur (S) to produce sulfur dioxide ($SO_2$) gas which is used in the preparation of sulfite cooking acids. Also referred to as *burners* or *rotary burners*.

**SULFUR DIOXIDE ($SO_2$):** A gas formed when sulfur (S) is burned in the acid plant of a pulp mill for use in making up sulfite types of cooking liquors.

**SULFUR DIOXIDE GAS STRENGTH:** In the acid plant of a sulfite pulp mill, it is the percentage of sulfur dioxide ($SO_2$) by volume in the dry gas from the sulfur burner, or in the total gas after dilution in a liquid sulfur dioxide system.

**SULFUR LOSSES:** The total sulfur lost during an alkaline pulping process, usually expressed as a percentage of the original total sulfur in the white liquor to the digesters or in pounds of air dry pulp produced.

**SULFUR MAKEUP:** The total sulfur chemicals, such as sodium sulfate, sulfur, etc., added to an alkaline pulping process and usually expressed in pounds per ton of air dry pulp produced.

**SULFUR RECOVERY INDEX:** A sulfate pulp mill chemical recovery performance value determined by dividing the sulfur in the smelt in the recovery furnace by the total sulfur entering the unit, and may be expressed as a percentage.

**SULFURIC ACID ($H_2SO_4$):** A mineral acid used in some chlorine dioxide ($ClO_2$) generating plants as a raw material to react with chlorates to produce chlorine dioxide gas. This gas is put into solution and used as a pulp bleaching agent. It is also used in other segments of the papermaking process and for cleaning purposes.

**SUMMERWOOD:** The slower growing portion of a tree that grows during the late summer and early autumn. *See* LATEWOOD.

**SUNDAY DRIVE:** A type of machine drive capable of being operated at slow speeds. It is usually associated with yankee dryers and used to keep certain pulping and papermaking machinery running at reduced speeds during periods of mill shutdowns, such as weekends and holidays, when it would be impractical to shut them down completely.

**SUPER MIXED PAPER:** A grade of paper consisting of a mixture of papers with a variety of qualities, usually packed in bales weighing less than 500 pounds with maximum outthrows less than 10 percent.

**SUPER NEWS:** Grades of newsprint paper used primarily for catalogs, directories, and advertising supplements. They possess such qualities as high finish, strength, and brightness brought about by calendering and increased chemical pulp content, resulting in a better printing surface than the standard newsprint.

**SUPER TINTED:** Paper that is colored by light addition of a dye solution to its surface during the calendering operation or by passing through a dye tub.

**SUPERBATCH:** A modified batch digester system (Sunds Defibrator) in which spent cooking liquors are collected in a series of accumulator vessels for use in subsequent batch cooks. *See* RAPID DISPLACEMENT HEATING (RDH).

**SUPERCALENDER:** An auxiliary piece of papermaking equipment used on some paper machines to obtain a denser paper with a higher finish than can be obtained on a calender. It uses paper or cotton-covered rolls arranged alternately with metal rolls through which the sheet is passed.

**SUPERCALENDERED (SC) PAPER:** Paper whose surface finish has a higher-than-normal glaze obtained by passing it between a series of alternately arranged metal and cotton or paper-covered rolls under pressure on a paper processing machine called a *supercalender.*

**SUPERCALENDERED GROUNDWOOD PAPER:** Groundwood-containing printing and writing papers that have been processed through a *supercalender* or series of supercalenders to improve gloss and printing performance. These papers have become competitive with more costly *lightweight coated papers (LWC)* in recent years. They are categorized into several sub-grades, such as SC-A, SC-B, SC-C, according to cost and quality.

**SUPERHEATED STEAM:** Steam at a pressure lower than the vapor pressure of water corresponding to the temperature of the steam, or steam at a temperature higher than the temperature at which water exerts a vapor pressure equal to the steam pressure.

**SUPERHEATER:** A heat exchanger inside a furnace which receives dry, saturated steam from the drum and raises its temperature.

**SUPERINTENDENT:** The person in charge of and responsible for specific operations in a pulp or paper mill.

**SUPPLY ROOM:** The storage area of a pulp and paper mill where materials such as machine parts, tools, etc., are kept for ready availability to maintain continuous operation.

**SURFACE COATED:** Any paper or paperboard sheet that has a coating material applied to one or both surfaces.

**SURFACE COLORED:** Paper or paperboard that is colored by the application of liquid dye to its surfaces when calendering the sheet or by passing it through a dye tub in a separate, off-machine operation.

**SURFACE COLORING:** The process of coloring paper and paperboard by the application of liquid dye to their surfaces when calendering the sheets or by passing them through a dye tub in a separate, off-machine operation. Also called *dipping* or *tub coloring.*

**SURFACE HARDENING:** A paper sheet defect caused by drying out the sheet too fast and hardening the outer fiber layer due to the resulting rapid evaporation.

**SURFACE LIFTING:** The raising of fibers and small pieces of coating from the paper sheet surface during the printing process, caused by the paper being too soft and/or the ink being too sticky.

**SURFACE MARK:** A paper defect that occurs after the calendering or supercalendering of the sheet on a paper machine due to an uneven surface on a winder or having it move at a different speed than the sheet.

**SURFACE PEELING:** A paper defect in which the surface is forming scale-like particles.

**SURFACE-SIZED PAPER:** Paper whose surface is treated with a sizing material after the sheet is formed on a paper machine or in a separate, off machine operation, occasionally after it has been internally sized. Sometimes called *top-sized paper. See also* TUB-SIZED PAPER.

**SURFACE SOFTNESS:** *See* SOFTNESS.

**SURFACE SPEED:** The outer peripheral speed of a pulpstone in a groundwood pulp mill, or rolls or drums on a paper machine, usually expressed in feet per minute.

**SURFACE STRENGTH:** The resistance of a paper sheet to the removal of fibers from its surface. It is measured by pulling a series of arbitrary grades of melted and then cooled wax sticks from its surface. The measurement procedure is called a *wax pick test.*

**SURFACE TENSION:** A property of liquids that causes the exposed surface to contract to the smallest possible area, in a drop-like formation.

**SURFACE WETTABILITY:** An affinity of paper for liquids, which determines its acceptability to be used as paper for writing purposes.

**SURFACTANT:** A material that enhances and stabilizes the ability of one liquid to disperse itself into another. Also known as *emulsifying agent* or *emulsifier.*

**SURGE:** A momentary increase, usually associated with pressure, level, flow, or other variables in vessels, storage chests, or other equipment in a process system.

**SURGE TANK:** A container or vessel attached to the process so that it will store gaseous, liquid, or solid materials to take up momentary increases in volume or pressure, and permit an intermittent operation to function without upsetting the continuity.

**SUSPENDED SOLIDS:** (1) Solid particles evenly distributed throughout a gaseous material, such as in power and recovery boiler flue exhausts. (2) Fibers and other colloidal solids held in uniform distribution throughout a mass by mechanical treatment in beaters or refiners. (3) The undissolved solid particles of organic and/or inorganic material in pulp and papermaking solutions, such as spent liquors, white water process liquids, and liquid wastes, that can be measured by filtering a representative sample and weighing the contents on the filter paper or by centrifuging it.

**SUSTAINED YIELD:** The yield that a forest produces by using a method of management designed so that the forest will produce continuously.

**SWEAT ROLL:** *See* COOLING ROLL.

**SWEETENER:** Long fiber or virgin pulps added to white water to improve its filtering properties in save-alls.

**SWELL:** The severe rise in a steam boiler drum level due to increased volume of steam bubbles occurring during a rapid increase in load.

**SWELLING:** An increase in the volume of fiber due to the absorption of water.

**SWING HAMMER-TYPE CHIP SHREDDER:** A machine used to shred chips in a wood preparation area of a pulp mill. It is equipped with rows of square-ended hammers that are pivoted to a steel disk held between spacing rings and form a rotor on a shaft.

**SWITCH:** A device usually used in an electrical circuit for making, breaking, or changing connections.

**SYNCHRONOUS MACHINE:** A machine in which the speed of normal operation is exactly proportional to the frequency of the current used to drive it.

**SYNCHRONOUS MOTOR:** A *synchronous machine* that converts electric power into mechanical power and has a speed strictly proportional to the frequency of the operating current, either single phase or polyphase.

**SYNERGISTIC EFFECT:** The cooperative action of separate substances producing a total effect greater than the sum of the effects of the substances acting independently. An example is when chlorine dioxide ($ClO_2$) solution is added to the chlorination stage of a pulp bleaching process.

**SYNTHETIC:** A general descriptive term applied to any man-made papermaking fibers and materials produced from chemicals or compounds, and products made from them.

**SYNTHETIC FIBERS:** Short filaments used in the manufacture of synthetic paper by extruding or spinning from synthetic resin materials.

**SYNTHETIC PAPER:** A sheet made from synthetic fibers or by extruding a film or foil of thermoplastic material. It does not contain any cellulosic-type fibers.

**SYNTHETIC RESINS:** Types of manufactured resins made from substances such as formaldehyde, phenol, urea, melamine, etc., and used extensively in paper mills for such purposes as coating adhesives and to impart greater wet and dry strength to a paper sheet.

**SYSTEMS ENGINEERING:** The analysis of pulp and paper mill problems, and the selection and implementation of all of the necessary equipment, devices, and schemes of operation to provide the most satisfactory solution.

**SZ:** *See* SOAKING ZONE.

# T

**T4S:** *See* TRIMMED FOUR SIDES.

**TABLE ROLL MARKS:** Basis weight variations on a paper web caused by waves on the wire (forming fabric) resulting from the supporting table rolls.

**TABLE ROLLS:** Cylindrical rolls used in the slower and narrower width paper machines to support the wet end wire (forming fabric) carrying the draining layer of stock to assist in the drainage of water and the retention and formation of the web. Sometimes referred to as *lube rolls*.

**TABLET BOARD:** A chip or news type of paperboard used for making the stiff backs for paper pads and tablets.

**TABLET NEWSPAPER:** A newsprint grade of paper used to make up pencil writing pads and tablets.

**TABLET PAPER:** A cheaper grade of hard-sized paper with a good writing surface. It is used in making pen and pencil pads, and tablets.

**TACHOMETER:** An instrument used by paper mill technicians and operators to measure speeds of rotation of equipment, such as pumps, ventilating fans, line shafts, and motors, and to determine linear travel of the paper machine with respect to its relative component parts or sections.

**TACK:** The sticky property of paper and paperboard adhesive and glue coating materials. Also used in reference to printing inks.

**TACKIES:** *See* STICKIES.

**TACKIFIER:** Any agent used in the preparation of adhesive paper-coating materials in order to increase or control its stickiness.

**TACKLE:** (1) That portion of a pulp beater and refiner which is replaceable. It consists of working surfaces made up of bars and knives that wear out with use or are changed to provide different designs to achieve certain desired actions on the pulp being processed. (2) An assembly of ropes and pulleys arranged to obtain mechanical advantage so that heavy loads can be hoisted and/or pulled around a pulp and paper mill.

**TACKY:** *See* AUTO ADHESION.

**TACKY SPECKLES:** Types of contaminant found in deinked pulp caused by plastic adhesive-like particles originating from such sources as heat-seal coatings, flying pasters, backing adhesive, etc.

**TAG:** Identifying marker, usually colored, inserted between the paper sheet while it is being rolled or put into stacks to indicate location of splices and defects, or the act of placing these markers. *See* SPLICE TAG.

**TAG BOARD:** A thin, strong, colored paperboard with a good printing surface used primarily for making tags for shipping identification.

**TAG STOCK:** A high-strength, durable paperboard made from a mixture of chemical pulps, mechanical pulp, rope, jute, and wastepapers. It possesses good writing and printing surfaces and is used in making a variety of identification tags.

**TAIL:** A strip cut in a newly formed sheet of paper on the tending side of the paper machine. It is used to transfer the sheet from the wet end and carry it through the presses and dryers.

**TAIL CUTTER:** An assembly mounted at the wet end of a paper machine incorporating a flexible hose and water jet that moves across the wet sheet and automatically cuts a tail so that it can be fed through the machine.

**TAILING SCREENS:** Screens in a pulp screening system that receive the rejected portions of raw pulp after they have gone through the primary screening operation. The accepted pulp is fed back to the primary screening operation and the rejected portion is fed to a rejects refiner before returning back to the primary screening operation.

**TAILINGS:** The rejected portions of raw pulp after it has gone through a primary screening operation to remove uncooked fiber bundles, coarse slivers, shives, knots, and any other extraneous matter.

**TALC:** A finely pulverized, hydrated magnesium silicate mineral used as a softening filler material for use in some grades of cheaper writing papers, envelope papers, papeteries, etc. *See also* SPANISH CHALK [$H_2Mg\ (SiO_3)_4$].

**TALL OIL:** A by-product made from the resins, fatty acids, and soap removed during the evaporation of sulfate black liquor. It is used to make chemicals for the manufacture of protective coatings, adhesives, detergents, paint, varnish, ink, lubricants, waxes, polishes, gasoline additives, agricultural products, etc.

**TALL OIL FATTY ACIDS (TOFA):** Fatty type acids formed during the process of manufacturing tall oil from digester relief gases in sulfate pulp mill.

**TALL OIL ROSIN (TOR):** A by-product of the sulfate pulping process obtained by the fractional distillation of crude tall oil, which is processed from the soap that separates out from the spent black liquor during the evaporation operation.

**TAMARACK:** A resinous evergreen species of trees used as a source of pulpwood by pulp mills.

**TANDEM COATING:** A paper-coating procedure used in a paper mill in which one coat is applied to one side and dried, followed by the application of the coating to the opposite side, and then the entire sheet is dried again.

**TANDEM THERMOMECHANICAL PULPING (TANDEM TMP):** A wood-pulping process in which chips are presteamed, followed by two stages of mechanical treatment in refiners at elevated temperature and pressure.

**TANGENTIAL FIRING:** A method of burning fuel in a furnace with an arrangement of burners mounted in the corners that fire tangentially to an imaginary circle in the firebox.

**TANK:** Any vessel, usually made of wood, metal, or cement, used in the pulp and papermaking process for storage of chemicals, pulp slurries, and other liquid solutions or suspensions.

**TANNING PAPER:** Paper that has been coated with an abrasive material and used to produce a smooth finish on leather surfaces by buffing it.

**TAPE PAPER:** Paper that has been cut into a narrow width and usually rolled up for further use or coating with various adhesive materials.

**TAPERED DUCT FLOW SPREADER:** A type of flow spreader in a paper machine wet end stock approach system. It is constructed so that there is a gradual change of its cross section from a circular pipe to a slot which runs the width of the paper machine.

**TAPERED ROLL:** The manner in which paper peels off the paper roll after a break during roll winding or rewinding, occurring diagonally from one side to the other.

**TAPIOCA STARCH:** A type of starch used by paper mills as a sizing material in papermaking.

**TAPPI:** Technical Association of the Pulp and Paper Industry. Headquarters: Atlanta, GA, (404) 446-1400.

**TAR PAPER:** Construction-type paper saturated or coated with tar or asphalt emulsion to make it impervious to water and the environment.

**TARE WEIGHT:** The weight of the container, box, wrapping, or holding means, which is subtracted from the gross weight of any packaged material to obtain the net weight of that material.

**TARGET BOARD:** Rigid-type paperboards suitable for making rifle and gun range targets.

**TARNISHPROOF BOARD:** An acid-free and reducible sulfur compound-free paperboard made in various weights. It is used to make containers for packing metallic articles to prevent rust and discoloration.

**TARNISHPROOF PAPER:** *See* ANTITARNISH PAPER.

**TARRED FELT:** A felt-like, thick construction paper saturated with coal tar or asphalt and used to waterproof buildings.

**TCMP:** *See* THERMO-CHEMIMECHANICAL PULP.

**TCZ:** *See* TEMPERATURE OF COOKING ZONE.

**TDS:** *See* TOTAL DISSOLVED SOLIDS.

**TEA:** *See* TENSILE ENERGY ABSORPTION.

**TEA BAG PAPER:** A thin, porous-type paper made from chemical pulps and bleached manila hemp fibers. It is used for converting into bags for extracting tea by immersion in hot water.

**TEA CARTRIDGE PAPER:** A high-strength, nonabsorbent type of paper made especially for converting into containers for packaging tea.

**TEAR FACTOR:** A tear-resistant related expression equal to the force in grams required to tear a strip of paper (under specific conditions) divided by the basis weight in grams per square meter multiplied by 100.

**TEAR OUTS:** Samples of paper that are torn from a sheet of paper for inspection and testing purposes.

**TEAR RATIO:** Paper sheet tear-resistance or tear-strength, expressed in points or grams per pound.

**TEAR RESISTANCE:** *See* TEAR STRENGTH.

**TEAR STRENGTH:** The resistance of a paper sheet to tearing. It is usually measured by the force required to tear a strip under standardized conditions. Sometimes called *tear resistance.*

**TEAR TEST:** *See* ELMENDORF TEST.

**TEASELS:** The barrel-shaped burrs located on the surface of a rotating drum that are used to raise the surface fibers on papermaking felt to make them soft and downy.

**TEE:** A pipeline fitting in the shape of the letter "T" used to connect a pipeline to another pipeline running at right angles to it.

**TELEGRAPH MANILA PAPER:** A sized grade of writing paper, usually yellow colored, made from chemical pulp or a mixture of chemical and mechanical pulp, and used in the manufacture of telegraph forms and writing pads.

**TELEPHONE DIRECTORY PAPER:** A lightweight grade of catalog or printing paper with good strength properties. It is made from a mixture of bleached chemical, mechanical, and reclaimed pulps for use in making telephone directories. Sometimes known simply as *directory paper.*

**TELESCOPED ROLL:** A paper roll defect term indicating that one end of the roll is cone-shaped because of slippage of the core. Also called *slipped roll.*

**TELLURIUM LEAD:** A lead alloy that is practically pure lead with a small percentage of tellurium added to improve the mechanical properties so it can be used for gas and acid lines in the acid plant of a sulfite chemical pulp mill.

**TEMPERATURE GRADIENT:** A common term used in reference to surface temperature profiles of dryers, normally in increasing value, indicated from wet to dry end.

**TEMPERATURE OF COOKING ZONE (TCZ):** A pulp mill expression normally used with continuous-type digesters. It refers to the temperature of the chips and liquor in that portion of the digester considered to be the cooking zone.

**TEMPERATURE OUT OF HEAT EXCHANGER (TOS):** A pulp mill expression normally used with continuous-type digesters referring to the temperature of liquor after it has been heated to cooking temperature in external steam heaters.

**TEMPLATE BOARD:** A dimensionally stable, hard-rolled paperboard usually made on a wet machine from wastepaper for use as a base for making patterns such as those used in machine shops.

**TENDER:** A descriptive term sometimes used to indicate that the referenced sheet, paperboard, or other items have deteriorated to or possess low strength qualities.

**TENDING AISLE:** The aisle between paper machines in a multi-machine paper mill that is free of drives and other mechanical equipment, allowing the machine to be more accessible for operation by the paper machine crew.

**TENDING SIDE:** *See* FRONT SIDE.

**TENSILE ENERGY ABSORPTION (TEA):** The amount of work necessary to pull the stretch out of a sheet of paper and finally rupture it. It is measured in foot-pounds per square foot in the testing of extensible paper used to manufacture multiwall bags.

**TENSILE ENERGY ABSORPTION TEST:** An evaluation procedure used on packaging materials to determine their durability to rough usage by measuring the energy absorbed when a sample is stressed to the point of rupture.

**TENSILE STRENGTH:** The resistant property of a sheet to pull or stress produced by tension. It is expressed as force per unit width of a sample tested to the point of rupture. *See also* BREAKING LENGTH.

**TENSILE STRENGTH TESTER:** A paper testing device used in paper mill testing laboratories to determine the tensile strength or the stretch of a paper sample by applying and measuring the tension force.

**TENSILE STRETCH:** The elongation of a sheet when put under tension. It usually refers to that force required to bring the sheet to the point of rupture.

**TENSIOMETER:** A device used at various locations along the dry end of a paper machine. It consists of a roll that rides on the top of the moving sheet of paper to feel or measure the tension so that it can be controlled.

**TENSION CONTROL SYSTEM:** A paper machine dry end control system that receives a tension measurement signal from a roll (*tensiometer*) riding on the top of the moving sheet of paper and adjusts the speed of the dryers to compensate for any change.

**TENSION WOOD:** Pulpwood that is usually obtained from the upper side of leaning or crooked hardwood trees where it has experienced abnormal growth.

**TERRA ALBA ($CaSO_4 \cdot 2H_2O$):** Ground gypsum used by paper mills as a filler material in papermaking to add whiteness to a sheet of paper.

**TERTIARY TREATMENT:** An extension of normal mill effluent waste and sewage treatment beyond the secondary or biological stages. It provides water of more potable quality by further chemical and mechanical treatment. *See also* ADVANCED WASTE TREATMENT.

**TERTIARY WALL:** The thin inner wall in the structure of papermaking fibers just inside the secondary wall. It is usually covered by a warty membrane and forms the central cavity or lumen.

**TEST:** A critical examination procedure conducted under standardized conditions, usually in a laboratory, to determine certain physical and/or chemical characteristics of pulp and paper mill raw materials, process materials, products, and by-products.

**TEST BOARD:** Types of paperboard that meet specific tests adopted by the packaging industry in order to qualify it for making shipping containers. *See also* RULE 41.

**TEST LINERBOARD:** Types of paperboard that meet specific tests adopted by the packaging industry to qualify for use as the outer facing layer for corrugated board, from which shipping containers are made. *See also* RULE 41.

**TEST SAMPLE:** A representative portion of raw materials, pulp, paper, and any material obtained in order to analyze, test, or make other measurements on it. Also referred to as a TEST SPECIMEN.

**TEST SHEET:** *See* HANDSHEET.

**TST SPECIMEN:** *See* TEST SAMPLE.

**TESTING PAPER:** Types of paper made with a special chemical treatment so they may be used to detect the presence of certain specific materials in solutions, gases, or vapors, by changing colors when exposed to them.

**TEXT PAPER:** A good quality, laid or woven book paper with a medium vellum-like finish. This type of paper is used in making deluxe editions and fancy stationery.

**TEXTILE PAPER:** Types of paper made from chemical and/or mechanical pulps and processed to be used by the textile industry, primarily for wrapping purposes.

**TEXTILE SPOOL PAPER:** A lined, lightweight, combination type of paperboard used to make spools utilized by the textile industry.

**TEXTILE TISSUE:** A light, high-strength tissue paper usually cut in narrow strips that are twisted into threads to be made into yarn in the manufacture of paper textiles.

**TEXTILE WRAPPING PAPER:** Any of a number of types of paper used by the textile industry to wrap and package cloth material. *See* TEXTILE PAPER.

**TEXTURE:** Sheet surface qualities that usually can be identified primarily by touch and sometimes by sight.

**THEORETICAL AIR:** The theoretical weight of air needed at a desired chemical recovery furnace condition to furnish the required supply of oxygen for optimum combustion of the liquor.

**THEORETICAL HEAD:** The total head of stock behind the slice of a paper machine headbox. It is calculated for a specific wire (forming fabric) speed, assuming a coefficient of spouting velocity discharge of one.

**THERM:** 100,000 Btu of heat.

**THERMAL COMPRESSION EVAPORATION:** A method of evaporation used on pulp mill spent liquors in which the steam removed from the boiling liquid being concentrated is compressed and fed back to the outside of the tubes of the evaporator. There it condenses and transfers its latent heat to the boiling liquid.

**THERMAL CONDUCTION:** The passage of heat through or along a mass, or between two bodies in direct contact.

**THERMAL CONDUCTIVITY:** The physical property of materials, such as paperboard used for insulating purposes, that is related to the rate of flow of heat from one side of a sheet through the other. It is expressed in the temperature gradient per unit area under constant conditions.

**THERMAL CONVECTION:** The conveying of heat from a heat source by contact with a moving fluid which is cooler than that of the heat source.

**THERMAL CYCLE:** The complete circuit that heat travels throughout a mill, beginning with the conversion of original fuel into heat used to generate steam from water, followed by the distribution of steam to locations in the mill where it is utilized, and the return of the condensate back to the boiler in the powerhouse.

**THERMAL EFFICIENCY:** The ability of a basic steam power plant to produce mechanical power by operating with a fluid carried through a closed cycle. The level of efficiency is calculated from the ratio of work output to heat input.

**THERMAL POLLUTION:** The undesirable increase of temperature that is harmful to the aquatic and biological life of natural receiving water bodies and streams by pulp and paper mill effluent discharges.

**THERMAL RADIATION:** Electromagnetic energy emitted by a mass as the result of excitation of its molecules by increasing temperature.

**THERMAL RECORDING PAPER:** Paper coated with a heat-sensitive material that will change color when heat is applied with a heated writing pointer.

**THERMAL SHOCK:** The effect of an abrupt temperature change on a device.

**THERMISTOR:** A semiconductor whose resistance is temperature-sensitive, permitting it to transmit a signal when the temperature changes.

**THERMO-CHEMIMECHANICAL PULP (TCMP):** *See* CHEMITHERMOMECHANICAL PULP (CTMP).

**THERMO COMPRESSOR:** An ejector or jet device that utilizes steam at a high pressure to entrain steam at low pressure and discharge the mixed streams at an intermediate pressure back to the process by reconverting velocity energy into pressure. These devices are used on dryer drainage systems on paper machines and on black liquor evaporators to improve their performance.

**THERMOCOUPLE:** A junction of dissimilar conducting materials used to measure temperatures in the pulp and papermaking process. Electrical current is produced around the junction and varies with temperature.

**THERMOMECHANICAL PULP (TMP):** Pulp made by presteaming chips and reducing them into their fiber components by an initial mechanical treatment in refiners under elevated temperature and pressure, with subsequent refining done at atmospheric pressure.

**THERMOMECHANICAL PULPING:** A wood-pulping process in which chips are presteamed before an initial refining treatment under elevated temperature and pressure, with subsequent refining treatment at atmospheric pressure.

**THERMOPLASTIC COATING:** Plastic paper-coating materials used in paper mills to coat paperboard to be used for milk cartons, etc. It has high molecular weight and can be melted to produce high viscosity fluid that solidifies and can be remelted repeatedly.

**THERMOREFINER MECHANICAL PULP (TRMP):** Wood pulp made by presteaming chips, followed by mechanical treatment in a refiner at atmospheric pressure.

**THERMOSETTING:** Plastic paper-coating materials that can be melted and solidified to an insoluble and infusible state. They cannot be remelted and are used to coat special purpose paper and paperboard.

**THICK BLACK LIQUOR:** Spent black liquor in a sulfate pulp mill after it is evaporated. It contains more than 50 percent total solids and is increased to 65 percent and above before incineration in the chemical recovery furnace.

**THICK PULP:** Any pulp slurry having a high consistency, generally between 10 and 16 percent.

**THICK STOCK:** A term for high consistency pulp slurry, varying slightly depending on the point of reference, but generally considered to fall between 10 and 16 percent.

**THICK STOCK VALVE:** A type of valve designed to operate on high consistency pulp slurries in the range of 10 to 16 percent.

**THICKENERS:** Types of filters used to remove water from a pulp slurry by passing it over stationary screens or by the use of screen-covered rotating cylinders.

**THICKNESS:** The dimension of a sheet of paper or paperboard usually measured between the two surfaces under specifically stated conditions, and expressed in thousandths of an inch as its *caliper.*

**THIN BLACK LIQUOR:** Spent black liquor in a sulfate pulp mill obtained after washing the brown stock from the digesters just before being evaporated. It contains less than 15 percent total solids.

**THIN PAPER:** Grades of lightweight papers of low caliper (less than those of facial, toilet, and hygienic type tissues).

**THINNING:** A pulp wood forest management procedure in which selective trees are harvested to provide optimum growth of those remaining.

**THIRD HAND:** A member of the paper mill machine crew who is a general utility person and assists the backtender around the calenders and winders.

**THIRD PRESS:** The third and usually last set of press rolls following the paper machine forming section, located just ahead of the dryer section.

**THIXOTROPIC:** The property of a material that makes it appear and act as a solid when undisturbed, but changes to a semiliquid when agitated. This occurs because its viscosity is a function of the shear rate to which it is subjected.

**THIXOTROPHY:** A phenomenon found in some stock slurries and paper making liquids in which there is a slow recovery on standing after there has been a drop of consistency through shearing forces. Also referred to as *shear thinning*.

**THIXOTROPY:** The property of a liquid material that allows its viscosity to increase when subject to constant shear rate, as exhibited by some paper coating color materials, under the actions of mixing and pumping equipment.

**THRASHER:** A machine used in a pulp mill, usually consisting of a revolving cylindrical drum or other means of beating old rags and papers to open them up so that loose dirt, dust, or other foreign particles can be shaken out and removed.

**THRASHING:** The use of a *thrasher* to remove loose dirt, dust, and other foreign particles from old rags and papers in pulp mills.

**THREAD BURR:** A pulpwood grindstone sharpening tool whose surface teeth are aligned in a continuous thread-like fashion as those on a pipe or screw.

**THREADING THE PAPER:** The passing of paper over the paper machine when starting through a new sheet. Once commonly done by hand on slower and smaller machines; now done mechanically by rope carriers.

**THREE-POCKET GRINDER:** A discontinuous type of pulpwood grinder in which the logs are fed against the grinding stone surface by three hydraulic rams operating through three compartments.

**THREE-ROLL PRESS:** A type of wet press used to remove water from a sheet of paper before it is dried on heated dryers. It consists of an arrangement of two suction-type rolls and a center roll combination.

**THREE-STAGE BLEACHING:** Once a very common multi-stage procedure for bleaching sulfite chemical pulp, in which the first stage was chlorination, followed by a caustic extraction stage and a hypochlorite stage.

**THREE-TERM CONTROLLER:** A process control device designed to have three modes of operation: proportional, integral, and derivative.

**THREE-WAY VALVE:** Type of valve used in pulp and paper mills designed with three flow openings so they can be used for splitting and blending flow streams.

**THROUGHPUT:** The volume of material passing through process equipment used in pulp and paper mills, also related to production.

**THROWING THE SEAM:** The act of twisting or "cocking" the seam of a paper machine felt when it is too wide for safe operation. After the felt is run on the machine for a while and settles down, it is widened and trimmed to fit properly.

**TICKER TAPE PAPER:** Paper once made specifically for converting into narrow-width rolls to be used by the stock market industry to continuously print stock quotations on tickers.

**TICKET BOARD:** A surface-colored, inferior grade of paperboard used to make a variety of tickets, such as for railroads, theaters, sporting events, etc.

**TICKET STOCK:** Any coated paperboard, usually colored and with a news pulp center layer, used to make a variety of tickets.

**TICKLER REFINER:** A type of stock refiner used in the stock preparation area for light mechanical treatment of fibers or as a trim refiner to give the fiber a touch-up mechanical treatment just ahead of the paper machine.

**TIE COAT:** A coating on the adhesive side of a paper backing or facing with a thin layer of adhesive material that serves as a bonding agent between the adhesive and the backing or facing.

**TIGHT EDGES:** *See* SHRUNKEN EDGES.

**TILE STOCK:** Any paper used to make high-quality wallpapers and paper moldings.

**TIME CARD:** A ticket used by pulp and paper mills to record the working time of employees for computing payrolls, usually by printing the arrival and departure times with time clocks.

**TIME CLOCK:** A clock used by pulp and paper mill employees to print work arrival and departure times for use in computing payrolls.

**TIME LAG:** The time during which a process variable undergoes an exponential change or delay in system output with respect to system input.

**TIME SHARING:** A process control device or system, such as a computer, used to work on two or more tasks, alternating the work from one task to another.

**TINFOIL PAPER:** (1) Paper used for decorative and some preservative purposes that has been coated with a mixture of an adhesive and ground tin powder. (2) A laminated paper sheet with an outer metallic layer resembling silver. Sometimes referred to as *silver paper.*

**TINNY:** The stiffness of a sheet that causes it to produce a ringing sound when shaken. This is produced by extended refining of the fibers. Similar to and sometimes called *crackle.*

**TINT:** Shade of colors obtained in preparing papermaking dye materials with white substances.

**TINTED:** A descriptive term for paper grades that have only been lightly colored.

**TIRE TRACKS:** A paper roll defect consisting of a peripheral ridging or indentation on the outside of a roll of paper, caused by the cumulative effects of uneven caliper across the web that was wound to make the roll. Sometimes called *bars, buckle, corrugation marks,* and *rope marks.*

**TIRE WRAPPING PAPER:** A dry or water-finished paper made from sulfate pulp, sometimes mixed with rope fiber, having one colored side. It is slit into small narrow rolls and used to wrap tires (automotive, tractors, trailers, bicycles, etc.) to protect them from mechanical damage during shipment and storage and to prevent harmful light (ultra-violet) deterioration of the rubber. Also called *automobile tire wrap.*

**TISSUE FILLER:** An anhydrous form of either natural or calcined calcium sulfate ($CaSO_4$), derived from ground gypsums, and used as a filler for tissue-type papers. *See also* ALABASTER, ANHYDRATE, PEARL FILLER.

**TISSUE MANILA PAPER:** Colored, thin lightweight manila paper made from hemp, jute, cotton, chemical wood pulp, or a mixture of them.

**TISSUE PAPER:** Thin, low weight, gauze-like types of paper made from virgin and/or reclaimed pulp, used to manufacture such items as sanitary products, wrapping material, protective packing paper, stock for waxing and twisting, etc.

**TITANIUM DIOXIDE ($TiO_2$):** A white manufactured oxide of titanium pigment having a high brightness. It is used in the paper mill as a filler or coating material to improve the brightness and opacity of the sheet of paper due to its inherent high refractive index.

**TITRATION:** A laboratory procedure used in pulp and paper mills to determine the quantity of substances in raw materials, process materials, and product solutions by reaction with measured amounts of standard chemical solutions.

**TMP:** *See* THERMOMECHANICAL PULP.

**TOBACCO PAPER:** A colored tobacco package wrapping paper made from sulfite chemical and mechanical pulps.

**TOCL:** *See* TOTAL ORGANIC CHLORINE.

**TOE-IN ANGLE**: *See* CANT ANGLE.

**TOFA:** *See* TALL OIL FATTY ACIDS.

**TOGGLE SWITCH:** A manually operated electric switch with a small projecting knob or arm that may be placed in either of two positions (on or off), and will remain in that position until changed. Other types are made for operation in more than two positions.

**TOILET TISSUE PAPER:** One of the most common lightweight grades of soft, clean, absorbent, sanitary paper.

**TOLERANCE:** (1) The deviation permitted in maintaining a specified dimension in the manufacture or between fitting of machine parts in operation, such as in the distance between plates in pulp refiners. (2) Allowable variations in specifications in basis weight, caliper, thickness, and other paper qualities, and the quantity shipped against that ordered in runs of paper on a paper machine.

**TONED PAPER:** Any grade of paper made specifically with pale creamy shades.

**TONER:** A water soluble dye. Also a heavy metal salt.

**TONS PER DAY (TPD):** An expression of throughput and production of pulp and paper in mills.

**TOOTH:** The rough surface finish characteristics of the types of paper to be used for drawing with pencil or crayon. Also expressed as *bite.*

**TOP BLOW:** Rapidly lowering the pressure in the dome of a batch digester for a pre-set time period to cause a boiling effect and to agitate the chip mass. This is done to facilitate a subsequent clean blow (removal of pulp) from the bottom.

**TOP FELT:** The felt on a double-felted paperboard machine press that contacts the upper surface of the sheet.

**TOP GLAZING:** The defect produced on the surface of a sheet of paper having high spots when it is calendered. This results in the higher spots having a higher gloss than the surrounding area, creating a poor appearance along with poor printability.

**TOP ROLL:** The roll in a paper machine wet press arrangement that comes in contact with the sheet of paper. It has a small-grained roughness or tiny craters to provide easy release of the wet web without pulling fibers away from the surface.

**TOP SHEET:** *See* TYMPAN PAPER.

**TOP SIDE:** (1) The paper surface that is opposite the paper machine wire (forming fabric) and comes in contact with the felts on the wet press. Sometimes referred to as the *felt side.* (2) The better quality surface of a paperboard made up of two different pulp stocks.

**TOP SIZED PAPER:** *See* SURFACE SIZED PAPER.

**TOP WEAR:** A paper machine operating problem consisting of undesirable wear on the upper surface of the wet end wire (forming fabric) due to an upward jam or dent, heavy riding deckle, improper positioning of the apron cloth, improper positioning of the slice, rough surface on the lump breaker, dandy, or outside return rolls, slippage between rolls and the wire, etc.

**TOPLINER:** The outer layer of a multi-ply paperboard.

**TOPPING TURBINE:** A steam turbine in a powerhouse that exhausts to the mill process steam system instead of to a condenser.

**TOPWIRE FORMER:** Former unit(s) normally located on top of a fourdrinier unit, in effect, adding a forming "side" to the top of the sheet.

**TOR:** *See* TALL OIL ROSIN.

**TORN DECKLE:** An edge made on paper by tearing the wet or dry web so that it resembles an edge made by the deckle of a paper machine.

**TORN SHEETS:** A paper finishing room expression referring to sheets of paper in a ream or stack that are of smaller size than the rest of the sheets due to tear-outs and defective paper that should have been removed during the sorting operation.

**TOS:** *See* TEMPERATURE OUT OF HEAT EXCHANGER.

**TOTAL AIR:** The total weight of air used by a power or chemical recovery furnace in order to supply the proper amount of oxygen for efficient combustion of the fuel or liquor. Depending on the design of the furnace, it may be divided into primary, secondary, or tertiary, and admitted at different locations.

**TOTAL ALKALI:** (1) The alkali present in sulfate pulping process cooking liquor ($NaOH + Na_2S + Na_2CO_3 + \frac{1}{2}Na_2SO_3$) expressed as grams per liter of $Na_2O$. (2) The alkali present in soda pulping process cooking liquor ($NaOH + Na_2CO_3$). Sometimes expressed in other equivalent units.

**TOTAL ALKALINITY:** The sodium sulfide ($Na_2S$) plus sodium hydroxide ($NaOH$) plus sodium carbonate ($Na_2CO_3$) plus sodium sulfite ($Na_2SO_3$) in sulfate pulp mill liquors, expressed in grams per liter of $Na_2O$.

**TOTAL CHEMICAL:** All of the sodium salts contained in sulfate- or soda-pulping cooking liquor, expressed as grams per liter of $Na_2O$, and is sometimes expressed in other equivalent units.

**TOTAL CHEMICAL LOSS IN ALKALINE PULPING:** A percentage determined by dividing the total chemical fed to alkaline pulping digesters into the total chemical in makeup chemicals, such as salt cake to the black liquor recovery boilers.

**TOTAL DISSOLVED SOLIDS (TDS):** The total amount of dissolved organic and/or inorganic materials in pulp and papermaking solutions, such as black liquor, white water, other process liquids, and liquid waste. It can be determined by evaporating a filtered sample to dryness and calculating the weight of the residual in pounds per gallon.

**TOTAL DYNAMIC HEAD:** The sum of the discharge head and the suction lift developed by pumps.

**TOTAL HEAD:** The pressure measured at the bottom of a pressurized storage vessel. It represents the sum of the pressure produced by the level of material contained within, plus any other applied pressure such as the level of the stock plus the pad pressure behind the slice of a paper machine headbox.

**TOTAL HEAT OF STEAM:** The heat necessary to convert 1 pound of water at 32°F to 1 pound of steam at the temperature of the boiling point.

**TOTAL IMMERSION TEST:** A paperboard mill test procedure used on sized heavy boards to determine their water resistance by completely immersing a sample in water for a specified period of time after weighing. The amount of water absorbed is measured by weighing the sample after immersion.

**TOTAL ORGANIC CHLORINE (TOCL):** The amount of organically bound chlorine that has carried over from the bleaching process into the waste treatment area. A specific test is made on samples of wastewater to determine TOCL content. *See also* ADSORBABLE ORGANIC HALIDE.

**TOTAL REDUCED SULFUR (TRS):** The total sulfur compounds existing in gaseous sulfate pulp mill emissions, condensate, and wastewaters that are the result of the reduction of sulfur by various existing components, typically hydrogen sulfide ($H_2S$), methyl mercaptan ($CH_3SH$), dimethyl sulfide ($CH_3SCH_3$), and dimethyl disulfide ($CH_3SSCH_3$).

**TOTAL SOLIDS:** The sum of dissolved and suspended constituents in pulp and paper mill process liquids, white water, chemicals, liquors, and waste effluents.

**TOTAL SULFUR DIOXIDE ($SO_2$):** An expression related to the strength of sulfite-pulping cooking liquor as determined by an iodometric titration testing procedure.

**TOTAL SUSPENDED SOLIDS (TSS):** The total amount of undissolved solid particles of organic and/or inorganic materials in pulp and papermaking solutions such as black liquor, white water, other process liquids, and liquid wastes. The amount of total suspended solids can be determined by filtering a sample and weighing the contents on the filter paper or by centrifuging it.

**TOTAL TITRATABLE ALKALI (TTA):** (1) The ($NaOH + Na_2CO_3 + Na_2S$) in the sulfate pulp making process liquors, expressed as grams per liter of $Na_2O$. (2) In the soda pulping process it is the ($NaOH + Na_2CO_3$). Sometimes expressed in other equivalent units.

**TOTAL TREE CHIPPING:** *See* WHOLE TREE CHIPPING.

**TOTAL VOLATILE SOLIDS (TVS):** The solids in pulp and paper process liquors, solutions, and slurries that will vaporize on heating.

**TOUCH PAPER:** Paper treated with selected chemical salts in order to pre-establish the way it will ignite and burn.

**TOUGH CHECK:** A high-strength type of paperboard made from chemical pulp, sometimes mixed with rope fibers, on a cylinder-type machine and used extensively for converting into tags, tickets, and other similar purposes.

**TOUR:** *See* SHIFT.

**TOWELLING PAPER:** A class of highly absorbent paper with fairly good strength properties. It is made from chemical and mechanical pulps especially for industrial, domestic, and institutional wiping purposes. The converted product is known as *paper towels*.

**TPD:** *See* TONS PER DAY.

**TRACHEID CELLS:** The fibrous, long, narrow cells of coniferous wood that make up the principal fiber structure of most papermaking pulps made from it. *See also* FIBER.

**TRACING PAPER:** A strong, high grade paper made from bleached chemical pulp and rag fibers, and treated mechanically in beaters and refiners until it can be formed into a low freeness transparent sheet. It is used for purposes allowing printed, written, and drawn matter to be viewed through a single thickness, and also for copy making.

**TRADEMARK PAPER:** Specially watermarked paper with a particular design that identifies its source and distinguishes it from other paper of the same grade.

**TRAILING BLADE COATER:** *See* BLADE COATER.

**TRAJECTORY FORMING:** A method of producing dry-formed paper by conveying fluffed fiber into a venturi section of a headbox where air carries it into the forming chamber and a vacuum is used to deposit it in a mat on a wire (forming fabric).

**TRANSDUCER:** An instrument that receives information in the form of one physical quantity and converts it to information in the form of the same or other physical quantity.

**TRANSFER PRESS:** A type of wet press consisting of an arrangement of one suction-type roll and plain roll combination. It is used to both remove water from the sheet before it is dried on heated dryers and to transfer the web from a top felt to a bottom felt (or remove the web from the felt).

**TRANSFORMER:** An electrical device used in pulp and paper mills to transfer energy in an *alternating current (A.C.)* system from one circuit to another with attendant increase or decrease in voltage. It consists essentially of two independent electrical circuits linked with a common magnetic circuit, thereby making it possible for energy at low voltage to be transformed into energy at high voltage or vice versa.

**TRANSIENT STATE:** A condition in which a process or control system is in a state of change.

**TRANSLUCENCY:** A characteristic of a material that scatters light passing through it in such a manner that objects behind it at a distance cannot be seen clearly, as with certain types of paper sheets.

**TRANSLUCENT PAPER:** A soft type of coated paperboard, made from chemical pulp, which is surface-coated and finished to a fairly high gloss. It is used as covers for pamphlets, booklets, brochures, etc.

**TRANSMITTANCE:** The property of a material that allows incident light to pass through it, such as found in sheets of certain types of paper.

**TRANSMITTER:** An instrument that responds to measured process variables by means of a sensing element and converts the measurement to a standardized transmission signal, which is a function of the measurement.

**TRANSPARENCY:** The ability of a material to allow light rays to pass through it in such a manner that objects behind it can be clearly seen, as with sheets of certain grades of paper.

**TRANSPARENT PAPER:** Types of usually thin papers through which objects held up against the back surface can be clearly seen.

**TRANSPORTATION LAG:** A delay caused by the time required for materials to move from one point to another.

**TRANSUDED MOISTURE:** The water that exists in the pores or interstices of the fiber in a paper sheet.

**TRASH IN ROLL:** Any extraneous material blown in or caught up into a roll of paper during its manufacture, such as dust, dirt, loose paper, etc. *See also* LOOSE PAPER.

**TRAVELING GRATE STOCKER:** A type of floor in a power furnace, usually coal burning, that conveys the fuel progressively through the furnace as it burns and discharges the ashes at the opposite end.

**TRAVELING WATER SCREEN:** A moving cleaning device used at the inlet of a pulp and paper mill open water source, such as a lake or river, to remove the debris from the water before it is treated and used in the process and for cooling purposes.

**TRAY:** A collector plate located under paper machine wet end wire (forming fabric) table rolls and press rolls to collect water that is squeezed or drained from the sheet and direct it into a storage chest for recirculation or filtering and reuse.

**TRAY CLARIFIER:** A type of multi-compartment tank used to separate settleable solids from green liquor in the recausticizing operation in a sulfate pulp mill. It consists of a flow box that distributes the raw green liquor to three or more tiers or trays that are scrapped with rotating arms to push the settleable solids toward the center where they fall and are discharged into a washer.

**TREE CROWN:** That part of a tree that is made up of leaves, branches, stems, and twigs.

**TREE SPECIES:** A distinct type of tree or a group of trees having one or several common characteristics that distinguish them from another tree or group of trees.

**TREE LENGTH:** A general reference to pulpwood that has not been cut into *shortwood* lengths (10 ft or less), and is transported to a pulp mill in lengths of 12 ft to 55 ft. Also called *random length* and *long log.*

**TRIAL RUN:** A small order made on a paper machine for the purpose of doing experimental work or to obtain samples on new grades for evaluation or customer approval.

**TRICKLING FILTER:** A synthetic or rock media filter system used in the treatment of some paper and paper mill wastewaters in which the biological growth removes waste organics from the wastewater as the wastewaters trickle through the media.

**TRIM:** (1) The dimension of the widest sheet of paper that can be made on a paper machine, not including the edges (which are normally cut off). (2) those edge portions of the sheet on the paper machine that are cut or trimmed off, or a reference to the process of cutting the edges off.

**TRIM CONVEYING SYSTEMS:** Paper broke and trim removal systems used after paper winding and finishing operations in a mill. They can be classified into four basic categories: trim handling systems (trim discharged into a repulper below the winder), high velocity air injector systems, combined shredder and transport fan systems, and vacuum systems.

**TRIM IN ROLL:** A paper roll defect caused by the winder slitter trim shavings inadvertently being wound up in the roll.

**TRIM LOSSES:** *See* TRIM WASTE.

**TRIM SQUIRT:** A pressurized jet of water used on the wet end of a paper machine to cut the sides of the wet web so that the desired width leaves the wire.

**TRIM WASTE:** Any wastepaper resulting from the trimming operation of the edges of a sheet of paper. Also called *trim losses.*

**TRIMMED PAPER:** Sheets of paper usually cut on a guillotine-type cutter on two or more sides so that the corner angles are precisely square.

**TRIMMED SPLICE:** A joint made in a torn continuous sheet of paper during processing in which the two ends are trimmed first.

**TRIMMED FOUR SIDES ($T_4S$):** A stack of paper (usually a ream package) that was cut to size by trimming the sheets on all four sides.

**TRIMMER:** A paper-cutting machine used in paper finishing operations. It consists of a long, heavy, mechanically driven knife used to cut stacks of paper sheets. Commonly known as a *guillotine.*

**TRIMMER STICK:** A wooden or plastic tool designed and used under the knife of paper trimmers in paper mills.

**TRIPLE PAPER:** Paper made up of three sheets.

**TRIPLEX PAPER:** A single sheet of paper made up of three plies built while it is being made or stuck together with an adhesive.

**TRIPPER:** A device installed on chip conveyors, positioned over pulp mill chip bins, and mounted on four flanged wheels which ride on tracks on each side of the conveyor belt. The conveyor belt is looped back over two pulleys mounted in a frame so that the belt has a slope that allows the chips to discharge from it into the chip bin.

**TRMP:** *See* THERMOREFINER MECHANICAL PULP.

**TROUBLESHOOT:** A mill maintenance expression meaning to search for the cause of a process and/or process equipment malfunction or breakdown so that it can be corrected to prevent recurrence.

**TROUGH:** A channel built in a "V" shape and used to transport logs by chain or water, and also to convey chips from one location to another.

**TRS:** *See* TOTAL REDUCED SULFUR.

**TRUCK:** (1) A variety of wheeled vehicles used in pulp and paper mills to move or carry paper and other loads from one location to another. (2) A type of automobile used to transport material to and from the mill site.

**TRUING:** A mechanical pulp mill term referring to the procedure of putting a pulpwood grindstone in perfect concentricity by dressing or burring with a diamond pointed burr.

**TRUNK BOARD:** A rigid paperboard made for the purpose of making luggage and storage trunks.

**TSS:** *See* TOTAL SUSPENDED SOLIDS.

**TTA:** *See* TOTAL TITRATABLE ALKALI.

**TUB COLORING:** The process of coloring paper or paperboard by application of a dye to the paper after it is formed and dried on the machine, and then passed through a dye tub or flooded nips of rolls. Also referred to as *dipping* or *surface coloring*.

**TUB SIZE PRESS:** A type of machine located between the dryers of a paper machine, used for applying sizing materials to the surface of formed and dried paper. It consists of an immersion tub or vat with two press rolls oriented so that the bottom one picks up liquid-sizing material from the vat and applies it to the paper web as it passes through the nip. One is usually made of rubber and the other of brass. Sometimes known as a *size tub*.

**TUB SIZED PAPER:** Paper whose surface is treated with a sizing material after the sheet is formed on the paper machine, usually by immersion in a vat or *tub size press* located between the dryer sections. Also referred to as *vat sized paper*.

**TUBE BOARD:** Types of unsized paperboard used as a base from which the material is cut into narrow strips for the manufacture of mailing tubes and similar containers.

**TUBE ROLLS:** A number of parallel rolls of steel or brass tubing designed to support the wire (forming fabric) on the wet end of a paper machine by forming a level table on which the fabric runs. Often called *table rolls*.

**TUGBOAT ANNIE:** A familiar nickname for a breaking-type pulper used to break up and defiberize baled pulp in the stock preparation operations of paper mills.

**TUMBLING BARREL BARKING DRUM:** *See* BARKING DRUM.

**TUNNEL DRYER:** A large insulated tube or box-like container used to dry coated and uncoated paper and paperboard by running it through from one end of the dryer to the other using a number of conveying methods while circulating hot air around it. Sometimes referred to as *flat dryers* and *ladder dryers.*

**TURBIDIMETER:** An instrument used for the measurement of light scattering properties of liquids in order to determine the amount of suspended particles contained therein.

**TURBIDITY:** A physical characteristic of liquid solutions caused by the presence of suspended and colloidal particles that scatter and absorb light rather than allowing it to be transmitted in a straight line through the solution.

**TURBINE METER:** A flow measuring device having a bladed rotor mounted in a specially designed passage through which the fluid stream is directed. The flowing fluid stream rotates the rotor whose speed of rotation is measured and calibrated in rate of flow units.

**TURBINE PUMP:** A pump whose construction resembles that of a reaction type of water turbine wheel with stationary diffusion vanes between the impeller and the free waterway of the casing.

**TURBULENT FLOW:** The flow of liquid such that the velocity at any fixed point in the fluid varies irregularly, causing an unsteady motion of the particles. The motion varies in no definite manner.

**TURMERIC PAPER:** Paper that has been treated with the juice of the turmeric root so that it can be used in test work to detect the alkaline content of solutions.

**TURN INSULATION:** A strong, high dielectric paper with a low power factor. It is commonly made on a cylinder-type paper machine from sulfate chemical pulp or manila rope fiber and properly treated with insulating liquids so it can be used as insulation on electrical wires and cables.

**TURN-UP:** The completion of a roll of paper from the paper machine and the starting of a new one.

**TURNED EDGE:** The folded-over outer side of paper in a stack of sheets that is held in a creased position by the other layers in the pile.

**TURNING:** The procedure of periodically cleaning off the accumulation of pitch and fines in the pores of a grinder pulpstone by running a diamond cut burr across its face.

**TURNKNIFE:** A long-life "Z" shaped knife for use on pulpwood chippers.

**TURNOVER:** A paper roll defect caused by any sheet tear, cut, or hole, resulting in a loose paper edge folding back during the winding operation.

**TURPENTINE:** A mixture of trepenes, principally pinene, obtained by steam distillation of pine gum recovered from the condensation of exhaust vapors from the cooking of softwoods by the sulfate process. Sometimes referred to as *sulfate turpentine.*

**TURPENTINE RECOVERY:** A chemical process used in some sulfate pulp mills to recover turpentine from digester relief gases by condensation, distillation, and purification procedures.

**TURPENTINE TEST:** A laboratory test used in paper mills in which the time of turpentine penetration in a paper sample is measured as an indication of the degree of grease resistance.

**TVS:** *See* TOTAL VOLATILE SOLIDS.

**TWIN MOTOR SHEETER:** Reference to type of sheeter on which two motors are directly coupled to pull the roll drum, which determines draw on the cutter or knife drum.

**TWIN SCREW FEEDER:** A chip mover having two screw-like elements. It can be positioned at any point under a chip bin to pull out chips and discharge them in a constant stream on a conveyor belt.

**TWINWIRE FORMER:** A type of multi-ply paper or paperboard forming unit having two wires (or fabrics) between which the sheet is formed.

**TWINWIRE MACHINE:** Any type of paper, paperboard, or pulp sheet forming machine that utilizes two wires (forming fabrics), or has a *twinwire former.*

**TWINWIRE PAPER:** Paper made on paper machines whose wet ends are fitted with two wires, i.e., they have a *twinwire former.*

**TWINE PAPER:** High-strength paper made from chemical pulp for the purpose of cutting into narrow strips that are wetted and twisted to form a string or twine product.

**TWISTING PAPER:** A high-strength kraft paper made especially to be twisted into cords, twines, and yarns. It is used to make paper fabrics, rugs, and reed-like material used in the automobile and furniture industries.

**TWITCHING:** The hauling of logs by skidding or dragging with a horse-drawn skid during the harvesting of pulpwood.

**TWO-ROLL PRESS:** A type of paper machine wet press used to remove water from a sheet of paper before it is dried on heated dryers. It consists of an arrangement of one suction-type roll and plain roll combination.

**TWO-SIDED PAPER:** Finished paper whose surfaces are readily distinguishable from each other because of different color and texture characteristics produced during the formation and processing operations.

**TWO-SIDEDNESS:** The visual difference between the top or felt side of a sheet of paper and the bottom or wire side.

**TWO SIDES COATED:** *See* COATED TWO SIDES ($C_2S$).

**TWO-STAGE ACID-ALKALINE PROCESS:** A type of multi-stage sulfite pulping process in which the first cooking stage is done with a bisulfite or acid sulfite liquor, followed by a second stage at a higher pH after sodium carbonate or caustic soda solution is added.

**TWO-STAGE ACID-NEUTRAL PROCESS:** A type of multi-stage sulfite pulping process in which the first cooking stage is done with an acid or bisulfite liquor, followed by a second stage in which a bisulfite or neutral sulfite liquor is used at a higher pH level.

**TWO-STAGE ALKALINE-ACID PROCESS:** A type of multi-stage sulfite pulping process in which the first cooking stage is done under alkaline conditions and contains either caustic or carbonate, followed by a second cooking stage in which sulfur dioxide ($SO_2$) is added to lower the pH.

**TWO-STAGE NEUTRAL-ACID PROCESS:** A type of multi-stage sulfite pulping process in which the first cooking stage is done with bisulfite or neutral sulfite liquor at about a pH value of 6, followed by a second cooking stage in which sulfurous acid or sulfur dioxide ($SO_2$) is added to lower the pH.

**TWO-STAGE PEROXIDE-HYDROSULFITE BLEACHING:** A multi-stage procedure for bleaching groundwood and semichemical pulps using a two-stage treatment with peroxide and hydrosulfite.

**TWO-TERM CONTROLLER:** A process control device that has been designed to have two modes of operation: proportional and integral, or integral and derivative.

**TWO-WAY VALVE:** A type of valve used in pulp and paper mills designed with two flow openings and used to control or shut off the flow of streams.

**TYMPAN PAPER:** A uniform caliper paper with a high finish, widely used in the printing industry on presses for backing between the platen and the printed sheet. Also called *top sheet*.

**TYPEWRITER PAPER:** Bond and writing papers specifically cut to size and packaged for convenient use on typewriters.

# U

**U-TUBE MANOMETER:** A liquid column gauge built in the shape of a "U" and used as a basic measurement device for gauging low and differential pressures.

**UF:** *See* ULTRAFILTRATION.

**UHLE BOX:** *See* VACUUM BOX.

**ULTIMATE ADHESION:** The force required to lift a strip of adhesive coated paper from a surface to which it had adhered for an extended length of time.

**ULTIMATE BIOCHEMICAL OXYGEN DEMAND:** The total quantity of oxygen required to completely satisfy the entire biochemical oxygen demand of pulp and paper mill wastewaters.

**ULTRA-HIGH-SOLIDS EVAPORATION:** Method of black liquor evaporation in which solids are concentrated above the conventional 65 to 70 percent level to facilitate more efficient burning in the recovery boiler, thereby reducing energy requirements.

**ULTRA HIGH YIELD PULPS:** Pulps made with a process in which 85 to 95 percent of the original fibrous material is converted into pulp, such as chemimechanical pulps.

**ULTRAFILTRATION (UF):** A filtration technique in which pulp mill spent liquor effluents are passed, under pressure produced by means of a pump, between semipermeable membranes that retain polluting, toxic, and colored substances and molecularly separate useful products such as lignins, wood, sugars, etc.

**ULTRA HIGH YIELD SULFITE PULP:** A type of papermaking pulp manufactured by cooking softwood chips in bisulfite cooking liquor, followed by atmospheric refining under conditions that produce very high yields.

**ULTRAMARINE:** A lightfast, bright blue pigment used for coating and tinting of high-grade rag content paper. It is made by the calcination of clay, sulfur, sodium sulfate, sodium carbonate, silica, and coal.

**ULTRASONIC FLOWMETER:** A flow measurement device that uses the time difference between upstream and downstream sound wave travel in a fluid moving through a pipeline as the basis of determining the velocity of the stream which, in turn, is calibrated into volume flow rate units.

**ULTRASONICS:** Sound waves in the frequency of about 20kHz. Sound at this frequency is sometimes used to detect chips, pulp, and other solid material levels in storage and holding tanks.

**ULTRAVIOLET LIGHT (UV):** Invisible light made up of rays that are shorter than visible violet light rays of the same spectrum. It is used by optical instruments for testing the fastness of paper and identifying certain foreign ingredients in paper.

**UNBEATEN PULP:** Pulp fiber that has not undergone any type of mechanical treatment with particular reference to beaters.

**UNBEATEN STOCK:** Stock slurry made up of pulp fibers that have not undergone any type of mechanical treatment with particular reference to beaters.

**UNBLEACHED PAPER:** Any type of paper made from pulp that was not treated in any bleaching process.

**UNBLEACHED PULP:** Pulp that has not been treated in a bleaching process and can be used as is in inferior grades of paper and paperboard.

**UNBOUND WATER:** Water in hygroscopic material, such as paper, that is in excess of the equilibrium water content, corresponding to the saturation water.

**UNCALENDERED PAPER:** Paper made on a paper machine and rolled up directly on a reel without going through the calender stack.

**UNCOATED:** A term used in conjunction with types of paper and paperboard that have been manufactured without the use of coating materials.

**UNCOATED FREE-SHEET:** Grades of paper made up of essentially all chemical pulp with no mechanical pulp and without any applied sheet coating material. It is used for printing and writing purposes.

**UNCOATED GROUNDWOOD:** A grade of paper that has properties similar to newsprint and is used as inexpensive printing paper for directories, catalogs, advertising circulars, periodicals, etc.

**UNCOATED UNBLEACHED KRAFT BOARD:** A grade of paperboard made from sulfate pulp without treatment.

**UNCOATED WEIGHT:** *See* RAW WEIGHT.

**UNDERCOOKED CHIPS:** Partially digested wood chips in the pulp slurry leaving a digester in a pulp mill. They are screened out and usually recycled back to the digester. Also called *underdigested chips.*

**UNDERDIGESTED CHIPS:** *See* UNDERCOOKED CHIPS.

**UNDERFIRE AIR:** Controlled, forced, or induced combustion air fed beneath the grate and passed through the bed of a solid fuel furnace.

**UNDERLAY BOARD AND PAPER:** Paperboard and paper (base material) from which pieces are cut to be used for placing under a cut or printing form in a printing press to bring it up to the proper printing level.

**UNDERLINER:** A layer of fiber in a multi-ply paperboard that is formed between the top surface layer and the filler layer.

**UNDERRUN:** A quantity of paper made for an order that is within the established tolerance specifications but less than the theoretical quantity for that order.

**UNDERSLICE SHOWERS:** A shower pipe system located under the slice of a paper machine, providing a layer of water on the wire between the slice and breast roll for the stock to ride on.

**UNDERTONE:** The color produced by transmitted light on thin pigmented films.

**UNDERWATER WOOD STORAGE:** An outdoor, on-site pulpwood storage system used by pulp mills in which the logs are immersed in a pond equipped with the necessary cranes and conveyors to get them into and out of the storage pond.

**UNEVEN DRAW:** Improper tension in the sheet of paper between the various sections of a paper machine.

**UNEVEN WINDING:** A random edge formed at the end of a paper roll during winding caused by the lateral movement of the core or sheet.

**UNFINISHED PAPER:** Paper that comes directly from a paper machine and goes through no additional surface treatment before it is used.

**UNGLAZED:** Types of uncoated paper of the same grade but without a high glossy polished finish, as compared with those having such a finish.

**UNIDIRECTIONAL WEBS:** A class of extremely soft and absorbent, nonwoven sheets produced like textile fabrics on garnets or cards in which they align themselves so that the greater strength results in the machine direction. They are used for making sanitary napkins, towels, wiping cloths, and medical items.

**UNIFLOW CYLINDER MACHINE:** A cylinder-type paper machine in which the stock flows in the vat in the same direction as the rotating mold.

**UNIFORM FORMATION:** The even distribution of fibers in a sheet of paper that produces uniform *translucency*.

**UNIFORMITY:** The sameness of a certain quality, character, or property of pulp, paper, or paperboard, such as brightness, strength, color, finish, distribution of fibers, formation, etc.

**UNION:** (1) A very common pipe fitting used in mills to join or couple the ends of the pipelines. It is designed especially to facilitate frequent connection and disconnection. (2) An organization of mill employees for the purpose of group representation in collective bargaining and other interests of the members with the mill management.

**UNION LABEL:** An identifying mark attached to packages of pulp, paper, paperboard, and by-products indicating they have been made by organized labor.

**UNIT OF PULPWOOD:** A total of 168 cubic feet of rough wood, bark, and air space. It is traditionally considered to be a stack of pulpwood logs that measures 6 ft x 4 ft x 5 ft-3 in.

**UNIT SLAKER CAUSTICIZER:** A combination green liquor gravity type slaking and causticizing unit used in the recausticizing operation in a pulp mill in which burned lime is mixed and slaked in one unit.

**UNLINED CORRUGATED PAPER:** A corrugated paperboard that has no facings and is used as a wrapping material for glassware and the lining of shipping cartons for fragile articles.

**UNOXIDIZED BLACK LIQUOR:** Spent black liquor in a sulfate pulp mill that has not been subjected to air or oxygen treatment prior to evaporation and combustion for environmental purposes. *See* BLACK LIQUOR OXIDATION.

**UNPROTECTED BUNDLES:** Packages of paper that are made up without the use of stiff heavy board top and bottom protectors for the sheets.

**UNREDUCED SALT CAKE:** Sodium sulfate ($Na_2SO_4$) passing through the chemical recovery furnace in a sulfate pulp mill and existing in the green liquor. It is usually expressed in grams per liter.

**UNREELING STAND:** *See* UNWINDING STAND.

**UNREFINED PULP:** Pulp fiber that has not undergone any type of mechanical treatment with particular reference to conical and disk refiners.

**UNREFINED STOCK:** Stock slurry made up of pulp fibers that have not undergone any type of mechanical treatment, with particular reference to conical and disk refiners.

**UNSEALED STOCK:** Ream stacks of paper that have not been wrapped.

**UNSIZED:** A descriptive term used to distinguish paper and paperboard made without the use of any type of sizing material.

**UNSLAKED LIME:** Dry lime (CaO) that will hydrate and form limewater [$Ca(OH)_2$] when mixed with water.

**UNTRIMMED PAPER:** Sheets of paper cut on slitters and rotary cutters but not squared off on guillotine-type cutters.

**UNWIND ADHESION:** The force needed to unwind a roll of adhesive coated paper under a certain set of conditions.

**UNWINDING STAND:** A holding stand fitted with bearings and used in paper mills to hold rolls of paper to be rewound on a winder or to be supercalendered. An adjustable brake is used to adjust the sheet tension as the roll size decreases while rewinding. Also referred to as an *unreeling stand*.

**UPFLOW BLEACHING:** Multi-stage pulp bleaching process in which the movement of the pulp and chemicals is from the bottom to the top of a vertical retention tower.

**UPFLOW CLARIFICATION:** A procedure used on wastewater from deinking of wastepapers in which the wastewater is introduced at the bottom of the treatment unit so that it is filtered through a blanket of suspended sludge.

**UPFLOW DIGESTER:** Vertical, continuous-type, pressurized cooking vessels in which the movement of chips is vertical from the bottom to the top.

**UPSIDE-DOWN PRESS:** *See* INVERSE PRESS.

**UPSTREAM:** Some thing or event that preceded or occurred before those under current consideration.

**UPTIME:** The amount of time a piece of equipment or process is in service and operating satisfactorily.

**UPWARD FLOW FLAT SCREEN:** A flat-type stock cleaning system in which the pulp slurry flows upwards through a series of screen plates, entering between plates and a vibrating diaphragm, causing the heavier dirt particles to settle below the screen so they can be carried away.

**UREA:** A chemical base used to make chemical substances, such as ureaformaldehyde resins. They are used in the manufacture of paper to impart wet strength to the sheet.

**UREA-FORMALDEHYDE RESINS:** Chemical substances used in the manufacture of paper and paperboard to enhance the physical strength of the sheet when wet.

**USED KRAFT BAGS:** A grade of wastepaper used as a source of fiber for making up certain types of paperboards. It consists of used bags made from sulfate paper in baled form.

**UV:** *See* ULTRAVIOLET LIGHT.

# V

**V-BELT:** A rubber composition connection belt made so that the top surface reduces to the bottom surface in the shape of a "V." It is used in V-shaped sheaves on various machine drives.

**V-PORTED VALVE:** A valve whose plug is cut with a V-shaped groove or opening.

**VACUUM:** The absence of or negative pressure created by the action of air, or artificially by the use of air pumps, or a space from which most of the air or gas has been removed, leaving its pressure less than atmospheric.

**VACUUM BOX:** A perforated, covered, narrow box extending across and over the area through which the paper machine wire (forming fabric) runs so that it will remove water from the forming paper web by the use of suction created by a pump connected to it. It is also used to remove water from wet felts (press fabrics) in the same manner. Sometimes referred to as a *suction box.* and a *uhle box.*

**VACUUM EVAPORATION:** Concentration of spent wash liquors in pulp mills by removing water with heat, usually in multiple-effect units maintained under negative pressure by an evacuation system.

**VACUUM FILTER:** Any type of slurry filter in which suction is employed to deposit and form a pad of solids on the surface of a separating material (screen) with the liquid flowing through it.

**VACUUM GAUGE:** An indicating device showing the degree of negative pressure that is created or exists in process equipment.

**VACUUM PICKUP:** The lifting of the paper web from the wire (forming fabric) of a paper machine in order to transfer it to the presses by using a suction-type roll that is lowered onto the wire between the couch and the drive roll, pressing a pickup felt against the sheet.

**VACUUM PUMP:** A pump designed and used to create and maintain a suction in mill process equipment, such as suction boxes, suction cylinders, and some types of filters.

**VACUUM SAVE-ALL:** A cylinder- or disk-type vacuum filter used in paper mills to reclaim fibers from white water, as well as for deckering, washing, or dewatering pulp stock.

**VACUUM TRANSFER PRESS:** Rolls used in paper machine presses for transferring the sheet from one press to another by the use of vacuum. The sheet is pressed against a pickup felt and becomes sandwiched between it and the press felt.

**VACUUM WASHER:** Pulp mill equipment designed to separate undesirable components in a pulp slurry from the acceptable fibers using an evacuated, partially immersed, screen-covered cylinder to pick up the fibers from the slurry to form a mat, which is removed for further processing.

**VALUE:** The level of a process variable being measured or controlled.

**VALUE OF COLOR:** The increments of paper colors related to their brightness or dullness with reference to a gray scale. It is usually measured with a photometer.

**VALVE:** A variable orifice used to shut off or adjust the flow of a fluid by direct manual operation with a handwheel or lever, or remote operation with a signal generated by a manual station or automatic control device.

**VALVE MANUFACTURERS ASSOCIATION OF AMERICA (VMA):** Headquarters: Washington, D.C., (202) 331-8105.

**VALVE POSITIONER:** A device mounted on remote-operated valves with direct connection to the stem. It is furnished with its own power supply in order to cause the valve plug to assume a position proportional to an instrument output signal by overcoming stem friction, valve plug and stem unbalance, and diaphragm and spring hysteresis.

**VALVELESS FILTER:** Type of cylindrical vacuum pulp washer frequently used in brownstock washing and bleaching processes where the operation depends on vacuum produced by suction channels in the filter drum alternately connected to drop legs, instead of by means of a vacuum pump.

**VANILLIN:** A product made from sulfite and sulfate lignin by heating with an alkali. It is used in the manufacture of flavoring materials and perfumes.

**VAPOR:** The gaseous phase of substances that normally are liquids at atmospheric temperature and pressure, such as digester blow gas, steam, and other similar mill effluents, which can be condensed by increasing pressure while holding the temperature constant or lowering the temperature at constant pressure.

**VAPOR ABSORPTION SYSTEM:** A piping system used in paper machines to remove vapor leaving paper, which collects in pockets formed by the dryer.

**VAPOR ABSORPTION UNIT:** A relatively high pressure, semi-high temperature air supply unit which is a part of the air system of a paper mill. It consists of crossed air nozzles extending across the width of the dryer face, and having either extended or depressed fittings directing air at the sheet so as to remove vapors formed in the drying process.

**VAPOR COMPRESSION (V-C) EVAPORATION:** A method of evaporation used on pulp mill spent wash liquors in which the steam and vapor removed from the boiling liquid being concentrated is compressed and fed back to the outside of the evaporator tubes, where it condenses and transfers its latent heat to the boiling liquid.

**VAPOR ECONOMIZER:** A heat reclaimer used on the vapor removed from the dry end of a paper machine.

**VAPOR PERMEABILITY:** The paper sheet property related to its ability to pass vapor through it, exclusive of water vapor or air, which is more closely related to *air permeability* and *porosity*, sometimes incorrectly associated with it.

**VAPOR PHASE INHIBITED (VPI) PAPER:** Paper used as wrapping material for metallic items in which the anticorrosion and antitarnish properties have been instilled by treatment with a vapor phase inhibitor.

**VAPOR PHASE PULPING:** A chip cooking process sometimes used in a sulfite pulp mill in which fully impregnated chips are separated from the excess impregnation liquor. Cooking subsequently takes place predominately in a vapor phase.

**VAPOR PLUME:** Stack effluent consisting of flue gases made visible by condensed water droplets or mist.

**VAPOR PRESSURE:** The pressure at which saturated vapor exists in equilibrium with the liquid at a specific temperature and is a function of temperature only. *See also* ABSOLUTE HUMIDITY.

**VAPOR PROOF PAPER:** A paper sheet or board through which vapors cannot pass. It usually consists of a multi-layer laminate with a vapor resistant coating or an interposed layer barrier material. It is commonly used a food packaging material.

**VAPOR RECOMPRESSION EVAPORATOR (VRE):** A procedure used in the concentration of pulp mill liquors by evaporation in which the vapor is compressed and returned to the shell side of the evaporator to be used as a heat source to evaporate more water.

**VAPOR RECOVERY SYSTEM:** A system used to collect various mill vapors, such as digester blow gases, from pulp mills and process them to prevent their emission into the atmosphere. This is accomplished by re-routing them to the drop leg of the chlorination filter in the bleach plant or to a furnace for incineration purposes.

**VAPORIZATION:** The change of a substance from the liquid to the gaseous state.

**VAPORIZERS:** Specially constructed, thermostatically controlled vessels used to convert liquid chemicals, such as chlorine, into their gaseous state so they can be used in processes such as the chlorination of pulp during bleaching and the manufacture of hypochlorite bleach liquor in a pulp mill bleach plant.

**VAPORSPHERE:** A kraft pulp mill term for a spherical vessel located on the roof of the digester house that is used to collect blow gases in order to reduce the odor released to the atmosphere by rerouting them to the drop leg of the chlorination filter in the bleach plant or to a furnace for combustion purposes.

**VARIABLE PITCH SHEAVE:** A grooved wheel that carries the connecting drive belt on a machine drive assembly. The pitch diameter can be changed by moving the driving motor base, thereby adjusting speed.

**VARIABLE SPEED WINDER DRIVE:** A type of drive on a paper winder that allows the unit to be started at a slow speed and gradually increased as the paper roll starts to build up.

**VARIABLES:** Quantities and conditions associated with the pulp and papermaking process, the value of which are subject to change with time.

**VARIATOR:** An electrical device used on rotary vacuum filters to adjust the rotational speed of the wire-covered cylinder on which the pulp mat is formed.

**VARNISH:** Gums or resin solutions in solvents such as alcohol and linseed oil used as a paper and paperboard coating film material.

**VARNISH LABEL PAPER:** Paper made from bleached chemical pulp specifically to be converted into labels that are subsequently varnished.

**VARNISHABILITY:** That property of a sheet of paper or paperboard related to smoothness and low absorbency, facilitating coating of the surface with gums or resin solutions.

**VARNISHING MACHINE:** Mechanical equipment used to apply an overall coat of gum or resin solution on sheets of paper and paperboard.

**VASCULAR PLANT:** Pulp and papermaking plant fiber source in which the formation of lignin is essentially unique in developing tissues that specialize in such functions as transportation of aqueous solutions and mechanical support.

**VAT:** Any holding container used to retain quantities of pulp slurries, process chemicals, and other liquids for specific purposes, such as picking up fibers from pulp slurries with a wire screen hand mold, or a screen-covered, rotating cylinder used to form a sheet or mat. Such a container is also used to hold dyes and sizing and coating materials through which a web of paper can be passed for coloring, sizing, and coating.

**VAT CIRCLE:** The curved wall and floor located below and around the cylinder mold in the vat of a cylinder-type paper machine.

**VAT COLORS:** *See* VAT DYES.

**VAT DYES:** Water-insoluble, pigment-type, lightfast dyes usually applied to the paper web by immersing in vats. Also called vat *colors.*

**VAT DYESTUFFS:** Water-insoluble paper tinting coloring materials that have very good fastness to light, acids, alkali, chlorine, etc. They are made from complex organic compounds, solubilized by sodium hydrosulfite reducing agents.

**VAT LINED:** A multi-ply paperboard made on a multi-cylinder paper machine, consisting of a series of vats with different types and/or colors of stock from which individual sheet layers are made and subsequently squeezed together in a press. *See* VAT MACHINE.

**VAT MACHINE:** A type of paper machine used primarily to make paperboards. It is characterized by the use of wire-covered cylinders or molds, rotating partially submerged in vats of dilute stock suspension from which the fibers are picked up to form a sheet on its surface. *See* CYLINDER MACHINE.

**VAT PAPER:** Paper usually made one sheet at a time by dipping special wire screen molds in vats of pulp stock and allowing water to drain away by gravity.

**VAT SIZED PAPER:** *See* TUB SIZED PAPER.

**VAT SYSTEM:** A continuously recirculating system on a cylinder-type paper machine in which the new stock from a regulating box is constantly added to the drainage water and from which the couched sheet is constantly removed by a pickup felt.

**VEGETABLE GUMS:** Paper machine wet end additive materials obtained from guar, locust bean, and deacetylated gum karaya, used to improve bursting strength, tensile strength, and the folding endurance of the finished sheet.

**VEGETABLE PARCHMENT:** A hard, dense sheet of grease-resistant paper, having a very high wet strength property. It is used extensively for greaseproof wrapping purposes. It is usually made by passing a sheet of paper through sulfuric acid which softens the outer portion of the cellulose fibers. An immediate neutralization with ammonia converts it into a hard coating.

**VEHICLE:** The binder and volatile thinner liquid portion of vapor finishing coaters.

**VEINED:** A descriptive reference to paper and paperboard containing some various colored fibers (or fibers of varying color intensity) that show up on the sheets as a granite or mottled effect.

**VELLUM:** A high-quality parchment sheet made from young calves skins. It is used for making up deluxe books, certificates, diplomas, stationery, and other especially high-grade documents.

**VELLUM FINISH:** A smooth, dull finish applied to book and stationery paper surfaces to simulate that of sheets originally made from young calves skins.

**VELLUM PAPER:** A creamy white, high-quality, rag content, parchment-like paper finished to resemble sheets used in deluxe bookbinding, diplomas, certificates, and special grades of stationery. Once produced from young calves skins. Sometimes incorrectly referred to as *vellum*.

**VELOCITY:** The rate of motion or speed of a body at any instant.

**VELOCITY FORMING:** The formation of a sheet from the headbox onto the wire (forming fabric) of a paper machine under conditions in which the head in the headbox is less than the wire speed and is referred to as DRAG.

**VELOCITY HEAD:** The head required to give water or other fluids its velocity of flow through a pipe.

**VELOCITY METER:** A fluid flow measurement instrument that employs a vaned wheel driven at approximately the same velocity as the flowing fluid.

**VELOUR COATING:** Tufts of cotton, wool, or rayon applied to a paper surface with an adhesive varnish or lacquer in order to produce a suede-like appearance.

**VELOUR PAPER:** Box-covering decorative paper usually made from chemical pulp, coated with an adhesive varnish or lacquer, and dusted with tufts of cotton, wool, or rayon to produce a suede-like appearance.

**VELVET FINISHED PAPER:** A paper with a smooth finish without any gloss, simulating velvet.

**VENT:** (1) To reduce pressure. (2) An opening in an enclosed vessel to provide relief of excessive contents.

**VENT STACK:** A flue or chimney mounted on storage or process vessels for the purpose of relieving foul or stagnant air or gases.

**VENT TANK:** A vessel into which excess material, air, or gases are relieved for possible treatment or return to the process.

**VENTED NIP PRESS:** A grooved bottom press roll whose purpose is to reduce the fluid pressure gradient within the nip of the press on a paper machine by providing an area of escape for the water pressed out of the paper and felt.

**VENTILATING HOOD:** A canopy located over process machinery, such as a paper machine dryer or a pulp washer, to catch and confine the vapors before they have a chance to disperse in the room.

**VENTILATION:** The process of introducing fresh air and expelling foul or stagnant air from areas such as in a paper machine room, bleach plant, and other process areas in a mill.

**VENTILATOR:** A device designed to introduce fresh air and expel foul or stagnant air from an area or piece of process equipment in a mill.

**VENTURI FLUME:** A liquid flow primary measurement device consisting of an open flume with a contracted throat. The contracted throat causes a drop in the hydraulic grade line. It is often used to measure mill water and waste.

**VENTURI METER:** An instrument used to measure fluid flow in pipelines. It consists of a venturi tube and flow registering device, with the difference in velocity heads between the entrance and the contracted throat being an indication of the rate of flow.

**VENTURI SCRUBBER:** A type of direct contact evaporator and fume collector also used to remove dust from alkaline pulp mill recovery furnace flue gas. It is characterized by a venturi section into which alkaline black liquor is sprayed into the throat where fine dust and fumes are carried with the fluid gases and are mixed with it, forming small droplets that go through an evaporation process. The droplets then agglomerate, and larger dust particles are separated from the mixture in the cyclone-type evaporator section.

**VENTURI TUBE:** A fluid flow primary measurement device that consists of a closed conduit or pipe containing a gradual contraction to a throat. This contraction causes a pressure head reduction by which the velocity may be determined.

**VERDOL PAPER:** A high-strength, rigid paperboard or pressboard made primarily from jute fibers. Also called *jacquard board.*

**VERTICAL BLEACHER:** A pulp mill term for a batch-type pulp bleaching vessel in which the height is much greater than the width and the flow of pulp is in the vertical direction.

**VERTICAL FILM-TYPE EVAPORATOR:** *See* FALLING FILM EVAPORATOR and LONG TUBE EVAPORATOR.

**VERTICAL SHAFT DISK REFINER:** A machine used to mechanically treat fibers in pulp and paper mills. It consists of a single rotating disk with a vertical shaft.

**VERTICAL SLICE:** A type of paper machine headbox outlet that regulates the flow of stock onto the wire (forming fabric) by the use of a long narrow steel strip (called the "top lip") that is moved in a vertical direction. It can also be bent in its own plane for local slice adjustment in order to change the profile of the sheets of paper being formed.

**VERTICAL SPLITTER:** A log-cutting machine used in the wood preparation area of pulp mills in which large diameter logs are placed in the upright position to be split into smaller sizes so that they will slide through the chipper chute openings and fit into pulpwood grinder pockets.

**VERTICAL STATIONARY DIGESTER:** An upright type of cylindrical pressurized vessel in which chips and cooking liquor are charged, sealed, and heated with steam to reduce them into pulp in a batch-wise procedure.

**VERTICAL UNION:** *See* INDUSTRIAL UNION.

**VESSEL:** (1) A tube-like structure in hardwoods made up of a series of coalesced cells that conduct water in the tree. (2) A storage or holding container used in the pulping and papermaking processes.

**VESSEL ELEMENT:** One of the cellular components of tube-like structures found in hardwoods for the primary purpose of conducting water throughout the tree.

**VIBRATING ROTARY SCREEN:** A type of pulp screening device in which the rotating screening chamber or drum is put under vibration.

**VIBRATING SCREEN:** A cleaning device made with a screening surface that is mechanically vibrated at high speed. It is used throughout the mill to screen chips, coal, and other solid, dry materials.

**VIBRATION:** A quivering or trembling motion experienced by process equipment during normal operation or produced on it by external units designed for that purpose.

**VIBRATION ANALYZER:** An instrument used to measure the acceleration, velocity, or amplitude of vibrations as a means of identifying faults and disturbances around the wet end of a paper machine.

**VIBRATOR:** A device mounted on solid material storage vessels, such as chip and chemical bins, to provide a vibrating motion to the vessel's sides to aid movement of the mass downward, usually toward a bottom discharge opening.

**VIBRATOR CONVEYOR:** Short-distance, solid material moving unit used in pulp and paper mills to feed chemicals and other dry materials to mixers. It is constructed of suspended metal troughs that are electrically or mechanically vibrated to move the material from one point to another.

**VIBRATORY CHIP SCREEN:** A type of screening system used to separate good chips going to a pulp mill from the sawdust and slivers. It consists of a two-bearing, counterbalanced, eccentric rotating shaft to impart vibrating action on the screen body.

**VIBRATORY DEKNOTTER:** A type of coarse screen designed to remove knots and larger pieces of trash from pulp coming from the digesters in a pulp mill just before the washing process. It consists of a suspended screen basket kept under vibration.

**VIBRATORY STOCK SCREEN:** A screening device used for trash removal from pulp slurries, especially wastepaper stock. The primary element is a basket with a bottom made up of screen plates, and is freely supported in a vat. Stock enters the basket and is shaken over the plates by the application of a vibrating action.

**VINYL FIBERS:** Synthetic resin fibers used in making specialty papers by blending them with other synthetic and natural cellulose fibers in order to achieve various specific physical and chemical properties.

**VIRGIN FIBER:** Papermaking fiber unused in the papermaking process.

**VIRGIN PULP:** Pulp consisting of unused fibers. It contains no secondary or recycled fibers.

**VIRGIN STOCK:** Paper machine stock furnishes made of various pulps, additives, dyes, chemicals, etc., that have never been used.

**VISCOMETER:** An instrument used to measure the deformation or resistance-to-flow property of liquids. Sometimes called a *viscosimeter.*

**VISCOSE:** A cellulose base chemical compound solution used to coat and impregnate paper, and as a wet strength agent. It is also used to make rayon thread by extruding it through small openings into a coagulating bath that regenerates the cellulose.

**VISCOSE PULP:** Chemically treated pulp in which the cellulose has been converted to cellulose xanthate and used to regenerate the cellulose fibers into synthetic materials such as rayon.

**VISCOSIMETER:** *See* VISCOMETER.

**VISCOSITY:** The resistance to flow or deformation of a liquid or of the solution produced when pulp fibers are dissolved in a suitable solvent, indicating the length of the cellulose molecule component it contains.

**VISE:** A gripping tool made up of two adjustable jaws, usually drawn together by a screw, lever, or cam, and used to hold materials to be worked on by a squeezing force.

**VISUAL EFFICIENCY:** The ratio of the luminosity of the light reflected from a paper or pulp sheet surface to the luminosity of the light reflected from a perfect, diffuse, reflecting body.

**VMA:** *See* VALVE MANUFACTURERS ASSOCIATION OF AMERICA.

**VOCs:** *See* VOLATILE ORGANIC COMPOUNDS.

**VOID FRACTION:** The volume of open area or void (*recesses*) divided by the total volume of a sheet of paper and paperboard.

**VOIDS IN PAPER:** Open areas in a sheet of paper. Same as *recesses.*

**VOLATILE:** The property of a material related to the relative ease of evaporation at low temperatures.

**VOLATILE ORGANIC COMPOUNDS (VOCs):** Sulfate pulp mill organic air pollutants, such as phenols and terpenes, that originate in the digester and in the black liquor evaporation stages.

**VOLT:** A unit of electromotive force in an electrical system equal to 1 ampere of current multiplied by 1 ohm of resistance. Expressed as $E = IR$.

**VOLTAGE:** An electrical term referring to the force required to send 1 ampere of current through 1 ohm of resistance in an electrical system. It is analogous to head or pressure in a hydraulic system.

**VOLUME RESISTIVITY:** The volume resistance between opposite faces of a cubic centimeter of material, usually expressed in ohms per centimeter. It is determined in paper by measuring the resistance of electrical current between electrodes placed in contact with opposing surfaces of a specimen.

**VOMIT STACKS:** Vent flues mounted on top of the blow pits in sulfite pulp mills.

**VORTEX BEATER:** A stock preparation machine designed to combine pulping with beating. It has been used to disintegrate dried pulps and wastepapers in the manufacture of paperboards and deinking operations.

**VORTEX CLEANER:** *See* CYCLONE CLEANER.

**VORTEX METER:** A flow sensing device used to measure the flow of gases, liquids, and slurries. It utilizes the phenomenon that a regular series of vortices occur when these materials flow past a specially shaped object immersed in the flowing stream, with the number of vortices per unit of time being proportional to the flow rate.

**VORTRAP CLEANER:** One of the first pressure-drop, vortex-type, pulp-cleaning devices in which the pulp was pumped under pressure tangentially to the side of a conical chamber, with the heavier particles of dirt settling to the bottom to be removed and the cleaned stock passing out through the top.

**VPI PAPER:** *See* VAPOR PHASE INHIBITOR PAPER.

**VRE:** *See* VAPOR RECOMPRESSION EVAPORATOR.

**VULCANIZED FIBER PAPER:** A dense, hard paper or paperboard with high strength properties, produced by gelatinizing the surface cellulose with zinc chloride ($ZnCl_2$) or sulfuric acid ($H_2SO_4$) and followed by washing and calendering. It is used for electrical insulation, luggage, building material, mechanical assemblies, etc.

**VULCANIZING:** A process to convert paper and paperboard into a dense, hard, high-strength paper by treating it with zinc chloride ($ZnCl_2$) or sulfuric acid ($H_2SO_4$) which gelatinizes the surface cellulose.

**VULCANIZING PAPER:** Paper made especially for treatment with zinc chloride ($ZnCl_2$) and sulfuric acid ($H_2SO_4$) to gelatinize the surface cellulose which converts the paper into a hard, dense, tough sheet.

# W

**WAD-BURN:** The build-up of stock after a sheet break in the press nip that generates enough friction heat on a subsequent break to scorch the felt.

**WADDING:** A single- or multi-ply, loosely matted fiber sheet made from chemical pulp and used in packaging, thermal, and acoustical applications, and as a cushioning and protective medium. It is also used as a sanitary absorbent material and sometimes referred to more specifically as *cellulose wadding.*

**WAFERBOARD:** A wood board formed from knife-cut, flat, wafer-like chips and adhesives. Also called *flakeboard* and *particle board.*

**WALL FRACTION:** The percentage of a papermaking fiber radius that consists of the fiber wall.

**WALLBOARD:** Type of building construction board manufactured by forming and binding together a combination of wood chips and pulp.

**WALLPAPER:** A stronger grade of groundwood paper especially suitable for clay coating, printing, and/or embossing to make a wall covering. Sometimes referred to as *hanging paper.*

**WALNUT:** A hardwood tree species used as a source of fiber for making pulp and paper.

**WAREHOUSE:** A large room or building in a pulp or paper mill complex used for storage purposes.

**WARM COLORS:** Paper colors in the red, yellow, and orange range that produce a warm psychological visual effect as opposed to those in the blue, green, and violet range that are called *cold colors.*

**WARM-UP PERIOD:** The time required after starting a piece of machinery, device, or process operation before it can be brought up to its rated performance.

**WARP WIRES:** The strands of wire (fabric) in a paper machine wire (forming fabric) that run parallel to the machine direction and cross the wires (fabric strands) that run in the cross machine direction called *shute wires* or *weft wires.*

**WARPED BOARD:** A defective paperboard that does not lay flat because of a wet liner, too much tension on double face linings, drag on single face web, too much heat on double backer, or excessive moisture at the single face or the double face side.

**WARPING:** A common paperboard defect appearing as wavy edges or curl. It is usually caused by conditions under which the edges expand more than the center portion of the sheet.

**WASH PRESS:** A high efficiency, compact pulp washing unit fed thin pulp slurries which it discharges at high consistency.

**WASH ROLL:** The first outside roll on the return part of the wire (forming fabric) on a paper machine where the stock carried around with the wire is washed off and removed.

**WASH TANK:** *See* BLOW PIT.

**WASHBOARD MARKS:** Corrugation-like paper roll defects that run at an angle to the axis. They are caused by soft areas due to uneven paper thickness in the cross machine direction of the sheet.

**WASHBOARDING:** A paperboard defect appearing as if the single face or double face liner follows the flute formation as a result of using excessive adhesive, improper glue roll speed, or excessive moisture in the liner.

**WASHER:** Pulp mill equipment designed to separate soluble, undesirable components in a pulp slurry from the acceptable fibers. It usually consists of some type of screening method combined with diffusion and displacement with wash liquid, utilizing vacuum, or the natural force of gravity.

**WASHER SHOWERS:** Water sprayed on the pulp mat formed on the pulp washer drum as it comes out of the washer vat in the washing and blending operation of a pulp mill.

**WASHER VAT:** That portion of a pulp washer that holds the slurry from which the wire-covered cylinder picks up the fibers to form a mat for washing and transporting.

**WASHING:** The process of separating soluble, undesirable impurity components of pulp slurries from the fibers. Normally this is done after cooking, and during and after the bleaching operation by the use of some method of screening combined with the use of fresh water and other liquids.

**WASHING DEINKING:** A process of removing ink from recycled furnishes using *washing* equipment rather than *flotation deinking* technology.

**WASHING DRUM:** A wire-cloth covered skeleton cylinder set in a beater washer to remove dirty water during the washing of rags.

**WASHING UP:** The cleaning of a paper machine wet end section and associated chests, pumps, and piping, along with the wire felts, press rolls, slice, headbox, etc. It is accomplished with water during a shutdown, usually between running various grades of paper of different colors.

**WASTE:** Any nonrecoverable loss in production of a pulp or paper mill due to improper materials and operations, malfunction of machinery, accidents, etc.

**WASTE HEAT BOILER:** A type of boiler installed behind alkaline chemical recovery incinerators to recover heat from gases in pulp mills.

**WASTE LIQUOR:** Used cooking liquor in a chemical pulp mill that is separated from the pulp after the cooking process and is not recoverable but discarded, a practice not normally followed today due to environmental pollution restrictions and the economic necessity to reduce chemical loss.

**WASTEPAPER:** All types of used paper or paper (discarded or not considered fit for a particular use) that provides a source of fiber for the manufacture of some papers, paperboards, and chipboards. Wastepaper is usually processed in a separate plant section of a mill that includes pulping/deinking (but not cooking), screening, cleaning, etc. Sometimes known as *old paper.*

**WASTEWATER:** Water carrying waste materials from a mill. It is actually a mixture of water, chemicals, and dissolved or suspended solids.

**WATER ABSORPTION:** The amount of water absorbed in a given time by one side of a unit area of paper or paperboard placed horizontally under a head of 1 cm of water. Sometimes referred to as a *cobb test.*

**WATER BACK:** Pressurized water used to move the pressure foot of a pulp grinder back to the position where the magazine can be reloaded with logs.

**WATER COOK:** A trial cook made in a newly constructed pulp digester using water only (with no chips or cooking liquor) for the purpose of checking out the installation and operation and to make adjustments to the equipment before actual production is started.

**WATER COOLED:** A descriptive term for machinery or components such as rolls, drives, or reactors to indicate that they are designed so that water, brine, or some other water solution can be circulated through them for cooling purposes to prevent damage from overheating.

**WATER CREPE:** A simulated, crinkly finish effect produced on a sheet of tissue or lightweight paper in an off-machine converting operation. *See* SECONDARY CREPE.

**WATER DROP TEST:** A test used on folding boxboard to measure the degree of sizing.

**WATER FELT:** The process of forming a felt mat on a screen by pouring a water suspension of fibers on it and allowing the water to drain through.

**WATER FINISHED (WF) PAPER:** Paper having a high-glazed finish produced by moistening the sheet with water or steam during the calendering operations in a paper mill.

**WATER GLASS:** A sizing material used in papermaking to provide water and ink penetration resistance in the sheet, to make an adhesive for producing corrugated paperboard, and in the lamination of solid fiberboard. Chemically named *sodium silicate* ($Na_2SiO_3$).

**WATER HARDNESS:** A measure of calcium and other salts in mill water, usually expressed as mg/liter of calcium carbonate ($CaCO_3$). It can cause scale deposits in power and recovery boilers, cooling systems, and other problems in the process.

**WATER JETS ON PAPER MACHINE:** Fine jets of water directed onto the paper machine wire (forming section) with enough force and so located as to cut the wet web to the correct width. One jet is sometimes made adjustable so it can be gradually fed across the wire to form a tail or leader strip on the wet sheet to facilitate feeding it to the wet presses and through the dryers.

**WATER LINES:** A surface design effect on a sheet of paper appearing as lines running parallel to the grain. It is produced by parallel wires laid around the surface of the dandy roll on the wet end of a paper machine.

**WATER LOG TRANSPORT:** A method of moving pulpwood logs to a pulp mill via waterways.

**WATER METER:** A device installed in a pressurized pipeline for measuring and registering the quantity of water passing through it.

**WATER POLLUTION:** The contamination of any receiving rivers, streams, lakes, etc., by pulp and paper mill liquid effluents, harmfully altering the physical, chemical, or biological properties of these natural bodies of water.

**WATER REPELLENCY:** The resistance of paper and paperboard to wetting by water.

**WATER RUN:** A trial run made on newly installed pulp and paper mill process equipment using only water. *See* WATER COOK.

**WATER RETENTION:** The ability of paper or paperboard to hold water.

**WATER SOFTENER:** An apparatus designed to remove the dissolved calcium and magnesium minerals that produce hardness from water to prevent scaling in power and recovery boilers and other problems.

**WATER SPRAYED LOGS:** Pulpwood logs stored under a water spray to protect them from fungi and insects.

**WATER STORED LOGS:** The storage of pulpwood logs in water at a pulp mill.

**WATER STREAKS:** A type of paper defect appearing as light streaks running in the machine direction. They are the result of fewer fibers and more water accumulating in these areas during the formation of the wet web on the wet end of the paper machine.

**WATER SUPPLY:** The primary source of natural water used in a pulp and paper mill, such as streams, rivers, lakes, and wells.

**WATER TREATMENT:** The processing of mill source waters from rivers, lakes, and streams to remove impurities by sedimentation, filtration, and the addition of chemicals such as alum, sodium carbonate, and chlorine. Such treatment is normally done to make the water suitable for use in pulping and papermaking processes.

**WATER TUBES:** Pipes aligned along power and recovery furnace walls that contain water for heating and conversion into steam. Collectively called *water wall.*

**WATER TURBINE:** A machine in which a wheel is turned by jets of water impinging on peripheral blades and used to run other machinery and electrical generators.

**WATER VAPOR PERMEABILITY:** The property of a sheet of paper that allows water vapor to pass through it.

**WATER VAPOR RESISTANT:** A descriptive reference to paper and paperboard coated with moisture-repellent, film-forming material, such as asphalt, resins, and wax, so that they and products made from them will resist the passage of water vapor.

**WATER WALL:** The bank of pipes (*water tubes*) that lines the walls of the power and recovery boiler and contains the water to be heated and converted into steam.

**WATERCOLOR PAPER:** A medium weight, well-sized paper made with a coarse grain surface so that it will be suitable for painting with water colors.

**WATERBOX:** A container used on a calender, designed to provide a means of applying water, starch, or similar materials directly to the surface of the sheet, usually on heavier types of paper and paperboards.

**WATERJET CUTTING:** The severing of materials such as paper, paperboard, plastics, etc., using a fine pressurized stream of water at a high velocity.

**WATERLEAF PAPER:** A soft, loosely felted, absorbent type of paper made with no sizing, thereby having the specific property of readily taking up water, ink, and other liquids.

**WATERMARK DANDY ROLL:** A wire-covered skeleton roll covered with a woven wire cloth and located on the top of a paper machine wet end wire (forming fabric). It has a raised surface pattern that impresses the wet web so that the design appears in the final dried sheet. *See also* WATERMARKS and DANDY ROLL.

**WATERMARK PAPER:** Paper made with a distinguishing design that is used for identification as to source, quality, etc.

**WATERMARKS:** Translucent marks or designs made in paper as the result of modifying the fiber orientation of the wet web. It is done by impressing the sheet with raised patterns on rolls so that the localized, displaced fibers produce more transparent areas in the final dried sheet. They are usually used for identification purposes. Also known simply as *marks. See also* WATERMARK DANDY ROLL.

**WATERPROOF PAPER:** A type of paper made water-repellent by impregnating or coating the sheet with certain substances, or laminating a number of sheets with a material such as asphalt that resists the penetration of water.

**WATERPROOFNESS:** A property associated with paper and paperboard whose furnish has been heavily sized, or the sheet coated or impregnated with a water-repellent substance to make it highly resistant to the absorption and penetration of water.

**WATT:** A unit of electrical energy that represents the product of ampere and voltage, when 1 ampere of current flowing at 1 volt produces 1 watt of energy.

**WAVE DANDY ROLL:** A type of dandy roll on the wet end of a paper machine that is covered with a woven wire cloth so that no wire marks are impressed on the paper.

**WAVE PAPER:** A general reference to paper made without any visible wire (forming fabric) marks on the bottom side of the sheet, except for a possible imprint left by the woven wire (fabric) on which it was made.

**WAVINESS:** A paper defect caused by variations in moisture content due to changing atmospheric conditions. It produces distorted edges along the grain of the sheet.

**WAX EMULSION:** Paraffin and microcrystalline wax solutions mixed with paper stock furnish as a sizing agent to make the sheet more resistant to moisture. It is produced by vigorous agitation after mixing with other emulsifying chemicals. *See also* PARAFFIN WAX and WAX SIZE.

**WAX HOLDOUT:** The effect obtained by treating paper and paperboard with fluorochemical size to prevent opacity loss, brightness loss, color change, and strike-through when a sheet is waxed.

**WAX PICK TEST:** A paper mill control test to determine adhesion properties of paper. It consists of melting a series of arbitrarily graded wax sticks, allowing them to cool on the paper surface, and then pulling them away to identify at what stickiness level the fibers will pull away. *See also* SURFACE STRENGTH.

**WAX SIZE:** An emulsified solution of waxes and/or fatty substances added to the furnish stock in a beater or applied to the surface after the sheet is made. This will then make it more resistant to moisture. *See also* WAX EMULSION.

**WAX SPOTS:** A type of defect found in paper made with wax emulsion sizing, in which undispersed particles of wax are visible in the sheet.

**WAXED:** A descriptive reference to any type of paper and paperboard to which paraffin wax has been applied either to the surfaces or by impregnation to make it more resistant to moisture.

**WAXED KRAFT:** Paper made from pulp manufactured by the sulfate cooking process and coated with paraffin or other type of wax.

**WAXED PAPER:** Paper used primarily for moisture-resistant wrapping purposes, especially with foods. The paper is coated or impregnated with wax by dipping it into a molten paraffin solution.

**WAXING PAPER:** Any type of paper made specifically for use as a base in the manufacture of wax paper.

**WAYBILL MANILA:** A yellow colored lightweight paper used for copy sheets by railroads. Also referred to as *yellow waybill tissue.*

**WBLO:** *See* WEAK BLACK LIQUOR OXIDATION.

**WEAK ACID:** The acid solution formed in the cooking liquor preparation process of a sulfite pulp mill by dissolving the excess sulfur dioxide gas from the absorption tower in a solution that is later fortified to produce strong cooking acid used to cook chips in the digester.

**WEAK BLACK LIQUOR:** Diluted, spent cooking liquor resulting from the washing of sulfate pulp from the blow tank in a pulp mill. This liquor is the washwater not yet concentrated by evaporation.

**WEAK BLACK LIQUOR OXIDATION (WBLO):** Treatment of spent sulfate black liquor with air or oxygen (for environmental purposes) immediately after washing it from the pulp and before it is concentrated to about 15 percent solids in evaporators. *See* BLACK LIQUOR OXIDATION.

**WEAK WASH:** A liquid stream in the kraft process that results from lime mud washing and is used mainly for dissolving smelt from the chemical recovery furnace.

**WEATHERING:** The natural degradation of pulpwood, pulp, paper, paperboard, etc., due to extremes of moisture, temperature, sunlight, etc.

**WEATHEROMETER:** Device used to determine the life of a film coating by subjection to ultraviolet light, water, etc.

**WEATHERPROOF PAPER:** Paper such as roofing felt that is treated to withstand extremes in weather conditions.

**WEAVE:** A defect in a paper roll that produces a side-to-side oscillating motion of the sheet during the winding operation in papermaking.

**WEB:** A continuous sheet of paper produced and rolled up at full width on the paper machine. It is not cut into sheets, but is used directly from rolls for converting into consumer goods.

**WEB CALENDERED PAPER:** Paper whose surface was finished or glazed by passing it through a calender in one continuous web.

**WEB CONTRACTION:** The shrinking of a paper web due to internal tensions that are created within it by the effect of drying or changes in relative humidity.

**WEB DRAW:** The tension developed in a sheet of paper between sections of the machine on which it is being made.

**WEB GRAB:** *See* AUTO ADHESION.

**WEB TENSION:** The tension that exists in a sheet being made on a paper machine caused by the pull of faster moving sections, such as calender stacks, reels, winders, etc., and by shrinkage in the dryer section.

**WEBBING:** *See* STRINGINESS.

**WEDDINGS:** A class of very high quality writing papers of medium to heavy basis weight, especially formulated for ease of folding, and used primarily for printing announcements of weddings or other important social events.

**WEDGES:** Wooden tools designed and used to drive into and separate previously slit rolls of paper in a paper mill.

**WEDGE SAMPLING:** A method of taking samples from baled pulp to make moisture content determinations. It consists of obtaining wedgeshaped specimens from representative bales in predetermined lots.

**WEEPS:** *See* BLADE BLEEDS.

**WEFT WIRES:** *See* SHUTE WIRES.

**WEFTLESS FELTS:** Felts made without cross directional fibers in order to improve cross machine directional flow in paper machine presses.

**WEIGHT OF PAPER:** The weight of a ream of paper, in pounds, normally consisting of 480, 500, or 1,000 sheets of specified size as stated in the standard definition of the basic size for the referred grade. The standard ream size varies with different grades of paper according to trade practices. The weight is sometimes converted and expressed as the weight of an arbitrary standard ream of 500 sheets, 24 x 36 inches.

**WEIGHT TOLERANCE:** The amount that the weight of paper or paperboard is allowed to vary and still be considered within specifications for a particular order in accordance with standards set by the industry.

**WEIGHTOMETER:** A scale designed for weighing materials being carried on a conveyor belt.

**WEIR:** A diversion dam or flow measuring device located in a liquid stream, and having a crest and side containment of known "V," trapezoidal, or rectangular shape. The flow is related to the upstream height of water above the crest.

**WELDING ENDS:** Pipeline and fitting connection ends that are designed for welding to another section or fitting.

**WELL-CLOSED SHEET:** Paper made with an even distribution of fibers throughout the sheet. The appearance is thus very smooth and highly uniform when observed with translucent light.

**WELL-SIZED:** *See* HARD-SIZED.

**WELTS:** Paper formation defects appearing as a series of ridges and valleys running along the machine direction in a sheet.

**WET BOTTOM:** A type of ash removal system in a power boiler using low ash fusion temperature coal. The ash is removed in a molten state, quenched in a slag tank, and pumped out as slurry.

**WET BREAK TEST:** A test used by some paper mills to measure the degree of sizing and wet strength of paper with test strips of paper being immersed in water or other suitable test liquids.

**WET BROKE:** Wet paper removed from a paper machine at the wet presses (prior to drying) because of machine breakdown, startup, shutdown, off-spec product, etc.

**WET BULB TEMPERATURE:** A temperature measurement made in the ambient air of a paper mill under saturated conditions and compared with actual temperature, called *dry bulb temperature*, to determine the relative humidity.

**WET COMPACTIBILITY:** The ability of papermaking fibers to collapse into a compact mass so that they will produce a denser sheet when formed into paper.

**WET DRAW:** The looseness or tightness of the wet web intrasection tension in the wet end of a paper machine.

**WET END:** The section of the head end of a paper machine, which includes the headbox, wire (forming fabric), and wet press sections, where the sheet is formed from the stock furnish and most of the water is removed before entering the dryer section. Sometimes called *wire end*.

**WET END ADDITIVES:** Any substances, such as slime preventives, defoamers, retention aids, drainage aids, fillers, etc., added to the paper machine furnish between the stock preparation operation and the wire (forming section) on a paper machine.

**WET END FINISH:** Any surface finish on a sheet of paper obtained by some operation on the web or treatment of the web at the wet end of a paper machine.

**WED-END BARRING:** Cyclic or random machine direction basis weight variations that occur at frequencies usually higher than 10 Hz. *See also* BARRING.

**WET FELT:** *See* PRESS FABRIC and FELT.

**WET FINISHED PAPER:** *See* WATER FINISHED.

**WET LAPS:** *See* LAPS.

**WET MACHINE:** A paper machine consisting essentially of a wire-covered cylinder rotating in a vat of pulp stock on which a mat of varying thickness is formed by drainage. These mats are removed either intermittently in thick sheets called laps, or continuously. A wet machine is used to manufacture certain types of paperboards.

**WET MACHINE BOARD:** A gray-colored, single-ply, glazed board made on a wet machine, and used primarily for binding hardcover books and in shoe manufacturing.

**WET NIP:** The line of contact between rolls that is normally wet, such as in the wet press section of a paper machine.

**WET PACK PAPER:** A special grade of paper made by sizing the surfaces of the sheet in a separate unit (off the paper machine) equipped with a sizing tub and dryers. Sometimes referred to as *wet soak paper.*

**WET PRESS:** One of a number of dewatering units comprising that portion of the paper machine called the *press section.*

**WET PULP:** *See* LAPS.

**WET ROLL:** A defective roll of paper that has inadvertently become damp or wet, causing the sheets to stick together and then dry with discolored separation of the sheet plies.

**WET RUB:** The ability of wet paper to withstand the lifting of fibers or scuffing when rubbed by or on another surface.

**WET SOAK PAPER:** *See* WET PACK PAPER.

**WET SPOT:** A paper defect consisting of any small area in a web that is wetter than the rest of the sheet. It may have been caused by any bruise or depression in a press roll, and produces a blemish when passing through a calender.

**WET STEAM:** Saturated steam that is accompanied by entrained water particles.

**WET STREAKS:** Paper defects appearing as darkened stripes oriented in the machine direction of the paper. They are caused by nonuniform drying across the sheet.

**WET STRENGTH:** The tensile strength of a sheet of paper when completely wet, sometimes calculated as a percentage of its dry strength.

**WET STRENGTH PAPER:** A paper in which the fiber constituents and/or the sheet were chemically treated to enhance resistance to tear, rupture, or falling apart after becoming saturated with liquids.

**WET STRENGTH RESINS:** Synthetic chemical substances, such as urea- and formaldehyde-based resins, used in the manufacture of paper and paperboard to improve the physical strength of a wet sheet.

**WET TENSILE STRENGTH:** The resistance of a paper sheet to pull or stress produced by applied tension after it has become saturated with liquids.

**WET WEIGHT:** The total weight of freshly formed pulp sheets, paper, and paperboard as they come off of a wet lap machine and the wet presses before entering any drying operation.

**WET WRINKLE:** A paper defect showing up as a wrinkle or blemish in the sheet that originated on the wet end portion of a paper machine. *See also* WRINKLE.

**WET-WAXED PAPER:** Waxed paper made by coating the sheet with molten wax followed by an immediate cooling over chilled rolls or with cold water.

**WETLAY PROCESS:** A method of sheet formation used in making paper. It implies that its fiber constituents are carried by water and deposited, with the water being drained from it to form a web in a moist or wet condition, as contrasted with the *air laid process* in which the formation is in a dry condition.

**WETNESS:** The capacity of fibers in a pulp slurry to absorb and hold water as developed by mechanical treatment of the pulp in refiners and beaters. Sometimes referred to as *freeness* and *slowness*. The phenomenon is known as *hydration*.

**WETNESS OF FELT:** The ability of paper machine wet press felts to accept water squeezed out of a sheet of paper at the nip; the wetter the felt, the less acceptance of water.

**WETTABILITY:** The affinity of paper and paperboard to liquids.

**WETTED:** A term used to indicate that a substance has a strong affinity to water.

**WETTING:** Any treatment that increases the quantity of water physically absorbed by fibers.

**WETTING AGENTS:** Materials used in paper and paperboard to improve their affinity to liquids and to make them more absorbent. It is also used in the preparation of adhesive coating materials used in the paper and paperboard industry.

**WETTING DOWN WOOD PILES:** A pulp mill practice of routinely hosing down pulpwood logs in storage piles to prevent deterioration.

**WETTING OUT A FELT:** An operation conducted on new wet felts for paper machines to assure there is a film of water over each of the thread fibers in the woven fabric before putting it in operation. It is accomplished by running the felt around with a large amount of shower water and detergent until all the intrafiber air is removed.

**WETTING TEST:** A test used to determine paper and paperboard surface wettability characteristics by observing the spread of water drops on a sheet sample.

**WF PAPER:** *See* WATER FINISHED PAPER.

**WHATMAN PAPER:** *See* BALSTON'S PAPER.

**WHEAT STRAW:** A fiber source used in the manufacture of less expensive grades of wrapping papers and corrugating paperboard.

**WHIPPING SPEED:** *See* CRITICAL SPEED.

**WHIRLIES:** Fixed, gallows-type cranes used to move materials in the pulp and paper industry. They incorporate power hoists that can be swung in a full circle.

**WHIRLING SPEED:** *See* CRITICAL SPEED.

**WHISKERING:** A defect occurring during the blade-coating of paper. It consists of a buildup of coating on the downstream side of the blade caused by the extrusion of semi-dry color under the blade.

**WHISKERS:** *See* BLADE BLEEDS.

**WHITE BROKE:** Paper trimmings or paper damaged as the result of a breakdown in the paper machine or finishing operations, or anywhere else in the process of paper manufacturing. This material contains no dyes or other coloring material.

**WHITE DEXTRINS:** Starches treated with acids and used in the preparation of adhesive coating materials in the manufacture of paper and paperboard.

**WHITE FIR:** A softwood tree species used as a source of fiber for making pulp and paper.

**WHITE LIQUOR:** Cooking liquor formed by refortifying green liquor in the causticizing operation of an alkaline-type pulp mill so that it contains the active chemicals that will reduce chips into their fiber components by dissolving the lignin cementing material during the digester operation, thereby producing pulp.

**WHITE LIQUOR CLARIFICATION:** The removal of calcium carbonate ($CaCO_3$) and other impurities from the causticizing liquor, usually by gravity sedimentation in units called clarifiers. This takes place in the liquor recausticizing process of a pulp mill in order to obtain a clear liquor for cooking wood.

**WHITE MUD:** *See* LIME MUD.

**WHITE PAPER:** (1) Any paper made from pulp stock whose natural color has been corrected by the addition of blue, yellow, and red dyestuff. (2) To a printer, it is any paper sheet devoid of printing material. (3) A general reference to high brightness, coated and uncoated printing and writing papers.

**WHITE PIGMENTS:** A class of pigment or pigment combinations, used in the formulation of paper and paperboard coating materials, that are characteristically white.

**WHITE PINE:** A softwood tree species used as a source of fiber for making pulp and paper.

**WHITE PITCH:** Sticky depositions with a whitish appearance, usually caused by some form of latex in the coating binder that occurs in paper mill recycling systems using broke with this type of coating binder.

**WHITE SHINER:** A glossy, white fiber that occurs in groundwood pulp or in chemical pulp and shows up later in the finished paper from which it is made.

**WHITE SIZE:** A rosin size used to impart water resistance characteristics to paper and paperboard. It is made up largely of unsaponified, emulsified, free rosin.

**WHITE SPRUCE:** A softwood tree species used as a source of fiber for making pulp and paper.

**WHITE WATER:** Mill waters that have a white, cloudy appearance due to a very fine dispersion of fibers picked up when separated from pulp suspension on paper machines, washers, thickeners, savealls, and other pulp-filtering equipment. It may also contain fine suspensions of sizing, dyestuffs, and filling materials, and is reused in the papermaking process or is refiltered to reclaim the suspended fibers. *See also* BACK WATER and REWATER.

**WHITE WATER PIT:** A tank, usually circular and lined with glazed tile, located at the side of the fourdrinier wire (forming) section of a paper machine or at the breast roll to receive white water drained from the table rolls and from the seal pit.

**WHITENER:** *See* BRIGHTENERS.

**WHITENESS:** The reflectivity of a sheet of pulp or paper for blue light or the attribute by which it approaches a theoretically perfect white. Commonly used interchangeably with *brightness. See* G.E. BRIGHTNESS and ELREPHO BRIGHTNESS.

**WHITING:** *See* PARIS WHITE.

**WHOLE STUFF:** Pulp that has been beaten in the stock preparation area of a rag paper mill.

**WHOLE TREE CHIPPING:** An operation, usually performed in the woodlands, consisting of converting the entire tree, including limbs, branches, bark, roots, etc., into chips to be used to make pulp in pulp mills. Also known as *complete tree chipping* and *total tree chipping.*

**WHOLE WOOD FIBER:** Fiber manufactured by the reduction of wood to a fibrous state without chemical action and with essentially minimum delignification so that the final product has about the same composition as the original wood, with yield approaching 100 percent.

**WICKING:** The tendency of liquid to move through a sheet of paper.

**WIDOW MAKER:** A pulpwood harvesting expression for a limb hanging on a tree or a tree that is lodged against another and is liable to fall at any time.

**WIDTH:** The cross-machine directional dimension of a paper machine or the shorter dimension of a sheet of paper.

**WILD FORMATION:** The disorganized orientation of nonuniform fiber distribution during the sheet forming operation. It is usually observed by translucent light. Also called *cloudy* and *snowstorm.*

**WILLOW:** A hardwood tree species used as a source of fiber for making pulp and paper.

**WINDBOX:** A plenum chamber attached to and surrounding the burners on a power boiler to distribute air furnished by a forced draft (FD) fan to the burner.

**WINDER:** A machine located after the dry end section of a paper machine that winds and sometimes slits the sheet into rolls as it comes off the dryers.

**WINDER CRACKS:** Puckers and wrinkles in paper, usually running in the machine direction, that break through the sheet when subjected to winder pressure.

**WINDER SHAFT:** A metal rod or tube around which paper or paperboard is wound into rolls for transporting and shipping.

**WINDING DRUM:** A revolving skeleton cylindrical frame around which paper emerging from the dryer section of a paper machine is wound to form rolls.

**WINDOW:** Reference to a beginner's fill-in job in a pulp and paper mill.

**WING DUSTER:** A type of machine used to remove dust from cut rags in a pulp mill rag preparation operation. It consists of a revolving drum with helically arranged wings inside a screen, revolving in the opposite direction and through which the cut rags to be cleaned are blown.

**WINGS:** Adjusting devices located on counterflow-type cylinder board machines. They control the levelness of paperboard made on the cylinder machine.

**WIPES:** Absorbent paper made into towels for cleaning purposes.

**WIRE:** A general reference to the *forming fabric* used on the wet end of a paper machine. These continuous screen belts, on which fibers in the pulp slurry form a sheet of paper as water drains away, are commonly made of woven plastic fibers (or monofilament strands), whereas they once were generally made of woven metal wires, such as bronze (some machine still use metal wires). The term may also be used to generally indicate the forming section of a fourdrinier type of paper machine. *See also* FOURDRINIER WIRE.

**WIRE CHANGING:** A common and essential operation performed in every paper mill, consisting of the removal of worn wires (forming fabrics) on a paper machine and the replacement with new ones.

**WIRE CLOTH:** The wire screen covering the rotating, filtering cylinder on pulp washers, savealls, and similar pulp and paper equipment.

**WIRE DRIVE ROLLS:** *See* WIRE TURNING ROLLS.

**WIRE END:** A paper mill expression referring to that portion of a paper machine in which the sheet-forming wire (forming fabric) is located. It is a part of the *wet end*, a term with which it sometimes is used interchangeably.

**WIRE GUIDES:** A mechanism attached to the paper machine wire (forming fabric) support frame in such a manner that it senses the sideways creep of the wire and automatically makes necessary adjustments to reposition it and prevent it from creeping too far in either direction.

**WIRE HOLE:** A paper defect appearing as a hole in the sheet caused by a hole in the wet end forming wire not retaining enough fiber to bridge the gap.

**WIRE LIFE:** The elapsed time from the initial installation of a paper machine wire (forming fabric) to the time it has to be removed and replaced due to wear and other conditions experienced under normal operation.

**WIRE LIFE EXTENDERS:** Chemicals added in small quantities to the white water system of the wet end of a paper machine for the purpose of increasing the length of time before the wire (forming fabric) has to be changed. These extenders retard corrosion and prevent chemical decomposition.

**WIRE LOADING:** The application of fillers to a paper sheet during the wet web formation on the wire (forming fabric) in the wet end of a paper machine.

**WIRE MARKS:** Small impressions produced on the bottom surface of a sheet of paper caused by the mesh of the wire screen (forming fabric) on which the wet web is formed in the wet end of a paper machine. If they are pronounced, they may be objectionable.

**WIRE PIT:** The large receiving chest located under the wire (forming section) at the wet end of a paper machine. It is used to collect drainage water resulting from sheet formation of the stock furnish. In addition, it provides a source of dilution water to the stock furnish that is being fed to the headbox of the paper machine.

**WIRE RETURN ROLLS:** Cylindrical rolls located on the return run of a fourdrinier paper machine wire whose functions are to drive, guide, and apply tension to the wire (forming fabric).

**WIRE ROLL DOCTORS:** Doctor blades mounted on wire (forming fabric) return rolls and paper machine wet end sections to prevent stock buildup that would damage the wire.

**WIRE SHAKE:** The side to side motion of paper machine wires (forming fabrics) running at slow speeds. It is caused by oscillating the table rolls on which the fabric is carried in order to align more fibers in the cross direction and improve sheet formation.

**WIRE SIDE:** The bottom side of a sheet of paper that comes in contact with the wire (forming fabric) as the web is being formed in the wet end of a paper machine.

**WIRE SPEED:** The rate of speed at which a paper machine wire (forming fabric) is moving or the maximum rate of speed it is designed to move. It is expressed in feet per minute, and is indicative of the paper production capabilities of the machine.

**WIRE SPOTS:** Any undesirable markings on a sheet of paper caused by defects in the wire (forming fabric) in the paper machine wet end.

**WIRE TURNING ROLLS:** Those rolls on the paper machine wet end whose functions include turning or helping to drive the wire (forming fabric). This function is sometimes performed by the couch roll or separate additional solid rolls. Also referred to as the *wire drive rolls*.

**WITH-THE-GRAIN:** An expression used to denote the dimension in paper that is parallel to the machine direction. It also describes the orientation of any association with the sheet such as fold, markings, defects, etc.

**WITHERITE:** A natural mineral, primarily barium carbonate ($BaCO_3$), from which paper-coating materials are made.

**WOOD:** That part of the stem of a plant, located between the bark and the pith, that is one of the primary sources for fiber used in the manufacture of pulp and paper.

**WOOD BROOMED ENDS:** A condition of pulpwood logs that results in the separation of fibers and causes a broom-like appearance.

**WOOD CELLULOSE:** The material that makes up the major part of the wood fiber. It has long, hair-like units measuring 2 to 4 millimeters in length, bound together by a glue-like substance called lignin.

**WOOD CHARGE:** The oven-dry weight of wood loaded into a batch-type digester. It is mixed with the proper amount of cooking and dilution liquors to make up one cook, and is usually expressed in tons.

**WOOD CHIPS:** *See* CHIPS.

**WOOD CHIP SIZE ANALYZER:** An instrument used to measure particle size distribution in a quantity of wood chips.

**WOOD CONVEYOR:** *See* LOG HAUL-UP.

**WOOD CORE:** A tube made of wood and used as the center around which paper rolls are wound.

**WOOD DENSITY:** The moisture-free weight per unit volume of wood, usually expressed in pounds per cubic foot.

**WOOD FEEDER:** The employee in a groundwood pulp mill whose duty is to fill the grinder magazines with pulpwood logs.

**WOOD FIBER:** An elongated, tapering, thick-walled, cellular unit that is the structural component of woody plants. Wood fibers are separated from each other during the pulping operation in a pulp mill and reassembled during the papermaking process in the paper mill.

**WOOD FLOUR:** *See* FINES.

**WOOD GRINDER:** A mechanical pulping machine that reduces wood logs into their fiber components through abrasion by pressing them against a revolving grindstone.

**WOOD GRINDING:** The mechanical pulping process in which wood logs are reduced into their fiber components through abrasion by pressing them against a revolving grindstone.

**WOOD HANDLING:** All operations conducted on pulpwood logs from the time the tree is cut down in the woodlands to its being converted into chips, including its transportation to the pulp mill site.

**WOOD PARENCHYMA:** The living elements or cells of wood, located in the sapwood, that contain starch, oil, and resins.

**WOOD PILE:** A method of outdoor storage of pulpwood logs in a pulp mill yard when they are to be held in reserve for chipping and converting into pulp. The logs are usually stacked in large mounds.

**WOOD PREPARATION:** Those operations conducted in a pulp mill to prepare wood for pulping, including removal of bark from logs, converting them into chips, and screening.

**WOOD PULP:** *See* PULP.

**WOOD PULP BOARD:** A paperboard made from mechanical, chemical, or wastepaper pulp, or all three.

**WOOD REFUSE:** Wood chip rejects from pulp-screening operations that are chipped and crushed (hog fuel), and bark that is used as a fuel, alone and in combination with coal.

**WOOD ROOM:** A room or building in a pulp mill where a system of machines and conveyors clean the wood, remove bark from logs for chipping, and screen the wood into sizes suitable for cooking in digesters.

**WOOD SPEAR:** *See* SPEAR POLES.

**WOOD SPECIES:** A method of classifying pulpwoods based on their biological features.

**WOOD YIELD:** A pulp mill expression indicating the quantity of pulp produced from a specific amount of wood. Usually expressed as pulp per unit volume (cord or cunit) of wood.

**WOODFREE:** *See* FREE SHEET.

**WOODYARD:** The area on a pulp mill site where pulpwood logs or chips are stored and processed (screened, sorted, etc.) prior to being used by the mill.

**WORD:** The basic size of a piece of information making up process computer languages used for communication purposes. It consists of bits or characters treated as a unit and capable of being stored in one computer location.

**WORMED ROLLS:** Rubber-covered paper machine wet end press rolls built with a worm-like surface design to provide a working action on the fibers of the felt passing over it, thereby keeping it in a spread and open condition.

**WOVE PAPER:** A reference made to papers produced by the use of woven type wires on the wet end of the paper machine that show no laid or chain lines.

**WRAP CURL:** A type of paper deformation that occurs across the grain of a sheet when it takes a set after being wrapped around a core in a roll for an extended length of time. Also called *roll curl* and *roll set.*

**WRAPPER:** *See* BODY WRAP.

**WRAPPED PAPER:** A paper machine dryer with paper wrapped around it after a break in the sheet.

**WRAPPED ROLL:** A paper machine dryer with paper wrapped around it after a break in the sheet.

**WRAPPER PULP SHEETS:** Larger size sheets of pulp used to cover a stack of pulp sheets to protect them from being soiled during transportation and/or storage.

**WRAPPING PAPER:** Paper with high strength and of different weights made especially to be used for wrapping purposes.

**WRAPPING STATION:** The area in a paper mill finishing department where stacks or rolls of paper are wrapped in preparation for shipping.

**WRINKLED ROLLS:** *See* BAGGY ROLL.

**WRINGER PRESS:** A type of press made up of solid rolls and used on paper machines to condition felts.

**WRINKLES:** Paper defects appearing as puckers or creases, caused during its manufacture. The defect would be called *wet wrinkle* if produced while the sheet is wet, usually at the wet end of the paper machine, and *dry wrinkle* if produced while the sheet is dry, usually at the dry end of the paper machine.

**WRITING PAPER:** Any grade of paper that is specially sized so that it can be satisfactorily used for pen and ink writing purposes.

**WRONG SIDE OF PAPER:** The bottom or wire (forming fabric) side of watermarked paper that is opposite from the upper or felt side, and from which the watermark cannot be properly viewed. *See also* RIGHT SIDE OF PAPER.

**WYE (Y)**: A type of pipeline fitting with three ports, located at other than 90° angles with each other. Same as *y (wye)*.

# XY

**X-PAPER:** A fairly strong paper made from sulfate pulp with a surface that readily accepts crayon, pencil, and other marks. It is used primarily to make patterns for designing and making clothes. Also called *black pattern paper.*

**XYLAN:** A gum-like substance usually extracted from the hulls of oats and used as a wet end additive in the manufacture of paper.

**XYLEM:** The part of a tree located between the bark and the pith which forms the mechanical structure that constitutes the wood from which the fiber is obtained for use in pulp and papermaking.

**Y (WYE):** A type of pipeline fitting with three ports located at other than 90° angles with each other. Same as *wye (y).*

**YANKEE DRYER:** A type of steam-heated paper dryer consisting of a large revolving drum equipped with a felt to hold the sheet in contact with its highly polished surface. It produces a highly glazed surface on that side of the paper. It is commonly used for drying tissue-type papers. Also called a *flying Dutchman.*

**YANKEE MACHINE:** A type of papermaking machine having one large dryer drum with a highly polished surface, primarily used to make machine glazed papers.

**YARD:** (1) The area on a pulp mill site where the pulpwood logs or chips are stored prior to being used by the mill. (2) A linear measurement equal to 3 feet.

**YELLOW STRAWBOARD:** Unbleached paperboard manufactured from pulp obtained by cooking straw.

**YELLOW STRAWPAPER:** Paper made entirely from unbleached pulp obtained by cooking straw.

**YELLOW WAYBILL TISSUE:** *See* WAYBILL MANILA.

**YELLOW WOVE:** A cheap blue woven paper.

**YELLOWING:** The discoloration of white paper, primarily due to the effect of aging. Also known as *brightness reversion.*

**YIELD:** (1) The ratio of the total amount of raw material entering a pulp and papermaking operation to the equivalent product output, usually expressed as a percentage. (2) A unit sometimes used (especially in Canada) to express *the finish* or *density* of a sheet of paper in points of 0.0001 inch. It is determined by multiplying the number of sheets in a standard bundle by the caliper in points of the finished sheet.

# Z

**Z-DIRECTIONAL STRENGTH:** The strength of paper and paperboard as related to the force and measured by means of tension applied perpendicular to the surfaces of the test sample.

**ZEIN:** A protein present in corn and used as a major component in making up some paper adhesives and coatings.

**ZEOLITE:** A natural alumino-silicate mineral having water-softening properties. It is now used as a general reference to all synthetic resins having the same property.

**ZEOLITE PROCESS:** An ion exchange process for softening water consisting of passing cold water through a bed of active zeolite.

**ZERO SPAN TENSILE TEST:** A test sometimes made on paper to indicate the amount of sheet strength due to fiber binding.

**ZETA POTENTIAL:** An electrokinetic potential existing between the fibers and surrounding liquid in a pulp slurry, and one of the forces that tends to hold the fibers in a stable suspension.

**ZIG-ZAG PAPER:** Paper that is embossed or watermarked with a design having short lines or sharp angles.

**ZINC CASEIN:** Casein paper and paperboard coating mixture that contains zinc compounds added for the purpose of imparting wet-rub resistance without introducing any discoloration.

**ZINC HYDROSULFITE ($ZnS_2O_4$):** A common chemical solution used to bleach groundwood pulp.

**ZINC OXIDE ($ZnO$):** A white pigment used in the formulation of paper and paperboard coating material.

**ZINC SULFIDE ($ZnS$):** A chemical material used to make loading substances or coating pigments for use in paper manufacturing.

**ZINC WHITE:** A common name for zinc oxide ($ZnO$) used as a filler in paper to provide opacity and color to the sheet.

# Appendix A
# Bibliography

A list of references is provided not only to acknowledge some of the sources of the information used, but also to provide readers requiring more in-depth knowledge a guide to help them find it. As can be expected, this does not comprise a complete list, but it is a representative cross section. With the aid of other references found in these sources one can delve more deeply into any area of interest in the pulp and paper industry.


Ainsworth, John H. *Paper—The Fifth Wonder*. Kaukauna, WI: Thomas Printing and Publishing Co. Ltd., 1959.

American Paper Institute, Inc., *The Dictionary of Paper*, 4th ed. Philadelphia: Winchell Co., 1980.

American Writing Co. *Handbook of Quality Standard Papers*. Holyoke, 1923.

Atkins, J.F., ed. *The Paper Machine Wet End Press Manual*. Atlanta: TAPPI Press, 1979.

Battista, O.A. *Synthetic Fibers in Papermaking*. New York: Wiley-Interscience, 1964.

Britt, Kenneth W., ed. *Handbook of Pulp and Paper Technology*. 1st and rev. eds. New York: Van Nostrand Reinhold Co. Inc., 1964, 1970.

Browning, B.L. *Analysis of Paper*. New York: Marcel Dekker, Inc., 1969.

Calkin, John B., ed. *Modern Pulp and Paper Making*. 3rd ed. New York: Van Nostrand Reinhold Co. Inc., 1957.

Casey, James P., ed. *Pulp and Paper, Chemistry and Chemical Technology*. Vols. 1, 2. New York: Wiley-Interscience, 1952.

———. *Pulp and Paper, Chemistry and Chemical Technology*. 2d ed. 3 vols. New York: Wiley-Interscience, 1957.

———. *Pulp and Paper, Chemistry and Chemical Technology*. Vol. 1. 3d ed. New York: Wiley-Interscience, 1980.

Clark, James d'A. *Pulp Technology and Treatment for Paper*. San Francisco: Miller Freeman Publications, Inc., 1978.

CPPA Printing and Graphic Arts Committee. *Glossary of Printing Terms for the Papermaker*. Montreal: Canadian Pulp and Paper Assn., Technical Section,1980.

Domtar Pulp and Paper Ltd. *Dictionary of Terms Used in the Paper, Printing, and Graphic Arts Industries*. Montreal, 1948.

English, J., et al. *Glossary of Terms for Pressure Sensitive (Self Adhesive) Labels*. Stamford: The Tag and Label Manufacturers Institute, 1981.

Fishwick, H.P., ed. *Paper Machine Steam and Condensate Systems*. 3d ed. Atlanta: TAPPI Press, 1982.

The Foxboro Co. *Glossary of Analog and Digital Process Control Terms*. Pub. 126. Foxboro, MA, 1967.

Gavelin, G. *Fourdrinier Papermaking-Stock Supply Through Presses*. New York: Lockwood Trade Journal Co., 1963.

———. *Science and Technology of Mechanical Pulp Manufacture*. New York: Lockwood Publishing Co. Inc., 1966.

———. *Drying of Paper and Paperboard.* New York: Lockwood Publishing Co., 1972.

Goerl, Stephen. *Papermaking in America.* New York: Bulkley Dunton Organization, 1945.

Gordon, Mary, ed. *Thesaurus of Forest Products Terms.* Madison: Forest Products Research Society, 1980.

Heiser, E.J. and J.L. Allswede, eds. *Blade Coating Defect Terminology.* Atlanta: TAPPI Press, 1971.

Hunter, Dard. *Papermaking—The History and Technique of an Ancient Craft.* 2d ed. New York: Alfred A. Knopf, 1967.

———. *Papermaking, The History and Technique of an Ancient Craft.* 3d ed. London: Dover Publications, 1978.

Joint Executive Committee on Vocational Education representing the pulp and paper industry of the U.S. and Canada. *The Manufacture of Pulp and Paper.* 3d ed. 5 vols. New York: McGraw-Hill Book Co., 1937.

Jones, H.R. *Pollution Control and Chemical Recovery in the Pulp and Paper Industry.* New York: Noyes Data Corporation, 1973.

Keyes, M.A., S.L. Seabury, and D.M. Gillespie. *Paper Oriented Glossary Covering Advanced Process Control Technology.* Atlanta: TAPPI, 1970.

Klemm, K.H. *Modern Methods of Mechanical Pulp Manufacture.* New York: Lockwood Trade Journal Co.Inc., 1958.

Lavigne, John R. *An Introduction to Paper Industry Instrumentation.* Rev. ed. San Francisco: Miller Freeman Publications, Inc. 1977

———. *Instrumentation Applications for the Pulp and Paper Industry.* San Francisco: Miller Freeman Publications, Inc., 1979.

Libby, E. Earl, ed. *Pulp and Paper Science and Technology.* Vols. 1, 2. New York: McGraw-Hill Book Co., 1962.

Lowe, Kenneth E. *Metrication for the Pulp and Paper Industry.* San Francisco: Miller Freman Publications, Inc., 1975.

———., et al. *Pulp and Paper Factbook.* San Francisco: Miller Freeman Publications, Inc., 1980.

MacDonald, R.G. and John N. Franklin, eds. *Pulp and Paper Manufacture.* 2d ed. 3 vols. New York: McGraw-Hill Book Company, 1969.

Manypenny, Gerry O. *Glossary of Packaging Terms.* 2d ed. New York: Packaging Institute, 1955.

Mardon, J., et al. *Paper Machine Crew Operating Manual.* New York: Lockwood Trade Journal Co., 1963.

McGill, Robert J. *Measurement and Control in Papermaking.* Bristol, England: Adam Hilger Ltd., 1980.

Mosher, R.H. and D.S. Davis, *Industrial and Specialty Papers.* 4 vols. New York: Chemical Publishing Co., 1968.

Mosinee Paper Mills Co. *Paper Terms and What They Mean To You.* Mosinee, WI.

Packaging Institute Inc. *Glossary of Packaging Terms.* 3d ed. New York, 1961.

*Pocket Encyclopedia of Paper and Graphic Arts Terms.* 2d ed. Kaukauna, WI: Thomas Printing and Publishing Company, Ltd., 1965.

*Pulp and Paper Manufacture.* 4 vols. New York: McGraw-Hill Book Co., 1950–55.

Pulp and Paper Research Institute of Canada. *Thesaurus of Pulp and Paper Terms.* Pointe Claire, P.Q., Canada, 1971.

Ranee, R.E., ed. *Handbook of Paper Science.* Vol. 1. London: Elsevier Scientific Publishing Company, 1980.

Rydholm, Sven A. *Pulping Process.* New York: Wiley-Interscience, 1965.

Saltman, D. *Paper Basics.* New York: Van Nostrand Reinhold Co., Inc., 1978.

Shearer, H.E., et al. *Glossary of Textile Terminology for the Paper Manufacturer.* CA Report No. 3. Atlanta: TAPPI, 1965.

Shieve, R.N. *Chemical Process Industries.* New York: McGraw-Hill Book Co., 1967.

Singh, R.P., ed. *The Bleaching of Pulp.* 3d ed. Atlanta: TAPPI Press, 1979.

Smith, David C. *History of Papermaking in the United States (1691-1969).* New York: Lockwood Publishing Co. Inc., 1970.

Stephensen, J. Newell, ed. *Pulp and Paper Manufacturers.* 4 vols. New York: McGraw-Hill Book Co., 1950.

Streit, F. *Paper Quality Control.* New York: Lockwood Publishing Co. Inc., 1968.

Sutermeister, Edwin. *The Story of Papermaking.* Boston: S.D. Warren Company, 1954.

———. *Chemistry of Pulp and Paper Making.* 3d ed. New York: John Wiley and Sons, 1945.

TAPPI Monograph Series. Atlanta: Technical Assn. of the Pulp and Paper Industry.

U.S. Environmental Protection Agency. *Environmental Pollution Control in the Pulp and Paper Industry.* Technical Transfer, 1967.

Van Derveer, Paul D. and Leonard E. Haas, eds. *International Glossary of Technical Terms for the Pulp and Paper Industry.* San Francisco: Miller Freeman Publications, Inc. 1976.

Voet, Andries. *Ink and Paper in the Printing Process.* New York: Wiley-Interscience, 1952.

Wenzl, Herman F.J. *Sulphite Pulping Technology.* New York: Lockwood Trade Journal Co., 1965.

———. *Kraft Pulping Theory and Practice.* New York: Lockwood Publishing Co., Inc., 1967.

Witham, G.S., Sr. *Modern Pulp and Papermaking: A Practical Treatise.* 2d ed. New York: Van Nostrand Reinhold Company, 1942.

**Magazines and Trade Journals**

*American Papermaker.* Atlanta: MacLean Hunter Inc.

*Canadian Papermaker.* Toronto: MacLean Hunter Ltd.

*International Papermaker.* Toronto: MacLean Hunter Ltd.

*Japan Pulp and Paper.* Tokyo: Shigyo Times Sha Ltd.

*Paper.* London: Benn Group.

*Paper Age.* Braintree, Mass.: Global Publications, Inc.

*Paper Industry.* Montgomery, AL: Hatton-Brown Publishers, Inc.

*Paper Southern Africa.* Cape Town: George Warman Publications (PTY) Ltd.

*Paper Technology.* London: British Paper and Board Makers Assn.

*PIMA.* Arlington Heights, IL: Paper Industry Management Assn.

*Pulp & Paper.* San Francisco: Miller Freeman Inc.

*Pulp & Paper Canada.* Westmount, Quebec: Southam Business Publications.

*Pulp & Paper International (PPI).* Brussels, Belgium: Miller Freeman Inc.

*Svensk Papperstidning.* Stockholm: Factu, Medlem av Foreningen Svensk Fackpress.

*TAPPI.* Atlanta: Technical Association of the Pulp and Paper Industry, Inc.

**Standards and Data Sheets**

British Paper and Board Makers Assn. *British Standards Year Book. Routine Test Methods.* London.

Canadian Pulp and Paper Assn. Standard Methods and Data Sheets. Montreal.

International Organization for Standardization. *ISO Standards.* Geneva.

Technical Association of the Pulp and Paper Industry. *TAPPI Standards* and *Technical Information Sheets*. Atlanta.

# Appendix B
# Conversion Tables

English units of measurement are still in common use in pulp and paper mills in the United States, particularly for nonelectrical quantities. However, the movement toward use of the International System of Units, commonly referred to as SI units or the metric system, is accelerating. Reprinted here are three tables giving common and metric units of measure for pulp and paper terminology.

Detailed conversion charts for measurement, temperature, and various paper and board specifications can be found in several sources, three of which are: *Metrication for the Pulp and Paper Industry*, by Kenneth E. Lowe, and my two books, *Instrumentation Applications for the Pulp and Paper Industry* and *An Introduction to Paper Industry Instrumentation*.

## PULP AND PAPER MILL METRIC UNITS

| | *North American practice* | *Recommended metric practice* |
|---|---|---|
| **Wood/woodyard** | | |
| Wood measurement | cord | cubic meter ($m^3$) |
| | cunit | cubic meter ($m^3$) |
| | weight (mass) | kilogram (kg) or metric ton (t) |
| Chip density | pound/cubic foot | kilogram/cubic meter ($kg/m^3$) |
| **Woodroom** | | |
| Chip size: length | inch | millimeter (mm) |
| thickness | inch | millimeter (mm) |
| Chipper disk diameter | foot | meter (m) |
| Rotation | rpm | rev/min |
| **Pulping** | | |
| Digester capacity | cubic feet | cubic meter ($m^3$) |
| Steam pressure | psi | *pascal (Pa) or kilogram-force/square centimeter ($kgf/cm^2$) |
| Temperature | °F | Celsius (°C) |
| Chemical concentrations | pound/cubic foot | gram per liter (g/l) or *kilogram/cubic meter ($kg/m^3$) |
| Liquor volume | gallon, cubic foot | *cubic meter ($m^3$) or liter (l) |
| Tall oil yield | pound/cord | kilogram/cubic meter of pulpwood |
| Turpentine yield | gallon/cord | liter/cubic meter of pulpwood |
| **Refining** | | |
| Peripheral/speed | feet/sec | meter/second (m/s) |
| Power rating | horsepower | kilowatt (kW) |
| Power consumption | hp-day/ton | *joule/metric ton (J/t) kilowatt-hour/metric ton (kW · h/t) |
| Plate clearance | thousandth of inch | millimeter (mm) |
| Closing pressure | psi | *pascal (Pa) or kilogram-force/square centimeter ($kgf/cm^2$) |

*continued*

**PULP AND PAPER MILL METRIC UNITS,** *continued*

| | *North American practice* | *Recommended metric practice* |
|---|---|---|
| **Stock preparation** | | |
| Stock flow | gallon per minute, cubic foot/minute | *cubic meter/minute ($m^3$/min) or liter/minute (l/min) |
| Consistency | percentage | percentage (%) |
| **Coating/chemical preparation** | | |
| Weighing | pound or ton | kilogram (kg) or metric ton (t) |
| Viscosity | centipoise | *pascal-second (Pa • s) or centipoise (cP) |
| Concentrations | pound/gallon | kilogram/cubic meter (kg/$m^3$) gram/liter (g/l) |
| Flows | gallon/minute | cubic meter/minute ($m^3$/min) or liter/minute (l/min) |
| **Paper/pulp machine** | | |
| Headbox level | feet | meter (m) |
| Wire and machine dimensions: width | inch | millimeter (mm) or *meter (m) |
| length | feet | meter (m) |
| Heads and vacuums | psi | *pascal (Pa) |
| | inch water or mercury | millimeter, Hg (mm Hg) millimeter, $H_2O$ (mm $H_2O$) |
| Lineal or nip pressure | pound/lineal inch (pli) | *kilonewton/meter (kN/m) kilogram-force/centimeter (kgf/cm) |
| Machine speed | feet/minute (fpm) | meter per minute (m/min) |
| Draws | feet/minute (fpm) | meter per minute (m/min) |
| Steam pressure | psi | *pascal (Pa) or kilogram-force/square centimeter (kgf/$cm^2$) |
| Trim sizes | inch | meter (m) or millimeter (mm)† |
| Roll size: diameter | inch | meter (m) or millimeter (mm) |
| width | inch | meter (m) or millimeter (mm) |
| Production | ton or pound | metric ton (t) or kilogram (kg) |
| **Power and recovery** | | |
| Recovery rating, black liquor solids | ton/day pound/hour | metric ton/day (t/day) kilogram/hour (kg/h) |
| Steam flow | pound/hour | kilogram/hour (kg/h) |
| Steam pressure | psi | *pascal (Pa) or kilogram-force/square centimeter (kgf/$cm^2$) |
| Electric current | ampere | ampere (A) |
| Electric potential | volt | volt (V) |
| Electrical resistance | ohm | ohm (Ω) |
| Power rating | horsepower | kilowatt (kW) or watt (W) |
| Energy consumption | kilowatt-hour | *joule (J) or kilowatt-hour (kW • h) |
| Temperature | °F | ° Celsius (°C) |
| Air flow: velocity | foot/min | meter/min (m/min) |
| volume | cubic foot | cubic meter ($m^3$) |
| Steam generating efficiency | Btu/kilowatt-hour | joule/kilowatt-hour (J/kW • h) |

*continued*

**PULP AND PAPER MILL METRIC UNITS,** *continued*

| | *North American practice* | *Recommended metric practice* |
|---|---|---|
| **Testing** | | |
| Basis weight (grammage) | pound/ream<br>pound/1,000 square feet | gram/square meter ($g/m^2$) |
| Ream weight | | kilogram/1,000 sheets |
| Film or coat weight | pound/ream<br>pound/1,000 $ft^2$ | gram/square meter ($g/m^2$) |

* Preferred unit of usage

† Note use of centimeter is discouraged in SI system

*Source*: Lowe, K.E. 1975. Metrication for the Pulp and Paper Industry. San Francisco: Miller Freeman Publications, Inc.

## PHYSICAL TESTING OF PULP AND PAPER

| *Test* | *Customary Units* × | *Conversion Factor* = | *SI Units* |
|---|---|---|---|
| Air Resistance, Gurley | s/100 ml | 1 | s/100 ml |
| Air Resistance, Bendtsen | ml/min | 1 | ml/min |
| Basis Weight | $g/m^2$ | 1 | $g/m^2$ |
| Beating Time, Valley | min | 1 | min |
| Beating Time, PFI | rev | 1 | rev |
| Brightness | % Absolute | 1 | % Absolute |
| Bulk | $cm^3/g$ | 1 | $cm^3/g$ |
| Burst Index (Burst Factor) | $\frac{g_f/cm^2}{g/m^2}$ | $9.8067 \times 10^{-2}$ | $kPa \cdot m^2/g$ |
| Burst | $lb_f/in^2$ | 6.8948 | kPa |
| Caliper | mm | 1 | mm |
| | inches | 25.4 | mm |
| Compressibility | % | 1 | % |
| Concora Crush | $lb_f$ | 4.4482 | N |
| Freeness, C.S.F. | ml | 1 | ml |
| Flat Crush | $lb_f/in^2$ | 6.8948 | kPa |
| Internal Bond (Scott) | $ft \cdot lb_f/in^2$ | 2.1015 | $kJ/m^2$ |
| Opacity | % | 1 | % |
| Recovery | % | 1 | % |
| Rejects | % | 1 | % |
| Ring Crush | $lb_f$ | 4.4482 | N |
| "S" Factor, CSF (48/100 fraction) | ml | 1 | ml |
| Screen Classification (each fraction) | % | 1 | % |
| Smoothness, Bendtsen | ml/min | 1 | ml/min |
| Smoothness, Sheffield | ml/min | 1 | ml/min |
| Softness, Gurley-Hill | s/100 ml | 1 | s/100 ml |

*continued*

**PHYSICAL TESTING OF PULP AND PAPER,** *continued*

| *Test* | *Customary Units* × | *Conversion Factor* = | *SI Units* |
|---|---|---|---|
| Specific Light Absorption Coefficient | | | |
| (Based on Basis Weight) | $cm^2/g$ | 1 | $cm^2/g$ |
| (Based on Thickness) | $mm^{-1}$ | 1 | $mm^{-1}$ |
| Specific Light Scattering Coefficient | | | |
| (Based on Basis Weight) | $cm^2/g$ | 1 | $cm^2/g$ |
| (Based on Thickness) | $mm^{-1}$ | 1 | $mm^{-1}$ |
| Stiffness (Gurley) | mgs | $9.8067 \times 10^{-3}$ | mN |
| Stiffness (Taber) | Taber units | 2.03 | mN |
| Stretch | % | 1 | % |
| TEA (Tensile Energy Absorption) | $J/m^2$ | 1 | $J/m^2$ |
| Tear Index (Tear Factor) | $100gs \cdot m^2/g$ | $9.8067 \times 10^{-2}$ | $mN \cdot m^2/g$ |
| Tear Strength | gs | 9.8067 | mN |
| Tensile Breaking Length | m | 1 | m |
| Tensile Strength (of Paper) | kgs/15mm | $6.5378 \times 10^{-1}$ | kN/m |
| | $lb_f$/in. | $1.7513 \times 10^{-1}$ | kN/m |
| Toughness Index | gs·cm | $9.8067 \times 10^{-2}$ | mJ |
| Visual Efficiency | % | 1 | % |
| Zero Span Tensile Breaking Length | m | 1 | m |

*Source*: *PAPRICAN Metric Practice Guide*. Pulp and Paper Research Institute of Canada.

## DERIVED METRIC UNITS USED FOR ELECTRICAL TERMS

| *Measurement* | *SI unit* | *Symbol* | *Definition* |
|---|---|---|---|
| Electric charge | coulomb | c | $c = A \cdot s$ The *coulomb* is the quantity of electricity transported in one second by a current of one ampere. |
| Electric potential difference | volt | V | $V = W/A$ The *volt* is the difference of potential between two points of a conducting wire carrying a constant current of one ampere, when the power dissipated between these points is equal to one watt. |
| Electric resistance | ohm | Ω | $\Omega = V/A$ The *ohm* is the resistance between two points of a conductor when a constant difference of potential of one volt, applied between these two points, produces in this conductor a current of one ampere, this conductor not being the source of any electromotive force. |
| Electric capacitance | farad | F | $F = c/V$ The *farad* is the capacitance of a capacitor between the plates of which there appears a difference of potential of one volt when it is charged by a quantity of electricity equal to one coulomb. |
| Electrical conductance | siemens | s | $s = A/V$ The *siemens* is the electrical conductance of a conductor in which a current of one ampere is produced by an electric potential difference of one volt. |
| Electrical inductance | henry | H | $H = Wb/A$ The *henry* is the inductance of a closed circuit in which an electromotive force of one volt is produced when the electric current in the circuit varies uniformly at the rate of one ampere per second. |
| Magnetic flux | weber | Wb | $Wb = V \cdot s$ The *weber* is the magnetic flux which, linking a circuit of one turn, produces in it an electromotive force of one volt as it is reduced to zero at a uniform rate in one second. |
| Magnetic flux density | tesla | T | $T = Wb/m^2$ The *tesla* is the magnetic flux density given by a magnetic flux of one weber per square meter. |

*Source*: Mechtly, E.A. 1973. *The International System of units.* Washington, D.C.: NASA, Scientific and Technical Information Office.

# *About the Author and Editor*

**John R. Lavigne** is an international authority on pulp and paper control systems engineering. Retired from the Pulp and Paper Division of Foxboro Co., he now provides consulting and technical writing services to the industry. He is a registered professional engineer and a well-known author and lecturer on the subject of industrial instrumentation.

His credits include feature articles for pulp and paper magazines, the presentation of many papers at industry conferences, chapters for technical books, and two books: *Introduction to Paper Industry Instrumentation* and *Instrumentation Applications for the Pulp and Paper Industry.*

A fellow of the Instrument Society of America and senior member of TAPPI and CPPA, he has received a number of awards from these organizations.

**Ken L. Patrick** joined *Pulp & Paper* magazine as southern editor in 1977 and became editor in 1982. Earlier he served as technical editor and communications supervisor for International Paper Co., in Mobile, Alabama, and as manager of public affairs at their mill in Natchez, Mississippi. He has a B.A. from the University of South Alabama.